Fluid Mechanics (Vol. 2)

Shiv Kumar

Fluid Mechanics (Vol. 2)

Basic Concepts and Principles

Fourth Edition

**Ane Books
Pvt. Ltd.**

Shiv Kumar
New Delhi, India

ISBN 978-3-030-99756-4   ISBN 978-3-030-99754-0   (eBook)
https://doi.org/10.1007/978-3-030-99754-0

Jointly published with ANE Books Pvt. Ltd.
In addition to this printed edition, there is a local printed edition of this work available via Ane Books in South Asia (India, Pakistan, Sri Lanka, Bangladesh, Nepal and Bhutan) and Africa (all countries in the African subcontinent).
ISBN of the Co-Publisher's edition: 978-9-384-72699-7.

This Springer imprint is published by the registered company Springer Nature Switzerland AG
The registered company address is: Gewerbestrasse 11, 6330 Cham, Switzerland

Dedicated
to

My Parents
My Wife Dr. Kusum and My Son Tanishq

Preface

This book has been written for the introductory course on Fluid Mechanics at the undergraduate level. This book fulfills the curriculum needs of UG students of Mechanical Engineering, Mechanical and Automation Engineering, Chemical Engineering, Electrical Engineering, Civil Engineering, Production Engineering, Automobile Engineering, aeronautical Engineering, Manufacturing Engineering, Tool Engineering and Mechatronics Engineering etc. Fluid Mechanics is dividing into two volumes. Fluid Mechanics Volume-II includesten chapters:

1. Laminar Flow (Viscous Flow), 2. Turbulent Flow, 3. Boundary Layer Theory, 4. Flow through Pipe, 5. Pipe Flow Measurement,6. Orifices and Mouthpieces, 7. Flow Past Submerged Bodies, 8. Flow through Open Channels, 9.Notches and Weirs, 10. Compressible Flows. Fluid Mechanics deals with the innovative use of the laws of Fluid Mechanics in solving the relevant technological problems. This introductory textbook aims to provide undergraduate engineering students with the knowledge (basics principles and fluid mechanics laws) they need to understand and analyze the fluid mechanics problems they are likely to encounter in practice.

The book is developed in the context of the author's simpler methodology to present even complex things. The most positive factor about the book is that it is concise, and everything is described from an elementary and tangible perspective.

The book presents the concepts in a very logical format with complete word descriptions. The subject matter is illustrated with a lot of examples. A great deal of attention is given to select the numerical problems and solving them. The theory and numerical problems at the end of each chapter also aim to enhance the creative capabilities of students. Ultimately as an introductory text for the undergraduate students, this book provides the background necessary for solving the complex problems in thermodynamics.

Writing this book made me think about a lot more than the material it covers. The methods I used in this book are primarily those that worked best for my students. The suggestions from the teachers and students for the further improvement of the text are welcome and will be implemented in the next edition. The readers are requested to bring out the error to the notice, which will be gratefully acknowledged.

Shiv Kumar

Acknowledgements

First of all, I would like to express my deep gratitude to God for giving me the strength and health for comleting this book. I am very thankful to my colleagues in the mechanical engineering department for their highly appreciable help and my students for their valuable suggestions.

I am also thankful to my publishers Shri Sunil Saxena and Shri Jai Raj Kapoor of Ane Books Pvt. Ltd. and the editorial group for their help and assistance.

A special thanks goes to my wife Dr. Kusum Lata for her help, support and strength to complete the book.

Shiv Kumar

Contents

Laminar Flow
(Viscous Flow or Flow with Low Reynolds Number)

1.1 INTRODUCTION

The flow of real fluid differs from that of ideal fluid in respect of viscous effects that take place near to the surface of the solid body. As ideal fluid is inviscid, the presence of a solid body does not affect the flow; the fluid is assumed to slip over the surface of solid body and no velocity gradient $\left(\dfrac{du}{dy}\right)$ exists over the surface of a solid body, and hence no shear stress $\left(\tau = \mu\dfrac{du}{dy}\right)$ acts on the layers. On the other hand, when real fluid flows over the surface of a solid body, the fluid particles contacting with surface get zero velocity. This flow characteristic of a real fluid at the surface is in accordance with the no-slip condition. These particles then retard the motion of particles in the adjoining fluid layer and so on. This change of velocity gradient is responsible for development of viscous shear resistance which opposes the motion.

The following velocity profile is shown in Fig. 1.1, when ideal and the real fluid flow through pipes and over the surface of solid body.

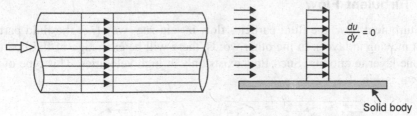

$\dfrac{du}{dy} = 0$

Solid body

(a) Velocity distribution in ideal fluid flow $\left(\dfrac{du}{dy}=0\right)$ through pipe and over the surface of solid body.

S. Kumar, *Fluid Mechanics (Vol. 2)*,
https://doi.org/10.1007/978-3-030-99754-0_1

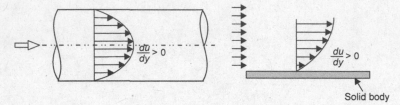

Solid body

(b) Velocity distribution in real fluid flow $\left(\dfrac{du}{dy}>0\right)$ through pipe and over the surface of solid body.

Fig. 1.1: Comparison of ideal and real fluid flows.

1.2 LAMINAR AND TURBULENT FLOW

The flow of a real fluid is divided into two types:

(i) Laminar flow.

(ii) Turbulent flow.

1.2.1 Laminar Flow

The laminar flow is a smooth regular flow in layers. Such flow exists only at low velocities. Fluid particles remain in motion in respective layers. In other words there will be no exchange of fluid particles from one layer to another. Thus, there will be no momentum transmission from one layer to another. The various particles in a layer describe definite straight line path as shown in Fig. 1.2.

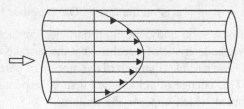

Fig. 1.2: Laminar flow through pipe.

1.2.2 Turbulent Flow

In the turbulent flow, the fluid particles flow in zig-zag way (i.e. the fluid particles are not moving in layer). In the other words, there will be exchange of fluid particles from one layer to another. Such flow exists only at high velocities. This type of flow is shown is Fig. 1.3.

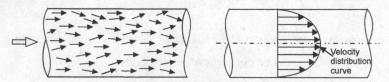

Velocity distribution curve

Fig. 1.3: Turbulent flow through pipe.

The type of flow, either laminar or turbulent, is described on the basis of Reynolds number (Re).

Reynolds number: $Re = \dfrac{\rho DV}{\mu} = \dfrac{DV}{\mu}$

where ρ = density of the fluid

D = diameter of the pipe

V = mean velocity of the fluid

μ = viscosity of fluid

ν = kinematic viscosity of the fluid

If $Re < 2000$, the flow through pipe is laminar and Re > 4000, the flow through pipe is turbulent.

1.3 REYNOLDS EXPERIMENT

Osborne Reynolds was a scientist who discovered that Reynolds number $\left(Re = \dfrac{\rho VD}{\mu} \right)$ is the criterion for determining the type of flow (either laminar or turbulent) in a circular pipe. The arrangement of the apparatus is shown in Fig. 1.4.

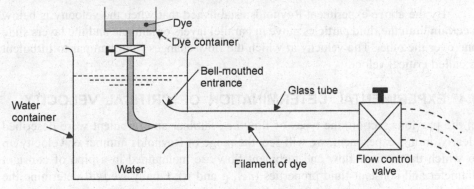

Reynolds apparatus for demonstration of laminar and turbulent flow.

(a) Straight line filament of dye shows laminar flow.

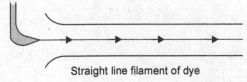

Straight line filament of dye

(b) Wavy filament of dye indicating onset of turbulence.

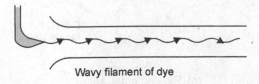

Wavy filament of dye

(*c*) Diffused filament of dye indicating turbulent flow.

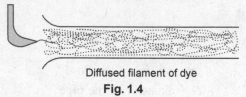

Diffused filament of dye

Fig. 1.4

The main parts of the apparatus are:

(*i*) A water container.

(*ii*) An arrangement to inject a fine filament of dye into the bell-mouthed entrance of a glass tube through which water flows, and

(*iii*) A valve to control the flow through the glass tube.

A fine filament of dye was introduced into the glass tube near the smoothly rounded bell-mouthed entrance. At low rate of flow the filament of dye appeared as a straight line parallel to the tube axis indicating laminar flow as shown in Fig. 1.4(*a*). As the valve was further opened resulting in high velocity of flow, the dye filament became wavy leading to its breakup and finally diffusing into the flowing water, Fig. 1.4. (*b*) and (*c*) shows the flow condition at increasing velocities indicating the onset of turbulence and then flow completely changing over to turbulence.

By the above experiment Reynolds established that when the velocity is below a certain limit the fluid particles move in parallel layers or laminae and the layers slide one over the other. The velocity at which the flow changes from laminar to turbulent is called critical velocity.

1.4 EXPERIMENTAL DETERMINATION OF CRITICAL VELOCITY

In the previous section, the types of flow *i.e.*, laminar and turbulent were described clearly. In this experiment, we will see, the range of Reynolds number or velocity up to which the laminar flow and turbulent flow are maintained in a pipe of constant diameter and constant fluid properties (*i.e.*, ρ and μ). Finally we will determine the critical value of velocity and Reynolds number.

The energy loss in certain length of straight circular pipe is determined by pressure drop measured by a differential manometer as shown is Fig. 1.5 (a).

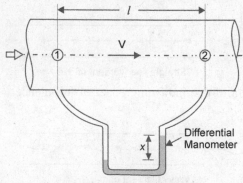

(*a*) Experimental setup.

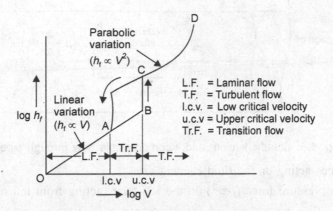

(b) Energy or head loss variation with velocity

Fig. 1.5

Then mean velocity of flow is calculated from the known volumetric rate of flow (Q) passing through the pipe. The energy or head loss (h_f) and corresponding mean velocity (V) may be plotted on a log-log graph paper. The head loss (h_f) is seen to increase linearly with velocity (V) till point B. The linear relationship between h_f and V indicates laminar flow while a higher velocities a nearly parabolic relationship $(h_f \mu V^2)$, the parabolic relationship between h_f and V indicates turbulent flow (CD). Between the two flow regimes there lies a transition zone in which flow is in the process of changing over from laminar to turbulent. As velocity is increased the head loss (h_f) varies with the mean velocity (V) according to the laws $(h_f \mu V, h_f \mu V^2)$ and is indicated by line $OABCD$, where for decreasing velocities it will be seen to follow the path $DCAO$. From this experiment, it may be deduced that the points A and B lying on the straight line OB define the lower and upper critical velocities respectively.

In flow through pipe, the values of Reynolds number at point A (i.e., lower critical velocity): $Re = 2000$ and at point B (i.e., upper critical velocity): $Re = 4000$, the transition from laminar to turbulent may occur at the value of Reynold's number between 2000 and 4000.

1.5 STEADY LAMINAR FLOW THROUGH A CIRCULAR PIPE

The laminar, incompressible and steady flow through pipe may be completely analysed by two laws:

  (i) Newton's law of viscosity and

  (ii) Newton's second law of motion.

  Consider a horizontal pipe of radius R and a concentric cylindrical fluid element of radius r and length dx as shown is Fig. 1.6.

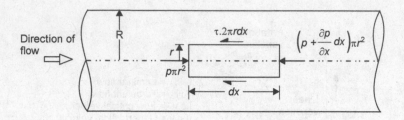

Fig. 1.6: Steady, laminar and incompressible flow through pipe.

The forces acting on the fluid element are:

1. The pressure force ($p\pi r^2$) on the left face (acting from left to right).

2. The pressure force $\left(p+\dfrac{\partial p}{\partial x}.dx\right)\pi r^2$ on the right face. (acting from right to left).

3. The shear force ($\tau.2\pi r dx$) acting on the surface of the fluid element opposing the motion (acting from right to left).

As the flow is steady, laminar and uniform, the total acceleration is zero.

According to Newton's second law of motion, the summation of all forces in the direction of flow is equal to product of mass and acceleration *i.e.,* the summation of all forces in the direction of flow must be zero ($\because$ acceleration is zero).

$$p\pi r^2 - \left(p+\frac{\partial p}{\partial x}dx\right)\pi r^2 - \tau.2\pi r\,dx = 0$$

$$-\frac{\partial p}{\partial x}dx\,\pi r^2 - \tau.2\pi r\,dx = 0$$

$$-\frac{\partial p}{\partial x}r - \tau.2 = 0$$

or
$$\tau = -\frac{\partial p}{\partial x}.\frac{r}{2} \qquad\qquad ...(1.5.1)$$

This equation gives the distribution of shear stress across a flow section, the pressure gradient $\dfrac{dp}{dx}$ depends only on the direction of flow *i.e.* x-direction, hence above Eq. (1.5.1) provides a linear relationship between shear stress τ and radius r, it is evident from the equation that the shear stress will be maximum at the pipe surface where $r=R$ $\left(i.e.,\ \tau_{max}=-\dfrac{\partial p}{\partial x}.\dfrac{R}{2}\right)$ and zero at the centre of pipe where $r=0$, as shown in Fig. 1.7.

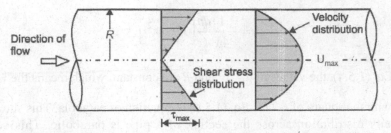

Fig. 1.7: Shear stress and velocity distribution across a section.

Velocity distribution:

The shear stress is related to the velocity gradient by Newton's law of viscosity.

$$\text{Shear stress} : \tau = \mu \frac{du}{dy}$$

Here, y is distance measured from the pipe wall,

$$y = R - r$$

Differentiating, $\qquad dy = -dr$ $\qquad\qquad\qquad \because R = \text{constant}$

$\therefore$ Shear stress in present case: $\tau = \mu \dfrac{du}{-dr}$

$$\tau = -\mu \frac{du}{dr} \qquad\qquad ...(1.5.2)$$

It is to be noted that the velocity gradient: $\dfrac{du}{dr}$ is negative *i.e.* as r increases, u decreases.

Equating Eqs. (1.5.2) and (1.5.1), we get

$$-\mu \frac{du}{dr} = -\frac{dp}{dx} \frac{r}{2}$$

$$\frac{du}{dr} = \frac{1}{2\mu} \frac{dp}{dx} r$$

Integrating *w.r.t.* r, we get

$$u = \frac{1}{2\mu} \frac{dp}{dx} \frac{r^2}{2} + C$$

$$u = \frac{1}{4\mu} \frac{dp}{dx} r^2 + C \qquad\qquad ...(1.5.3.)$$

where C is the constant of integration and its value is obtained from the boundary condition that at $r = R$, $u = 0$

$$0 = \frac{1}{4\mu} \frac{dp}{dx} R^2 + C$$

or $\qquad\qquad C = -\dfrac{1}{4\mu} \dfrac{dp}{dx} R^2$

Substituting the value of C in Eq. (1.5.3), we get

$$u = \frac{1}{4\mu} \frac{dp}{dx} r^2 - \frac{1}{4\mu} \frac{dp}{dx} R^2$$

$$u = -\frac{1}{4\mu}\frac{dp}{dx}\left[R^2 - r^2\right] \qquad ...(1.5.4)$$

In Eq. (1.5.4), the values of μ, $\frac{dp}{dx}$ and R are constant, which means the velocity u varies with the square of r. This Eq. (1.5.4) is equation of parabola. This shows that the velocity distribution across the section of a pipe is parabolic. This velocity distribution is shown is Fig. 1.7

[**Note :** Pressure variation along the x-direction *i.e.,* along the direction of flow, $\frac{dp}{dx}$ is always negative because of continuous decrease of pressure in the direction of flow].

Ratio of maximum to average velocity:

The velocity is maximum, when $r = 0$ in Eq. (1.5.4). Thus maximum velocity, U_{max} is obtained as

$$U_{max} = -\frac{1}{4\mu}\frac{dp}{dx}R^2 \qquad ...(1.5.5)$$

The average velocity: $\quad \bar{u} = \dfrac{\text{Discharge through pipe}:Q}{\text{Cross-sectional area of pipe}:\pi R^2}$

$$\bar{u} = \frac{Q}{\pi R^2}$$

Consider a small circular ring element of radius r and thickness dr as shown in Fig. 1.8.

The fluid flowing per second through small elementary ring: dQ

Fig. 1.8: Discharge flow through pipe

dQ = velocity at radius r × area of a small ring element.

$dQ = u \times 2\pi r dr$

Substituting the value of u from Eq. (1.5.4) in above equation, we get

$$dQ = -\frac{1}{4\mu}\frac{dp}{dx}\ [R^2 - r^2]\ 2prdr$$

Net discharge flow through pipe: Q

$$Q = \int_0^R dQ$$

$$= \int_{0}^{R} -\frac{1}{4\mu}\frac{dp}{dx}[R^2 - r^2]2\pi r\, dr$$

$$= -\frac{1}{4\mu}\frac{dp}{dx}2\pi\int_{0}^{R}(R^2 - r^2)r\, dr$$

$$= -\frac{1}{4\mu}\frac{dp}{dx}2\pi\left[\frac{R^2 r^2}{2} - \frac{r^4}{4}\right]_{0}^{R}$$

$$= -\frac{1}{4\mu}\frac{dp}{dx}2\pi\left[\frac{R^4}{2} - \frac{R^4}{4}\right]$$

$$= -\frac{1}{4\mu}\frac{dp}{dx}2\pi\times\frac{R^4}{4}$$

$$Q = -\frac{\pi}{8\mu}\frac{dp}{dx}R^4$$

$\therefore$ Average velocity: $\bar{u} = \dfrac{Q}{\pi R^2}$

$$\bar{u} = \frac{-\dfrac{\pi}{8\mu}\dfrac{dp}{dx}R^4}{\pi R^2}$$

$$\bar{u} = -\frac{1}{8\mu}\frac{dp}{dx}R^2 \qquad\qquad\qquad ...(1.5.6)$$

Dividing Eq. (1.5.5) by Eq. (1.5.6), we get

$$\frac{U_{max}}{\bar{u}} = \frac{-\dfrac{1}{4\mu}\dfrac{dp}{dx}R^2}{-\dfrac{1}{8\mu}\dfrac{dp}{dx}R^2} = 2$$

$\therefore$ Maximum velocity: $U_{max} = 2$ times average velocity: $\bar{u}$

$$U_{max} = 2\bar{u}$$

Drop of pressure head for a given length (*l*) of a pipe:

Recalling the average velocity through pipe from Eq. (1.5.6)

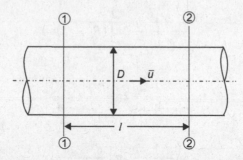

Fig. 1.9: Drop of Pressure head in a pipe

$$\bar{u} = -\frac{1}{8\mu}\frac{dp}{dx}R^2$$

or

$$-\frac{dp}{dx} = \frac{8\mu\bar{u}}{R^2}$$

The term $\left(-\frac{dp}{dx}\right)$ represents pressure drop per unit length through pipe and may be written as

$$-\frac{\Delta p}{l} = \frac{-(p_2-p_1)}{l} = \frac{p_1-p_2}{l}$$

$$\therefore \qquad \frac{p_1-p_2}{l} = \frac{8\mu\bar{u}}{R^2}$$

$$p_1 - p_2 = \frac{8\mu\bar{u}l}{\left(\frac{D}{2}\right)^2} \qquad\qquad \because R = \frac{D}{2}$$

$$= \frac{32\mu\bar{u}l}{D^2}$$

Dividing by ρg on the both sides, we get

$$\frac{p_1-p_2}{\rho g} = \frac{32\mu\bar{u}l}{\rho g D^2}$$

$$\boxed{h_f = \frac{32\mu\bar{u}l}{\rho g D^2}} \qquad\qquad\qquad ...\,(1.5.7)$$

where

$$h_f = \frac{p_1-p_2}{\rho g}, \text{ loss of pressure head}$$

$$\mu = \text{viscosity of fluid}$$

$\bar{u}$ = average velocity of flow

ρ = density of fluid

Equation (1.5.7) is called **Hagen-Poiseuille equation**.

We know that the average velocity : $\bar{u} = \dfrac{Q}{\text{Area}} = \dfrac{Q}{\dfrac{\pi}{4}D^2} = \dfrac{4Q}{\pi D^2}$

Substituting $\bar{u} = \dfrac{4Q}{\pi D^2}$ in Eq.. (1.5.7), we get

$$h_f = \frac{128 Q \mu l}{\pi \rho g D^4} \qquad \qquad ...(1.5.8)$$

Equation (1.5.8) is also called **Hagen-Poiseuille equation**.

Note: Hagen–Poiseuille Eqs. (1.5.7) and (1.5.8) are only applicable for steady, laminar and incompressible flow through pipe.

If the pressure loss (or head loss) is known, the required pumping power to overcome the pressure loss is determined from

$$\begin{aligned} P_{\text{pump}, L} &\sim Q\Delta p \sim Q\rho g h_f \\ &= mg\,h_f \end{aligned} \qquad [\because\ m = \rho Q]$$

where Q = volume flow rate

m = mass flow rate

The average velocity for laminar flow in horizontal pipe is, from Eq. (1.5.7)

$$\bar{u} = \frac{\rho g h_f D^2}{32\mu l} = \frac{\Delta p D^2}{32\mu l} \qquad \because\ \Delta p = \rho g h_f$$

Then the volume flow rate for laminar flow through a horizontal pipe of diameter D and length l, from Eq. (1.5.8)

$$Q = \frac{\pi \rho g\, h_f\, D^4}{128\mu l} = \frac{\pi \Delta p\, D^4}{128\mu l} \qquad \qquad ...(1.5.9)$$

This equation is known as Poiseuille's law and this flow is called Hagen-Poiseuille flow in honour of the works of G. Hagen (1797-1884) and J. Poiseuille (1799-1869) on the subject. It is to be noted from Eq. (1.5.9) that for a specified volume flow rate, the pressure drops and thus the required pumping power is directly proportional to the length of the pipe and viscosity of the fluid, but it is inversely proportional to the fourth power of the diameter (or radius) of the pipe. Therefore, the pumping power requirement for a piping system can be reduced by a factor of 16 by doubling the pipe diameter as shown in Fig. 1.10.

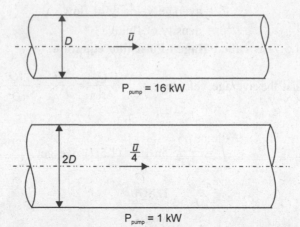

Fig. 1.10: Pumping power requirement can be reduced by a
factor 16 by doubling the pipe diameter

Of course the benefits of the reduction in power costs must be weighed against the
increased cost of construction due to using a larger-diameter pipe.

1.5.1 Comparison between Hagen-Poiseuille Equation and Darcy's Formula

Hagen-Poiseuille Equation	Darcy's Formula
Pressure loss: $\Delta p = \dfrac{32\mu\bar{u}l}{D^2}$	Pressure loss: $\Delta p = \dfrac{4f\,l\rho\bar{u}^2}{2D}$
Head loss: $h_f = \dfrac{32\mu\bar{u}l}{\rho gD^2}$ $= \dfrac{32\nu\bar{u}l}{gD^2}$ $\because \quad \nu = \dfrac{\mu}{\rho}$	Head loss: $h_f = \dfrac{4f\,l\bar{u}^2}{2gD}$
1. The above equations are valid only for fully developed laminar flow in horizontal circular pipes but not for inclined pipes.	1. The above equations are valid for both laminar and turbulent flows, in both circular and non-circular pipes, smooth or rough surface, horizontal or inclined pipes.
2. In these equations, pressure or head loss only due to viscous effect of the liquid.	2. In these equations, pressure loss or head loss is due to friction.

For laminar flow, by equating Hagen-Poiseuille equation and Darcy's formula, we get

$$\frac{32\mu\bar{u}l}{\rho g D^2} = \frac{4fl\bar{u}^2}{2gD}$$

$$\frac{16\mu}{\rho D \bar{u}} = f$$

or
$$f = \frac{16}{\dfrac{\rho \bar{u} D}{\mu}}$$

$$\boxed{f = \frac{16}{Re}}$$ ...(1.5.10)

where $Re = \dfrac{\rho \bar{u} D}{\mu}$, Reynolds number

The above Eq. (1.5.10) gives a relationship between coefficient of friction f and the Reynolds number Re, for laminar flow through a circular pipe.

Problem 1.1: An oil of viscosity 5 poise flows in a 5.0 mm diameter pipe, discharge rate being 5.5 liter/s. If the specific gravity of oil is 0.87, state whether flow is laminar or turbulent.

Solution: Given data:

Viscosity: $\mu = 5$ poise $= \dfrac{5}{10}$ Ns/m^2 = 0.5 Ns/m^2

Diameter of pipe: $D = 50$ mm = 0.05 m

Discharge: $Q = 5.5$ liter/s $= \dfrac{5.5}{1000}$ m^3/s = 0.005 m^3/s

Specific gravity of oil: S = 0.87

∴   Density of oil: $\rho = S \times 1000$ kg/m^3 = 0.87 × 1000 kg/m^3 = 870 kg/m^3

Average velocity: $\bar{u} = \dfrac{Q}{\dfrac{\pi}{4}D^2} = \dfrac{4Q}{\pi D^2} = \dfrac{4 \times 0.0055}{3.14 \times (0.05)^2} = 2.8$ m/s

Reynolds number: $Re = \dfrac{\rho \bar{u} D}{\mu} = \dfrac{870 \times 2.8 \times 0.05}{0.5} = 243.6$

Reynolds number (Re) is less than 2000.

Hence, **the flow in pipe is laminar.**

Problem 1.2: An oil of specific weight 8930 N/m^3 and kinematics viscosity 0.0002 m^2/s is pumed through a 150 mm diameter 300 m long pipe at the rate of 200 kN/h. Show that the flow is viscous and find the power required.

Solution: Given data:

Specific weight: $w = 8930$ N/m^3

Kinematic viscosity: $v = 0.0002$ m^2/s

Diameter: $D = 150$ mm $= 0.15$ m

Length: $l = 300$ m

Weight flow rate: $\dot{w} = 200$ kN/h $= 200 \times 10^3$ N/h

$$= \frac{200 \times 10^3}{60 \times 60} \text{ N/s} = 55.55 \text{ N/s}$$

Discharge: $Q = mv$

$$= \frac{m}{\rho} = \frac{mg}{\rho g} \qquad \qquad \because \text{ Weight flow rate: } \dot{w} = mg$$

$$= \frac{\dot{w}}{w}$$

Specific weight: $w = \rho g$

$$Q = \frac{55.55}{8930} \text{ m}^3/\text{s} = 0.00622 \text{ m}^3/\text{s}$$

also $\qquad Q = A\bar{u}$

$$0.00622 = \frac{\pi}{4} D^2 \bar{u}$$

$$0.00622 \times 4 = 3.14 \times (0.15)^2 \times \bar{u}$$

Or average velocity: $\bar{u} = 0.352$ m/s

Reynolds number: $Re = \dfrac{\bar{u}D}{v} = \dfrac{0.352 \times 0.15}{0.0002} = 264$

Since, Reynolds number is less than 2000, the flow is viscous.

According to Hagen-Poiseuille equation, the loss of pressure head:

$$h_f = \frac{32 \mu \bar{u} l}{\rho g D^2}$$

$$= \frac{32 v \bar{u} l}{g D^2} \qquad \qquad \because v = \frac{\mu}{\rho}$$

$$= \frac{32 \times 0.0002 \times 0.352 \times 300}{9.81 \times (0.15)^2} = 3.06 \text{ m}$$

$\therefore \qquad$ Power required $= mg\, h_f = \dot{W} h_f \qquad \qquad \because \dot{W} = mg$

$$= 55.55 \times 3.06 = 169.98 \text{ Nm/s or J/s} = \mathbf{169.98 \text{ W}}$$

Problem 1.3: An oil of viscosity 0.95 poise and specific gravity 0.92 is flowing through a horizontal pipe of diameter 120 mm and length 20 m. Find the difference of pressure of the two ends of the pipe, if 120 kg of the oil is collected in a tank in 40 seconds.

Solution: Given data:

Viscosity of oil: $\mu = 0.95$ poise $= \dfrac{0.95}{10}$ Ns/m^2 $= 0.095$ Ns/m^2

Specific gravity: $S = 0.92$

$\therefore$ Density of oil: $\rho = S \times 1000$ kg/m$^3 = 0.92 \times 1000$ kg/m$^3 = 920$ kg/m^3

Diameter of pipe: $D = 120$ mm $= 0.12$ m

Length of pipe: $l = 20$ m

Mass of oil collected: $M = 120$ kg in time : $t = 40$s

$\therefore$ Mass flow rate: $m = \dfrac{M}{t} = \dfrac{120}{40}$ kg/s $= 3$ kg/s

Also mass flow rate: $m = \rho A \bar{u}$ by continuity equation.

$$3 = 920 \times \frac{\pi}{4}(0.12)^2 \times \bar{u}$$

$$3 = 10.399 \ \bar{u}$$

or $$\bar{u} = \frac{3}{10.399} = 0.2884 \,\text{m/s}$$

Reynolds number: $Re = \dfrac{\rho \bar{u} D}{\mu}$

$$= \frac{920 \times 0.2884 \times 0.12}{0.095} = 335.15$$

As the Reynolds number is less than 2000. Hence, the flow is laminar.

Loss of head between two ends: $h_f = \dfrac{32\mu\bar{u}l}{\rho g D^2}$

also $$h_f = \frac{p_1 - p_2}{\rho g}$$

$\therefore$ $$\frac{p_1 - p_2}{\rho g} = \frac{32\mu\bar{u}l}{\rho g D^2}$$

$$p_1 - p_2 = \frac{32\mu\bar{u}l}{D^2} = \frac{32 \times 0.095 \times 0.2884 \times 20}{(0.12)^2}$$

$$= 1217.688 \text{ N/m}^2 \text{ or Pa} = \mathbf{1.217 \text{ kPa}}$$

Problem 1.4:   A fluid of viscosity 8 poise and specific gravity 1.2 is flowing through a circular pipe of diameter 100 mm. The maximum shear stress at the pipe wall is 210 N/m^2. Find:

(*i*) The pressure gradient,

(*ii*) The average velocity, and

(*iii*) Reynolds number of the flow.   (*GGSIP University, Delhi. Dec. 2006*)

Solution: Given data:

Viscosity: $\mu = 8$ poise $= \dfrac{8}{10}$ Ns/m^2 $= 0.8$ Ns/m^2

Specific gravity: $S = 1.2$

∴ Density: $\rho = S \times 1000$ kg/m^3 $= 1.2 \times 1000 = 1200$ kg/m^3

Diameter of pipe: $D = 100$ mm $= 0.1$m

∴ Radius of pipe: $R = \dfrac{D}{2} = \dfrac{0.1}{2} = 0.05$ m

Maximum shear stress: $\tau_{max} = 210$ N/m^2

(*i*) **Pressure gradient:** $\dfrac{\partial p}{\partial x}$

The maximum shear stress: $\tau_{max} = -\dfrac{\partial p}{\partial x}\dfrac{R}{2}$

$$210 = -\dfrac{\partial p}{\partial x} \times \dfrac{0.05}{2}$$

or $\dfrac{\partial p}{\partial x} = -\,\textbf{8400 N/m}^2$ **per** *m*

(*ii*) **Average velocity:** $\bar{u}$

We know that the expression of average velocity:

$$\bar{u} = -\dfrac{1}{8\mu}\dfrac{dp}{dx}R^2$$

$$= -\dfrac{1}{8 \times 0.8} \times (-\,8400) \times (0.05)^2$$

$$= \textbf{3.28 m/s}$$

(*iii*) **Reynolds number:** *Re*

$$Re = \dfrac{\rho D\bar{u}}{\mu} = \dfrac{12 \times 0.1 \times 3.28}{0.8} = \textbf{492}$$

Problem 1.5: An oil of specific gravity 0.9 is flowing through a pipe of diameter 110 mm. The viscosity of oil is 10 poise and velocity at the centre is 2 m/s. Find :

(*i*) Pressure gradient in the direction of flow,

(*ii*) Shear stress at the pipe wall,

(*iii*) Reynolds number, and

(*iv*) Velocity at a distance of 30 mm from the wall.

(*GGSIP University, Delhi. Dec. 2007*)

Solution: Given data:

Specific gravity of oil: $S = 0.9$

∴ Density of oil: $\rho = S \times 1000$ kg/m^3 $= 0.9 \times 1000 = 900$ kg/m^3

Diameter of pipe: $D = 110$ mm $= 0.11$ m

Radius of pipe: $R = \dfrac{D}{2} = \dfrac{0.11}{2} = 0.055$ m

Viscosity of oil: $\quad \mu = 10 \text{ poise} = \dfrac{10}{10} \text{ Ns/m}^2 = 1 \text{ Ns/m}^2$

Velocity at the centre *i.e.*, maximum velocity: $U_{max} = 2 \text{ m/s}$

(*i*) **Pressure gradient in the direction of flow:** $\dfrac{\partial p}{\partial x}$

We know that the expression of maximum velocity: U_{max}

$$U_{max} = -\frac{1}{4\mu}\,\frac{\partial p}{\partial x}\,R^2$$

$$2 = -\frac{1}{4 \times 1} \times \frac{\partial p}{\partial x} \times (0.055)^2$$

or $\qquad\qquad\qquad \dfrac{\partial p}{\partial x} = -2644.62 \text{ N/m}^3$

(*ii*) **Shear stress at the pipe wall:** $\tau_0 = \tau_{max}$

$$\tau_{max} = -\frac{\partial p}{\partial x} \cdot \frac{R}{2}$$

$$= -(-2644.62) \times \frac{0.055}{2} = 72.727 \text{ N/m}^2$$

(*iii*) **Reynolds number:** $\quad Re = \dfrac{\rho \bar{u} D}{\mu}$

$$\because\ U_{max} = 2\bar{u} \qquad \text{or} \qquad \bar{u} = \frac{U_{max}}{2}$$

$$= \frac{\rho U_{max} D}{2\mu} = \frac{900 \times 2 \times 0.11}{2 \times 1} = 99$$

(*iv*) **Velocity at a distance of 30 mm from the wall:** u

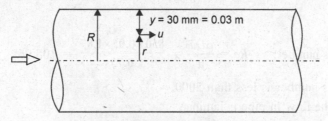

Fig. 8.11: Schematic for Problem 1.5

$$u = \frac{-1}{4\mu}\,\frac{\partial p}{\partial x}\,(R^2 - r^2)$$

where $\qquad\qquad r = R - y$

$$= 0.055 - 0.03 = 0.025 \text{ m}$$

$\therefore \qquad\qquad u = -\dfrac{1}{4 \times 1} \times [(0.055)^2 \times (0.25)^2]$

$$= 661.155 \times [3.025 \times 10^{-3} - 6.25 \times 10^{-4}]$$

$$= \mathbf{1.586 \text{ m/s}}$$

Problem 1.6: An oil having viscosity of 7.5 poise of specific gravity 0.85 flows through a horizontal 50 mm diameter pipe with a pressure drop of 18 kN/m² per metre length of pipe. Determine (i) the flow rate of oil and centre line velocity (ii) power required to maintained the flow in 100 m length of pipe (iii) velocity and shear stress at 8 mm from the wall.

Solution: Given data:

Viscosity of oil: $\mu = 7.5$ poise $= \dfrac{7.5}{10}$ Ns/m² $= 0.75$ Ns/m²

Specific gravity: $S = 0.85$

∴ Density of oil: $\rho = S \times 1000$ kg/m³ $= 0.85 \times 1000$ kg/m³ $= 850$ kg/m³

Diameter of pipe: $D = 50$ mm $= 0.05$ m

∴ Radius of pipe: $R = \dfrac{D}{2} = \dfrac{0.05}{2} = 0.025$ m

Pressure drop: $p_1 - p_2 = 18$ kN/m² per m length of pipe

i.e., $\dfrac{p_1 - p_2}{l} = 18$ kN/m³ $= 18 \times 10^3$ N/m³

Pressure drop for laminar flow through pipe is,

$$p_1 - p_2 = \frac{32\mu\bar{u}l}{D^2}$$

$$\frac{p_1 - p_2}{l} = \frac{32\mu\bar{u}}{D^2}$$

$$18 \times 10^3 = \frac{32 \times 0.75 \times \bar{u}}{(0.05)^2}$$

or $\bar{u} = 1.87$ m/s

Reynolds number: $Re = \dfrac{\rho D \bar{u}}{\mu} = \dfrac{850 \times 0.05 \times 1.87}{0.75} = 105.96$

Reynolds number is less than 2000.

Hence, the flow in pipe is laminar.

(i) **Flow rate:** $Q = A\bar{u} = \dfrac{\pi}{4}D^2\bar{u}$

$$= \frac{3.14}{4} \times (0.05)^2 \times 1.87 = \mathbf{0.003669 \ m^3/s}$$

The maximum velocity occurs at centre line of the pipe and it equals twice the average flow velocity.

$$U_{max} = 2\bar{u} = 2 \times 1.87 = \mathbf{3.74 \ m/s}$$

(ii) **Power required to maintain the flow in 100 m length of pipe: P**

$$P = mgh_f$$

$$= \rho Q g h_f \qquad\qquad \because m = \rho Q$$

where $\qquad h_f = \dfrac{32\mu \bar{u} l}{\rho g D^2} \qquad$ [Hagen–Poiseuille equation]

$$\therefore \qquad P = \dfrac{\rho Q g \times 32 \mu \bar{u} l}{\rho g D^2}$$

$$= \dfrac{32 \mu \bar{u} Q l}{D^2} \qquad\qquad \because l = 100 \ m$$

$$= \dfrac{32 \times 0.75 \times 1.87 \times 0.003669 \times 100}{(0.05)^2} = \mathbf{6586.58 \ W}$$

(*iii*) Velocity at radius r is given by

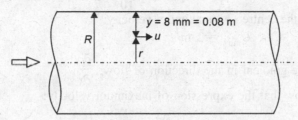

Fig. 1.12: Schematic for Problem 1.6

$$u = -\dfrac{1}{4\mu} \ \dfrac{dp}{dx} \ (R^2 - r^2)$$

Corresponding to 8 mm from the wall,

$$r = R - y$$
$$= 0.025 - 0.008 = 0.017 \ \mathrm{m}$$

and $\qquad \dfrac{p_1 - p_2}{l} = 18 \times 10^3 \ \mathrm{N/m^3}$

or $\qquad \dfrac{p_2 - p_1}{l} = -18 \times 10^3 \ \mathrm{N/m^3}$

or $\qquad \dfrac{dp}{dx} = -18 \times 10^3 \ \mathrm{N/m^3}$

$\therefore \qquad u = -\dfrac{1}{4 \times 0.75} \times (-18 \times 10^3) \times [(0.025)^2 - (0.017)^2]$
$$= 2.01 \ \mathrm{m/s}$$

and Shear stress: $\quad \tau = -\dfrac{dp}{dx}\dfrac{r}{2}$

$$= -(-18 \times 10^3) \times \dfrac{0.017}{2} = \mathbf{153 \ N/m^2}$$

Problem 1.7: An oil of specific gravity 0.9 and viscosity 10 poise is flowing through a pipe of diameter 110 mm. The velocity at the centre is 2 m/s. Find:

(i) Pressure gradient in the direction of flow.

(ii) Shear stress at the pipe wall.

Solution: Given data:

Specific gravity of oil: $S = 0.9$

$\therefore$ Density of oil: $\rho = S \times 1000$ kg/m^3 $= 0.9 \times 1000$ kg/m^3 $= 900$ kg/m^3

Diameter of pipe: $D = 110$ mm $= 0.11$ m

Radius of pipe: $R = \dfrac{D}{2} = \dfrac{0.11}{2} = 0.055$ m

Viscosity of oil: $\mu = 10$ poise $= \dfrac{10}{10}$ Ns/m^2 $= 1$ Ns/m^2

Velocity at the centre $i.e.$, maximum velocity:

$$U_{max} = 2 \text{ m/s}$$

(i) Pressure gradient in the direction of flow: $\dfrac{\partial p}{\partial x}$

We know that the expression of maximum velocity:

$$U_{max} = -\frac{1}{4\mu} \frac{\partial p}{\partial x} R^2$$

$$2 = -\frac{1}{4 \times 1} \times \frac{\partial p}{\partial x} \times (0.055)^2$$

or $\dfrac{\partial p}{\partial x} = \mathbf{-2644.62 \ N/m^2}$

(ii) Shear stress at the pipe wall: $\tau_0 = \tau_{max}$

$$\tau_{max} = -\frac{\partial p}{\partial x} \frac{R}{2} = -(-2644.62) \times \frac{0.055}{2} = \mathbf{72.727 \ N/m^2}$$

Problem 1.8: A laminar flow is taking place in a pipe of diameter 250 mm. The maximum velocity is 2 m/s. Find the mean velocity and the radius at which this occurs. Also find the velocity at 40 mm from the wall of the pipe.

Solution: Given data:

Diameter of pipe: $D = 250$ mm $= 0.250$ m

$\therefore$ Radius of pipe: $R = \dfrac{D}{2} = \dfrac{0.250}{2} = 0.125$ m

Maximum velocity: $U_{max} = 2$ m/s

(i) **Mean velocity:** $\bar{u}$

We have $\bar{u} = \dfrac{U_{max}}{2} = \dfrac{2}{2} = \mathbf{1 \ m/s}$

(ii) **Radius at which $\bar{u}$ occurs:** r

The velocity u at any radius r is given

$$u = -\frac{1}{4\mu} \frac{dp}{dx} [R^2 - r^2]$$

$$u = -\frac{1}{4\mu} \frac{dp}{dx} R^2 \left[1 - \frac{r^2}{R^2} \right]$$

$$u = U_{max} \left[1 - \left(\frac{r}{R} \right)^2 \right]$$

where $\qquad U_{max} = -\frac{1}{4\mu} \frac{dp}{dx} R^2$

Now the radius r at which $u = \bar{u} = 1$ m/s

$$\therefore \qquad 1 = 2 \left[1 - \left(\frac{r}{0.125} \right)^2 \right]$$

or $\qquad 1 - \left(\frac{r}{0.125} \right)^2 = \frac{1}{2}$

or $\qquad \left(\frac{r}{0.125} \right)^2 = 1 - \frac{1}{2} = \frac{1}{2} = 0.5$

or $\qquad \frac{r}{0.125} = \sqrt{0.5}$

$$r = 0.125 \times \sqrt{0.5} = 0.08838 \text{ m} = \textbf{88.38 mm}$$

(*iii*) **Velocity at 40 mm from the wall:** u

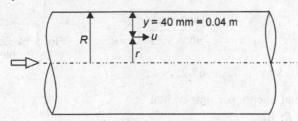

Fig. 1.13: Schematic for Problem 1.8

$$r = R - y = 0.125 - 0.04 = 0.085 \text{ m}$$

The velocity at radius: $\qquad r = 0.085$ m

OR

The velocity at 40 mm from pipe wall is

$$u = U_{Max} \left[1 - \left(\frac{r}{R} \right)^2 \right] = 2 \times \left[1 - \left(\frac{0.085}{0.125} \right)^2 \right]$$

$$= \textbf{1.075 m/s}$$

Problem 1.9: An oil of dynamic viscosity 1.5 poise and specific gravity 0.9 a laminar flows through a 20 mm diameter vertical pipe. Two pressure gauges have been fixed at 20 m apart. The pressure gauge fixed at higher level roads 600 kPa. Find the direction of flow and rate of flow through the pipe.

Solution: Given data:

Dynamic viscosity: $\mu = 1.5$ poise $= \dfrac{1.5}{10}$ Ns/m^2 = 0.15 Ns/m^2

Specific gravity: $S = 0.9$

$\therefore$ Density of oil: $\rho = S \times 1000$ kg/m^3 = $0.9 \times 1000 = 900$ kg/m^3

Diameter of pipe : $D = 20$ mm $= 0.02$ m

Length of pipe between two gauges: $l = 20$ m

Pressure at higher level gauge: $p_B = 200$ kPa $= 200 \times 10^3$ N/m^2

Pressure at lower level gauge: $p_A = 600$ kPa $= 600 \times 10^3$ N/m^2

Total energy per unit weight at lower level A: E_A

$$E_A = \frac{p_A}{\rho g} + \frac{V_A^2}{2g} + z_A$$

Assuming datum at level A,

$\therefore$ $z_A = 0$

Diameter of pipe is same and hence, velocity at A and B will be same.

     i.e., $V_A = V_B = V$

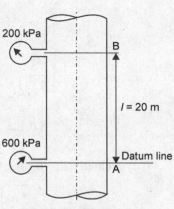

$$E_A = \frac{p_A}{\rho g} + \frac{V^2}{2g}$$

$$= \frac{600 \times 10^3}{900 \times 9.81} + \frac{V^2}{2g}$$

Fig. 1.14: Schematic for Problem 1.9

$$E_A = 67.95 + \frac{V^2}{2g} \qquad \qquad \qquad ...(i)$$

Similarly, total energy per unit weight at higher level B: E_B

$$E_B = \frac{p_B}{\rho g} + \frac{V^2}{2g} + z_B$$

$$= \frac{200 \times 10^3}{900 \times 9.81} + \frac{V^2}{2g} + 20 \qquad \qquad \because z_B = l = 20 \text{ m}$$

$$= 22.65 + \frac{V^2}{2g} + 20 = 42.65 + \frac{V^2}{2g} \qquad ...(ii)$$

It is clear from Eqs. (*i*) and (*ii*), $E_A > E_B$. Hence, the **flow takes place from A to B**.
Note: Fluid always flows from higher energy level to lower energy level.
Rate of flow: Q
Loss of head: $h_f = E_A - E_B$

$$= 67.95 + \frac{V^2}{2g} - 42.65 - \frac{V^2}{2g} = 25.3 \text{ m}$$

Keep in mind, the Hagen-Poiseuille's equation is only applicable for horizontal pipe flow. So, in present case (*i.e.*, vertical pipe), we used Darcy-Weisbach's formula.

$$h_f = \frac{4\,fl\bar{u}^2}{2gD}$$

(Darcy-Weisbach's formula)

where $\qquad f = \dfrac{16}{\mathrm{Re}}$, Darcy's coefficient of friction for laminar flow

$$= \frac{16}{\dfrac{\rho\bar{u}D}{\mu}} = \frac{16\mu}{\rho\bar{u}D}$$

$\therefore \qquad\qquad h_f = \dfrac{4\times16\mu}{\rho\bar{u}D}\times\dfrac{l\bar{u}^2}{2gD}$

$$h_f = \frac{32\mu l\bar{u}}{\rho g D^2}$$

$$25.3 = \frac{32\times0.15\times20\times\bar{u}}{900\times9.81\times(0.02)^2}$$

or $\qquad\qquad \bar{u} = 0.93$ m/s

∴ Rate of flow: $\qquad Q$ = cross-sectional area of pipe × average velocity

$$= \frac{\pi}{4}D^2\times\bar{u} = \frac{3.14}{4}\times(0.02)^2\times0.93$$

$$= 0.000292 \text{ m}^3/\text{s} = 0.000292 \times 1000 \text{ litre/s} = \textbf{0.292 litre/s}$$

1.6 FLOW BETWEEN PARALLEL PLATES

Consider two parallel plates, bottom being stationary while the top one moving with a constant velocity U. Let small fluid element of dimensions be dx (length), dy (thickness) as shown in Fig. 1.15. Let width of the plates be unity not shown in Fig. 1.15.

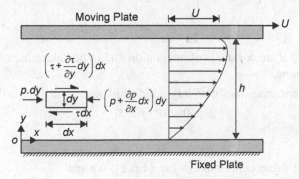

Fig. 1.15: Flow between two pallet plates.

According to the Newton's second law of motion for steady, laminar and incompressible flow is summation of all forces in the direction of flow = $m\,a_x$

$$p.dy - \left(p+\frac{\partial p}{\partial x}\cdot dx\right)dy + \left(\tau+\frac{\partial \tau}{\partial y}\cdot dy\right)dx - \tau\cdot dx = 0 \quad \because a_x = 0$$

$$-\frac{\partial p}{\partial x}dxdy + \frac{\partial \tau}{\partial y}dydx = 0$$

or
$$\frac{\partial p}{\partial x} = \frac{\partial \tau}{\partial y}$$

Here $p = f(x)$ and $\tau = f(y)$ only.

Hence partial derivative can be changed to total derivative.

$\therefore$
$$\frac{dp}{dx} = \frac{d\tau}{dy} \qquad\qquad\qquad\qquad ... (1.6.1)$$

In case of laminar flow, the shear stress is given by the Newton's law of viscosity.

Shear stress: $\tau = \mu\dfrac{du}{dy}$

Substituting the value of τ in Eq. (1.6.1), we get

$$\frac{dp}{dx} = \frac{d}{dy}\left(\mu\frac{du}{dy}\right)$$

or
$$\frac{dp}{dx} = \mu\frac{d^2u}{dy^2} \qquad\qquad\qquad \because \mu = \text{constant}$$

or
$$\frac{d^2u}{dy^2} = \frac{1}{\mu}\frac{dp}{dx}$$

On integrating above equation, we get

$$\frac{du}{dy} = \frac{1}{\mu}\frac{dp}{dx}y + A \qquad \text{At particular section, } \frac{dp}{dx} = c$$

Again integrating, we get

$$u = \frac{1}{\mu}\frac{dp}{dx}\frac{y^2}{2} + Ay + B$$

or
$$u = \frac{1}{2\mu}\frac{dp}{dx}y^2 + Ay + B \qquad\qquad ...(1.6.2)$$

where A and B are constants of integration and to be determined from the known boundary conditions.

In the present case, the boundary conditions are:

$$u = 0 \quad \text{at} \quad y = 0$$

and
$$u = U \quad \text{at} \quad y = h$$

Substituting these conditions in Eq. (1.6.2), we get

$$0 = B \quad \text{at} \quad u = 0, y = 0$$

or
$$\boxed{B = 0}$$

and at
$$u = U \quad \text{at} \quad y = h;$$

$$U = \frac{1}{2\mu}\frac{dp}{dx}h^2 + Ah + B$$

or $$Ah = U - \frac{1}{2\mu} \frac{dp}{dx} h^2 \qquad\qquad \because B = 0$$

or $$\boxed{A = \frac{U}{h} - \frac{1}{2\mu} \frac{dp}{dx} h}$$

Substituting the values of constants A and B in Eq. (1.6.2), we get

$$u = \frac{1}{2\mu} \frac{dp}{dx} y^2 + \frac{y}{h} U - \frac{1}{2\mu} hy \frac{dp}{dx}$$

$$\boxed{u = \frac{y}{h} U - \frac{1}{2\mu} \frac{dp}{dx} (hy - y^2)} \qquad\qquad \text{... (1.6.3)}$$

The flow between two parallel plates, one is fixed and the other is moving is known as **Couette flow.**

Rearranging Eq. (1.6.3), we get

$$u = U \frac{y}{h} - \frac{h^2}{2\mu} \frac{dp}{dx} \frac{y}{h} \left(1 - \frac{y}{h}\right)$$

or $$\frac{u}{U} = \frac{y}{h} - \frac{h^2}{2\mu U} \frac{dp}{dx} \frac{y}{h} \left(1 - \frac{y}{h}\right)$$

$$\frac{u}{U} = \frac{y}{h} + \alpha \frac{y}{h} \left(1 - \frac{y}{h}\right) \qquad\qquad \text{...(1.6.4)}$$

where $$\alpha = \frac{h^2}{2\mu U} \left(-\frac{dp}{dx}\right)$$ is the dimensionless pressure gradient

For $\alpha > 0$, the pressure is decreasing in the direction of flow, the velocity is positive over the whole width between the plates.

For $\alpha > 0$, the pressure is increasing in the direction of flow and the reverse flow begins to occur near the fixed plate as the value α becomes less than -1.

The reverse flow near the fixed plate is due to the dragging action of the faster neighbouring layers on the fluid close to the moving plate is not enough to overcome the influence of the adverse pressure gradient.

when $\qquad\qquad \alpha = 0$, Eq. (1.6.4) becomes

$$\frac{u}{U} = \frac{y}{h}$$

This pure shearing flow is called simple Couette flow.

The velocity distribution of the Couette flow is shown in Fig. 1.16 is a function of the distance from the fixed plate for various dimensionless pressure gradient (α).

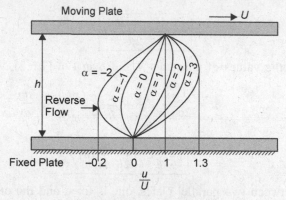

Fig. 1.16: Couette Flow

Average velocity and shear stress distribution.

Consider a small fixed element of cross-section with unit width and height dy; Discharge per unit width through the element:

$$dq = \text{velocity} \times \text{cross-sectional of the fluid element}$$
$$= u \, . \, dy \, .1$$
$$= u \, dy$$

Total discharge per unit width: q

$$q = \int_0^h dq$$

$$= \int_0^h u\,dy = \int_0^h \left[\frac{U}{h} y - \frac{1}{2\mu}\frac{dp}{dx}(hy - y^2) \right] dy$$

$$= \frac{U}{h}\int_0^h y\,dy - \frac{h}{2\mu}\frac{dp}{dx}\int_0^h y\,dy - \frac{1}{2\mu}\frac{dp}{dx}\int_0^h y^2\,dy$$

$$= \frac{U}{h}\left[\frac{y^2}{2} \right]_0^h - \frac{h}{2\mu}\frac{dp}{dx}\left[\frac{y^2}{2} \right]_0^h - \frac{1}{2\mu}\frac{dp}{dx}\left[\frac{y^3}{3} \right]_0^h$$

$$= \frac{U}{h}\frac{h^2}{2} - \frac{h}{2\mu}\frac{dp}{dx}\cdot\frac{h^2}{2} - \frac{1}{2\mu}\frac{dp}{dx}\frac{h^3}{3}$$

$$= \frac{Uh}{2} - \frac{h^3}{4\mu}\frac{dp}{dx} - \frac{h^3}{6}\frac{dp}{dx}$$

$$= \frac{Uh}{2} - \frac{h^3}{12\mu}\frac{dp}{dx}$$

Average velocity: $\quad \bar{u} = \dfrac{\text{Discharge: } q}{\text{Cross-sectional area of fluid flow}}$

$$= \frac{\dfrac{Uh}{2} - \dfrac{h^3}{12\mu}\dfrac{dp}{dx}}{h \times 1}$$

$$\boxed{\bar{u} = \frac{U}{2} - \frac{h^2}{12\mu}\frac{dp}{dx}} \qquad \ldots(1.6.5)$$

The shear stress distribution may be computed by using the Newton's law of viscosity.

$$\tau = \mu \frac{du}{dy}$$

$$= \mu \frac{d}{dy}\left[\frac{U}{h}y - \frac{1}{2\mu}\frac{dp}{dx}(hy - y^2) \right]$$

$$\boxed{\tau = \mu \frac{U}{h} - \frac{1}{2}\frac{dp}{dx}(y - 2y)} \qquad \ldots (1.6.6)$$

Pressure head loss for a given length (*l*):

Recalling average velocity through parallel plates from Eq. (1.6.5)

$$\bar{u} = \frac{U}{2} - \frac{h^2}{12\mu}\frac{dp}{dx}$$

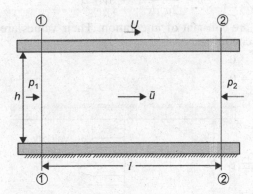

Fig. 1.17: Pressure head loss in Couette flow

or

$$-\frac{h^2}{12\mu}\frac{dp}{dx} = \bar{u} - \frac{U}{2}$$

$$-\frac{dp}{dx} = \frac{12\mu\bar{u}}{h^2} - \frac{6\mu U}{h^2}$$

The term $\left(-\dfrac{dp}{dx}\right)$ represents pressure drop per unit length of pole and may be written as

$$-\frac{\Delta p}{l} = -\frac{(p_2 - p_1)}{l} = \frac{p_1 - p_2}{l}$$

$$\frac{p_1 - p_2}{l} = \frac{12\mu\bar{u}}{h^2} - \frac{6\mu U}{h^2}$$

or

$$p_1 - p_2 = \frac{12\mu\bar{u}l}{h^2} - \frac{6\mu U}{h^2}l$$

Dividing by ρg above in both sides, we get

$$\frac{p_1 - p_2}{\rho g} = \frac{12\mu\bar{u}l}{\rho g h^2} - \frac{6\mu U l}{\rho g h^2}$$

$$h_f = \frac{12\mu\bar{u}l}{\rho g h^2} - \frac{6\mu U l}{\rho g h^2}$$

where $\quad h_f = \dfrac{p_1 - p_2}{\rho g}$, drop of pressure head due to friction.

$$\boxed{h_f = \frac{6\mu l}{\rho g h^2}\left[2\bar{u} - U\right]}$$

1.6.1 Both Plates are Fixed

Recalling the general Eq. (1.6.2) for velocity distribution:

$$u = \frac{1}{2\mu}\frac{dp}{dx}y^2 + Ay + B \qquad ...(1.6.7)$$

where A and B are constant of integration. Their values are obtained from the two below boundary conditions:

(*i*) at $y = 0$, $u = 0$

(*ii*) at $y = h$, $u = 0$

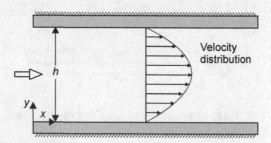

Fig. 1.18: Both Plates are fixed.

Substitution of $y = 0$, $u = 0$ in Eq. (1.6.7), we get

$$0 = 0 + 0 + B$$

or

$$\boxed{B = 0}$$

and substitution of $y = h$, $u = 0$ in Eq. (1.6.7), we get

$$0 = \frac{1}{2\mu}\frac{dp}{dx}h^2 + Ah + 0$$

or

$$A = -\frac{1}{2\mu}\frac{dp}{dx}h$$

Substituting the values of A and B in Eq. (1.6.7), we get

$$u = \frac{1}{2\mu}\frac{dp}{dx}y^2 + y\left(-\frac{1}{2\mu}\frac{dp}{dx}\right)h + 0$$

$$u = \frac{1}{2\mu}\frac{dp}{dx}y^2 - \frac{1}{2\mu}\frac{dp}{dx}h.y$$

$$u = -\frac{1}{2\mu}\frac{dp}{dx}\left[hy - y^2\right]$$ ... (1.6.8)

In the above equation μ, $\frac{dp}{dx}$ and h are constants. It means u varies with the square of y. Here Eq. (1.6.8) is a equation of a parabola. Hence velocity distribution across a section of the two parallel fixed plates is parabolic. This velocity distribution is shown in Fig. 1.18.

Ratio of Maximum Velocity to average velocity:

The velocity of the fluid is maximum at the centre between two fixed plate, putting the value of $y = \frac{h}{2}$ in Eq. (1.6.8), we get

$$U_{max} = -\frac{1}{2\mu}\frac{dp}{dx}\left[\frac{h^2}{2} - \frac{h^2}{4}\right]$$

$$= -\frac{1}{2\mu}\frac{dp}{dx}\cdot\frac{h^2}{4}$$

$$U_{max} = -\frac{1}{8\mu}\frac{dp}{dx}h^2$$ ...(1.6.9)

Average velocity: $\bar{u} = \dfrac{\text{Discharge}}{\text{Cross-sectional area of fluid flow}}$

Consider rate of flow of fluid through the element strip of thickness dy and unit width.

∴ The rate of flow through small element strip:

$$dq = \text{velocity at a distance } y \times \text{cross-sectional area of strip}$$

$$= -\frac{1}{2\mu}\frac{dp}{dx}[hy - y^2] \times dy \times 1$$

Net discharge flow through fixed plates:

$$q = \int_o^h dq = \int_o^h -\frac{1}{2\mu}\frac{dp}{dx}(hy - y^2)dy$$

$$= -\frac{1}{2\mu}\frac{dp}{dx}\left[\frac{hy^2}{2} - \frac{y^3}{3}\right]_0^h$$

$$= -\frac{1}{2\mu}\frac{dp}{dx}\left[\frac{h^3}{2} - \frac{h^3}{3}\right]$$

$$= \frac{1}{2\mu}\frac{dp}{dx}\cdot\frac{h^3}{6} = -\frac{1}{12\mu}\frac{dp}{dx}.h^3$$

$\therefore$ Average velocity: $\bar{u} = \dfrac{q}{A} = \dfrac{-\dfrac{1}{12\mu}\dfrac{dp}{dx}h^3}{h\times 1}$

$$\boxed{\bar{u} = -\frac{1}{12\mu}\frac{dp}{dx}h^2} \qquad\qquad ... (1.6.10)$$

Dividing Eq. (1.6.9) by Eq. (1.6.10), we get

$$\frac{U_{max}}{\bar{u}} = \frac{-\dfrac{1}{8\mu}\dfrac{dp}{dx}h^2}{-\dfrac{1}{12\mu}\dfrac{dp}{dx}.h^2} = \frac{12}{8} = \frac{3}{2} = 1.5$$

$$\boxed{\frac{U_{max}}{\bar{u}} = \frac{3}{2} = 1.5}$$

Maximum velocity: $U_{max} = \dfrac{3}{2}$ times average velocity: $\bar{u}$

i.e., $\boxed{U_{max} = \dfrac{3}{2}\bar{u} = 1.5\,\bar{u}}$

Shear Stress Distribution:

In case of fluid flowing through fixed plates, the shear stress is given by the Newton's law of viscosity.

Shear stress: $\tau = \mu\dfrac{du}{dy}$

Substituting the value of u from Eq. (8.6.8) in above equation, we get

Shear stress: $\tau = \dfrac{d}{dy}\left[-\dfrac{1}{2\mu}\dfrac{dp}{dx}(hy - y^2)\right]$

$$= \mu\times\left[-\frac{1}{2\mu}\frac{dp}{dx}(h - 2y)\right]$$

$$\tau = -\frac{1}{2}\frac{dp}{dx}\left[(h - 2y)\right] \qquad\qquad ...(1.6.11)$$

In above Eq. (1.6.11), $\dfrac{dp}{dx}$ and h are constants.

Hence shear stress τ varies linearly with y. The shear stress distribution is shown in Fig. (1.19). The shear stress is maximum when $y = 0$ and h *i.e.,* at the walls of the fixed plates.

$\therefore$ Maximum shear stress: $\tau_{max} = -\dfrac{1}{2}\dfrac{dp}{dx} \cdot h$

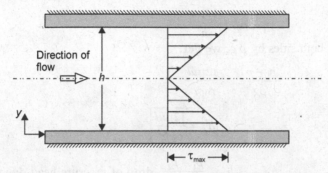

Fig. 1.19: Shear stress distribution across a section of parallel fixed plates.

and shear stress is zero at the centre line between the two *i.e.,* $y = \dfrac{h}{2}$.

Pressure head loss for a given length:

Recalling the average velocity through parallel fixed plates equation (1.6.10)

$$\bar{u} = -\frac{1}{12\mu}\frac{dp}{dx}\,h^2$$

or

$$-\frac{dp}{dx} = \frac{12\mu\bar{u}}{h^2}$$

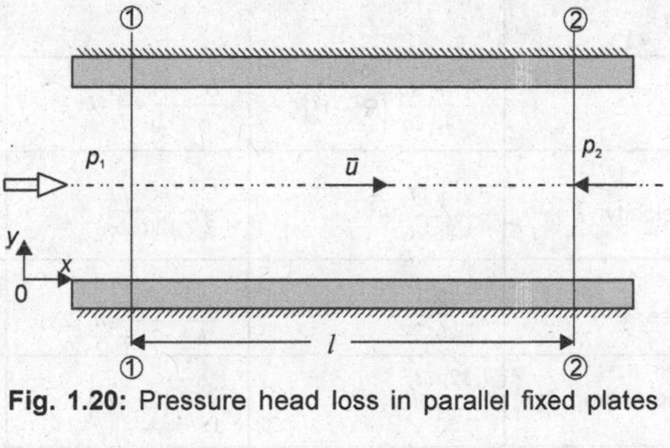

Fig. 1.20: Pressure head loss in parallel fixed plates

The term $\left(-\dfrac{dp}{dx}\right)$ represents pressure drop per unit length of plate and may be written as

$$-\frac{\Delta p}{l} = \frac{-(p_2 - p_1)}{l} = \frac{p_1 - p_2}{l}$$

$$\therefore \qquad \frac{p_1 - p_2}{l} = \frac{12\mu\overline{u}}{h^2}$$

$$p_1 - p_2 = \frac{12\mu\overline{u}l}{h^2}$$

Dividing both sides by ρg, we get

$$\frac{p_1 - p_2}{\rho g} = \frac{12\mu\overline{u}l}{\rho gh^2}$$

$$\boxed{h_f = \frac{12\mu\overline{u}l}{\rho gh^2}}$$

where $\qquad h_f = \dfrac{p_1 - p_2}{\rho g}$, drop of pressure head due to friction.

Table 1.1

S. No.	Variable	Flow through Pipe	Flow through Plates	
			One plate is fixed are and other moving	Both plates are fixed
1.	Velocity: u	$-\dfrac{1}{4\mu}\dfrac{dp}{dx}[R^2 - r^2]$	$\dfrac{Uy}{h} - \dfrac{1}{2\mu}\dfrac{dp}{dx}(hy - y^2)$	$-\dfrac{1}{2\mu}\dfrac{dp}{dx}(hy - y^2)$
2.	Average velocity: $\overline{u}$	$-\dfrac{1}{8\mu}\dfrac{dp}{dx}R^2$	$\dfrac{U}{2} - \dfrac{h^2}{12\mu}\dfrac{dp}{dx}$	$-\dfrac{1}{12\mu}\dfrac{dp}{dx}h^2$
3.	Shear stress: τ	$-\dfrac{dp}{dx}\dfrac{r}{2}$	$\dfrac{\mu U}{h} - \dfrac{1}{2}\dfrac{dp}{dx}(h - 2y)$	$-\dfrac{1}{2}\dfrac{dp}{dx}[h - 2y]$
4.	Pressure head drop: h_f	$\dfrac{32\mu\overline{u}l}{\rho gD^2}$	$\dfrac{6\mu l}{\rho gh^2}[2\overline{u} - U]$	$\dfrac{12\mu\overline{u}l}{\rho gh^2}$

Problem 1.10: An oil of viscosity 0.2 poise flowing between two fixed plates 1 m wide maintained 12 mm apart. The velocity midway between the plates is 2 m/s. Find

(*i*) the pressure gradient along flow,

(*ii*) the average velocity, and

(*iii*) the rate of flow.

Solution: Given data:

Viscosity of oil: $\qquad$ $\mu = 0.2$ poise $= \dfrac{0.2}{10}$ Ns/m^2 = 0.02 Ns/m^2

Width of plates: $\qquad$ $b = 1$ m

Distance between plates: $h = 12$ mm $= 0.012$ m

Velocity midway between the plates: $U_{max} = 2$ m/s

(i) **Pressure gradient:** $\dfrac{dp}{dx}$

As we know, $\qquad$ $U_{max} = -\dfrac{1}{8\mu}\dfrac{dp}{dx}h^2$

$$2 = -\dfrac{1}{8 \times 0.02}\dfrac{dp}{dx} \times (0.012)^2$$

or $\qquad\qquad\qquad \dfrac{dp}{dx} = -\ \textbf{2222.22 N/m}^3$

(ii) **Average velocity:** $\bar{u}$

$$U_{max} = 1.5\,\bar{u}$$

or $\qquad\qquad\qquad \bar{u} = \dfrac{U_{max}}{1.5} = \dfrac{2}{1.5} = \textbf{1.33 m/s}$

(iii) **Rate of flow:** Q

$\qquad\qquad Q$ = cross-sectional area of flow × average velocity

$\qquad\qquad\quad = b.\ h \times u$

$\qquad\qquad\quad = 1 \times 0.012 \times 1.33 = 0.01596$ m^3/s = **15.96 litre/s**

Problem 1.11: An oil of viscosity 18 poise flows between two horizontal fixed parallel plates which are kept at distance 150 mm apart. The maximum velocity of flow is 1.5 m/s. Find

$\quad$ *(i)* the pressure gradient,

$\quad$ *(ii)* the shear stress at the two horizontal parallel plates, and

$\quad$ *(iii)* the discharge per unit width for laminar flow of oil.

Solution: Given data:

Viscosity of oil: $\qquad\qquad\qquad\qquad\qquad \mu = 18$ poise $= \dfrac{18}{10}$ Ns/m^2 = 1.8 Ns/m^2

Distance between two plates: $\qquad\quad h = 150$ mm $= 0.15$ m

Maximum velocity: $\qquad\qquad\qquad U_{max} = 1.5$ m/s

(i) **Pressure gradient:** $\dfrac{dp}{dx}$

We have maximum velocity: $U_{max} = -\dfrac{1}{8\mu}\dfrac{dp}{dx}h^2$

$$1.5 = -\frac{1}{8 \times 1.8} \times \frac{dp}{dx} \times (0.15)^2$$

or $$\frac{dp}{dx} = -960 \text{ N/m}^3$$

(ii) **Shear stress at the wall:** τ_0

Maximum shear stress: $\tau_{max} = \tau_0 = -\frac{1}{2}\frac{dp}{dx}h = -\frac{1}{2} \times (-960) \times (0.15)^2$

$$= 10.8 \text{ N/m}^2$$

(iii) **Discharge per meter width:** Q

Q = average velocity × cross-sectional area of flow

$$= \bar{u} \times h \times b$$

$$= \frac{2}{3}U_{Max} \times h \times 1 \qquad \because b = 1\text{m}$$

$$= \frac{2}{3} \times 1.5 \times 0.15 = \mathbf{0.15 \text{ m}^3/\text{s}}$$

Problem 1.12: An oil of viscosity 8 poise flows between two parallel fixed plates which are kept at a distance of 60 mm apart. Find the rate of flow of oil between the plates if the drop of pressure in a length of 1.2 m be 3 kPa. The width of the plates is 150 mm.

Solution: Given data:

Viscosity: $\mu = 8 \text{ poise} = \frac{8}{10}\text{Ns/m}^2 = 0.8 \text{ Ns/m}^2$

Distance between plates: $h = 60 \text{ mm} = 0.06 \text{ m}$

Drop of pressure: $p_1 - p_2 = 3 \text{ kPa} = 3 \times 10^3 \text{ Pa or N/m}^2$

Length: $l = 1.2 \text{ m}$

Width: $b = 150 \text{ mm} = 0.15 \text{ m}$

We have,

Loss of head: $h_f = \frac{p_1 - p_2}{\rho g}$

also $h_f = \frac{12\mu\bar{u}l}{\rho g h^2}$

∴ $\frac{p_1 - p_2}{\rho g} = \frac{12\mu\bar{u}l}{\rho g h^2}$

$$P_1 - P_2 = \frac{12\mu\bar{u}l}{h^2}$$

$$3 \times 10^3 = \frac{12 \times 0.8 \times \bar{u} \times 1.2}{(0.06)^2}$$

or $\bar{u} = 0.9375 \text{ m/s}$

Rate of flow: Q

$$Q = \text{average velocity} \times \text{cross-sectional area of flow}$$
$$= \bar{u} \times b \times h = 0.9375 \times 0.15 \times 0.06$$
$$= 8.43 \times 10^{-3} \text{ m}^3/\text{s} = \textbf{8.43 liters/s}$$

Problem 1.13: There is a horizontal crack 30 mm wide and 3 mm deep in a wall of thickness 100 mm. Water leaks through the crack. Find the rate of leakage of water through the crack if the difference of pressure between the two ends of the crack is 298 N/m². Take the viscosity of water is 0.01 poise.

Solution: Given data:

Width of crack: $\qquad b = 30 \text{ mm} = 0.03 \text{ m}$

Depth of crack: $\qquad h = 3 \text{ mm} = 0.003 \text{ m}$

Length of crack: $\qquad l = 100 \text{ mm} = 0.1 \text{ m}$

Pressure difference: $\quad p_1 - p_2 = 295 \text{ N/m}^2$

Viscosity: $\qquad\qquad \mu = 0.01 \text{ poise} = \dfrac{0.01}{10} \text{ Ns/m}^2 = 0.001 \text{ Ns/m}^2$

Loss of head: $\qquad h_f = \dfrac{p_1 - p_2}{\rho g} = \dfrac{298}{1000 \times 9.81} = 0.03037 \text{ m}$

also $\qquad\qquad h_f = \dfrac{12 \, \mu \, \bar{u} \, l}{\rho \, g \, h^2}$

$$0.03037 = \dfrac{12 \times 0.001 \times \bar{u} \times 0.1}{1000 \times 9.81 \times (0.003)^2}$$

or $\qquad\qquad \bar{u} = 2.23 \text{ m/s}$

$$\text{Rate of leakage} = \text{average velocity} \times \text{cross-sectional area of crack}$$
$$= \bar{u} \times bh = 2.23 \times 0.03 \times 0.003 \text{ m}^3/\text{s} = 2 \times 10^{-4} \text{ m}^3/\text{s}$$
$$= \textbf{0.20 litre/s}$$

1.7 MOMENTUM CORRECTION FACTOR

Momentum correction factor is defined as the ratio of the momentum of the flow per second based on actual velocity a is cross a section to the momentum of the flow per second based on average velocity across the same section. It is denoted by β.

Mathematically,

Momentum correction factor:

$$\beta = \frac{\text{Momentum per second based on actual velocity}}{\text{Momentum per second based on average velocity}}$$

Momentum correction factor:

$$\beta = \frac{4}{3} \text{ for fully developed laminar flow through pipe.}$$
$$= 1.01 \text{ to } 1.04 \text{ for fully developed turbulent}$$
$$\text{flow through pipe.}$$

Problem 1.14: Show that the momentum factor for laminar flow through a circular pipe is $\dfrac{4}{3}$.

Solution:

We know that the velocity distribution through a circular pipe for laminar flow at any radius r is given by

$$u = -\frac{1}{4\mu}\,\frac{dp}{dx}\,(R^2 - r^2) \qquad\qquad ...\ (i)$$

Consider a small fluid element ring at radius r and width dr

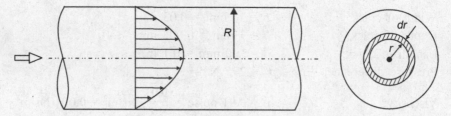

Fig. 1.21: Schematic for Problem 1.14

$\therefore$ Cross-sectional area of small ring: $dA = 2\pi r dr$

$$dQ = u \times 2\pi r dr = 2\pi u r dr$$

Mass flow rate through small ring = $\rho dQ = \rho \times 2\,\pi u r dr = 2\pi\rho u\ r dr$

Momentum of the fluid through a small ring per second:

= mass flow rate × velocity

$= 2\,\pi\rho\ u r\ dr \times u = 2\,\pi\rho\ u^2 r\ dr$

Substituting the value of u from Eq. (i), we get

$$= 2\pi\rho \left[-\frac{1}{4\mu}\frac{dp}{dx}(R^2 - r^2) \right]^2 r dr$$

$$= 2\pi\rho\ \frac{1}{16\mu^2}\left(\frac{dp}{dx}\right)^2 (R^4 + r^4 - 2R^2 r^2)\ r dr$$

Total actual momentum of the fluid per second

$$= \int_0^R 2\pi\rho \times \frac{1}{16\mu^2}\left(\frac{dp}{dx}\right)^2 (R^4 + r^4 - 2R^2 r^2)\ r dr$$

$$= \frac{\pi\rho}{8\mu^2}\left(\frac{dp}{dx}\right)^2 \int_0^R (R^4 r + r^5 - 2R^2 r^3)\ dr$$

$$= \frac{\pi\rho}{8\mu^2}\left(\frac{dp}{dx}\right)^2 \left[\frac{R^4 r^2}{2} + \frac{r^6}{6} - \frac{2R^2 r^4}{4} \right]_0^R$$

$$= \frac{\pi\rho}{8\mu^2}\left(\frac{dp}{dx}\right)^2 \left[\frac{R^4 r^2}{2} + \frac{R^6}{6} - \frac{R^2 r^4}{4} \right]$$

$$= \frac{\pi\rho}{8\mu^2} \left(\frac{dp}{dx}\right)^2 \left[\frac{R^6}{2} + \frac{R^6}{6} - \frac{R^6}{2}\right]$$

$$= \frac{\pi\rho}{8\mu^2} \left(\frac{dp}{dx}\right)^2 \times \frac{R^6}{6}$$

$$= \frac{\pi\rho}{48\mu^2} \left(\frac{dp}{dx}\right)^2 R^6 \qquad\qquad ...(ii)$$

Now we know the expression of average velocity: $\overline{u}$

$$\overline{u} = -\frac{1}{8\mu} \frac{dp}{dx} R^2$$

∴ Momentum of the fluid per second based on average velocity

= mass flow rate × average velocity

$$= \rho \, A \, \overline{u} \times \overline{u} = \rho A \overline{u}^2$$

$$= \rho \times \pi R^2 \times \left[-\frac{1}{8\mu} \frac{dp}{dx} R^2\right]^2$$

$$= \rho \; \pi R^2 \times \frac{1}{16\mu^2} \left(\frac{dp}{dx}\right)^2 R^4$$

$$= \frac{\pi\rho}{64\mu^2} \left(\frac{dp}{dx}\right)^2 R^6 \qquad\qquad ... (iii)$$

∴ Momentum correction factor:

$$\beta = \frac{(\text{Momentum/second})_{act}}{(\text{Momentum/second})_{avg}}$$

$$= \frac{\text{Eq.} (ii)}{\text{Eq.} (iii)}$$

$$= \frac{\dfrac{\pi\rho}{48\mu^2} \left(\dfrac{dp}{dx}\right)^2 R^6}{\dfrac{\pi\rho}{64\mu^2} \left(\dfrac{dp}{dx}\right)^2 R^6} = \frac{64}{48} = \frac{4}{3}$$

1.8 KINETIC ENERGY CORRECTION FACTOR

Kinetic energy correction factor is defined as the ratio of the kinetic energy of flow per second based on actual velocity across a section to the kinetic energy of flow per second based on average velocity across the same section. It is denoted by α

Mathematically,

Kinetic energy correction factor: $\alpha = \dfrac{KE_{act}/\text{second}}{KE_{avg}/\text{second}}$

where
$$KE_{act}/s = \int_0^A \frac{1}{2} \, dm \, u^2$$

$$= \int_0^A \frac{1}{2} \rho \, u \, dA \, u^2 \qquad\qquad \because dm = \rho u \, dA$$

$$= \int_0^A \frac{1}{2} \rho u^3 \, dA = \frac{\rho}{2} \int_0^A u^3 \, dA$$

and
$$KE_{avg/s} = \frac{1}{2} m \bar{u}^2 = \frac{1}{2} \rho \bar{u} A \bar{u}^2 \qquad\qquad \because m = \rho A \bar{u}$$

$$= \frac{\rho}{2} A \bar{u}^3$$

$\therefore$
$$\alpha = \frac{\dfrac{\rho}{2} \displaystyle\int_0^A u^3 \, dA}{\dfrac{\rho}{2} A \bar{u}^3} = \frac{1}{A \bar{u}^3} \int_0^A u^3 \, dA$$

Kinetic energy correction factor:

α = 2 for fully developed laminar flow through pipe.

= 1.04 to 1.11 for fully developed turbulent flow
through pipe.

Problem 1.15: Show that the kinetic energy factor for fully laminar flow through a circular pipe is 2.

Solution: We know that velocity distribution through a circular pipe for laminar flow at any radius r is given by

$$u = -\frac{1}{4\mu} \frac{dp}{dx} (R^2 - r^2) \qquad\qquad ...(i)$$

Consider a small fluid element ring at radius r and width dr.

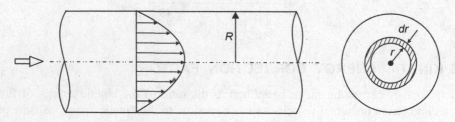

Fig. 1.22: Schematic for Problem 1.15

$\therefore$ Cross-sectional area of small ring: $\qquad dA = 2\pi \, r \, dr$

Rate of fluid flow through small ring: $\qquad dQ = u \times dA$

$$= u \times 2\pi \, r \, dr = 2\pi \, ur \, dr$$

Mass flow rate through small ring $\qquad = \rho dQ = \rho \times 2\pi\ ur\ dr$

$$= 2\pi\ \rho\ ur\ dr$$

Kinetic energy of the fluid flowing through a small ring per second

$$= \frac{1}{2} \times \text{mass flow rate} \times (\text{velocity})^2$$

$$= \frac{1}{2} \times 2\pi\ ur\ dr \times u^2 = \pi\ \rho\ u^3 r\ dr$$

Substituting the value of u from Eq. (i), we get

$$= \pi\ \rho \left[-\frac{1}{4\mu}\frac{dp}{dx}(R^2 - r^2) \right]^3 r dr$$

$$= \pi\ \rho \left(-\frac{1}{4\mu} \right)^3 \left(\frac{dp}{dx} \right)^3 [R^2 - r^2]^3\ r\ dr$$

$$= -\pi\ \rho \times \frac{1}{64\mu^3} \left(\frac{dp}{dx} \right)^3 [R^6 - r^6 + 3R^2 r^2 (R^2 - r^2)]\ r dr$$

$$= -\frac{\pi\rho}{64\mu^3} \left(\frac{dp}{dx} \right)^3 [R^6 - r^6 + 3R^4 r^2 - 3R^2 r^4]\ r dr$$

$$= -\frac{\pi\rho}{64\mu^3} \left(\frac{dp}{dx} \right)^3 [R^6 r - r^7 + 3R^4 r^3 - 3R^2 r^5]\ dr$$

$\therefore$ Total actual kinetic energy of flow per second

$$= \int_0^R -\frac{\pi\rho}{64\mu^3} \left(\frac{dp}{dx} \right)^3 [R^6 r - r^7 + 3\ R^4 r^3 - 3\ R^2 r^5]\ dr$$

$$= -\frac{\pi\rho}{64\mu^3} \left(\frac{dp}{dx} \right)^3 \int_0^R [R^6 r - r^7 + 3\ R^4 r^3 - 3\ R^2 r^5]\ dr$$

$$= -\frac{\pi\rho}{64\mu^3} \left(\frac{dp}{dx} \right)^3 \left[\frac{R^6 r^2}{2} - \frac{r^8}{8} + \frac{3R^4 r^4}{4} - \frac{3R^2 r^6}{6} \right]_0^R$$

$$= -\frac{\pi\rho}{64\mu^3} \left(\frac{dp}{dx} \right)^3 \left[\frac{R^8}{2} - \frac{R^8}{8} + \frac{3R^8}{4} - \frac{R^8}{2} \right]$$

$$= -\frac{\pi\rho}{64\mu^3} \left(\frac{dp}{dx} \right)^3 \left(\frac{R^8}{2} \right)$$

$$= -\frac{\pi\rho}{512\mu^3} \left(\frac{dp}{dx} \right)^3 R^8 \qquad\qquad ...(ii)$$

Now we know the expression of average velocity: $\bar{u}$

$$\overline{u} = \frac{1}{8\mu} \frac{dp}{dx} R^2$$

$\therefore$ Kinetic energy of the fluid flowing through pipe per second based on average velocity

$$= \times \text{ mass flow rate } \times \text{ (average velocity)}^2$$

$$= \frac{1}{2} \times \rho A \overline{u} \times \overline{u}^2 = \frac{1}{2} \rho A \overline{u}^3$$

$$= \frac{1}{2} \rho \pi R^2 \times \left[-\frac{1}{8\mu} \frac{dp}{dx} R^2 \right]^3 \qquad \qquad \because A = \pi R^2$$

$$= \frac{1}{2} \rho \pi R^2 \left(-\frac{1}{8\mu} \right)^3 \left(\frac{dp}{dx} \right)^3 R^6$$

$$= -\frac{\rho \pi}{1024 \overline{\mu}^3} \left(\frac{dp}{dx} \right)^3 R^8 \qquad \qquad \ldots (iii)$$

$\therefore$ Kinetic energy correction factor: α

$$\alpha = \frac{KE_{act} / \text{second}}{KE_{avg} / \text{second}}$$

$$= \frac{\text{Eq.}(ii)}{\text{Eq.}(iii)}$$

$$= \frac{-\dfrac{\pi \rho}{512 \mu^3} \left(\dfrac{dp}{dx} \right)^3 R^8}{-\dfrac{\pi \rho}{1024 \mu^3} \left(\dfrac{dp}{dx} \right)^3 R^8} = \frac{1024}{512} = \mathbf{2}$$

1.9 POWER ABSORBED IN VISCOUS RESISTANCE

Oil used as lubricant in the bearings is example of viscous flow. In order to minimize the frictional loss a thin layer of oil is maintained between the fixed and rotating parts of the bearings. Viscosity is a very important property of the lubricating oil since its load-carrying capacity at higher pressure and temperature is proportional to the viscosity. If the viscosity of the oil is too low, a liquid film cannot be maintained between the rotating and fixed parts. On the other hand, if the viscosity is too high, it will offer more resistances to the rotating parts. So, it is very essential to use correct viscosity of oil for lubrication. The power required to overcome the viscous resistance in the following three types of bearing:

    (i) Journal bearing,

   (ii) Foot-step bearing, and

   (iii) Collar bearing.

1.9.1 Journal Bearing

Let d = diameter of shaft

t = thickness of oil film

l = length of the bearing

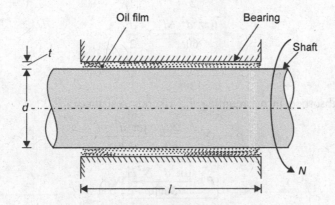

Fig. 1.23: Journal Bearing.

N = speed of shaft in rpm

ω = angular velocity of the shaft in rad/s

μ = viscosity of the oil film

We know that the angular velocity of the shaft: ω

$$\omega = \frac{2\pi N}{60}$$

$\therefore$ Tangential velocity of the shaft: $u = \omega R = \omega \dfrac{d}{2}$

$$= \frac{2\pi N}{60} \frac{d}{2} = \frac{\pi dN}{60}$$

According to Newton's law of viscosity,

Shear stress: $\tau = \mu \dfrac{du}{dy}$

where $\dfrac{du}{dy}$ = velocity gradient

$$= \frac{u-0}{t} = \frac{u}{t}$$

$\therefore$ Shear stress: $\tau = \mu \dfrac{u}{t} = \dfrac{\mu}{t} \dfrac{\pi dN}{60} = \dfrac{\mu \pi dN}{60t}$

Shear force: F = shear stress × surface of the shaft contact with oil (*i.e.*, wetted area of the shaft)

$$= \frac{\mu \pi dN}{60t} \times \pi dl = \frac{\mu \pi^2 \ d^2 \ Nl}{60t}$$

Torque required to overcome the viscous resistance (*i.e.*, shear force):

$$T = \text{shear force} \times \frac{d}{2}$$

$$= F \times \frac{d}{2}$$

$$= \frac{\mu\pi^2 d^2 Nl}{60t} \times \frac{d}{2}$$

$$= \frac{\mu\pi^2 d^3 Nl}{120t}$$

Power absorbed in overcoming the viscous resistance: P

$$P = \omega T = \frac{2\pi N}{60} \times \frac{\mu\pi^2 d^3 Nl}{120t}$$

$$\boxed{P = \frac{\mu\pi^3 d^3 N^2 l}{3600t}} \ \mathbf{W}$$

P in W

when m in Ns/m^2

N in rpm and d, t, l in m.

1.9.2 Foot-Step Bearing

In this types of bearing, the shaft is vertical rotating. An oil film is maintained between the bottom surface of the shaft and bearing as shown in Fig. 1.24.

Let N = speed of the shaft in rpm

  t = thickness of oil film

  R = radius of the shaft

Let a small element ring of radius r and thickness dr.

Area of small element ring: $dA = 2\pi r dr$

Now stear stress on the ring:

$$d\tau = \mu\frac{du}{dy} = \mu\frac{u}{t}$$

where u is the tangential velocity of the shaft at radius r:

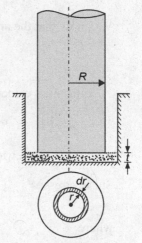

Fig. 1.24: Foot-step bearing.

$$\therefore \qquad u = \omega r = \frac{2\pi N}{60} r$$

The shear force on the ring: $dF = d\tau \times$ area of the ring

$$= d\tau \times 2\pi r \, dr = \mu\frac{u}{t} \times 2\pi r dr$$

$$= \frac{u}{t} \times \frac{2\pi Nr}{60} \times 2\pi r \, dr = \frac{\mu}{15t}\pi^2 N r^2 dr$$

$\therefore$ Torque required to overcome the viscous resistance:

$$dT = dF \times r = \frac{\mu}{15t} \pi^2 N r^2 dr$$

Total torque required to overcome the viscous resistance:

$$\int_0^T dT = \int_0^R \frac{\mu}{15t} \pi^2 N r^3 dr$$

$$T = \frac{\mu}{15t} \pi^2 N \int_0^R r^3 dr = \frac{\mu}{15t} p^2 N \left[\frac{r^4}{4} \right]_0^R$$

$$T = \frac{\mu}{15t} \pi^2 N \frac{R^4}{4} = \frac{\mu}{60t} \pi^2 N R^4$$

$\therefore$ Power absorbed in overcoming the viscous resistance: P

$$P = \frac{2\pi NT}{60} \; W$$

Here P in W

when T in Nm

and N in rpm

Substituting the value of T, we get

$$P = \frac{2\pi N}{60} \times \frac{\mu}{60t} \pi^2 NR^4$$

$$\boxed{P = \frac{\mu \pi^3 N^2 R^4}{1800t} \; W}$$

Here P in W

when μ in Ns/m^2

R and t in m

N in rpm

1.9.3 Collar Bearing

The collar bearing is shown in Fig. 1.25 where an oil film of uniform thickness separates the face of the collar from the surface of the bearing.

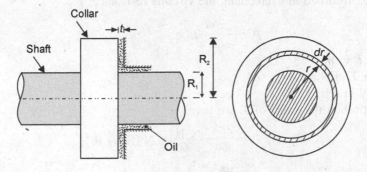

Fig. 1.25: Collar bearing.

The collar bearing is employed to take up the axial thrust of a rotating shaft.

Let $\qquad N$ = speed of the shaft in rpm

$\qquad R_1$ = radius of the shaft

$\qquad R_2$ = radius of the collar

$\qquad t$ = thickness of oil film

Consider an elementary circular ring of radius r and width dr of the bearing surface.

$\therefore$ Area of elementary ring: $dA = 2\pi\, r\, dr$

Now shear stress on the ring: $d\tau = \mu \dfrac{du}{dy} = \mu \dfrac{u}{t}$

where u is the tangential velocity of the shaft at radius r:

$$\therefore \qquad\qquad u = \omega r = \frac{2\pi N}{60} \times r = \frac{\pi\, Nr}{30}$$

$$\therefore \qquad\qquad d\tau = \frac{\mu}{t} \times \frac{\pi Nr}{30} = \frac{\pi\mu\, Nr}{30t}$$

The shear force on the ring: $dF = d\tau \times 2\pi r\, dr$

$$dF = \frac{\pi\mu\, Nr}{30t} \times 2\pi r\, dr = \frac{\mu\pi^2\, Nr^2 dr}{15t}$$

Torque required to overcome the viscous resistance: $dT = dF \times r$

$$= \frac{\mu\pi^2\, Nr^3 dr}{15t}$$

Total torque required to overcome the viscous resistance:

$$\int_0^T dT = \int_{R_1}^{R_2} \frac{\mu}{15t}\, \pi^2 N r^3\, dr$$

$$T = \frac{\mu}{15t}\, \pi^2 N \int_{R_1}^{R_2} r^3\, dr = \frac{\mu}{15t}\, \pi^2 N \left[\frac{r^4}{4} \right]_{R_1}^{R_2}$$

$$= \frac{\mu}{15t}\, \pi^2 N \left[\frac{R_2^4 - R_1^4}{4} \right] = \frac{\mu}{60t}\, \pi^2 N \left[R_2^4 - R_1^4 \right]$$

Power absorbed in overcoming the viscous resistance: P

$$P = \frac{2\pi NT}{60}\, W$$

Here P in W

when T in Nm

and N in rpm

Substituting the value of T, we get

$$P = \frac{2\pi N}{60} \times \frac{\mu}{60t}\, \pi^2 N\, [R_2^4 - R_1^4]$$

$$P = \frac{\mu\pi^3\, N^2}{1800t}\, [R_2^4 - R_1^4]\; W$$

Here P in W
when μ in Ns/m^2
N in rpm
R_1, R_2 and t in m.

Problem 1.16: A shaft of 100 mm diameter rotates in journal bearing at 60 rpm in a 200 mm long bearing. Taking that the two surfaces are uniformly separated by a distance of 0.5 mm and taking linear velocity distribution in the lubricating oil having dynamic viscosity of 0.004 N s/m², find the power absorbed in the bearing.

Solution: Given data:

Diameter of shaft: $d = 100$ mm
Speed: $N = 60$ rpm
Length of bearing: $l = 200$ mm $= 0.2$ m
Thickness of the oil film: $t = 0.5$ mm $= 0.5 \times 10^{-3}$ m
Viscosity: $\mu = 0.004$ Ns/m²

We know that the power absorbed in the journal bearing:

$$P = \frac{\mu \pi^3 d^3 N^2 l}{3600 t} \text{ W}$$

$$= \frac{0.004 \times (3.14)^3 \times (0.1)^3 \times (60)^2 \times 0.2}{3600 \times 0.5 \times 10^{-3}} \text{ W} = \mathbf{0.0495\ W}$$

Problem 1.17: A journal bearing has a length of 300 mm and supports a shaft of 150 mm diameter rotating at 120 rpm. If the internal diameter of the bearing is 150.25 mm, find the torque required to keep the shaft in rotation at the speed mentioned above. Find also the power required. The viscosity of the oil film is 0.25 Ns/m².

Solution: Given data:

Length of bearing: $l = 300$ mm $= 0.3$ m
Diameter of shaft: $d = 150$ mm $= 0.15$ m
Diameter of bearing: $D = 150.25$ mm $= 0.15025$ m
Speed of shaft: $N = 120$ rpm

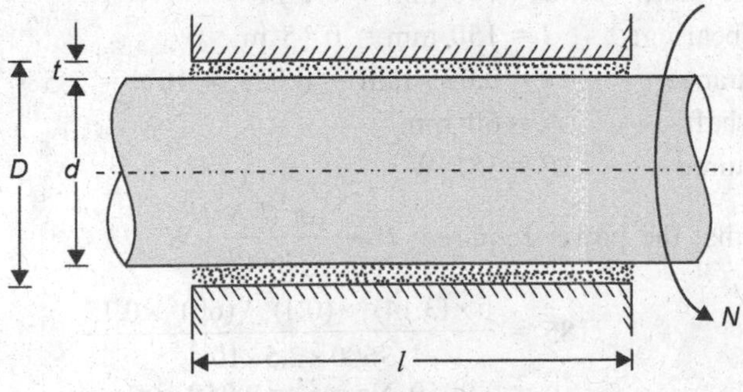

Fig. 1.26: Schematic for Problem 1.17

Thickness of oil film:

$$t = \frac{D-d}{2} = \frac{0.15025 - 0.15}{2} = 1.25 \times 10^{-4} \text{ m}$$

Viscosity of oil film:

$$\mu = 0.25 \text{ Ns/m}^2$$

Tangential viscosity of shaft:

$$u = \frac{\pi d N}{60} = \frac{3.14 \times 0.15 \times 120}{60} = 0.942 \text{ m/s}$$

$\therefore$ Shear stress:
$$\tau = \mu \frac{du}{dy} = \mu \frac{u}{t} = \frac{0.25 \times 0.942}{1.25 \times 10^{-4}} = 1884 \text{ N/m}^2$$

Shear force:
$$F = \tau \times \pi d l$$
$$= 1884 \times 3.14 \times 0.15 \times 0.3 = 266.20 \text{ N}$$

Torque required:
$$T = F \frac{d}{2} = 266.20 \times \frac{0.15}{2} = 19.965 \text{ Nm}$$

Power required:
$$P = \frac{2\pi N T}{60} = \frac{2 \times 3.14 \times 120 \times 19.965}{60} W = \textbf{250.76 W}$$

OR

Torque required:
$$T = \frac{\mu \pi^2 d^3 N l}{120 t}$$
$$= \frac{0.25 \times (3.14)^2 \times (0.15)^3 \times 120 \times 0.3}{120 \times 1.25 \times 10^{-4}} = 19.965 \text{ Nm}$$

Power required:
$$P = \frac{\mu \pi^3 d^3 N^2 l}{3600 t} W = \frac{0.25 \times (3.14)^2 \times (0.15)^3 \times 120 \times 0.3}{3600 \times 1.25 \times 10^{-4}}$$
$$= \textbf{250.76 W}$$

Problem 1.18: A shaft 100 mm in diameter runs in a bearing of length 150 mm with a radial clearance of 0.025 mm at 60 *rpm*. Find viscosity of the oil, if the power required to overcome the viscous resistance is 185 W.

Solution: Given data:

Diameter of shaft: $d = 100$ mm $= 0.1$ m
Length of bearing: $l = 150$ mm $= 0.15$ m
Radial clearance: $t = 0.025$ mm $= 0.025 \times 10^{-3} = 2.5 \times 10^{-5}$ m
Speed of shaft: $N = 60$ rpm
Power required: $P = 185$ W

We know that the power required: $P = \dfrac{\mu \pi^3 d^3 N^2 l}{3600 t} W$

$$185 = \frac{\mu \times (3.14)^3 \times (0.1)^3 \times (60)^2 \times 0.15}{3600 \times 2.5 \times 10^{-5}}$$

or $\mu = 0.9959$ Ns/m^2 = **9.959 poise**

Problem 1.19: A vertical shaft of diameter 100 mm rotates at 650 rpm. The lower end of the shaft rests on a foot-step bearing. The end of the shaft and surface of the bearing are both flat and are separated by an oil film of thickness of 0.4 mm. The viscosity of the oil is given 1.5 poise. Find the torque and power required.

Solution: Given data:

Diameter of shaft: $d = 100$ mm $= 0.1$ m

Radius of shaft: $R = \dfrac{d}{2} = \dfrac{0.1}{2} = 0.05$ m

Speed of shaft: $N = 650$ rpm

Thickness of oil film: $t = 0.4$ mm $= 0.4 \times 10^{-3}$ m

Viscosity of oil: $\mu = 1.5$ poise $= \dfrac{1.5}{10}$ Ns/m^2 $= 0.15$ Ns/m^2

We know that the torque required for a food–step bearing: $T = \dfrac{\mu}{60t} \pi^2 \, N \, R^2$

$$= \frac{0.15 \times (3.14)^2 \times 650 \times (0.05)^2}{60 \times 0.4 \times 10^{-3}} = \mathbf{100.13 \ Nm}$$

Power required: $P = \dfrac{2\pi NT}{60} = \dfrac{2 \times 3.14^2 \times 650 \times 100.13}{60} = 6812.17$ W

$$= \mathbf{6.81 \ kW}$$

Problem 1.20: The external and internal diameters of a collar bearing are 200 mm and 150 mm respectively. An oil film 0.30 mm thick is maintained between the collar surface and the bearing. Find the power lost in overcoming the viscous resistance of the oil when the shaft is running at 240 rpm. Take $\mu = 0.10$ Ns/m^2.

Solution: Given data:

External diameter of collar: $D = 200$ mm $= 0.2$ m

$\therefore$ External radius of collar: $R_2 = \dfrac{D}{2} = \dfrac{0.2}{2} = 0.1$ m

Internal diameter of collar: $d = 150$ mm $= 0.15$ m

$\therefore$ Internal radius of collar: $R_1 = \dfrac{d}{2} = \dfrac{0.15}{2} = 0.075$ m

Thickness of oil: $t = 0.30$ mm $= 0.3 \times 10^{-3}$ m

Speed of shaft: $N = 240$ rpm

Viscosity: $\mu = 0.10$ Ns/m^2

Power lost in overcoming the viscous resistance of oil in collar bearing: P

$$P = \frac{\mu \pi^3 N^2}{1800t} [R_2^4 - R_1^4]$$

$$= \frac{0.10 \times (3.14)^3 \times (240)^2}{1800 \times 0.3 \times 10^{-3}} [(0.1)^4 - (0.075)^4]$$

$$= \mathbf{22.57 \ W}$$

1.10 DASH-POT MECHANISM: MOVEMENT OF A PISTON IN A DASH-POT

Dash-pot mechanism is used to damp the mechanical vibrations of the machine element that communicates with the piston. It consists of a piston that moves within a cylinder, the diameter of the cylinder is slightly greater than the diameter of the piston. The cylinder is filled with a highly viscous fluid. When piston moves downwards under the influence of load W, oil moves upwards through the clearance between cylinder and piston. When piston moves upwards, the oil moves downwards through the clearance. The flow of oil offers viscous resistance to the movement of piston. This helps to absorb the mechanical vibrations of the machine element that communicates with the piston.

Let

D = diameter of the piston
l = length of the piston
W = load exerted by the machine member
μ = viscosity of oil
V = velocity of the piston
$\bar{u}$ = average velocity of oil flow in the clearance
b = clearance between the piston and cylinder
dp = pressure difference across the two ends of the piston.

$$dp = \frac{W}{\frac{\pi}{4}D^2} = \frac{4W}{\pi D^2} \qquad \qquad ...(i)$$

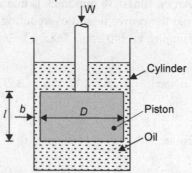

Fig. 1.27: Dash-Pot Mechanism.

The flow of oil through clearance is a similar case to the laminar flow between two parallel plates. The pressure difference for parallel plate for length l is given by

$$dp = \frac{12\mu\bar{u}l}{b^2} \qquad \qquad ...(ii)$$

Equating Eqs. (i) and (ii), we get

$$\frac{4W}{\pi D^2} = \frac{12\mu\bar{u}l}{b^2}$$

or

$$\mu = \frac{Wb^2}{3\pi\,\mu l\,D^2} \qquad \qquad ...(iii)$$

Rate of flow of oil in dash-pot

= velocity of the piston × cross-sectional area of dash pot.

$$= V \times \frac{\pi}{4} D^2$$

$$= \frac{\pi}{4} D^2 V \qquad \qquad \qquad ...(iv)$$

Rate of flow of oil through clearance = average velocity of oil through clearance × cross-sectional area of flow

$$= \bar{u} \times \pi D b \qquad \qquad \qquad ...(v)$$

According to continuity equation, rate of flow through clearance must be equal to rate of flow through dash-pot

Equating Eqs. (v) and (iv), we get

$$\bar{u} \, \pi D b = \frac{\pi}{4} D^2 V$$

$$\bar{u} = \frac{DV}{4b} \qquad \qquad \qquad ...(vi)$$

Equating Eqs. (iii) and (vi), we get

$$\frac{Wb^2}{3\pi\mu l D^2} = \frac{DV}{4b}$$

or

$$\boxed{\mu = \frac{4Wb^3}{3\pi l D^3 V}}$$

Velocity of the piston: $V = \dfrac{4Wb^3}{3\pi l D^3 \mu}$

Problem 1.21: An oil dash pot consists of a piston moving in a cylinder having oil. This arrangement is used to damp out the vibrations. The pistons falls with uniform speed and covers 50 mm in 100 seconds. If an additional weight of 1.35 N is placed on the top of the piston, it falls through 5 mm in 86 seconds with uniform speed. The diameter of the piston is 75 mm and its length is 100 mm. The clearance between the piston and the cylinder is 1.20 mm which is uniform throughout. Find the viscosity of oil.

Solution: Given data:

Distance moved by piston due to own weight = 50 mm = 0.05 m

Time taken = 100 s

Additional weight: $w = 1.35$ N

Time taken to cover distance of 50 mm due to additional weight (w),

= 86 s

Diameter of the piston: D = 75 mm = 0.075 m

Length of the piston: l = 100 mm = 0.1 m

Clearance: b = 1.20 mm = 0.0012 m

Velocity of piston without additional weight:

$$V = \frac{\text{Distance moved}}{\text{Time taken}} = \frac{0.05}{100} = 5 \times 10^{-4} \text{ m/s}$$

Velocity of piston with additional weight:

$$V' = \frac{\text{Distance moved}}{\text{Time taken}} = \frac{0.05}{86} = 5.81 \times 10^{-4} \text{ m/s}$$

We know the expression of viscosity in dash-pot:

$$\mu = \frac{4Wb^3}{3\pi l D^3 V} \quad \text{for own weight of piston } W$$

$$= \frac{4[W+w]b^3}{3\pi l D^3 V'} \quad \text{for additional weight } (W + W)$$

$$\frac{4Wb^3}{3\pi l D^3 V} = \frac{4[W+w]b^3}{3\pi l D^3 V'}$$

$$\frac{W}{V} = \frac{W+w}{V'}$$

$$\frac{W}{5\times10^{-4}} = \frac{W+1.35}{5.81\times10^{-4}}$$

$$\frac{W}{5} = \frac{W+1.35}{5.81}$$

$$5.81\, W = 5W + 1.35 \times 5$$

$$5.81\, W - 5\,W = 6.75$$

$$0.81\, W = 6.75$$

or

$$W = 8.33\ N$$

We know that the viscosity of oil: $\mu = \dfrac{4Wb^3}{3\pi l D^3 V} = \dfrac{4\times8.33\times(0.0012)^3}{3\times3.14\times0.1\times(0.075)^3\times5\times10^{-4}}$

$$= 0.289 \text{ Ns/m}^2 = \textbf{2.89 poise}$$

1.11 STOKES' LAW

If a sphere is placed in a flow of a highly viscous fluid at low velocity, the drag force acting on a sphere and drag coefficient are given by

Drag force:
$$F_D = \bar{C}_f A \frac{\rho U^2}{2}$$

where U = velocity of free stream.

ρ = density of fluid

If fluid is stationary, sphere is moving with constant velocity V. The motion of the sphere is resisted by the drag force, which acts in the direction opposite to motion. As the velocity of the body increases, so does the drag force. This continues until all the forces balance each other and the net force acting on the body is zero.

Drag force acts on sphere: $F_D = \bar{C}_f A \dfrac{\rho V^2}{2}$

The drag coefficient $(\bar{C}_f)$ exhibits different behaviour in the low (creeping), moderate (laminar), and high (turbulent) regions of the Reynolds number. The inertia effects are negligible in low Reynolds number flows $(Re < 1)$, called creeping flows, and the fluid crops around the body smoothly. The drag coefficient in this case is inversely proportional to the Reynolds number, and for sphere it is determined to be

$$\bar{C}_f = \frac{24}{Re} \qquad \text{for sphere when } Re \leq 1$$

Then the drag force acting on spherical object at low Reynolds number becomes

$$F_D = \frac{24A}{Re}\frac{\rho V^2}{2} \qquad \left| \because A = \frac{\pi}{4}d^2, \, Re = \frac{\rho V d}{\mu} \right.$$

$$= \frac{24}{\dfrac{\rho V d}{\mu}} \times \frac{\pi}{4}d^2 \times \frac{\rho V^2}{2}$$

$$F_D = 3\pi\mu V d \qquad\qquad \text{...(1.11.1)}$$

Equation (8.11.1) is known as **Stokes' law**. This relation shows that for a very low Reynolds number, the drag force acting on spherical objects is proportional to the diameter (d), the velocity (V), and the viscosity of the fluid. This relation is often applicable to dust particles in the air and suspended solid particles in water.

1.12 MEASUREMENT OF VISCOSITY

The device used to determine the viscosity of fluid is called viscometer or viscosimeter. The devices which are used to measure the viscosity are based on the principle of existence of fully established laminar flow. Some of the common devices for the measurement of viscosity may be classified as:

(1) Capillary tube viscometer.

(2) Rotating cylinder viscometer.

(3) Falling sphere viscometer.

(4) Industrial viscometers.

1.12.1 Capillary Tube Viscometer

A capillary tube viscometer consists of a horizontal capillary tube through which the given fluid is made to flow. A large tank in which the liquid whose viscosity is to be determined is filled, and one end of a capillary tube is connected horizontally to the tank and other end open to atmosphere. The pressure in the capillary at a distance 'l' upstream the open end is recorded by means of a vertical piezometer. It is necessary

to leave some distance the flow at the entry to the capillary tube is of developing
nature, the flow of liquid must be laminar through considering length l of capillary tube
as shown in Fig. 1.28.

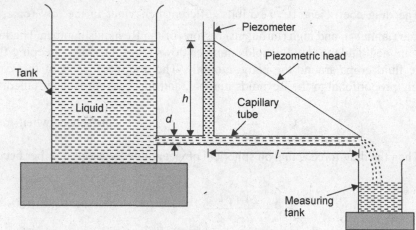

Fig. 1.28: Capillary tube viscometer.

The level of the liquid in tank is maintained constant so as to ensure steady flow
through the tube. The rate of discharge through the tube is estimated by the quantity
of fluid collected in measuring tank for a given interval of time using a stop watch and
head loss h over length l of the capillary tube determined by using piezometer.

Using Hagen–Poiseuille equation

Head loss: $h_f = \dfrac{32\,\mu \bar{u} l}{\rho g d^2}$

where $h_f = h$, pressure head loss over the fully developed laminar
                        flow through length l of the tube

                        $\bar{u}$ = average velocity of flow.

                        $Q = \dfrac{\pi}{4} d^2 \bar{u}$

or $\bar{u} = \dfrac{4Q}{\pi d^2}$

$\therefore$ $h = \dfrac{128\,\mu Q l}{\pi \rho g d^4}$

or Viscosity of liquid: $\boldsymbol{\mu = \dfrac{\pi \rho g h d^4}{128\, l Q}}$

Here viscosity: μ in Ns/m^2
when ρ in kg/m^3
$g = 9.81$ m/s^2
h, d, l in m
Q in m^3/s

Demerits of using of this viscometer

It is difficult to install piezometer in a fine capillary tube capable of providing laminar flow. Further, it may be difficult to measure accurately diameter of such a tube. A small error in measuring the diameter will cause an appreciable error in the viscosity because viscosity μ is proportional to fourth power of diameter, *i.e.*, $\mu \propto d^4$

1.12.2 Rotating Cylinder Viscometer

This viscometer consists of two concentric cylinders of radii R_1 and R_2 as shown in Fig. 1.29, where R_1 and R_2 are the radii of inner stationary cylinder and outer rotating cylinder respectively. The space between two cylinders is filled with liquid of which the viscosity is required to be determined. The torque applied on the outer cylinder is transferred to the inner cylinder through the shear resistance or viscosity of the liquid which is absorbed by torsional spring and measured on the dial means of the pointer as shown in Fig. 1.29.

Let the outer cylinder is rotated at a constant angular speed of ω.

Peripheral (*i.e.*, tangential) speed of the outer cylinder: $u_2 = \omega R_2$

Peripheral velocity of liquid layer in contact with outer cylinder will be equal to the peripheral velocity of outer cylinder.

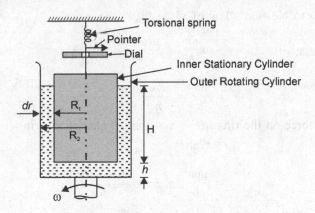

Fig. 1.29: Rotating cylinder viscometer.

∴ Velocity of liquid layer with outer cylinder: u_2

$$u_2 = \omega R_2$$

and velocity of liquid layer with inner cylinder: $u_1 = 0$ because inner cylinder is stationary.

∴ Velocity gradient over the radial distance $(R_2 - R_1)$: $\dfrac{du}{dr}$

$$\frac{du}{dr} = \frac{u_2 - u_1}{R_2 - R_1} = \frac{\omega R_2 - 0}{R_2 - R_1}$$

or
$$= \frac{\omega R_2}{R_2 - R_1}$$

According to Newton's law of viscosity,

Shear stress: $\quad\quad\quad\quad\quad \tau = \mu\dfrac{du}{dr} = \dfrac{\mu\omega R_2}{R_2 - R_1}$

$\therefore$ Shear force: $\quad\quad\quad F =$ shear stress $\times$ wetted area of inner cylinder

$$= \tau \times 2\pi R_1 H$$

$$= \dfrac{\mu\omega R_2}{R_2 - R_1} \times 2\pi R_1 H$$

Torque acting on the inner cylinder due to shearing action on the fluid: T_1

$$T_1 = \text{shear force} \times \text{radius } (R_1)$$

$$= \dfrac{\mu\omega R_2}{R_2 - R_1} v 2\pi R_1 H \ R_1 = \dfrac{2\pi\mu\omega R_2 R_1^2 H}{R_2 - R_1}$$

The torque transferred to the inner cylinder through its bottom surface, by the bottom surface of the outer cylinder, will be determined as follows:

Consider an elementary ring of the bottom of the outer cylinder of radius r and thickness dr circumferential velocity at radius, $r : u$

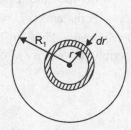

$$u = \omega r$$

According to Newton's law of viscosity,

Shear stress: $\quad\quad\quad\quad\quad \tau = \mu\dfrac{du}{dy} = \dfrac{\mu(u-0)}{h}$

$$= \dfrac{\mu u}{h} = \dfrac{\mu\omega r}{h} \quad\quad\quad\quad\quad\quad [\because u = \omega r]$$

$\therefore$ Shear force on the ring: $dF = \tau \times$ area of elementary ring

$$= \tau 2\pi r dr$$

$$= \dfrac{\mu\omega r}{h} 2\pi r dr = \dfrac{2\pi\mu\omega r^2 dr}{h}$$

$\therefore$ Torque acting on the elementary ring: $dT_2 = dF \times r$

$$= \dfrac{2\pi\mu\omega r^2 dr}{h} \times r = \dfrac{2\pi\mu\omega r^3 dr}{h}$$

$\therefore$ Total torque: $\quad \displaystyle\int dT_2 = \int_0^{R_1} \dfrac{2\pi\mu\omega r^3 dr}{h}$

$$T_2 = \dfrac{2\pi\mu\omega}{h} \int_0^{R_1} r^3 dr$$

$$T_2 = \dfrac{2\pi\mu\omega}{h} \left[\dfrac{r^4}{4}\right]_0^{R_1} = \dfrac{2\pi\mu\omega}{h} \times \dfrac{R_1^4}{4}$$

$$T_2 = \frac{2\pi\mu\omega R_1^4}{2h}$$

Total torque acting on the inner cylinder: T

$$T = T_1 + T_2$$

$$= \frac{2\pi\mu\omega R_2 R_1^2 H}{(R_2 - R_1)} + \frac{\pi\mu\omega R_1^4}{2h}$$

$$= \frac{(2\pi\mu\omega R_2 R_1^2 H) \times 2h + \pi\mu\omega R_1^4 (R_2 - R_1)}{2h(R_2 - R_1)}$$

$$= \frac{4\pi\mu\omega R_2 R_1^2 Hh + \pi\mu\omega R_1^4 (R_2 - R_1)}{2h(R_2 - R_1)}$$

$$T = \frac{\pi\omega R_1^2 \mu [4HhR_2 + R_1^2 (R_2 - R_1)]}{2h(R_2 - R_1)}$$

or   Dynamic viscosity of the liquid: μ

$$\mu = \frac{2h(R_2 - R_1)T}{\pi\omega R_1^2 [4HhR_2 + R_1^2 (R_2 - R_1)]}$$

where T = torque measured on the dial by means of torsion spring. Dynamic viscosity μ is determined when other parameter

R_1 and R_2 = radii of inner and outer cylinder

h = clearance at the bottom of cylinders

H = height of the stationary cylinder submerged in liquid

ω = angular velocity of the other cylinder; are known.

1.12.3 Falling Sphere Viscometer

This viscometer consists of a long vertical transparent cylindrical tank, which is filled with the liquid where viscosity is to be measured.

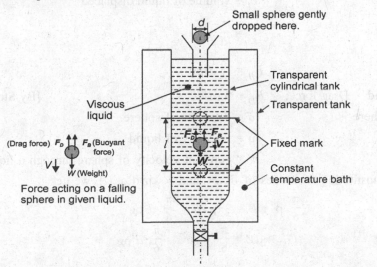

Fig. 1.30: Falling sphere viscometer.

This transparent tank is surrounded by another transparent tank to keep the temperature of the liquid in the cylindrical tank constant.

The experiment consists of gently dropping a small sphere on the surface of the liquid as shown in Fig. 8.30. After the sphere has attained a constant velocity (V), the time taken (t) by the sphere in covering a known distance 'l' marked between two points.

Thus the constant velocity of sphere: $V = \dfrac{l}{t}$ while moving downward through the viscous fluid, three forces are acting on the ball:

(*i*) Its own weight: $W\downarrow$ [By law of gravitation]

(*ii*) Drag force: $F_D\uparrow$ [By Stokes' law]

(*iii*) Buoyant force: $F_B\uparrow$ [By Archimede's principle]

For equilibrium:

Weight of sphere = drag force + buoyant force

$$\boxed{W = F_D + F_B} \qquad \qquad ...(i)$$

We know, Weight of sphere: $W = mg$ [Mass = density × volume]

$$W = \rho_s \times \text{volume} \times g \qquad \left[\text{Volume of sphere is } \frac{\pi}{6}d^3 \right]$$

$$W = \frac{\pi}{6} \rho_s g d^3$$

where ρ_s = density of sphere

                       d = diameter of sphere

and   Buoyant force: $F_B = \rho V_l g$

where ρ = density of liquid

                      V_l = volume of liquid displaced

$$= \frac{\pi}{6} d^3$$

$$F_B = \frac{\pi}{6} d^3 \rho g$$

and   Drag force: $F_D = 3\pi\mu V d.$ [By Stokes' law]

where d = diameter of sphere

                 μ = viscosity of liquid

                 V = constant velocity of sphere through a liquid

Substituting the values of W, F_B and F_D in Eq. (*i*), we get

$$\frac{\pi}{6} \rho_s g d^3 = 3\pi\mu V d + \frac{\pi}{6} d^3 \rho g$$

or $3\pi\mu V = \dfrac{\pi}{6} \rho_s g d^2 - \dfrac{\pi}{6} d^2 \rho g$

$$3\mu V = \frac{g d^2}{6} [\rho_s - \rho]$$

or Dynamic viscosity: $\mu = \dfrac{g d^2}{18 V}[\rho_s - \rho]$

By using of above equation, the dynamic viscosity of a liquid can be measured if ρ_s, ρ, d and V are density of sphere, density of liquid, diameter of sphere and constant velocity of sphere through the liquid are known.

1.12.4 Industrial Viscometers

These viscometer require the measurement of time taken by a certain quantity of the liquid to flow through a short capillary tube. The coefficient of viscosity is then obtained by comparing with the coefficient of a liquid whose viscosity is known or by the use of conversion factor.

Viscometers of different designs but based on the same principle are used in different countries. USA favours the use of Saybolt viscometer while in the United Kingdom (UK) Redwood viscometer is more commonly used. Both the methods are described as following:

Saybolt Viscometer

This viscometer is shown as Fig.1.31, which consists of a tank at the bottom of which a short capillary tube is fitted. In this tank the liquid whose viscosity is to be measured is filled. This tank is surrounded by another tank, called constant temperature bath. The liquid is allowed to gravity flow through the capillary tube at constant temperature. The time taken by 60 cc (*i.e.*, 60 ml or 0.06 litre) of liquid to flow through the capillary tube is noted down. From the time t required for the gravity flow of 60 cc of the given liquid, known as Saybolt Universal Seconds (SUS), the kinematic viscosity of the liquid can be determined from the following equations.

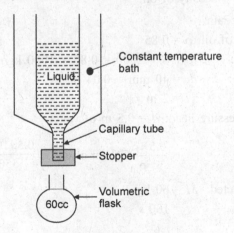

Fig. 1.31: Saybolt viscometer.

Kinematic viscosity: $\quad \nu = \left(0.22t - \dfrac{195}{t}\right)$ stokes $\quad$ for $t > 50$ and < 100 SUS

$$= \left[0.22t - \dfrac{135}{t}\right] \text{ stokes } \quad \text{for } t > 100 \text{ SUS}$$

where t is time noted in seconds, known as Saybolt Universal Seconds (SUS)

ν = kinematic viscosity in stokes

and Dynamic viscosity: $\mu = \nu\rho$

where ρ = density of the liquid at same temperature

Redwood Viscometer

In the Redwood viscometer, the time t required for the gravity flow of 50 cc of the given liquid known as Redwood Seconds (RWS). The kinematic viscosity of the liquid can be determined from the following equations.

[Same Fig. 1.31, only 50 cc volumetric flask is used instead of 60 cc]

Kinematic viscosity: $\nu = \left[0.00246 - \dfrac{0.9}{t}\right]$ **stokes** for $t > 40$ and < 85 RWS

$= \left[0.00246 - \dfrac{0.85}{t}\right]$ **stokes** for $t > 85$ and and < 200 RWS

where t is time noted in seconds, known as Redwood seconds (RWS).

ν = kinematic viscosity in stokes.

and Dynamic viscosity: $\mu = \nu\rho$

where ρ = density of the liquid at same temperature.

Problem 1.22:   The specific gravity of an oil of 0.85 is flowing through capillary tube of diameter 40 mm. The difference of pressure head between two points 2 m apart is 0.5 m of water. The mass of oil collected in a measuring tank is 60 kg in 100 seconds. Find the viscosity of oil.

Solution: Given data:

Specific gravity of oil: $\rho = 0.85$

$\therefore$ Density of oil: $\rho = S \times 1000 = 0.85 \times 1000 \text{ kg/m}^3 = 850 \text{ kg/m}^3$

Diameter of tube: d = 40 mm = 0.04 m

Length of tube: $l = 2$ m

Difference of pressure head: $h = 0.5$ m of water

$$= \frac{0.5 \times \rho_{\text{water}}}{\rho_{\text{oil}}} \text{ m of oil} = \frac{0.5 \times 1000}{550} = 0.588 \text{ m of oil}$$

Mass of oil collected: $M = 60$ kg

Time taken: $t = 100 \ s$

$\therefore$ Mass flow rate: $m = \dfrac{M}{t} = \dfrac{60}{100} = 0.6$ kg/s

Discharge: $Q = mv$

$$= \frac{m}{\rho} = \frac{0.6}{850} = 7.05 \times 10^{-4} \text{ m}^3/\text{s}$$

According to Hagen-Poiseuille equation,

Head loss: $h_f = \dfrac{128\mu Q l}{\pi \rho g d^4}$

or   Viscosity: $m = \dfrac{\pi \rho g\, h_f\, d^4}{128 Q l} = \dfrac{3.14 \times 850 \times 9.81 \times 0.588 (0.04)^4}{128 \times 7.05 \times 10^{-4} \times 2}$ $\because h_f = h$

                         $= 0.2183 \text{ Ns/m}^2 = 0.2183 \times 10 \text{ poise} = \mathbf{2.183 \text{ poise}}$

Problem 1.23: A capillary tube of diameter 30 mm and length 2 m is used for measuring viscosity of a liquid. The difference of pressure between two ends of the tube is 40 kPa and viscosity of liquid is 0.8 poise. Find the rate of flow of liquid through the tube.

Solution: Given data:

Diameter of tube: $d = 30$ mm $= 0.03$ m

Length of tube: $l = 2$ m

Difference of pressure: $p_1 - p_2 = 40$ kPa $= 40 \times 10^3$ Pa or N/m^2

Viscosity of liquid: $\mu = 0.8$ poise $= 0.08$ Ns/m^2

According of Hagen–Poiseuille equation,

Loss of head: $h_f = \dfrac{128\mu Q l}{\pi \rho g d^4}$

also $h_f = \dfrac{p_1 - p_2}{\rho g}$

$\therefore$ $\dfrac{p_1 - p_2}{\rho g} = \dfrac{12\mu Q \rho}{\pi \rho g d^4}$

or $p_1 - p_2 = \dfrac{128\mu Q l}{\pi d^4}$

                           $40 \times 10^3 = \dfrac{128 \times 0.08 \times Q \times 2}{3.14 \times (0.03)^4}$

or $Q = 4.96 \times 10^{-3}$ m^3/s $= 4.96 \times 10^{-3} \times 1000$ litre/s

                           $= \mathbf{4.96 \text{ litre/s}}$

Problem 1.24: A capillary tube of diameter 4 mm and length 150 mm is used for measuring viscosity of a liquid. The pressure difference between the two ends of the tube is 0.7848 N/cm^2 and the viscosity of liquid is 0.2 poise. Find the rate of flow of liquid through the tube.

Solution: Given data:

Diameter of tube: $d = 4$ mm $= 0.004$ m

Length of tube: $l = 150$ mm $= 0.15$ m

Difference of pressure: $p_1 - p_2 = 0.7848$ N/cm^2 $= 0.7848 \times 10^4$ N/m^2

Viscosity of a liquid: $\mu = 0.2$ poise $= \dfrac{0.2}{10}$ Ns/m^2 $= 0.02$ Ns/m^2

According to Hagen-Poiseuille equation:

Loss of head : $h_f = \dfrac{128 \mu Q l}{\pi \rho g d^4}$

also $h_f = \dfrac{p_1 - p_2}{\rho g}$

$\therefore$ $\dfrac{p_1 - p_2}{\rho g} = \dfrac{128 \mu Q l}{\pi \rho g d^4}$

or $p_1 - p_2 = \dfrac{128 \mu Q l}{\pi d^4}$

$$0.7848 \times 10^4 = \dfrac{128 \times 0.02 \times Q \times 0.15}{3.14 \times (0.004)^4}$$

or $Q = \mathbf{1.6428 \times 10^{-5}}$ **m^3/s**

Problem 1.25: Find the viscosity of an oil for the following data:
Diameter of inner and outer cylinder and 250 mm and 255 mm respectively.
Height of liquid in the cylinder = 320 mm
Clearance at the bottom of two cylinders = 6 mm
Speed of outer cylinder = 350 rpm
Reading of the torsion meter = 5.2 Nm

Solution: Given data:
Diameter of inner cylinder:

$$D_1 = 250 \text{ mm} = 0.25 \text{ m}$$

$\therefore$ Radius of inner cylinder:

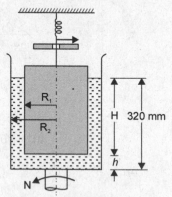

$$R_1 = \dfrac{D_1}{2} = \dfrac{0.25}{2}$$
$$= 0.125 \text{ m}$$

Diameter of outer cylinder:

Fig. 1.32: Schematic for Problem 1.25

$$D_2 = 255 \text{ mm} = 0.255 \text{ m}$$

$\therefore$ Radius of outer cylinder:

$$R_2 = \dfrac{D_2}{2} = \dfrac{0.255}{2}$$
$$= 0.1275 \text{ m}$$

Height of liquid from the bottom of outer cylinder = 320 mm = 0.32 m

Clearance at the bottom of two cylinders: h = 6 mm = 0.006 m

$\therefore$ Height of inner cylinder immersed in liquid $H = 0.32 - h = 0.32 - 0.006 = 0.314$ m

Speed of outer cylinder: $N = 350$ rpm

$\therefore$ Angular speed: $\omega = \dfrac{2\pi N}{60} = \dfrac{2 \times 3.14 \times 350}{60} = 36.63$ rad/s

Reading of the torsion meter: $T = 5.2$ Nm

By using rotating cylinder viscometer,

Viscosity of oil: $\mu = \dfrac{2h(R_2 - R_1)T}{\pi \omega R_1^2 \,[4HhR_2 + R_1^2(R_2 - R_1)]}$

$= \dfrac{2 \times 0.006 \times (0.1275 - 0.125) \times 5.2}{3.14 \times 36.63 \times (0.125)^2 . [4 \times 0.314 \times 0.006 \times 0.1275 + 0.125^2\,(0.1275 - 0.125)]}$

$= \dfrac{1.56 \times 10^{-4}}{1.797 \times [9.608 \times 10^{-4} + 3.906 \times 10^{-5}]}$

$= \dfrac{1.56 \times 10^{-4}}{1.797 \times [9.608 \times 10^{-4} + 0.3906 \times 10^{-4}]}$

$= \dfrac{1.56 \times 10^{-4}}{17.967 \times 10^{-4}} = 0.08682$ Ns/m^2 = **0.8682 poise**

Problem 1.26: A sphere of diameter 2.5 mm falls 200 mm in 25 seconds in a viscous liquid. The density of the sphere is 7000 kg/m^3 and the liquid is 900 kg/m^3. Find the dynamic viscosity of the liquid.

Solution: Given data:

Diameter of sphere: $d = 2.5$ mm = 0.0025 m

Distance travelled by sphere: $l = 200$ mm = 0.2 mm

Time taken: $t = 25$ s

$\therefore$ Velocity of sphere: $V = \dfrac{l}{t} = \dfrac{0.2}{25} = 0.008$ m/s

Density of sphere: $\rho_s = 7000$ kg/m^3

Density of liquid: $\rho = 900$ kg/m^3

According to falling sphere viscometer,

Dynamic viscosity: $\mu = \dfrac{g\rho^2}{18V}\,[\rho_s - \rho] = \dfrac{9.81 \times (0.0025)^2}{18 \times 0.008}\,[7000 - 900]$

$= 2.597$ Ns/m^2 = 2.597×10 poise = **25.97 poise**

Problem 1.27: Find the viscosity of an oil of specific gravity 0.8, when a gas bubble of diameter 12 mm rises steadily through the oil at a velocity of 0.02 m/s. Neglect the weight of the bubble.

Solution: Given data:

   Specific gravity of oil: $S = 0.8$

   $\therefore$ Density of oil: $\rho = S \times 1000 \text{ kg/m}^3 = 0.8 \times 100 = 800 \text{ kg/m}^3$

   Diameter of gas bubble: $d = 12$ mm $= 0.012$ m

   Velocity of bubble: $V = 0.02$ m/s

   According of falling sphere viscometer,

   Viscosity: $\mu = \dfrac{gd^2}{18 V} [\rho_s - \rho]$ when $\rho_s > \rho$

   $= \dfrac{gd^2}{18 V} [\rho - \rho_s]$ when $\rho_s < \rho$

   $= \dfrac{9.81 \times (0.012)^2}{18 \times 0.02} \times [800 - 0]$

$\because$ Weight of bubble neglected, so $\rho_s = 0$

$= 3.139 \text{ Ns/m}^2 = \mathbf{31.39 \text{ poise}}$

1.13 NAVIER-STOKES EQUATIONS OF MOTION

Navier-Stokes equation based on the law of conservation of momentum or the momentum acting on a fluid states that the sum of the forces acting on a fluid mass is equal to the change in momentum of flow per unit time in the direction of flow.

   If $\vec{F_B}$ is the body force per unit volume (*e.g.* $\vec{F_B} = \rho\vec{g}$) and F_s is the surface force per unit volume, then according to momentum equation

$$\rho \frac{D\vec{V}}{Dt} = \vec{F_B} + \vec{F_S} \qquad \qquad ...(1)$$

   where the operator: $\dfrac{D}{Dt} = \dfrac{\partial}{\partial t} + u \dfrac{\partial}{\partial x} + v \dfrac{\partial}{\partial y} + w \dfrac{\partial}{\partial z}$

   is sometimes called the Eulerian derivative which implies the rate of change of a certain fluid particle due to both non-stationary effects and convective effects in the direction of fluid particle motion.

$\vec{F_B} = F_{Bx}i + F_{By}j + F_{Bz}k,$ body force vector per unit volume.

$\vec{F_S} = F_{Sx}i + F_{Sy}j + F_{Sz}k,$ surface force vector per unit volume

$\vec{V} = ui + vj + wk$, velocity vector.

   Eq. (1), written along x, y and z-directions.

   along x-direction,

$$\frac{\rho Du}{Dt} = F_{Bx} + F_{Sx} \qquad \qquad ...(2)$$

   along y-direction,

$$\frac{\rho Dv}{Dt} = F_{By} + F_{Sy} \qquad\qquad ...(3)$$

and along z-direction,

$$\frac{\rho Dv}{Dt} = F_{Bz} + F_{Sz} \qquad\qquad ...(4)$$

Consider an elementary parallelepiped of dimension dx, dy and dz along x, y and z-axes respectively. All normal and shear stresses are shown on elementary parallelepiped in Fig. 1.33.

Net force in the x-direction due to stress;

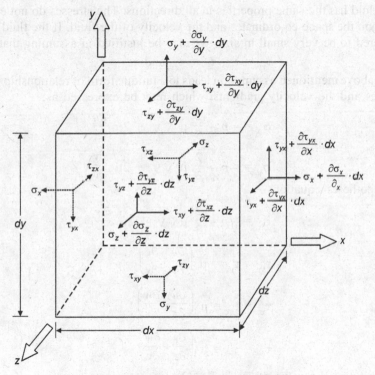

Fig. 1.33: Elementary Parallel piped

$$= \left(\sigma_x + \frac{\partial \sigma_x}{\partial x} \cdot dx - \sigma_x\right) dy\, dz + \left(\tau_{xy} + \frac{\partial \tau_{xy}}{\partial y} dy\right) dx\, dz$$

$$- \tau_{xy} dx\, dz + \left(\tau_{xz} + \frac{\partial \tau_{xz}}{\partial z} dz - \tau_{xz}\right) dx.dy$$

$$= \frac{\partial \sigma_x}{\partial x} dx.dx.dz + \frac{\partial \tau_{xy}}{\partial y} dx.dy.dz + \frac{\partial}{\partial z}\tau_{xz}.dx.dy.dz$$

$$= \left(\frac{\partial \sigma_x}{\partial x} + \frac{\partial \sigma_{xy}}{\partial y} + \frac{\partial \tau_{xz}}{\partial z}\right) dx.dy.dz$$

Net surface force per unit volume along x-direction

$$F_{sx} = \frac{\partial \sigma_x}{\partial x} + \frac{\partial \tau_{xy}}{\partial y} + \frac{\partial \tau_{xz}}{\partial z}$$

where σx is normal stress along x-axis and plane $\perp ar$ to x-axis.

τ_{xy} is shear stress in x-direction, plane $\perp ar$ to y-axis (or plane parallel to x-axis).

τ_{xz} is shear stress in x-direction, plane $\perp ar$ to z-axis (or plane parallel to z-axis).

Experimental investigations have shown that the stresses in newtonian fluids are related linearly to the derivatives of velocities and that most fluids are isotropic *i.e.,* the fluid properties are not dependent on the direction in space. In other words, an isotropic fluid has the same properties in all directions. The stresses do not explicitly depend upon the space co-ordinates and the velocity of the fluid. If the fluid element is considered to be very small in size, we may be justified in assuming that

$$\tau_{xy} = \tau_{yx}, \tau_{yz} = \tau_{zy} \text{ and } \tau_{zx} = \tau_{xz}$$

The above mentioned assumption leads to a unique form of relationship between the stresses and the velocity gradients, which may be expressed as:

$$\sigma_x = -p - \frac{2}{3} \mu \, \nabla \vec{V} + 2\mu \, \frac{\partial u}{\partial x}$$

$$\sigma_y = -p - \frac{2}{3} \mu \, \nabla \vec{V} + 2\mu \, \frac{\partial v}{\partial y}$$

Stokes hypotheses equations.

$$\sigma_x = -p - \frac{2}{3} \mu \nabla \vec{V} + 2\mu \, \frac{\partial w}{\partial z}$$

$$\tau_{xy} = \tau_{yx} = \mu \left(\frac{\partial u}{\partial y} + \frac{\partial v}{\partial x} \right)$$

$$\tau_{xz} = \tau_{zx} = \mu \left(\frac{\partial u}{\partial z} + \frac{\partial w}{\partial x} \right)$$

and

$$\tau_{yz} = \tau_{zy} = \mu \left(\frac{\partial v}{\partial z} + \frac{\partial w}{\partial y} \right)$$

Substituting the value of F_{sx} in Eq. (2), we get

$$\rho \frac{Du}{Dt} = F_{Bx} + \frac{\partial \sigma_x}{\partial x} + \frac{\partial \tau_{xy}}{\partial y} + \frac{\partial \tau_{xz}}{\partial z}$$

Similarly along y and z-directions

$$\rho \frac{Dv}{Dt} = F_{By} + \frac{\partial \tau_{yx}}{\partial x} + \frac{\partial \sigma_y}{\partial y} + \frac{\partial \tau_{yz}}{\partial z}$$

and

$$r \frac{Dw}{Dt} = F_{Bz} + \frac{\partial \tau_{zx}}{\partial x} + \frac{\partial \tau_{zy}}{\partial y} + \frac{\partial \sigma_z}{\partial z}$$

Substituting the values of stresses in above equations, we get

$$\rho \frac{Du}{Dt} = F_{Bx} - \frac{\partial p}{\partial x} + \frac{\partial}{\partial x} \left[2\mu \frac{\partial u}{\partial x} - \frac{2}{3} \mu \nabla \vec{V} \right] + \frac{\partial}{\partial y} \left[\mu \frac{\partial u}{\partial y} + \frac{\partial v}{\partial x} \right]$$

$$+ \frac{\partial}{\partial z}\left[\mu\left(\frac{\partial u}{\partial z}+\frac{\partial w}{\partial x}\right)\right] \qquad \qquad ...(5)$$

$$\rho \frac{Dv}{Dt} = F_{By} - \frac{\partial p}{\partial y} + \frac{\partial}{\partial x}\left[\mu\left(\frac{\partial u}{\partial x}-\frac{\partial v}{\partial x}\right)\right]$$

$$+ \frac{\partial}{\partial y}\left[2\mu\frac{\partial v}{\partial y}-\frac{2}{3}\mu\nabla\vec{V}\right] + \frac{\partial}{\partial z}\left[\mu\left(\frac{\partial v}{\partial z}+\frac{\partial w}{\partial y}\right)\right] ...(6)$$

$$\rho \frac{Dw}{Dt} = F_{Bz} - \frac{\partial p}{\partial y} + \frac{\partial}{\partial x}\left[\mu\left(\frac{\partial u}{\partial z}-\frac{\partial w}{\partial x}\right)\right]$$

$$+ \frac{\partial}{\partial y}\left[\mu\left(\frac{\partial v}{\partial z}+\frac{\partial w}{\partial y}\right)\right] + \frac{\partial}{\partial z}\left[2\mu\frac{\partial w}{\partial z}-\frac{2}{3}\mu\nabla\vec{V}\right] ...(7)$$

The above differential Eqs. (5), (6) and (7) are known as the **Navier-Stokes equations** along x, y and z-directions respectively.

These Navier-Stokes equations of motions for a newtonian fluid of varying density and viscosity in a gravitational field. (*i.e.*, N.S. equations of motion of a viscous and compressible fluid]

If the viscosity is assumed to be constant, these equations may be simplified and rearranged as:

$$\rho \frac{Du}{Dt} = F_{Bx} - \frac{\partial p}{\partial x} + 2\mu\frac{\partial^2 u}{\partial x^2} - \frac{2}{3}\mu\frac{\partial}{\partial x}(D.\vec{V})$$

$$+ \mu\frac{\partial^2 u}{\partial y^2} + \mu\frac{\partial^2 v}{\partial y\partial x} + \frac{\mu\partial^2 w}{\partial z\partial x} + \frac{\mu\partial^2 u}{\partial z^2}$$

$$\rho \frac{Du}{Dt} = F_{Bx} - \frac{\partial p}{\partial x} + \frac{\mu\partial^2 u}{\partial x^2} + \frac{\mu\partial^2 u}{\partial y^2} + \frac{\mu\partial^2 u}{\partial z^2}$$

$$+ \frac{\mu\partial^2 u}{\partial x^2} + \frac{\mu\partial^2 v}{\partial y\partial x} + \mu\frac{\partial^2 w}{\partial z\partial x} - \frac{2}{3}\mu\frac{\partial}{\partial x}(\nabla.\vec{V})$$

$$\rho \frac{Du}{Dt} = F_{Bx} - \frac{\partial p}{\partial x} + \mu\left[\frac{\partial^2 u}{\partial x^2}+\frac{\partial^2 u}{\partial y^2}+\frac{\partial^2 u}{\partial z^2}\right]$$

$$+ \mu\frac{\partial}{\partial x}\left[\frac{\partial u}{\partial x}+\frac{\partial v}{\partial y}+\frac{\partial w}{\partial z}\right] - \frac{2}{3}\mu\frac{\partial}{\partial x}(\nabla.\vec{V})$$

$$\rho \frac{Du}{Dt} = F_{Bx} - \frac{\partial p}{\partial x} + \mu\left[\frac{\partial^2 u}{\partial x^2}+\frac{\partial^2 u}{\partial y^2}+\frac{\partial^2 u}{\partial z^2}\right]$$

$$+ \mu\frac{\partial}{\partial x} - \frac{2}{3}\mu\frac{\partial}{\partial x}(\nabla.\vec{V})$$

$$\rho \frac{Du}{Dt} = F_{Bx} - \frac{\partial p}{\partial x} + \mu\left[\frac{\partial^2 u}{\partial x^2}+\frac{\partial^2 u}{\partial y^2}+\frac{\partial^2 u}{\partial z^2}\right] + \frac{1}{3}\mu\frac{\partial}{\partial x}(\nabla.\vec{V})$$

Similarly for y and z-directions, we get

$$\rho \frac{Dv}{Dt} = F_{By} - \frac{\partial p}{\partial y} + \mu \left[\frac{\partial^2 v}{\partial x^2} + \frac{\partial^2 v}{\partial y^2} + \frac{\partial^2 v}{\partial z^2} \right] + \frac{1}{3} \mu \frac{\partial}{\partial y} (\nabla.\vec{V}))$$

$$\rho \frac{Dw}{Dt} = F_{Bz} - \frac{\partial p}{\partial z} + \mu \left[\frac{\partial^2 w}{\partial x^2} + \frac{\partial^2 w}{\partial y^2} + \frac{\partial^2 w}{\partial z^2} \right] + \frac{1}{3} \mu \frac{\partial}{\partial z} (\nabla.\vec{V})$$

For incompressible fluid where both the density ρ and viscosity coefficient μ are constant. Then $\nabla.\vec{V}$ becomes zero.

Thus

$$\rho \frac{Du}{Dt} = F_{Bx} - \frac{\partial p}{\partial x} + \mu \left[\frac{\partial^2 u}{\partial x^2} + \frac{\partial^2 u}{\partial y^2} + \frac{\partial^2 u}{\partial z^2} \right]$$

$$\rho \frac{Du}{Dt} = F_{Bx} - \frac{\partial p}{\partial x} + \mu \nabla^2 u \qquad \qquad ...(8)$$

Similarly for y and z-directions, we get

$$\rho \frac{Dv}{Dt} = F_{By} - \frac{\partial p}{\partial y} + \mu \nabla^2 v \qquad \qquad ...(9)$$

$$\rho \frac{Dw}{Dt} = F_{Bz} - \frac{\partial p}{\partial z} + \mu \nabla^2 w \qquad \qquad ...(10)$$

Eqs. (8), (9) and (10) are **Navier-Stokes equations** of motion of a viscous in compressible fluid in cartesian co-ordinates.

Navier-Stokes equation in vector form.

$$\boxed{\rho \frac{D\vec{V}}{Dt} = \vec{F}_B - \operatorname{grad} p + \mu \nabla^2.\vec{V}}$$

where $\vec{F}_B$ = body force per unit volume.

$$\frac{D\vec{V}}{Dt} = \frac{\overline{F_B}}{\rho} - \frac{1}{\rho} \operatorname{grad} p + \frac{\mu}{\rho} \nabla^2.\vec{V}$$

$$\boxed{\frac{D\vec{V}}{Dt} = \vec{F}_B - \frac{1}{\rho} \operatorname{grad} p + \nu \nabla^2.\vec{V}} \qquad ... (11)$$

where $\overline{F_B}$ = body fore per unit volume,

$\nu = \dfrac{\mu}{\rho}$, kinematic viscosity.

Equation (11) in x, y and z-directions as

$$\frac{Du}{Dt} = F_{Bx} - \frac{1}{\rho} \frac{\partial p}{\partial x} + \nu \nabla^2 u$$

$$\frac{Dv}{Dt} = F_{By} - \frac{1}{\rho} \frac{\partial p}{\partial y} + \nu \nabla^2 v$$

$$\frac{Dw}{Dt} = F_{Bz} - \frac{1}{\rho} \frac{\partial p}{\partial z} + \nu \nabla^2 w$$

Problem 1.28: A viscous liquid of density ρ and viscosity μ flows down a wide include plate under the influence of gravity. The angle of inclination of the plate with horizontal is θ. For all practical purposes the flow is parallel to the plate and a fully developed. The velocity is not a function of the distance down the plate. The depth of liquid normal to the plate is h. The viscosity of the air in contact with the upper surface of the liquid may be neglected. For steady laminar two-dimensional flow of the incompressible liquid, determine.

(i) the velocity distribution normal to the flow.

(ii) the shear stress at the plate boundary

(iii) the average velocity,

(iv) the discharge per unit width of the plate

(v) the free surface velocity of flow.

Solution: For steady two dimensional flow of an incomplete viscous liquid, the Navier-Strokes equations of motion in x and y-directions are

$$\frac{u\partial u}{\partial x} + \frac{v\partial u}{\partial y} = F_{Bx} - \frac{1}{\rho}\frac{\partial p}{\partial x} + v\nabla^2 u \qquad \ldots(i)$$

and

$$\frac{u\partial v}{\partial x} + \frac{v\partial v}{\partial y} = F_{Bx} - \frac{1}{\rho}\frac{\partial p}{\partial y} + v\nabla^2 v \qquad \ldots(ii)$$

Since the steady flow is in the x-direction and is fully developed the force, $u = f(y)$ and $v = 0$. On the liquid-air interface, the pressure is everywhere atmospheric and since the depth of flow is constant throughout, therefore, $\dfrac{\partial p}{\partial x} = 0$.

Under these conditions, the Navier-Strokes equations are reduced to

$$4\,\frac{\partial u}{\partial x} = F_{Bx} - v\frac{\partial^2 u}{\partial y^2} \qquad \ldots(iii)$$

$$0 = F_{By} - \frac{1}{\rho}\frac{\partial p}{\partial y} \qquad \ldots(iv)$$

$$\frac{\partial u}{\partial x} = 0 \qquad\qquad \because u = f(y)$$

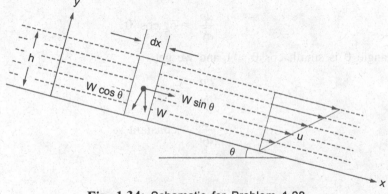

Fig. 1.34: Schematic for Problem 1.28

Consider a length dx and unit width of flow. The weight of the liquid contain in small volume is

$$W = \rho g \, dx.h$$

Body force in the x-direction = $W \sin \theta = \rho g \, dxh \sin \theta$

Body force per unit mass of liquid in the text x-direction,

$$F_{Bx} = \frac{\rho g \, dx \, h \sin \theta}{\rho dx h} = g \sin \theta$$

Similarly the body force per unit mass of liquid in the y-direction,

$$F_{By} = \frac{\rho g \, dx \, h \cos \theta}{\rho dx \, h} = g \cos \theta$$

Substituting the values of $\dfrac{\partial u}{\partial x} = 0$, and $F_{Bx} = g \sin \theta$ in Eq. (iii), we get

$$0 = g \sin \theta + v \frac{d^2 u}{dy^2}$$

or $\qquad v \dfrac{d^2 u}{dy^2} = -g \sin \theta \qquad\qquad \therefore v = \dfrac{\mu}{\rho}$

$$\frac{u}{\rho} \frac{d^2 u}{dy^2} = -g \sin \theta$$

or $\qquad \dfrac{d^2 u}{dy^2} = -\dfrac{\rho g}{\mu} \sin \theta \qquad\qquad ...(v)$

Substituting the value of $F_{BY} = g \cos \theta$ in eq. (iv), we get

$$0 = g \cos \theta - \frac{1}{\rho} \frac{dp}{dy}$$

or $\qquad \dfrac{1}{\rho} \dfrac{dp}{dy} = g \cos \theta$

or $\qquad \dfrac{dp}{dy} = \rho g \cos \theta$

If angle θ is small, $\cos \theta \approx 1$ and we get

$$\frac{dp}{dy} = \rho g$$

$$\frac{dp}{dy} = \text{constant} \qquad\qquad ...(vi)$$

Equation (*vi*) indicates the hydrostatic variation of pressure in the *y*-direction. On integrating Eq. (*v*), we get

$$\frac{du}{dy} = -\frac{\rho g}{\mu} \sin\theta \times y + A$$

Again integrating, we get

$$u = \frac{-\rho g}{\mu} \sin\theta \times \frac{y^2}{2} + Ay + B \quad ...(vii)$$

where A and B are constants of integration and the values of A and B are to be determined from the boundary conditions.

At
$$y = h,$$
$$\tau_{yx} = 0$$
$$\mu\frac{du}{dy} = 0$$

or
$$\frac{du}{dy} = 0$$

∴
$$0 = -\frac{\rho g \sin\theta}{\mu} y + A$$

or
$$\boxed{A = \frac{\rho g\, h \sin\theta}{\mu}}$$

At $y = 0$, $u = 0$

we get from Eq (*vii*)

$$0 = 0 + 0 + B$$

Or
$$\boxed{B = 0}$$

Substituting the values of A and B in Eq. (*vii*), we get

$$u = \frac{\rho g y^2}{2\mu}\sin\theta + \frac{\rho g h y \sin\theta}{\mu}$$

$$u = \frac{\rho g \sin\theta}{2\mu}(2\,hy - y^2)$$

(*i*) The velocity distribution normal to the flow,

$$u = \frac{\rho g \sin\theta}{2\mu}(2\,hy - y^2)$$

(*ii*) The shear stress at the plate boundary,

$$(\tau_{yz})_{y=0} = \mu\left(\frac{du}{dy}\right)_{y=0}$$

$$= \mu\left[\frac{\rho g \sin\theta}{2\mu}(2\,h - 2y)\right]_{y=0}$$

$$= \mu \left[\frac{\rho g \sin\theta}{2\mu} (2h - 0) \right]$$

$$(\tau_{yz})_{y=0} = \rho g h \sin\theta$$

(iii) The average velocity of flow,

$$\bar{u} = \frac{1}{h} \int_0^h u \, dy$$

$$= \frac{1}{h} \int_0^h \frac{\rho g \sin\theta}{2\mu} (2hy - y^2) \, dy$$

$$= \frac{1}{h} \frac{\rho g \sin\theta}{2\mu} \left[\frac{2hy^2}{2} - \frac{y^3}{3} \right]_0^h$$

$$= \frac{\rho g \sin\theta}{2h\mu} \left[\frac{2h^3}{2} - \frac{h^3}{3} - 0 \right]$$

$$= \frac{\rho g \sin\theta}{2h\mu} \left[h^3 - \frac{h^3}{3} \right]$$

$$= \frac{\rho g \sin\theta}{2h\mu} \times \frac{2}{3} h^3$$

$$\bar{u} = \frac{\rho g h^2 \sin\theta}{3\mu}$$

(iv) The discharge per unit width of plate

$$= \text{Area of flow} \times \text{average velocity}$$
$$= h \times 1 \times \bar{u}$$

$$= h \times \frac{\rho g h^2 \sin\theta}{3\mu}$$

$$= \frac{\rho g h^3 \sin\theta}{3\mu} \ \text{m}^2/\text{s}$$

$$= \frac{g h^3 \sin\theta}{3\nu} \ \text{m}^2/\text{s} \qquad \because \nu = \frac{\mu}{\rho}$$

(v) The free surface velocity of flows: U

At $y = h$, $\qquad\qquad U = (u)_{y=h}$

$$= \frac{\rho g \sin\theta}{2\mu} \ (2h^2 - h^2)$$

$$= \frac{\rho g \sin\theta \times h^2}{2\mu} = \frac{\rho g h^2 \sin\theta}{2\mu}$$

Problem 1.29: A thin film of lubricating oil (μ = 0.9 Pa.s, ρ = 1260 kg/m^3) is flowing uniformly down an inclined plane θ = 30°. Calculate the free surface velocity and the flow rate per unit width of the plate, if the depth is 5 mm.

Solution: Given Data

$$\mu = 0.9 \text{ Pa.s} = 0.9 \text{ Ns/m}^2$$
$$\rho = 1260 \text{ kg/m}^3$$
$$\theta = 30°$$

Depth of flow : h = 5 mm = 0.005 m

Free surface velocity of flow is given by

$$U = \frac{\rho g h^2 \sin \theta}{2\mu}$$

$$= \frac{1260 \times 9.81 \times (0.005)^2 \times \sin 30°}{2 \times 0.9}$$

$$= 0.0858 \text{ m/s} = \textbf{85.8 mm/s}$$

Flow rate per unit width $= \dfrac{\rho g h^3 \sin \theta}{3\mu}$

$$= \frac{1260 \times 9.81 \times (0.005)^3 \sin 30°}{3 \times 0.9}$$

$$= \textbf{2.86} \times \textbf{10}^{-4} \textbf{ m}^2\textbf{/s}$$

1.14 FLUIDIZATION

When liquid or gas is passed at low velocity through the horizontal bed of solid particles in upward direction, the particles do not move, and the pressure drops. If the fluid velocity is steadily increased, the pressure drop and drag on individual particles increase, and the particles start to move and become suspended in the fluid. The terms fluidization and fluidized bed are used to describe the condition of fully suspended particles, since the suspension behaves as a dense fluid. If the bed is tilted, the top surface remains horizontal and large objects will either float or sink in the bed depending on their density relative to the suspension. The fluidized solids can be drained from the bed through pipes and valves just as a liquid can, and this fluidity is one of the main advantages of the use of the fluidization for handling solids.

1.14.1 Conditions for Fluidization

Consider a vertical tube filled with fine solid particles. The tube is open at the top and has a porous plate at the bottom to support the bed of solid particles and to distribute the flow uniformly over the entire cross section. Air is admitted below the distributor plate at a low flow rate and passes upward through the bed without causing any particle motion. If the particles are quite small, flow in the channels between the particles will be laminar and the pressure drop across the bed will be proportional to the superficial velocity V_0. As the velocity is gradually increased, the pressure drop increases, but the particles do not move and the bed height remains the same. At a certain velocity, the pressure drop across the bed counterbalances the force due gravity on the particles or the weight of the bed, and any further increase in velocity causes the particles to move. This is point A on the graph. Sometime the bed expands slightly with grains still in contact since just a slight increase in porosity or void fraction can offset an increase of serval percentage in V_o and keep Δp constant. With a further increase in velocity, the particles become separated enough to move about in the bed, and true fluidization starts at point B.

Once the bed is fluidized, the pressure drop across the bed stays constant, but bed height continues to increase with increasing flow. The bed can be operated at quite high velocities with very little or no loss of solids, since the superficial velocity needed to support a bed of particles in much less than the terminal velocity for individual particles.

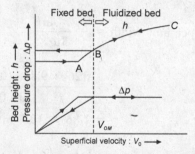

Fig. 1.35: Pressure drop and bed height versus superficial velocity for a bed.

If the flow rate to the fluidized bed is gradually reduced, the pressure drop remains constant, and the bed height decreases, following the line BC which was observed for increasing velocities. However, the final bed height may be greater than the initial value for the fixed bed, since solids dumped in a tube tend to pack more tightly than solids, slowly settling from a fluidized state. The pressure drop at low velocity is then less than that in the original fixed bed. On the starting up again, the pressure drop offsets the weight of the bed at point B, and this point, rather than point B, should be fluidization velocity V_{oM}. To determine V_{oM}, the considered to give the minimum bed should be fluidized vigorously and allowed to settle with the gas turned off, and the flow rate increased gradually until the bed starts to expand. More reproducible values of V_{oM} can sometimes be obtained from the intersection the fixed bed and the fluidized bed.

1.14.2 Types of Fluidization

The fluidization has the following two types as:

    (*i*) Particulate fluidization.

    (*ii*) Aggregative or bubbling fluidization

(i) **Particulate fluidization:**

When fluidizing sand with water, the particles move farther apart and their motion becomes more vigorous as the velocity is increased, but the average bed density at a given velocity is the same in all sections of the bed. This is called particulate fluidization and is characterized by a large but uniform expansion of the bed of high velocities.

(ii) **Aggregative or bubbling fluidization:**

Beds of solids fluidized with air usually exhibit what is called aggregative or bubbling fluidization. At superficial velocities greater than V_{om}, most of the gas passes through bed as bubbles or voids which are almost free of solids, and only a small fraction of the gas flows in the channels between the particles. The particles move erratically and are supported by the fluid, but in the space between bubbles, the void fraction is about the same as that at incipient fluidization. The non-uniform nature of the bed was first attributed to aggregation of the particles, and the term aggregation fluidization wall applied, but there is no evidence that the particles stick together, and the term bubbling fluidization is a better description of the phenomenon. The bubbles that form, behave much as air bubbles in water or bubbles of vapour in boiling liquid.

SUMMARY

1. The viscous flow is a smooth regular flow in layers. Such flow exists only at low velocities.

2. Reynolds number:
$$Re = \frac{\rho DV}{\mu} \text{ for flow through pipe}$$
$$= \frac{DV}{\nu}$$

If $Re < 2000$, the flow through pipe is laminar.
and $Re > 4000$, the flow through pipe is turbulent.

3. Viscous flow though pipe:

(a) Shear stress:
$$\tau = -\frac{\partial p}{\partial x} \frac{r}{2} \text{ shear stress distribution}$$
$$\tau = f(r)$$

(b) Velocity:
$$u = -\frac{1}{4\mu} \frac{dp}{dx} [R^2 - r^2] \text{ velocity distribution}$$
$$u = f(r)$$

(c) Maximum Velocity:
$$U_{max} = -\frac{1}{4\mu} \frac{dp}{dx} R^2$$

(d) Average velocity:
$$\bar{u} = -\frac{1}{8\mu} \frac{dp}{dx} R^2$$

Contd...

(e) $U_{max} = 2\bar{u}$

(f) Hagen–Poiseuille equation,

Loss of pressure head: $h_f = \dfrac{32\mu\bar{u}l}{\rho g D^2}$

also $\qquad\qquad\qquad h_f = \dfrac{p_1 - p_2}{\rho g}$

Average velocity: $\qquad \bar{u} = \dfrac{Q}{\dfrac{\pi}{4}D^2} = \dfrac{4Q}{\pi D^2}$

$\therefore \qquad\qquad\qquad\qquad h_f = \dfrac{128 Q \mu l}{\pi \rho g D^4}$

4. **Viscous flow between parallel plates.**

Case I : One plate is fixed and the other is moving: Couette flow.

(a) Velocity: $\qquad\qquad u = \dfrac{y}{h}U - \dfrac{1}{2\mu}\dfrac{dp}{dx}(yh - y^2)$

$\qquad\qquad\qquad\qquad u = f(y)$

velocity distribution

(b) Average velocity: $\qquad \bar{u} = \dfrac{U}{2} - \dfrac{h^2}{12\mu}\dfrac{dp}{dx}$

(c) Shear stress: $\qquad\qquad \tau = \mu\dfrac{U}{h} - \dfrac{1}{2}\dfrac{dp}{dx}[h - 2y]$

$\qquad\qquad\qquad\qquad \tau = f(y)$ shear stress distribution

(d) Loss of pressure head: $h_f = \dfrac{6\mu l}{\rho g h^2}[2\bar{u} - U]$

Case II : Both plates are fixed:

(a) Velocity: $\qquad\qquad u = -\dfrac{1}{2\mu}\dfrac{dp}{dx}[hy - y^2]$

$\qquad\qquad\qquad\qquad u = f(y)$ velocity distribution

(b) Maximum velocity: $U_{max} = -\dfrac{1}{8\mu}\dfrac{dp}{dx}h^2$

(c) Average velocity: $\qquad \bar{u} = -\dfrac{1}{12\mu}\dfrac{dp}{dx}h^2$

(d) $U_{max} = 1.5\bar{u}$

(e) Shear stress: $\qquad\qquad \tau = -\dfrac{1}{2}\dfrac{dp}{dx}[h - 2y]$

Contd...

$$\tau = f(y) \text{ shear stress distribution}$$

(f) Maximum shear stress: $\tau_{max} = -\dfrac{1}{2} \dfrac{dp}{dx} h$

(g) Loss of pressure head: $h_f = \dfrac{12\mu \bar{u} l}{\rho g h^2}$

5. Momentum correction Factor: β

$$\beta = \dfrac{\text{Momentum per second based on actual velocity}}{\text{Momentum per second based on average velocity}}$$

$\qquad\qquad = \dfrac{4}{3}$ for laminar flow through pipe.

$\qquad\qquad = 1.01$ to 1.04 for turbulent flow through pipe.

6. Kinetic energy correction factor: α

$$\alpha = \dfrac{KE_{act}/\text{second}}{KE_{avg}/\text{second}}$$

$\qquad\qquad = 2$ for laminar flow through pipe

$\qquad\qquad = 1.04$ to 1.11 for turbulent flow through pipe.

7. Power absorbed in overcoming the viscous resistance in journal bearing is given by

$$P = \dfrac{\mu \pi^3 d^3 N^2 l}{3600 t} \text{ watt}$$

where
$\qquad\qquad d = $ diameter of shaft in m
$\qquad\qquad N = $ Speed of shaft in rpm
$\qquad\qquad l = $ length of bearing in m
$\qquad\qquad t = $ thickness of oil film in m
$\qquad\qquad \mu = $ viscosity of oil in Ns/m^2

8. Power absorbed in overcoming the viscous resistance in foot-step bearing is given by

$$P = \dfrac{\mu \pi^3 N^2 R^4}{1800 t} \text{ watt}$$

μ in N$_3$/m^2, N in *rpm*, R and *t* in *m*

9. Power absorbed in overcoming the viscous resistance in collar bearing is given by

$$P = \dfrac{\mu \pi^2 N^2 [R_2^4 - R_1^4]}{1800 t} \text{ watt}$$

$\qquad\qquad \mu = $ viscosity of oil in Ns/m^2
$\qquad\qquad N = $ speed of shaft in *rpm*

Contd...

R_1 = radius of shaft in m

R_2 = radius of Collar in m

T = thickness of oil film in m

10. Velocity of the piston in a dash-pot: V

$$V = \frac{4Wb^3}{3\pi \, lD^3\mu} \, \text{m/s}$$

where

W = load exerted by the machine member in N

b = clearance between the piston and cylinder in m

l = length of the piston in m

D = diameter of the piston in m

μ = viscosity of oil in Ns/m^2

11. Stokes law:

Drag force act on sphere: $F_D = 3\pi\mu \, Vd$.

where

V = velocity of sphere in stationary fluid

μ = viscosity of fluid

d = diameter of sphere.

12. Measurement of viscosity: The devices used to determine the viscosity of fluid is called viscometer or viscosimeter. Some of the common devices for the measurement of viscosity may be classified as:

(*i*) Capillary tube viscometer:

Dynamic viscosity: $\mu = \dfrac{\pi \, \rho g \, hd^4}{128 \, lQ}$

where

ρ = density of liquid

h = pressure head loss

d = diameter of tube

l = length of tube

Q = discharge flow through pipe

(*ii*) Rotating cylinder viscometer:

Dynamic viscosity: $\mu = \dfrac{2h(R_2 - R_1)T}{\pi\omega R_1^2 \, [4Hh \, R_2 + R_1^2(R_2 - R_1)]}$

where

h = clearance at the bottom of the cylinders

R_1, R_2 = radii of inner and out cylinder

ω = angular velocity of the outer cylinder

T = torque measured on the dial by means of torsion spring.

Contd...

(iii) Falling sphere viscometer:

Dynamic viscosity: $\mu = \dfrac{gd^2}{18v}[\rho_s - \rho]$

where $\quad d$ = diameter of sphere

V = velocity of sphere

ρ_s = density of sphere

ρ = density of liquid

(iv) Industrial viscometers:

(a) Saybolt viscometer:

Kinematic viscosity: $v = \left[0.22t - \dfrac{195}{t}\right]$ stokes for $t > 50$ and < 100 SUS

$= \left[0.22t - \dfrac{135}{t}\right]$ stokes for $t > 100$ SUS

where the t is time noted in seconds, known as Saybolt Universal Seconds (SUS) and dynamic viscosity: $\mu = v\rho$

where ρ = density of the liquid at the same temperature

(b) Redwood viscometer,

Kinematic viscosity: $v = \left[0.00246 - \dfrac{0.9}{t}\right]$ stokes

for $t > 40$ and < 85 RWS

$= \left[0.00246 - \dfrac{0.85}{t}\right]$ stokes

for $t > 85$ and < 200 RWS

where t is time noted in seconds, known as Redwood Second (RWS) and dynamic viscosity: $\quad \mu = v\rho$

where ρ = density of the liquid at same temperature.

ASSIGNMENT - 1

1. What do you mean by viscous flow?
2. Differentiate between laminar and turbulent flow.
3. Describe Reynolds experiment to demonstrate the two types of flow.
4. What do you mean by critical velocity of a fluid through a circular pipe?
5. What is the relation between shear stress and the velocity gradient?

6. Show for viscous flow through a circular pipe and the velocity distribution across the section is parabolic. Show that the mean velocity is equal to one half of the maximum velocity.

7. Prove that the loss of pressure head for the viscous flow through a circular pipe is given by

$$h_f = \frac{32\mu \bar{u}l}{\rho g D^2}$$

where
$\bar{u}$ = average velocity of flow

μ = viscosity of flowing fluid

L = length of pipe

D = diameter of pipe and

ρ = density of fluid

(GGSIP University Delhi, 2006)

8. What is a Couette flow?   *(GGSIP University, Delhi, Dec. 2005)*

9. Derive an expression for velocity distribution for laminar flow through a circular pipe. Show that the Darcy friction coefficient is equal to 16/Re where Re is the Reynolds number.   *(GGSIP University, Delhi, Dec. 2008)*

10. Prove that for a steady laminar flow between two fixed parallel plates, the velocity distribution across a section is parabolic and the average velocity is 2/3rd of the maximum velocity.

11. Prove that the maximum velocity of flow between two fixed parallel plates and for laminar flow, is equal to 1.5 times of the average velocity of flow.

(GGSIP University, Delhi, Dec. 2005)

12. What do you mean by momentum correction factor and kinetic energy correction factor?

13. Show that the momentum correction factor is 4/3 for viscous flow through a pipe.

14. Show that the kinetic energy correction factor is 2 for viscous flow through a pipe.

15. What are the various types of bearings?

16. Prove that power absorbed in overcoming viscous resistance in foot-step bearing is given by

$$P = \frac{\mu \pi^3 N^2 R^4}{1800t} \text{ watt}$$

where
μ = viscosity of liquid in Ns/m^2

N = speed of shaft in rpm

R = radius of shaft in m

t = clearance between shaft and

foot-step bearing in m

17. Prove that power absorbed in overcoming viscous resistance in collar bearing is given by

$$P = \frac{\mu \pi^3 N^2}{1800t} [R^4_2 - R^4_1] \text{ watt}$$

where

N = speed of shaft in rpm
R_1 = radius of shaft in m
R_2 = radius of collar in m
t = thickness of oil film in m

18. Show that the velocity of the piston in a dash pot is given by

$$V = \frac{4Wb^3}{3\pi l D^3 \mu} \text{ with the usual notations}$$

19. What are the different methods of determining the co-efficient of viscosity of a liquid? Describe with neat sketch any one method in detail.

20. What are the differents methods of measurement of viscosity of a liquid? Describe any one method in detail.

21. Describe the rotating cylinder method of determining the coefficient of viscosity of a liquid. (*GGSIP University, Delhi, Dec. 2005*)

22. Derive an expression for Navier-Stokes equations of motion of a viscous incompressible fluid in cartesian co-ordinates.

23. The Navier-Stokes equations for incompressible flow in vector notation is given by

$$\rho \frac{D\vec{V}}{Dt} = \vec{F}_B - \text{grad } p + \mu \nabla^2 \vec{V}$$

Where

$\vec{V}$ = velocity vector
$\vec{F}_B$ = body force vector, and
∇^2 = laplace operator

Derive the above equations in cartesian co-ordinates

24. What do mean by fluidization?

ASSIGNMENT - 2

1. An oil of viscosity 4.5 poise flows in a 40 mm diameter pipe, discharge rate being 5 litre/s. If the specific gravity of oil is 0.85, state whether flow is laminar or turbulent. **Ans.** Laminar

2. An oil of specific weight 8910 N/m^3 and kinematic viscosity of 1.5 stoke is pumped through a 160 mm diameter and 250 mm long pipe at the rate of 210 kN/h. Show that the flow is viscous and find the power required.

Ans. The flow is viscous, 90.77 W

3. A crude oil of viscosity 0.97 poise and specific gravity 0.9 is flowing through a horizontal circular pipe of diameter 100 mm and length 10 m. Find the difference of pressure at the two ends of the pipe, if 100 kg of the oil is collected in a tank in 30 seconds.   **Ans.** 1.462 kPa

4. A fluid of viscosity 7 poise and specific gravity 1.3 is flowing through a circular pipe of diameter 100 mm. The maximum shear stress at the pipe wall is given as 196.2 N/m^2, find (*i*) the pressure gradient (*ii*) the average velocity and (*iii*) Reynolds number of flow.

Ans. (*i*) − 7848 N/m^2 per m (*ii*) 3.5 m/s (*iii*) 650

5. A laminar flow is taking place in a pipe of diameter of 200 mm. The maximum velocity is 1.5 m/s. Find the mean velocity and the radius at which this occurs. Find also the velocity at 40 mm from the wall of the pipe.

Ans. 0.75 m/s, r = 70. 71 mm, 0.96 m/s

6. An oil of viscosity 20 poise flows between two horizontal fixed parallel plates which are kept at a distance 100 mm apart. The maximum velocity of flow is 2 m/s. Find

  (*i*) the pressure gradient,

 (*ii*) the shear stress at the two horizontal parallel plates, and

(*iii*) the discharge per unit width for laminar flow of oil.

Ans. (*i*) − 3200 N/m^3 (*ii*) 160 N/m^2 (*iii*) 0.1333 m^3/s

7. A shaft of diameter of 50 mm rotates in a journal bearing having a diameter of 50.15 mm and length 100 mm. The angular space between the shaft and the bearing is filled with oil having viscosity of 0.9 poise. Find the power absorbed in the bearing when the speed of the shaft is 60 rpm.

Ans. 46.43 W

8. A shaft 100 mm diameter runs in a bearing of length 200 mm with a radial clearance of 0.025 mm at 30 rpm. Find the viscosity of oil, if the power required to overcome the viscous resistance is 184 W.   **Ans.** 29.71 poise

9. Find the power required to rotate a circular disc of diameter 200 mm at 250 rpm. The circular disc has a clearance of 0.5 mm from the bottom flat plate and clearance contains oil of viscosity 1.5 poise.   **Ans.** 32.249 W

10. The external and internal diameter of a collar bearing are 200 mm and 150 mm respectively. Between the collar surface and bearing, an oil film of thickness 0.25 mm and of viscosity 1 poise is maintained. Find the torque and power lost in overcoming the viscous resistance of oil when shaft is running at 500 rpm.   **Ans.** 2.246 Nm, 117.54 W

11. The specific gravity of an oil of 0.9 is flowing through capillary tube of diameter 50 mm. The difference of pressure head between two points 2.5 m apart is 0.6 m of oil. The mass of oil collected in a measuring tank is 70 kg in 100 seconds. Find the viscosity of oil.   **Ans.** 4.177 poise

12. A capillary tube of diameter 20 mm and length 1.5 m is used for measuring viscosity of a liquid. The difference of pressure between the two ends of the tube is 50 kPa and viscosity of liquid is 0.2 poise. Find the rate of flow of liquid through the tube.   **Ans.** 6.54 litre/s

13. A sphere of diameter 2 mm falls 200 mm in 20 seconds in a viscous liquid. The density of the sphere is 800 kg/m^3 and a liquid is 850 kg/m^3. Find the viscosity of the liquid.   **Ans.** 15.58 poise

14. A sphere of diameter 1 mm falls through 335 mm in 100 seconds in a viscous fluid. If the relative densities of the sphere and the liquid are 7 and 0.96 respectively, find the dynamic viscosity of the liquid.   **Ans.** 9.82 poise

Hint:

$$d = 1 \text{ mm} = 0.001 \text{ m}$$

$$l = 335 \text{ mm} = 0.335$$

$$t = 100 \text{ s}$$

$$V = \frac{l}{t} = \frac{335}{100} = 0.00335 \text{ m/s}$$

Relative density of sphere $= 7$

∴   Density of sphere: $\rho_s = 7 \times 1000 \text{ kg/m}^3 = 7000 \text{ kg/m}^3$

Relative density of liquid $= 0.96$

∴   Density of liquid: $\rho = 0.96 \times 1000 \text{ kg/m}^3 = 960 \text{ kg/m}^3$

∴

$$\mu = \frac{8d^2}{18V} [\rho_s - \rho] = \frac{8 \times (0.001)^2}{18 \times 0.00335}[7000 - 960]$$

$$= 0.9826 \text{ Ns/m}^2 = \textbf{9.82 poise}$$

15. Find the viscosity of a liquid for the following given data:

Diameters of inner and outer cylinder are 200 mm and 205 mm respectively

Height of liquid in the cylinder $= 300$ mm

Clearance at the bottom of two cylinders $= 5$ mm

Speed of outer cylinder $= 400$ rpm

Reading of the torsion meter $= 4.5$ Nm   **Ans.** 1.474 poise

□□□

Turbulent Flow

2.1 INTRODUCTION

In the previous chapter with the help of Reynolds experiment, we explained the types of flow *i.e.,* laminar or turbulent. In practical application most fluid flows are turbulent in nature. This term (turbulent) denotes a motion in which an irregular fluctuation (mixing, or eddying motion) is superimposed on the main stream. This chapter deals with the fundamental concepts of turbulence, various theories of turbulence and the derivation of equation for velocity distribution and frictional resistance in turbulent flow through pipes.

2.2 TYPES OF VELOCITIES IN A TURBULENT FLOW

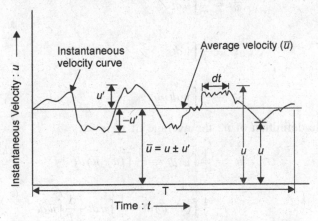

Fig. 2.1: Variation of velocity with time in turbulent flow.

© The Author(s) 2023
S. Kumar, *Fluid Mechanics (Vol. 2),*
https://doi.org/10.1007/978-3-030-99754-0_2

Turbulent flow is three-dimensional flow. The velocity at any point in a turbulent flow fluctuates both in magnitude and direction in all the three axes. H.L. Dryden first measured the velocity fluctuating using a hot wire anemometer. Figure 2.1 shows the u-component of instantaneous velocity varying with the t. Similar variation can be shown for v and w components.

In turbulent flows, three types of velocities involved are:

(i) Instantaneous velocity

(ii) Fluctuation velocity

(iii) Average velocity.

(i) **Instantaneous Velocity**: The velocity of turbulent flow at any instant known as instantaneous velocity. It is denoted by u, v and w along x, y and z-axes respectively.

(ii) **Fluctuation Velocity**: The variation of the velocity in turbulent flow with respect to average velocity is called fluctuation velocity. It is denoted by u', v' and w' along x, y and z-axes respectively.

(iii) **Average Velocity**: The velocity noted at a point over a certain period of time and there after time averaged, is called as average velocity. It is denoted by $\bar{u}$, $\bar{v}$ and $\bar{w}$ along x, y and z-axes respectively. Average velocity is also known as temporal velocity. By definition, the average velocity at any quantity, $a = f(x, y, z, t)$ may be denoted by $\bar{a}$ and expressed as:

$$\bar{a} = \frac{1}{T}\int_0^T a\,dt$$

Average velocity components may be expressed as

$$\bar{u} = \frac{1}{T}\int_0^T u\,dt$$

$$\bar{v} = \frac{1}{T}\int_0^T v\,dt$$

and

$$\bar{w} = \frac{1}{T}\int_0^T w\,dt$$

and by the definition of u', the average of u' is

$$\bar{u'} = \frac{1}{T}\int_0^T u'\,dt = \frac{1}{T}\int_0^T (u-\bar{u})\,dt \qquad \because\ u' = u - \bar{u}$$

$$= \frac{1}{T}\int_0^T (u-\bar{u})\,dt = \frac{1}{T}\int_0^T u\,dt - \frac{1}{T}\int_0^T \bar{u}\,dt$$

$$= \bar{u} - \bar{u} \qquad \because\ \frac{1}{T}\int_0^T \bar{u}\,dt = \bar{u}$$

$$= 0$$

Similarly, $\qquad \bar{v'} = \bar{w'} = 0$

where T is the period of sampling which must be sufficiently large to include an adequate number of fluctuations.

2.2.1 Relation between Various Velocities

The relationship between the instantaneous velocity, fluctuation velocity and time average velocity is shown in Fig. 9.1.

Time average velocity along x-axis : $\bar{u} = u \pm u'$

Similarly along y-axis: $\bar{v} = v \pm v'$

and along z-axis: $\bar{w} = w \pm w'$

RMS (root-mean square) value of turbulent fluctuation of velocity:

$$\bar{u'} = \sqrt{\frac{1}{T}\int_0^T u'^2\, dt} = \sqrt{\bar{u'^2}}$$

where $\bar{u'}$ is RMS value on x-axis.

Similarly $\bar{v'}$ is RMS value of turbulent fluctuation of velocity in y-axis:

$$\bar{v'} = \sqrt{\frac{1}{T}\int_0^T v'^2\, dt} = \sqrt{\bar{v'^2}}$$

and $\bar{w'}$ is RMS value of turbulent fluctuation of velocity:

$$\bar{w'} = \sqrt{\frac{1}{T}\int_0^T w'^2\, dt} = \sqrt{\bar{w'^2}}$$

2.2.2 Degree or Level of Turbulence

It is the square root of the arithmetic mean of the mean squares values of velocities in x, y and z axes. It is denoted by D.

Degree or level of turbulence:

$$D = \sqrt{\frac{1}{3}\left(\bar{u'^2} + \bar{v'^2} + \bar{w'^2}\right)}$$

2.2.3 Intensity of Turbulence

It is the ratio of the degree of turbulence (D) and the free-stream velocity (U). It is denoted by I.

Mathematically,

Intensity of turbulence:

$$I = \frac{\text{Degree of turbulence}: D}{\text{Free-stream velocity}: U}$$

$$= \frac{1}{U}\sqrt{\frac{1}{3}\left(\bar{u'^2} + \bar{v'^2} + \bar{w'^2}\right)}$$

An isotropic flow is a turbulent flow for which the average fluctuation velocity is same in all three coordinate directions.

i.e., $$\bar{u'^2} = \bar{v'^2} = \bar{w'^2}$$

In this case the longitudinal velocity u' alone can be used for the turbulence intensity; thus

$$I = \sqrt{\frac{\overline{u'^2}}{U}}$$

The drag measurement at spheres in different wind tunnels show great dependence of the critical flow Reynolds number (Re_c) or turbulence intensity (I); Re_c increases greatly with decreasing I. The turbulence intensity in older wind tunnels was about 0.01.

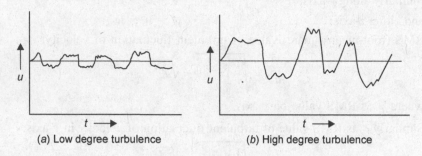

(*a*) Low degree turbulence   (*b*) High degree turbulence

Fig. 9.2: Turbulence

2.3 CLASSIFICATION OF TURBULENCE

Turbulence may be generated in various ways other than just flowing through a pipe. In general, it can result either from contact of the flowing fluid with solid boundaries or from contact between two layers of fluid flow moving at different velocities.

There are two types of turbulence (*i*) Wall turbulence, (*ii*) Free turbulence.

(*i*) *Wall Turbulence*: The turbulence generated by the viscous effect of the flowing fluid due to the presence of a solid boundary is called wall turbulence. For example: flow through pipes, flow through open channels and flow over solid bodies.

(*ii*) *Free Turbulence*: If turbulence is generated, in the absence of a solid boundary, by contact between two layers of fluid flow at different velocities, it is called free turbulence. For example: turbulence in jet mixing region.

2.4 SHEAR STRESS IN TURBULENT FLOW

The mechanism of turbulence is very complex because of three-dimensional velocity fluctuation occurring rapidly, mixing of fluid particles causing creation of eddies and transfer of momentum. The following numbers of semi-empirical theories and formulae were developed to find the shear stress in turbulent flow as:

(*i*) Reynolds theory.

(*ii*) Boussinesq eddy-viscous theory.

(*iii*) Prandtl's mixing length theory, and

(*iv*) Von-Karman's theory.

2.4.1 Reynolds Theory

Osborne Reynolds in 1894 developed an expression for turbulent shear stress between two layers of a fluid due to the transverse momentum exchange in turbulent mixing process.

Consider an arbitrary plane AB between two layers of fluid, separated by a distance dy (or l) as shown in Fig. 2.3. At the point of intersection of the plane AB with velocity profile.

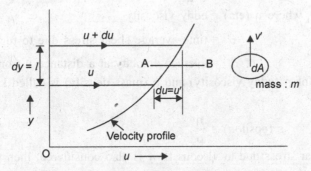

Fig. 2.3: Transverse momentum exchange in turbulence.

Let $\dfrac{du}{dy}$ = velocity gradient,

m = mass of small fluid element,

dA = area of the small fluid element,

v' = fluctuating component of velocity in the direction of y due to turbulence.

Mass per second moving across AB $= \rho dAv'$

Turbulent shear force = rate of change of momentum.

$\quad\quad\quad\quad\quad\quad$ = mass per second × change in velocity

$\quad\quad\quad\quad\quad\quad$ $= \rho dAv' du$

where $du = u'$, the longitudinal turbulent fluctuation.

∴ Turbulent shear force $= \rho dAv'(-u')$

where −ve sign shows v' increases then u' decreases.

$\quad\quad\quad\quad\quad\quad$ $= -\rho dA u' u'$

Turbulent shear stress: $\tau_t = \dfrac{\text{Turbulent shear force}}{\text{Area}}$

$\quad\quad\quad\quad\quad\quad$ $= \dfrac{-\rho dAv'u'}{dA} = -\rho v'u'$ $\quad\quad\quad\quad$ ...(2.4.1)

As u' and v' are varying and hence τ_t will also vary. Hence to find turbulent shear stress, the time average on both sides of the Eq. (2.4.1) is taken

∴ $\quad\quad\quad\quad\quad\quad \overline{\tau_t} = -\rho \overline{v'u'}$ $\quad\quad\quad\quad\quad\quad$ ...(2.4.2)

The turbulent shear stress in Eq. (2.4.2) is known as **'apparent'** or **Reynolds stress.**

2.4.2 Boussinesq Eddy-viscous Theory

According to Boussinesq (1877), in a turbulent flow, apart from the viscous shear stress, there is shear stress due to eddies formation. Time average shear stress in turbulent flow is

$$\overline{\tau_t} = \eta \frac{d\overline{u}}{dy}$$

where η (eta) = eddy viscosity

$\overline{\tau_t}$ = time average shear stress due to turbulence

$\overline{u}$ = average velocity at a distance y from boundary.

The ratio of η (eddy viscosity) and ρ (mass density) is called kinematic eddy viscosity (ε).

i.e., ε (epsilon) = $\dfrac{\eta}{\rho}$

If the shear stress due to viscous flow is also considered, then the total shear stress becomes:

$$\tau = \overline{\tau_v} + \overline{\tau_t}$$

$$= \mu \frac{d\overline{u}}{dy} + \eta \frac{d\overline{u}}{dy} = (\mu + \eta)\frac{d\overline{u}}{dy}$$

Actually η varies widely in magnitude, not only with general direction of flow but also from point to point in a given flow. While μ represents a fluid property independent of the type of flow, η depends upon the fluid density as well as upon the intensity of turbulent fluctuations.

In laminar flow there being no transverse fluctuation, $\eta = 0$. For other cases the value of η may be several thousand times the value of μ.

Dynamic viscosity: μ = kinematic viscosity (ν) × mass density (ρ)

i.e., $\mu = \nu\rho$

and   Eddy viscosity: η = kinematic eddy viscosity (ε) × mass density (ρ).

i.e., $\eta = \varepsilon\rho$

SI units of μ and η is Ns/m^2 and ν and ε is m^2/s.

Kinematic eddy viscosity (ε) may be treated as a measure of the transporting capacity of the turbulent mixing process.

2.4.3 Prandtl's Mixing Length Theory

Recalling Reynolds shear stress in turbulent flow,

$$\overline{\tau_t} = -\rho \overline{v'\,u'} \qquad\qquad\qquad ...(2.4.3)$$

$\overline{\tau_t}$ can only be calculated if the value of $u'v'$ is known. But it is very difficult to measure $\overline{u}\,\overline{v}$. To overcome this problem, Prandtl in 1926, presented mixing length hypothesis which can be used to express turbulent shear stress in terms of measurable quantities.

According to Prandtl, mixing length (l) is defined as the average distance between two layers in the transverse direction such that small mass of fluid particles from one layer could reach the other layer in such a way that the momentum of the particles in the direction of x is same. He also assumed that the velocity fluctuation in the x-direction u' is related to mixing length l as (see Fig. 2.3)

$$u' = l \frac{du}{dy}$$

and v' the fluctuation component of velocity in y-direction is of the same order of magnitude as u' and hence

$$v' = l \frac{du}{dy} = u'$$

$$\overline{u'} = l \frac{d\overline{u}}{dy} = \overline{v'}$$

$$\therefore \qquad \overline{u'v'} = l \frac{d\overline{u}}{dy} \times l \frac{d\overline{u}}{dy} = l^2 \left(\frac{d\overline{u}}{dy} \right)^2$$

Substituting the value of $\overline{u'v'}$ in Eq. (2.4.3), we get, the expression for shear stress in turbulent flow due to Prandtl as:

$$\overline{\tau_t} = \rho l^2 \left(\frac{d\overline{u}}{dy} \right)^2$$

Thus, the total shear stress at any point in turbulent flow is sum of shear stress due to viscous shear and turbulent shear and can be written as:

$$\tau = \overline{\tau_v} + \overline{\tau_t}$$

$$\boxed{\tau = \mu \frac{d\overline{u}}{dy} + \rho l^2 \left(\frac{d\overline{u}}{dy} \right)^2}$$

But the viscous shear stress is negligible except near the boundary. Prandtl's mixing length (l) is influenced by Reynolds number and transverse distance from boundary (y). The simplest relation between l and y is given by:

$$\boxed{l = ky}$$

where k is a proportionality constant which must be determined from experiments. In region close to the boundary the value of $k = 0.4$.

2.4.4 Von-Karman's Theory

Von-Karman (1936) developed a mathematical expression for mixing length (l) of Prandtl, in terms of the derivatives of the time average velocity, Von-Karman found that:

$$l = \kappa \frac{d\overline{u}/dy}{d^2\overline{u}/dy^2}$$

So that turbulent shear stress:

$$\overline{\tau_t} = \rho l^2 \left(\frac{d\overline{u}}{dy}\right)^2 = \rho\kappa^2 \frac{\left(d\overline{u}/dy\right)^2}{\left(d^2\overline{u}/dy^2\right)^2} \times \left(\frac{d\overline{u}}{dy}\right)^2$$

$$\overline{\tau_t} = \frac{\rho\kappa^2 \left(d\overline{u}/dy\right)^2}{\left(d^2\overline{u}/dy^2\right)^2}$$

where κ (kappa) is a universal constant known as Karman's mixing length constant (determined from experiments) and is independent of the boundary geometry, roughness and Reynolds number.

2.5 VELOCITY DISTRIBUTION LAW IN TURBULENT FLOW

The flow decelerates near a solid boundary due to viscous effect by the formation of boundary layer. The velocity of fluid particles decreases towards the solid boundary due to the action of viscous shear and becomes equal to zero at the solid boundary.

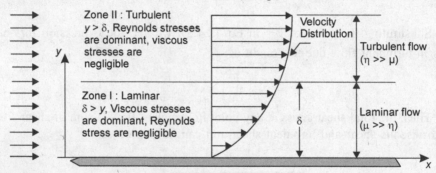

Fig. 2.4: Comparative role of Reynolds stresses and viscous stresses in affecting flow phenomena near a smooth boundary.

According to Nikurades and Karman, turbulent flow over a solid boundary may be divided into three zones. The first zone consist of a thin layer near a solid boundary in which the viscous shear stress predominates while the Reynolds stress is negligible. This zone is the region of laminar sub-layer. The second zone immediately above the laminar sub-layer is called the 'buffer-zone' in which the viscous shear stress is equal to the Reynolds stresses. In buffer zone, the flow changes from laminar to turbulent flow. The third zone above the buffer zone in which the Reynolds stresses are predominant while the viscous shear stress is negligible. This zone lies between turbulent boundary layer and buffer zone. This zone is called the turbulent zone.

For convenience, the flow is usually divided into two zones. The first zone near the solid boundary accommodating the laminar sub-layer is called the zone of laminar layer. In this layer, the viscous shear stresses are predominant over the Raynolds stress *i.e.,* dynamic viscosity is greater than eddy viscosity *i.e.,* $\mu \gg \eta$.

The second zone represents the remaining portion up to turbulent boundary layer, is called turbulent layer. In this layer, the Reynolds stresses are predominant over viscous shear stress *i.e.,* the eddy viscosity: $\eta \gg \mu$, dynamic viscosity. These two zones are shown in Fig. 2.4.

2.5.1 Velocity Distribution in Laminar Region: Zone-I

The shear stress very close to the boundary may be taken as equal to the surface shear stress (τ at $y = 0$).

According to Newton's law viscosity,

$$\tau_0 = \mu \frac{du}{dy}$$

$$\tau_0 = \mu \frac{u}{y} \qquad \qquad \because \frac{du}{dy} = \frac{u}{y} \text{ in laminar region}$$

Dividing by ρ on both sides, we get

$$\frac{\tau_0}{\rho} = \frac{\mu}{\rho} \frac{u}{y}$$

Here $\sqrt{\dfrac{\tau_0}{\rho}}$ has the dimension $\sqrt{\dfrac{ML^{-1}T^{-2}}{ML^{-3}}} = \sqrt{\dfrac{L^2}{T^2}} = \dfrac{L}{T}$

This is dimension of velocity and is generally known as shear velocity u_*.

i.e.

$$\sqrt{\frac{\tau_0}{\rho}} = u_*$$

or

$$\frac{\tau_0}{\rho} = u_*^2$$

Substituting $\dfrac{\tau_0}{\rho} = u_*^2$ in above equation, we get.

$$u_*^2 = \frac{\mu}{\rho} \frac{u}{y}$$

$$u_*^2 = \nu \frac{u}{y} \qquad \qquad \because \frac{\mu}{\rho} = \nu \ : \text{kinematic viscosity}$$

or

$$\frac{u_* y}{\nu} = \frac{u}{u_*}$$

or

$$\frac{u}{u_*} = \frac{u_* y}{\nu} \qquad \qquad \qquad \ldots(2.5.1)$$

Above Eq. (2.5.1) shows a linear variation between u and y in the laminar flow.

2.5.2 Velocity Distribution in the Turbulent Region: Zone-II

According to Prandtl's mixing length theory, the time-average turbulent shear stress:

$$\bar{\tau} = \rho l^2 \left(\frac{d\bar{u}}{dy} \right)^2 \qquad \qquad \ldots(2.5.2)$$

and Prandtl also assumed that mixing length l is linear variation to distance from the boundary of the flow y.

i.e.,

$$l \propto y$$

$$l = ky$$

where k is the Karman's constant to be determined from experiments.

Substituting the value of l in Eq. (2.5.2), we get,

$$\tau = \rho k^2 y^2 \left(\frac{d\overline{u}}{dy} \right)^2$$

Thus, turbulent shear stress near the boundary.

$$\tau_0 = \rho k^2 y^2 \left(\frac{d\overline{u}}{dy} \right)^2$$

$$\frac{1}{k^2 y^2} \frac{\tau_0}{\rho} = \left(\frac{d\overline{u}}{dy} \right)^2$$

or

$$\frac{d\overline{u}}{dy} = \frac{1}{ky} \sqrt{\frac{\tau_0}{\rho}}$$

$$\frac{d\overline{u}}{dy} = \frac{1}{ky} u_* \qquad \qquad \because \sqrt{\frac{\tau_0}{\rho}} = u_*, \text{ shear velocity}$$

$$d\overline{u} = \frac{u_*}{k} \frac{dy}{y}$$

On integration above equation, we get

$$\overline{u} = \frac{u_*}{k} \log_e y + c \qquad \qquad ...(2.5.3)$$

where c is the constant of integration and determined from the boundary condition.

At distance y' from the solid boundary, $u = 0$.

$$0 = \frac{u_*}{k} \log_e y' + c$$

or

$$c = -\frac{u_*}{k} \log_e y'$$

Substituting the value of c in Eq. (2.5.3), we get

$$u = \frac{u_*}{k} \log_e y - \frac{u_*}{k} \log_e y'$$

$$u = \frac{u_*}{k} \left[\log_e y - \log_e y' \right]$$

$$\frac{u}{u_*} = \frac{1}{k} \log_e \left(\frac{y}{y'} \right)$$

Nikuradse's experiments have revealed that, $k = 0.4$ for regions very close to the boundary.

$$\frac{u}{u_*} = \frac{1}{0.4}\log_e\left(\frac{y}{y'}\right)$$

$$\frac{u}{u_*} = 2.5\log_e\left(\frac{y}{y'}\right) \qquad\qquad ...(2.5.4)$$

Case-I: For smooth pipes: Nikuradse's experiments on smooth pipes yielded a value of

$$y' = \frac{0.111\,v}{u_*}$$

where v = kinematic viscosity of fluid.

substituting the value of y' in Eq. (2.5.4), we get

$$\frac{u}{u_*} = 2.5\log_e\left(\frac{y}{0.111\,v/u_*}\right)$$

$$= 2.5\log_e\left(\frac{yu_*}{0.111\,v}\right)$$

$$= 2.5\log_e\left(\frac{yu_*}{v}\right) - 2.5\log_e(0.111)$$

$$= 2.5\log_e\left(\frac{yu_*}{v}\right) - 2.5\,(-2.198)$$

$$= 2.5\log_e\left(\frac{yu_*}{v}\right) + 5.50$$

$$= 2.5 \times 2.3\log_{10}\left(\frac{yu_*}{v}\right) + 5.50$$

$$\therefore \qquad \frac{u}{u_*} = 5.75\log_{10}\left(\frac{yu_*}{v}\right) + 5.50 \qquad\qquad ...(2.5.5)$$

The above Eq. (2.5.5) is known as the Karman-Prandtl equation for turbulent flow near smooth boundaries.

Case-II: For rough pipes.

In this case, the roughness projections are larger than the thickness of laminar sub-layer *i.e.,* $k > \delta'$.

Nikuradse's experiments on rough pipes yielded as value of $y' = \dfrac{k}{30}$

Substituting the value of y' in Eq. (2.5.4), we get

$$\frac{u}{u_*} = 2.5\log_e\left(\frac{y}{k/30}\right) = 2.5\log_e\left(\frac{y \times 30}{k}\right)$$

$$= 2.5\log_e\left(\frac{y}{k}\right) + 2.5\log_e 30$$

$$= 2.5 \times 2.3 \log_{10}\left(\frac{y}{k}\right) + 2.5 \times 3.4$$

$$\boxed{\frac{u}{u_*} = \mathbf{5.75} \log_{10}\left(\frac{y}{k}\right) + \mathbf{8.5}}$$

...(2.5.6)

The above Eq. (2.5.6) is known as the Karman-Prandtl equation for turbulent flow near rough boundaries.

2.5.3 Relation between u_{max} and u

Recalling equation (2.5.3),

$$u = \frac{u_*}{k} \log_e y + C$$

...(2.5.7)

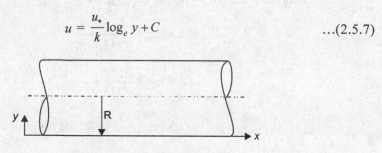

Fig. 2.5: Flow through pipe

The velocity of fluid particles maximum at the centre of pipe,

at $y = R$, $u = u_{max}$

$\therefore$

$$u_{max} = \frac{u_*}{k} \log_e R + C$$

or

$$c = u_{max} - \frac{u_*}{k} \log_e R$$

Substituting the value of c in Eq. (2.5.7), we get

$$u = \frac{u_*}{k} \log_e y + u_{max} - \frac{u_*}{k} \log_e R$$

$$u = u_{max} + \frac{u_*}{k}\left[\log_e y - \log_e R\right]$$

Karman constant: $k = 0.4$

$\therefore$

$$u = u_{max} + \frac{u_*}{0.4} \log_e\left(\frac{y}{R}\right)$$

$$u = u_{max} + 2.5 u_* \log_e\left(\frac{y}{R}\right)$$

...(2.5.8)

The above Eq. (2.5.8) is known as Prandtl's universal velocity distribution equation for turbulent flow in pipes. This equation is applicable to smooth as well as rough pipe boundaries.

Eq. (8) is also written as:

$$u_{max} - u = -2.5u_* \log_e \left(\frac{y}{R}\right) = 2.5u_* \log_e \left(\frac{R}{y}\right)$$

Dividing by u_* on both sides, we get

$$\frac{u_{max} - u}{u_*} = 2.5 \log_e \left(\frac{R}{y}\right) = 2.5 \times 2.3 \log_{10} \left(\frac{R}{y}\right)$$

$$\boxed{\frac{u_{max} - u}{u_*} = 5.75 \log_{10} \left(\frac{R}{y}\right)} \qquad \ldots(2.5.9)$$

Eq. (2.5.9) the difference between the maximum velocity (u_{max}) and local velocity (u) at any point *i.e.*, ($u_{max} - u$) is known as **velocity defect.**

2.6 HYDRODYNAMICALLY SMOOTH AND ROUGH BOUNDARIES

For laminar flow, all rough pipes irrespective of their roughness size and pattern offer the same resistance as that offered by a smooth pipe under similar condition of flow. As a matter of fact, there is no surface which may be regarded as perfectly smooth.

Let k be the average height of the irregularities projecting from the surface of a boundary as shown in Fig. 2.6.

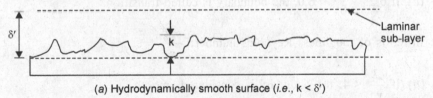

(a) Hydrodynamically smooth surface (*i.e.*, k < δ′)

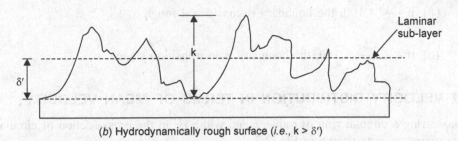

(b) Hydrodynamically rough surface (*i.e.*, k > δ′)

Fig. 2.6: Smooth and rough boundaries and δ′is the thickness of laminar sub-layer.

2.6.1 Hydrodynamically Smooth Boundary

If the irregularities of any actual surface are such that the surface projections are completely covered by the laminar sub-layer *i.e.*, the average height k of the irregularities projecting from the surface of a boundary is much less than δ′, the thickness of laminar sub-layer as shown in Fig. 2.6(a). This type of boundary is called **hydrodynamically smooth boundary**. This is because, outside the laminar sub-layer, the flow is turbulent and eddies of various size present in turbulent flow try to penetrate the laminar sub-layer and reach the surface of the boundary. But due to great thickness of laminar sub-layer the eddies are unable to reach in the irregularities of the surface and hence the boundary behaves as a smooth boundary.

2.6.2 Hydrodynamically Rough Boundary

If the irregularities of the surface are above the laminar sub-layer and eddies present in turbulent zone will come in contact with the irregularities of the surface and lot of energy will be lost. Such a boundary is called **hydrodynamically rough boundary** as shown in Fig. 2.6(*b*). The thickness of laminar sub-layer depends upon Reynolds number, if the Reynolds number of the flow is increased then the thickness of laminar sub-layer will decrease. In case of hydrodynamically rough boundary, the average height k of the irregularities projecting from the surface of a boundary is much greater than δ', the thickness of laminar sub-layer.

From Nikuradse's experiment:

(*a*) if $\dfrac{k}{\delta'}$ < 0.25, the boundary is called smooth boundary

(*b*) if $\dfrac{k}{\delta'}$ > 6.0, the boundary is called rough boundary

(*c*) if $0.25 < \dfrac{k}{\delta'} < 6.0$, the boundary is called transition.

In terms of roughness Reynolds number: $\dfrac{u_* k}{v}$

(*a*) if $\dfrac{u_* k}{v}$ < 4, the boundary is considered smooth

(*b*) if $\dfrac{u_* k}{v}$ > 100, the boundary is considered rough

(*c*) if $4 < \dfrac{u_* k}{v} < 100$, the boundary is in transition stage.

2.7 VELOCITY DISTRIBUTION IN TERMS OF MEAN VELOCITY

Considering a circular ring of radius r and width dr in the cross-section of circular pipe of radius R as shown in Fig. 2.7.

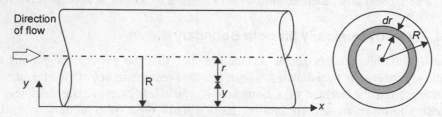

Fig. 2.7: Average velocity for turbulent flow.

Let u is velocity of fluid at radius r and area of small circular ring: $dA = 2\pi r dr$

Discharge through small circular ring:

$$dQ = dA.u = 2\pi r dr u$$

Net discharge through pipe:

$$\int dQ = \int_0^R 2\pi r u dr$$

$$Q = 2\pi \int_0^R r u\, dr \qquad\qquad\qquad ...(2.7.1)$$

Case-I: For smooth pipe: Considering the pipe boundary to be smooth and using the corresponding velocity distribution Eq. (2.5.5), we get

$$\frac{u}{u_*} = 5.75 \log_{10}\left(\frac{u_* y}{v}\right) + 5.50 \qquad\qquad ...(2.7.2)$$

Here $\qquad\qquad y = R - r$

$\therefore \qquad\qquad \dfrac{u}{u_*} = 5.75 \log_{10}\dfrac{u_*(R-r)}{v} + 5.50$

or $\qquad\qquad u = u_*\left[5.75\log_{10}\dfrac{u_*(R-r)}{v} + 5.50\right]$

Substituting the value of u in above Eq. (2.7.1), we get

$$Q = 2\pi\int_0^R r u_*\left[5.75\log_{10}\frac{u_*(R-r)}{v} + 5.50\right]dr$$

Integrating by parts, we get

$$Q = \pi R^2 u_* (5.75 \log_{10}\frac{u_* R}{v} + 1.75$$

The mean velocity is given by

$$\bar{u} = \frac{Q}{\pi R^2} = \frac{\pi R^2 u_* (5.75\log_{10}\dfrac{u_* R}{v} + 1.75}{\pi R^2}$$

$$\frac{\bar{u}}{u_*} = 5.75\log_{10}\frac{u_* R}{v} + 1.75 \qquad\qquad ...(2.7.3)$$

Subtracting Eq. (2.7.3) from Eq. (2.7.2), we get,

$$\boldsymbol{\frac{u - \bar{u}}{u_*} = 5.75\log_{10}\frac{y}{R} + 3.75}$$

Case-II: For rough pipe:

Considering the pipe boundary to be rough and using the corresponding velocity distribution Eq. (2.5.6), we get

$$\frac{u}{u_*} = 5.75\log_{10}\left(\frac{y}{k}\right) + 8.5 \qquad\qquad ...(2.7.4)$$

Here $\qquad\qquad y = R - r$

$$\therefore \qquad \frac{u}{u_*} = 5.75 \log_{10}\left(\frac{R-r}{k}\right) + 8.5$$

or
$$u = u_*\left[5.75 \log_{10}\frac{(R-r)}{k} + 8.5\right]$$

Substituting the value of u in Eq. (9.7.1), we get

$$Q = 2\pi \int_0^R r u_*\left[5.75 \log_{10}\frac{(R-r)}{k} + 8.5\right] dr$$

Integrating by parts, we get

$$Q = \pi R^2 u_*\left(5.75 \log_{10}\frac{R}{k} + 4.75\right)$$

The mean velocity is given by

$$\bar{u} = \frac{Q}{\pi R^2} = \frac{\pi R^2 u_*\left(5.75 \log_{10}\frac{R}{k} + 4.75\right)}{\pi R^2}$$

$$\bar{u} = u_*\left(5.75 \log_{10}\frac{R}{k} + 4.75\right) \qquad \qquad ...(2.7.5)$$

Subtracting Eq. (2.7.5) from Eq. (2.7.4), we get

$$\boxed{\frac{u - \bar{u}}{u_*} = 5.75 \log_{10}\frac{y}{R} + 3.75} \qquad \qquad ...(2.7.6)$$

Thus when expressed in terms of mean velocity, Karman-Prandtl velocity distribution equation for both smooth and rough boundaries becomes same.

2.8 POWER LAW FOR VELOCITY DISTRIBUTION IN SMOOTH PIPES

The velocity distribution formulae developed in the previous articles are logarithmic in nature. These formulae are not so convenient to use as compare to exponential ones. Nikuradse carried out experiments for wide range of Reynolds number $4 \times 10^3 \le \text{Re} < 2 \times 10^6$. The dimensionless velocity distribution law in smooth pipes, based on his work may be represented by an empirical equation in exponent form as:

$$\frac{u}{u_{max}} = \left(\frac{y}{R}\right)^n \qquad \qquad ...(2.8.1)$$

where exponent n is dependent on the Reynolds number Re. The value of exponent n decreases with increasing Reynolds number Re.

For Reynolds number:

$$\text{Re} = 4 \times 10^3, \qquad n = \frac{1}{6}$$

$$\text{Re} = 1.1 \times 10^5, \qquad n = \frac{1}{7}$$

$$\text{Re} \geq 2 \times 10^6, \qquad n = \frac{1}{10}$$

If $n = \frac{1}{7}$, the velocity distribution law becomes as

$$\frac{u}{u_{max}} = \left(\frac{y}{R}\right)^{1/7} \quad \text{for Re} \leq 10^5 \qquad \ldots(2.8.2)$$

Equation (2.8.2) is known as **one-seventh power law of velocity distribution for smooth pipes.** This equation is also known as the Blasius one-seventh power velocity distribution law.

2.9 DETERMINATION OF COEFFICIENT OF FRICTION *f*

(a) For laminar flow:

Coefficient of friction: $f = \dfrac{16}{\text{Re}}$ $\qquad \ldots(2.9.1)$

Equation (2.9.1) is valid for laminar flow, Re < 2000.

(b) For turbulent flow through smooth pipes.

(i) Coefficient of friction: $f = \dfrac{0.0791}{\text{Re}^{1/4}}$

where $4000 \leq \text{Re} < 10^5$.

(ii) $\dfrac{1}{\sqrt{4f}} = 2.03 \log_{10}\left(\text{Re}\sqrt{4f}\right) - 0.9$

This equation is valid up to Re = 4×10^6

(iii) Nikuradse's experimental result for coefficient of friction *f*

$$\frac{1}{\sqrt{4f}} = 2 \log_{10}\left(\text{Re}\sqrt{4f}\right) - 0.8 \qquad \ldots(2.9.2)$$

This equation is valid up to Re = 4×10^7.

But the Eq. (2.9.2) is solved by hit and trial method. The value of coefficient of friction can alternately be obtained as:

$$f = 0.0008 + \frac{0.05525}{(\text{Re})^{0.237}} \qquad \ldots(2.9.3)$$

(c) For turbulent flow through rough pipes:

Niuradse's experiment result gave the following relation for coefficient of friction *f*:

$$\frac{1}{\sqrt{4f}} = 2 \log_{10}\left(\frac{R}{k}\right) + 1.74 \qquad \ldots(2.9.4)$$

(d) For turbulent flow through commercial pipes:

(i) Smooth pipes:

$$\frac{1}{\sqrt{4f}} = 2 \log_{10}\left(\frac{\text{Re}\sqrt{4f}}{R/k}\right) - 0.8 \qquad \ldots(2.9.5)$$

(*ii*) Rough pipes:

$$\frac{1}{\sqrt{4f}} = 2\log_{10}\left(\frac{R}{k}\right) + 1.74 \qquad \qquad ...(2.9.6)$$

Problem 2.1: A pipeline carrying water has surface irregularities of average height 0.10 mm. If the shear stress developed is 7.85 N/m². Determine whether the pipe surface acts as smooth, rough, or in transition. The kinematic viscosity of water is 0.95×10^{-2} stokes.

Solution: Given data:

Average height of irregularities:
$$k = 0.10 \text{ mm} = 0.10 \times 10^{-3} \text{ m.}$$
Shear stress developed: $\tau_0 = 7.85$ N/m²
Kinematic viscosity: $v = 0.95 \times 10^{-2}$ stokes
$$= 0.95 \times 10^{-2} \times 10^{-4} \text{ m}^2/\text{s} = 0.95 \times 10^{-6} \text{ m}^2/\text{s}$$
We know that the shear velocity:

$$u_* = \sqrt{\frac{\tau_0}{\rho}} = \sqrt{\frac{7.95}{1000}} = 0.089 \text{ m/s}$$

Roughness Reynolds number $= \dfrac{u_* k}{v} = \dfrac{0.089 \times 0.10 \times 10^{-3}}{0.95 \times 10^{-6}} = 9.36$

Since the roughness Reynolds number $\dfrac{u_* k}{v}$ lies between 4 and 100 and hence, the pipe surface acts in transition.

Problem 2.2: A rough pipe is of diameter 100 mm. The velocity at a point 40 mm from wall is 35% more than the velocity at a point 10 mm from pipe wall. Find the average height of the roughness.

Solution: Given data:

Diameter of rough pipe:
$$D = 100 \text{ mm} = 0.1 \text{ m}$$
Let velocity of flow at 10 mm from pipe wall $= u$
Then, the velocity of flow at 40 mm from pipe wall $= u + 0.35u = 1.35 \ u$
We know that the velocity distribution for rough pipe from equation (9.5.6) is

$$\frac{u}{u_*} = 5.75\log_{10}\left(\frac{y}{k}\right) + 8.5$$

where k = average height of the roughness.
at $y = 10$ mm,

$$\frac{u}{u_*} = 5.75\log_{10}\left(\frac{10}{k}\right) + 8.5 \qquad \qquad ...(i)$$

at $y = 40$ mm,

$$\frac{1.354}{u_*} = 5.75 \log_{10}\left(\frac{40}{k}\right) + 8.5 \qquad \dots (ii)$$

Dividing Eq. (i) by (ii), we get

$$\frac{1}{1.35} = \frac{5.75 \log_{10}\left(\frac{10}{k}\right) + 8.5}{5.75 \log_{10}\left(\frac{40}{k}\right) + 8.5}$$

$$5.75 \log_{10}\left(\frac{40}{k}\right) + 8.5 = 1.35 \times 5.75 \log_{10}\left(\frac{10}{k}\right) + 1.35 \times 8.5$$

$$5.75 \log_{10}\left(\frac{40}{k}\right) + 8.5 = 7.76 \log_{10}\left(\frac{40}{k}\right) + 11.47$$

$$5.75 \log_{10} 40 - 5.75 \log_{10} k + 8.5 = 7.76 \log_{10} 40 - 7.76 \log_{10} k + 11.47$$

$$9.21 - 5.75 \log_{10} k + 8.5 = 12.43 - 7.76 \log_{10} k + 11.47$$

$$17.71 - 5.75 \log_{10} k = 23.9 - 7.76 \log_{10} k$$

or $\qquad 7.76 \log_{10} k = 23.9 - 17.71 = 6.19$

or $\qquad \log_{10} k = \dfrac{6.19}{7.76} = 0.7976$

or $\qquad k = 6.27 \text{ mm.}$

Problem 2.3: Find the wall shearing stress in a pipe of diameter 200 mm which carries water. The velocities at the pipe centre and 60 mm from the pipe centre are 3 m/s and 2 m/s respectively. The flow in pipe is given as turbulent.

Solution: Given data:

Diameter of pipe: $D = 200 \text{ mm} = 0.2 \text{ m}$

$\therefore$ Radius of pipe: $\qquad R = \dfrac{D}{2} = \dfrac{0.2}{2} = 0.1 \text{m}$

Velocity at centre: $\qquad u_{max} = 3 \text{ m/s}$

Velocity at 60 mm from centre $\quad = 2 \text{ m/s}$

i.e., Velocity at $\qquad r = 60 \text{ mm} : u = 2 \text{ m/s}$

$$y = R - r = 0.1 - 0.06 = 0.04 \text{ m}$$

For turbulent flow, the velocity distribution in terms of maximum velocity (u_{max}) is given by Eq. (9.5.9) as

$$\frac{u_{max} - u}{u_*} = 5.75 \log_{10}\left(\frac{R}{y}\right)$$

$$\frac{3 - 2}{u_*} = 5.75 \log_{10}\left(\frac{0.1}{0.04}\right)$$

$$\frac{1}{u_*} = 2.288$$

or $u_* = 0.437$ m/s

also $u_* = \sqrt{\dfrac{\tau_0}{\rho}}$

$$0.437 = \sqrt{\dfrac{\tau_0}{1000}} \qquad \because \ \rho = 1000 \ \text{kg/m}^3 \ \text{for water}$$

Squaring both sides, we get

$$(0.437)^2 = \dfrac{\tau_0}{1000}$$

or $\tau_0 = \mathbf{190.96 \ N/m^2}$

Problem 2.4:   For a turbulent flow in a pipe of diameter 300 mm, find the discharge when the centerline velocity is 2 m/s and the velocity of a point 100 mm from the centre as measured by pitot-tube is 1.6 m/s.

Solution : Given data :

Diameter of pipe : $D = 300$ mm $= 0.3$ m

$\therefore$ Radius of pipe : $R = \dfrac{D}{2} = \dfrac{0.3}{2} = 0.15$ m

Velocity at a centre : $u_{max} = 2$ m/s

Velocity at a point 100 mm from the centre is 1.6 m/s

i.e., $r = 100$ mm $= 0.1$ m

$u = 1.6$ m/s

$\therefore$ $y = R - r = 0.15 - 0.1 = 0.05$ m

We know that the velocity in terms of maximum velocity

$$\dfrac{u_{max} - u}{u_*} = 5.75 \log_{10} \left(\dfrac{R}{y} \right)$$

$$\dfrac{2 - 1.6}{u_*} = 5.75 \log_{10} \left(\dfrac{0.15}{0.05} \right)$$

$$\dfrac{0.4}{u_*} = 2.743$$

or $u_* = \dfrac{0.4}{2.743} = 0.1458$ m/s

Also we know that the relation between velocity at any point and average velocity

$$\dfrac{u - \bar{u}}{u_*} = 5.75 \log_{10} \dfrac{y}{R} + 3.75$$

at $y = R, \qquad u = u_{max}$

$$\therefore \qquad \frac{u_{max}-\bar{u}}{u_*} = 5.75\log_{10}\frac{R}{R}+3.75$$

$$\frac{2-\bar{u}}{0.1458} = 5.75 \times 0 + 3.75$$

$$\frac{2-\bar{u}}{0.1458} = 3.75$$

$$2-\bar{u} = 3.75 \times 0.1458$$

$$2-\bar{u} = 0.5467$$

or $\qquad \bar{u} = 2 - 0.5467 = 1.4533$ m/s

Discharge: $\qquad Q =$ cross-sectional area × average velocity

$$= \frac{\pi}{4}D^2 \times \bar{u} = \frac{3.14}{4}\times(0.3)^2\times1.4533 = \mathbf{0.10267\ m^3/s}$$

Problem 2.5: For turbulent flow in a pipe of diameter 300 mm, find the discharge when the centreline velocity is 2 m/s and the velocity at a point 60 mm from the centre is 1.7 m/s.

Solution: Given data:

Diameter of pipe: $\qquad D = 300$ mm $= 0.3$ m

$\therefore$ Radius of pipe: $\qquad R = \dfrac{D}{2}=\dfrac{0.3}{2}=0.15$ m

Velocity at centre: $\quad u_{max} = 2$ m/s
Velocity at a point 60 mm from the centre is 1.7 m/s

i.e., $\qquad r = 60$ mm $= 0.06$ m

$\qquad u = 1.7$ m/s

$\therefore \qquad y = R - r = 0.15 - 0.06 = 0.09$ m

We know that the velocity in terms of maximum velocity from Eq. (9.5.9) as:

$$\frac{u_{max}-u}{u_*} = 5.75\log_{10}\left(\frac{R}{y}\right)$$

$$\frac{2-1.7}{u_*} = 5.75\log_{10}\left(\frac{0.15}{0.09}\right)$$

$$\frac{0.3}{u_*} = 1.27$$

or $\qquad u_* = \dfrac{0.3}{1.27} = 0.236$ m/s

Also we know that the relation between velocity at any point and average velocity from Eq. (2.7.6), we get

$$\frac{u - \bar{u}}{u_*} = 5.75 \log_{10} \frac{y}{R} + 3.75$$

at

$$y = R, \ u = u_{max}$$

∴

$$\frac{u_{max} - \bar{u}}{u_*} = 5.75 \log_{10} \frac{R}{R} + 3.75$$

$$\frac{2 - \bar{u}}{0.236} = 5.75 \times 0 + 3.75$$

or

$$\frac{2 - \bar{u}}{0.236} = 3.75$$

or

$$2 - \bar{u} = 3.75 \times 0.236$$

$$2 - \bar{u} = 0.885$$

or

$$\bar{u} = 2 - 0.885 = 1.115$$

∴ Discharge: Q = cross-sectional area × average velocity

$$= \frac{\pi}{4} D^2 \times \bar{u} = \frac{3.14}{4} \times (0.3)^2 \times 1.115$$

$$= 0.07877 \ m^3/s = \textbf{78.77 litre/s}$$

Problem 2.6: An oil of specific gravity 0.9 is flowing through a rough pipe of diameter 400 mm and length 4 km at the rate of 0.4 m³/s. Find the power required to maintain this flow. Take the average height of roughness as $k = 0.5$ mm.

Solution: Given data:

Specific gravity of oil: $S = 0.9$

∴ Density of oil: $\rho = S \times \rho_{water} = 0.9 \times 1000 = 900 \ kg/m^3$

Diameter of rough pipe: $D = 400 \ mm = 0.4 \ m$

∴ Radius of pipe: $R = \dfrac{D}{2} = \dfrac{0.4}{2} = 0.2 \ m$

Length of pipe: $l = 4 \ km = 4 \times 1000 \ m = 4000 \ m$

Discharge: $Q = 0.4 \ m^3/s$

Average height of roughness:

$$k = 0.5 \ mm = 0.5 \times 10^{-3} \ m$$

For a rough pipe, the value of coefficient of friction f is given by the Eq. (9.9.6), we get

$$\frac{1}{\sqrt{4f}} = 2 \log_{10} \left(\frac{R}{k} \right) + 1.74$$

$$\frac{1}{\sqrt{4f}} = 2 \log_{10} \left(\frac{0.2}{0.5 \times 10^{-3}} \right) + 1.74$$

$$\frac{1}{\sqrt{4f}} = 5.20 + 1.74$$

$$\frac{1}{\sqrt{4f}} = 6.94$$

or $\qquad \sqrt{4f} = \dfrac{1}{6.94}$

Squaring both sides, we get

$$4f = \frac{1}{(6.94)^2}$$

or $\qquad f = 0.00519$

The average velocity: $\qquad \bar{u} = \dfrac{\text{Discharge}:Q}{\text{Cross-sectional area}:A}$

$$\bar{u} = \frac{Q}{\dfrac{\pi}{4}D^2} = \frac{4Q}{\pi D^2} = \frac{4 \times 0.4}{3.14 \times (0.4)^2} = 3.18 \text{ m/s}$$

Head loss due to friction: $\quad h_f = \dfrac{4f l \bar{u}^2}{2gd} = \dfrac{4 \times 0.00519 \times 4000 \times (3.18)^2}{2 \times 9.81 \times 0.4} - 106.99 \text{ m}$

∴   Power required: $\qquad P = \rho\, Q\, g\, h_f = 900 \times 0.6 \times 9.81 \times 106.99$ watt

$\qquad\qquad\qquad\qquad = 566768.82$ W = **566.768 kW**

Problem 2.7:   A smooth pipe of diameter 400 mm and length 1000 m carries water at the rate of 40 litre/s. Find the head lost due to friction, wall shear stress, maximum velocity and thickness of laminar sub-layer. Take the kinematic viscosity of water as 0.018 stokes.

Solution: Given data:

Diameter of pipe: $\qquad\qquad\qquad D = 400$ mm $= 0.4$ m

∴ Radius of pipe: $\qquad\qquad\quad R = \dfrac{D}{2} = \dfrac{0.4}{2} = 0.2$ m

Length of pipe: $\qquad\qquad\qquad l = 1000$ m

Discharge: $\qquad\qquad\qquad\qquad Q = 40$ litre/s $= 0.04$ m³/s

Kinematic viscosity: $\qquad\qquad v = 0.018$ stokes $= 0.018 \times 10^{-4}$ m²/s

We know that the discharge: $Q = A\bar{u}$

or $\qquad\qquad \bar{u} = \dfrac{Q}{A} = \dfrac{Q}{\dfrac{\pi}{4}D^2} = \dfrac{4Q}{\pi D^2} = \dfrac{4 \times 0.04}{3.14 \times (0.4)^2} = 0.3184$ m/s

∴ Reynolds number: $\qquad \text{Re} = \dfrac{\bar{u}D}{v} = \dfrac{0.3184 \times 0.4}{0.018 \times 10^{-4}} = 70755.55$

As Re > 4000, the flow is turbulent.

The coefficient of friction: $f = \dfrac{0.0791}{Re^{1/4}}$

where $4000 \leq Re < 10^5$

$$f = \dfrac{0.0791}{(70755.55)^{1/4}} = 0.00485$$

(i) Head loss due to friction: $h_f = \dfrac{4 f l \bar{u}^2}{2gD} = \dfrac{4 \times 0.00485 \times 1000 \times (0.3184)^2}{2 \times 9.81 \times 0.4}$

$$= 0.25 \text{ m}$$

(ii) Wall shear stress: $\tau_0 = \dfrac{f \rho \bar{u}^2}{2} = \dfrac{0.00485 \times 1000 \times (0.3184)^2}{2}$

$$= 0.2458 \text{ N/m}^2$$

(iii) The maximum velocity (u_{max}) for smooth pipe is given by eq. (9.5.5) as in which $u = u_{max}$ at $y = R$

$$\therefore \qquad \dfrac{u_{max}}{u_*} = 5.75 \log_{10} \dfrac{u_* R}{v} + 5.55$$

where $\qquad u_* = \sqrt{\dfrac{\tau_0}{\rho}} = \sqrt{\dfrac{0.2458}{1000}} = 0.0156 \text{ m/s}$

$$\therefore \qquad \dfrac{u_{max}}{0.0156} = 5.75 \log_{10} \left(\dfrac{0.0156 \times 0.2}{0.015 \times 10^{-4}} \right) + 5.55$$

$$\dfrac{u_{max}}{0.0156} = 18.62 + 5.55$$

or $\qquad u_{max} = 0.377 \text{ N/m}^2$

(iv) Thinkness of laminar sub-layer is given by:

$$\delta' = \dfrac{11.6v}{u_*} = \dfrac{11.6 \times 0.018 \times 10^{-4}}{0.0156}$$

$$= 0.001338 \text{ m} = 1.338 \text{ mm}$$

Problem 2.8:   The universal velocity distribution for the turbulent flow in a smooth pipe is given by the equation $\dfrac{u}{u^*} = 5.5 + 2.5 \log_e \dfrac{y u_*}{v}$. Calculate the ratio of the mean velocity to maximum velocity and the radius at which the local velocity is equal to the mean velocity.

Solution: The universal velocity distributions for the turbulent flow in a smooth pipe is given by

$$\dfrac{u}{u^*} = 5.5 + 2.5 \log_e \dfrac{y u^*}{v} \qquad\qquad ...(i)$$

where $y = R - r$

$$\therefore \qquad \dfrac{u}{u^*} = 5.5 + 2.5 \log_e \dfrac{(R-r) u^*}{v}$$

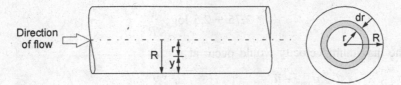

Fig. 2.8: Schematic for Problem 2.8

Discharge through a small circular ring,

$$\delta Q = dA \times u$$

$$= 2\pi r \, dr \times u_* \left[5.5 + 2.5 \log_e \frac{(R-r)}{\nu} u_* \right]$$

Net discharge through pipe,

$$\int_0^Q dQ = \int_0^R 2\pi u_* r \left[5.5 + 2.5 \log_e \frac{(R-r)}{\nu} u_* \right]$$

$$Q = 2\pi \mu_* \int_0^R r \left[5.5 + 2.5 \log_e \frac{(R-r)}{\nu} u_* \right]$$

$$= \pi R^2 \mu_* \left(1.75 + 2.5 \log_e \frac{R u_*}{\nu} \right)$$

The average velocity is given by

$$\bar{u} = \frac{Q}{\pi R^2}$$

$$= \frac{\pi R^2 \mu_* \left(1.75 + 2.5 \log_e \dfrac{R u_*}{\nu} \right)}{\pi R^2}$$

$$= u_* \left(1.75 + 2.5 \log_e \frac{R u_*}{\nu} \right)$$

or $$\frac{\bar{u}}{u_*} = 1.75 + 2.5 \log_e \frac{R u_*}{\nu} \qquad \qquad \ldots(ii)$$

Subtracting Eq. (ii) from Eq. (i), we get

$$\frac{u}{u_*} - \frac{\bar{u}}{u_*} = 5.5 + 2.5 \log_e \frac{y u_*}{\nu} - 1.75 - 2.5 \log_e \frac{R u_*}{\nu}$$

$$\frac{u - \bar{u}}{u_*} = 3.75 + 2.5 \log_e \frac{y u_*}{\nu} \times \frac{\nu}{R u_*}$$

$$= 3.75 + 2.5 \log_e \frac{y}{R} \qquad \qquad ...(iii)$$

The maximum velocity would occur at $y = R$

$$\therefore \qquad \frac{u_{max} - \bar{u}}{u_*} = 3.75 + 2.5 \log_e \frac{R}{R}$$

$$= 3.75$$

or $\qquad \qquad u_{max} - \bar{u} = 3.75 \; u_*$

or $\qquad \qquad u_{max} = \bar{u} + 3.75 \; u_*$

$$u_{max} = \bar{u} \left(1 + \frac{3.75 \; u_*}{\bar{u}} \right)$$

or $\qquad \qquad \dfrac{u_{max}}{\bar{u}} = 1 + \dfrac{3.75\,u_*}{u}$

Give condition,

  Local velocity at a point = Average velocity

$$u = \bar{u}$$

Substituting $u = \bar{u}$ in Eq. (iii), we get

$$\frac{\bar{u} - \bar{u}}{u_*} = 3.75 + 2.5 \log_e \frac{y}{R}$$

$$0 = 3.75 + 2.5 \log_e \frac{y}{R}$$

or $\qquad \qquad 2.5 \log_e \dfrac{y}{R} = -3.75$

or $\qquad \qquad \log_e \dfrac{y}{R} = \dfrac{-3.75}{2.5} = -1.5$

or $\qquad \qquad \dfrac{y}{R} = e^{-1.5} = 0.2231$

or $\qquad \qquad y = 0.2231 \; R$

also $\qquad \qquad y = R - r$

$\therefore \qquad \qquad 0.2231 \; R = R - r$

or $\qquad \qquad r = R - 0.2231 \; R = \mathbf{0.7769} \; \boldsymbol{R}$

Problem 2.9:   Find the distance from the pipe wall at which the local velocity is equal to the average velocity for turbulent flow in pipe. Also find the distance from the centre of the pipe.

Solution: Given condition:

Local velocity at a point = average velocity

i.e. $u = \bar{u}$

For a smooth or rough pipe, the difference of velocity at any point and average velocity is given by Eq. (9.7.6) as

$$\frac{u - \bar{u}}{u_*} = 5.75\log_{10}\left(\frac{y}{R}\right) + 3.75$$

Substituting the given condition *i.e.*, $u = \bar{u}$, we get

$$\frac{\bar{u} - \bar{u}}{u_*} = 5.75\log_{10}\left(\frac{y}{R}\right) + 3.75$$

$$0 = 5.75\log_{10}\left(\frac{y}{R}\right) + 3.75$$

or $$5.75\log_{10}\left(\frac{y}{R}\right) = -3.75$$

or $$\log_{10}\left(\frac{y}{R}\right) = \frac{-3.75}{5.75} = -0.6521$$

or $$\frac{y}{R} = 0.2227$$

or $$y = \textbf{0.2227 R}$$

also $y = R - r$

where r is distance from the centre of the pipe.

∴ $0.2227\ R = R - r$

or $r = R - 0.2227\ R$

$$r = \textbf{0.7773 R}$$

2.10 THERMAL (HOT-WIRE AND HOT FILM) ANEMOMETERS

Thermal anemometers are used for determining velocity at any point in fluid flow. Thermal anemometers consist of an electrically heated sensor and make use of thermal effect to measure velocity. Thermal anemometer has an extremely small sensor, so they can be used to measure velocity at any point in the flow without disturbing the flow. Thousands of velocity measurements can be taken per second so details of fluctuations in turbulent flow can be studied.

Depending on the sensor there are two types of thermal anemometers:

(*i*) Hot wire anemometer.

(*ii*) Hot film anemometer.

(*i*) **Hot wire anemometer:** If the sensing element is wire, the thermal anemometer is called hot wire anemometer. The hot wire anemometer consists of a very fine sensing wire having few microns of diameter (about 8 μm) and few millimeters (1 mm) of length. This sensing wire is made of platinum, tungsten, or platinum-iridium alloys, and is mounted on non-conducting structure.

(ii) **Hot film anemometer:** If the sensing element is a thin metallic film (less than 0.1 μm thick), the thermal anemometer is called hot film anemometer. This sensing element is mounted on a relatively thick ceramic support. Since the fine wire sensor of hot wire anemometer can easily break if the fluid contains excessive amounts of particulate matter. In such cases the hot-film anemometer is highly useful. But it is not suitable for studying the fine details of turbulent flow.

The two commercial setups of hot-wire anemometers are available of a constant temperature anemometer (CTA) and constant current anemometer (CCA). The operating principle of constant temperature is most common and is explained with the schematic in Fig. 2.9.

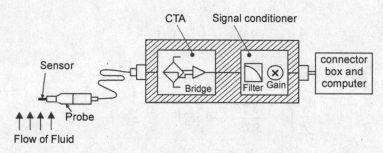

Fig. 2.9: Thermal anemometer system.

The sensor is heated to a specified temperature (normally about 200°C). When sensor is placed in a flowing fluid, it tends to cool due to exchange of heat between the wire and surrounding flowing fluid. The electronic control temperature by varying the electric current. The rate of heat transfer will increase with increasing fluid flow. So large voltage is applied across the sensor to maintain it at a constant temperature. Thus a close correlation exists between the flow velocity and voltage *i.e.,* flow velocity can be determined by electric current passing through the sensor.

The sensor is maintained at a constant temperature during the whole operation, thus its thermal energy remains constant. So the conservation of energy principle requires that the electrical Joule heating: $Q = I^2 R_w = \dfrac{E^2}{R_W}$ of sensor must be equal to total heat loss (Q_T) from the sensor.

So, using proper relations, the energy balance can be expressed by King's law as:
$$E^2 = a + bV^n \qquad \qquad ...(2.10.1)$$
where E is the voltage.

a, b and n are constants for a given probe, whose values will be given by the manufacture. Once the voltage E is measured, by using relation in Eq. (2.10.1) gives the flow velocity V directly.

2.11 LASER DOPPLER VELOCIMETRY

Laser Doppler Velocimetry (LDV), also called Laser Doppler Anemometry (LDA), is an optical technique to measure flow velocity at any desired point in fluid flow without disturbing the flow. LDA technique involves no sensing wire inserted into the flow so it is a non-intrusive technique. It can accurately measure velocity at a very small volume, and thus it can also be used to study the details of flow at a locality, including turbulent fluctuations, and it can be traversed through the entire fluid flow without any kind of intrusion.

The LDA technique was developed in mid-1960's and was highly adopted due to its high accuracy in measuring velocity for both gas and liquid flows. It can measure all three velocity components (*i.e.,* u, v and w along x, y and z-axes respectively). As compared to hot wire anemometer, LDA is quite costly. It requires a sufficient transparently between the laser source, the target location in the flow, and photodetector, and requirement for careful alignment of emitted and reflected beam for accuracy.

2.11.1 Operating Principle

The operating principle of LDA is based on sending a highly coherent monochromatic light beam toward the target area and determining the change in frequency of reflected radiation due to Doppler effect and relating this change in frequency to flow velocity of the fluid at the target area.

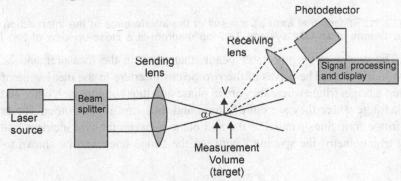

Fig. 2.10: Dual Beam Lasser Doppler Anemometry.

The basis setup of LDA to measure single velocity component is shown in Fig. 2.10. The heart of LDA systems is a laser power source which is usually a helium-neon or argon-ion laser with a power output of 10 mW to 20 W. Lasers are preferred over other light source because laser beams are highly focused and highly coherent. The laser beam is firstly splitted into two parallel beams of equal intensity by a beam splitter. Both beams are then passed through at a point in the flow (the target). The small fluid volume where these two beams intersect in the region where the velocity is measured and this region is called measurement volume or focal volume or target area. The laser light is scattered by particles in the measurement volume. The light scattered in a certain direction is collected by a receiving lens and is passed through a photodetector that converts the fluctuating in light intensity into fluctuating in voltage signal. Finally a signal processor determines the frequency of the voltage signal and thus velocity of the flow.

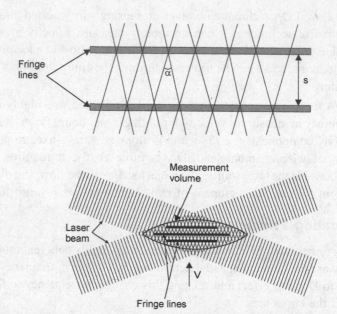

Fig. 2.11: Fringes that form as a result of the interference at the intersection of two laser beams of an LDA system. The top diagram is a close-up view of two fringes.

The waves of the two laser beams that cross in the measurement volume are shown in Fig. 2.11. The waves of the two beams interfere in the measurement volume, creating a bright fringe where they are in phase and thus support each other and creating a dark fringe where they are out of phase and thus cancel each other. The bright and dark fringe form lines parallel to the mid plane between the two incident laser beams. Using trigonometry, the spacing s between the fringe lines, can be shown to be:

$$s = \frac{\lambda}{2\sin\frac{\alpha}{2}}$$

where λ is the wavelength of laser and α is the angle between the two laser beams. When a particle traverses these fringe lines at a velocity V, the frequency of the scattered fringe lines is:

$$f = \frac{V}{s} = \frac{2V\sin\alpha/2}{\lambda}$$

This fundamental relation shows that the flow velocity to be proportional to the frequency and is known as the LDV equation.

SUMMARY

1. **Three types of velocities involved in turbulent flows:**
   (*i*) Instantaneous velocity
   (*ii*) Fluctuation velocity
   (*iii*) Average velocity.

2. **Degree or level of turbulence:** It is the square root of the arithmetic mean of the mean squares values of velocities in x, y and z-axes. It is denoted by letter D.

   Degree or level of turbulence: $D = \sqrt{\dfrac{1}{3}\left(\overline{u'^2} + \overline{v'^2} + \overline{w'^2}\right)}$

3. **Intensity of turbulence:** It is the ratio between the degree of turbulence (D) and the free-stream velocity (U). It is denoted by letter I.

   Intensity of turbulence: $\quad I = \dfrac{1}{U}\sqrt{\dfrac{1}{3}\left(\overline{u'^2} + \overline{v'^2} + \overline{w'^2}\right)}$

4. **Classification of turbulence:**
   (*i*) **Wall turbulence:** If turbulence generated by the viscous effect of the flowing fluid due to the presence of a solid boundary it is called the wall turbulence.
   (*ii*) **Free turbulence:** If turbulence generated, in the absence of a solid boundary, by contact between two layers of fluid flow at different velocities is called free turbulence.

5. **Turbulent shear stress :** $\overline{\tau_t}$
   (*i*) $\overline{\tau_t} = -\rho \overline{v'u'}$ (Reynolds theory)

   (*ii*) $\overline{\tau_t} = \eta \dfrac{d\overline{u}}{dy}$ (Boussinesq eddy-viscous theory)

   where η = eddy viscosity

   If the shear stress due to viscous flow is consider, then the total shear stress becomes as

   $$\tau = \overline{\tau_v} + \overline{\tau_t}$$

   where $\overline{\tau_v}$ = shear stress due to viscosity

   $\quad\quad \overline{\tau_t}$ = shear stress due to turbulence.

   $\therefore \quad\quad \tau = \mu \dfrac{d\overline{u}}{dy} + \eta \dfrac{d\overline{u}}{dy} = (\mu + \eta)\dfrac{d\overline{u}}{dy}$

   (*iii*) $\overline{\tau_t} = \rho l^2 \left(\dfrac{d\overline{u}}{dy}\right)^2$ (Prandtl's mixing length theory)

Contd...

where l = mixing length

$$\tau = \mu \frac{d\bar{u}}{dy} + \rho l^2 \left(\frac{d\bar{u}}{dy}\right)^2$$

(iv) $\bar{\tau}_t = \dfrac{\rho k^2 (d\bar{u}/dy)^2}{(d^2\bar{u}/dy^2)^2}$ (Von-Karman's theory)

6. The velocity distribution in the turbulent flow for pipes is given by the expression

$$\frac{u_{max} - u}{u_*} = 5.75 \log_{10}\left(\frac{R}{y}\right)$$

where u_{max} = maximum velocity at the centre-line

u = local velocity

$$u_* = \sqrt{\frac{\tau_0}{\rho}} \text{, shear velocity}$$

R = radius of pipe

y = distance from the pipe wall.

7. Hydrodynamically smooth and rough boundaries:

If the average height k of the irregularities projecting from the surface of a boundary is much less than d¢, the thickness of laminar sub-layer, such a boundary is called hydrodynamically smooth boundary.

If the irregularities of the surface are above the laminar sub-layer and eddies present in turbulent zero will come in contact with the irregularities of the surface, such a boundary is called hydrodynamically rough boundaries.

From Nikuradse's experiment

(i) if $\dfrac{k}{\delta'} < 0.25$, the boundary is called smooth boundary.

(ii) if $\dfrac{k}{\delta'} > 6.0$, the boundary is called rough boundary.

(iii) if $0.25 < \dfrac{k}{\delta'} < 6.0$, the boundary is called transition.

where k = average height of the irregularities of the surface.

δ' = thickness of the laminar sub-layer boundary.

In term of roughness Reynolds number: $\dfrac{u_* k}{v}$

(i) if $\dfrac{u_* k}{v} < u$, the boundary is considered smooth.

Contd...

(ii) if $\dfrac{u_* k}{v} > 100$, the boundary is considered rough.

(iii) if $4 < \dfrac{u_* k}{v} < 100$, the boundary is in transition stage.

8. Difference of local velocity and average velocity for smooth and rough pipes is

$$\frac{u - \bar{u}}{u_*} = 5.75 \log_{10}\left(\frac{y}{R}\right) + 3.75$$

9. Power law for velocity distribution in smooth pipes

$$\frac{u}{u_{max}} = \left(\frac{y}{R}\right)^n$$

where exponent n is dependent on the Reynolds number Re. The value of exponent n decreases with increasing Reynolds number Re.

For Reynolds number:

$$\text{Re} = 4 \times 10^3, \qquad n = \frac{1}{6}$$

$$\text{Re} = 1.1 \times 10^5, \qquad n = \frac{1}{7}$$

$$\text{Re} \geq 2 \times 10^6, \qquad n = \frac{1}{10}$$

10. **Determination of coefficient of friction:** f

(i) $f = \dfrac{16}{\text{Re}}$ for laminar flow *i.e.,* Re < 2000.

(ii) $f = \dfrac{0.0791}{\text{Re}^{1/4}}$ for turbulent flow *i.e.,* $4000 \leq \text{Re} < 10^5$

(iii) $\dfrac{1}{4f} = 2\log_{10}\left(R\sqrt{4f}\right) - 0.8$ for Re $\equiv 4 \times 10^7$

$$f = 0.0008 + \frac{0.05525}{\text{Re}^{0.237}} \qquad \text{for } 10^5 < \text{Re} \pounds\ 4 \times 10^7$$

(iv) $\dfrac{1}{\sqrt{4f}} = 2\log_{10}\left(\dfrac{R}{k}\right) + 1.74$ for rough pipe.

11. **For turbulent flow through commercial pipes:**

(i) Smooth pipes: $\dfrac{1}{\sqrt{4f}} = 2\log_{10}\left(\dfrac{\text{Re}\sqrt{4f}}{R/k}\right) - 0.8$

(ii) Rough pipes: $\dfrac{1}{\sqrt{4f}} = 2\log_{10}\left(\dfrac{R}{k}\right) + 1.74$

Contd...

12. **Thermal (hot-wire and hot film) anemometers:** Thermal anemometers are used for determining velocity at any point in fluid flow. Depending on the sensor there are two types of thermal anemometers:

(*i*) Hot-wire anemometer

(*ii*) Hot-film anemometer.

13. **Laser doppler velocimetry:** Laser Doppler Velocimetry (LDV), also called laser Doppler anemometry (LDA), is an optical technique to measure flow velocity at any desired point in fluid flow without disturbing the flow.

ASSIGNMENT - 1

1. What do you mean by turbulent flow?
2. Define the following terms:
    (*i*) Instantaneous velocity
    (*ii*) Fluctuation velocity
    (*iii*) Average velocity.
3. Explain briefly hydrodynamically smooth and rough boundaries.
                                    (*GGSIP University, Delhi, Dec. 2008*)
4. Define the terms: degree of turbulence and intensity of turbulence.
5. Distinguish between wall turbulence and free turbulence.
6. What is meant by turbulence? How does it affect the flow properties?
7. Derive an expression for the velocity distribution for turbulent flow in smooth pipes.   (*GGSIP University, Delhi, Dec. 2005*)
8. Show that velocity distribution for turbulent flow through rough pipe is given

by
$$\frac{u}{u_*} = 5.75 \log_{10}\left(\frac{y}{k}\right) + 8.5$$

where
u = shear velocity
y = distance from pipe wall
k = roughness factor.

9. Explain the Prandtl's mixing length theory for turbulent shear stress.
                                    (*GGSIP University, Delhi, Dec. 2005*)
10. What are the semi-empirical theories of turbulence? Explain the concept of mixing length introduced by Prandtl and state the relationship that exists between the turbulent shearing stress and mixing length.
                                    (*GGSIP University, Delhi, Dec. 2008*)
11. Prove that the difference of local velocity and average velocity for turbulent flow through rough or smooth pipes is given by:

$$\frac{u - \overline{u}}{u_*} = 5.75 \log_{10}\left(\frac{y}{R}\right) + 3.75$$

ASSIGNMENT - 2

1. A pipeline carrying water has average height of irregularities projecting from the surface of the boundary of the pipe as 0.15 mm. What type of boundary is it? The shear stress developed is 4.9 N/m². The kinematic viscosity of water is 0.02 stokes. **Ans.** Boundary is in transition

2. A rough pipe is of diameter 50 mm. The velocity at a point 30 mm from wall is 30% more than the velocity at a point 10 cm from pipe wall. Find the average height of the roughness. **Ans.** 77.25 mm

3. Find the wall shearing stress in a pipe of diameter 100 mm which carries water. The velocity at the pipe centre and 30 mm from the pipe centre are 2 m/s and 1.5 m/s respectively. The flow in pipe is given as turbulent.
   Ans. 47.75 N/m²

4. For turbulent flow in a pipe of diameter 300 mm, determine the discharge when the maximum velocity is 2 m/s and the velocity at point 100 mm from the centre is 1.6 m/s. **Ans.** 102.67 litre/s

5. An oil of specific gravity of 0.85 is flowing through a rough pipe of diameter 500 mm and length 4000 m at the rate of 0.5 m³/s. Find the power required to maintain this flow. Take the average height of roughness as $k = 0.40$ mm.
   Ans. 204.96 kW

6. A smooth pipeline 75 mm in diameter and 500 m long conveys water at the rate of 75 litre/s. Find the loss of head, wall shear stress, centreline velocity and thickness of laminar sub-layer. Take kinematic viscosity as .0195 stokes.
   Ans. 31m, 6.3 N/m², 1.89 m/s, 0.285 mm

7. For turbulent flow in a pipe of diameter 200 mm, find the discharge when the centreline velocity is 1.5 m/s and velocity at a point 50 mm from the wall as measured by pitot tube is 1.35 m/s. Also find the coefficient of friction and the average height of roughness projections.
   Ans. 0.037 m³/s, 0.0109, 2.99 mm

8. Water is flowing through a rough pipe of diameter 600 mm at the rate of 500 litre/s. The wall roughness is 3 mm. Find the power lost for 1000 m length of pipe. **Ans.** 39.57 kW

Boundary Layer Theory

3.1 INTRODUCTION

If a body is placed in the flow of ideal fluid, the fluid particles slip on the surface of the body in the direction of flow of the fluid due to the property of ideal fluid (*i.e.*, zero viscosity), no friction forces are exerted on the fluid in opposite direction of flow of the fluid. On the other hand if a body is placed in the flow of real fluid, the fluid particles adhere on the surface of the body and attain zero velocity, because the particles of real fluid do not slip on the surface of the body due to viscosity of the real fluid (*i.e.*, in case of real fluid viscosity must be taken into consideration). Now as the fluid particles on the surface of body have zero velocity (*i.e.*, there is no relative velocity between the fluid particles and surface of body), they will provide a resistance to adjacent layer and hence decrease its velocity. Similarly this layer acts as resistance to the next adjacent layer, which in turn decreases its velocity up to some extent and this process continues. In this way a thin layer of fluid developed close to the body surface and in which velocity gradient

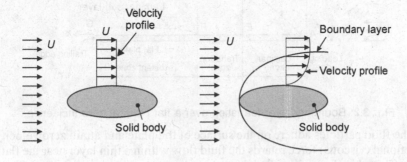

(a) Ideal fluid flow over solid body
(No boundary layer formed)

(b) Real fluid flow over solid body
(Boundary layer formed)

Fig. 3.1: Concept of boundary layer

© The Author(s) 2023
S. Kumar, *Fluid Mechanics (Vol. 2)*,
https://doi.org/10.1007/978-3-030-99754-0_3

$\left(\dfrac{du}{dy}\right)$ exists normal to the surface of the body. This thin layer across the narrow

region of the solid body is called boundary layer or friction layer. The theory dealing with the boundary layer flow is called boundary layer theory. In other words, we can say that in the boundary layer, the velocity of fluid increases from zero velocity on the surface of the body to free stream velocity (U) of the fluid in the direction normal to the body.

According to boundary layer theory, fluid flow on the surface of the body can be divided into two unequal regions:

   (i) A very thin layer of the fluid, called boundary layer, at which viscosity must

         be taken into account as there exists a velocity gradient $\left(\dfrac{du}{dy}\right)$.

   (ii) The region outside the boundary layer where the viscosity can be neglected

         and no velocity gradient $\left(\dfrac{du}{dy}\right)$ exists.

3.2 BOUNDARY LAYER FORMATION OVER A FLAT PLATE

Consider the flow of a fluid having free stream velocity (U) over a smooth thin flat plate placed at zero incidence (*i.e.*, flow direction makes zero angle with the plate surface) as shown in Fig. 3.2.

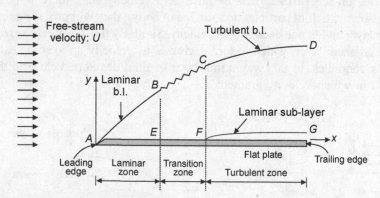

Fig. 3.2: Boundary layer formation over a flat plate at zero incidence.

The fluid particles adhere on the surface of the plate and attain zero velocity and that frictional (viscous) force retards the fluid flow within a thin layer near the flat plate surface. This thin layer is called boundary layer. Inside the boundary layer, the following types of variations take place:

   (i) The velocity of fluid increases from zero velocity on the plate surface to
         free-stream velocity (U) of the fluid in the direction normal to the plate

surface (*i.e.*, velocity gradient $\left(\dfrac{du}{dy}\right)$ exists normal to the plate surface).

(*ii*) The boundary layer thickness (δ) increases continuously from the leading edge of the plate to the downstream direction.

Let *ABCD* is boundary layer formation over a flat plate as shown in Fig. 3.2. This boundary layer is classified into three stages as:

1. Laminar boundary layer.
2. Transition boundary layer.
3. Turbulent boundary layer.

3.2.1 Laminar Boundary Layer

The initial stage of boundary layer development exhibits characteristics of laminar flow *i.e.*, the fluid particles at the leading edge of plate moves orderly in laminas parallel to the flat plate surface as shown in Fig. 3.3. In this figure *AB* is called laminar boundary layer. The length of the plate from the leading edge up to which the laminar boundary layer exists is called laminar zone.

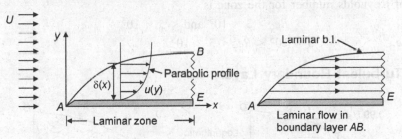

Fig. 3.3: Laminar boundary layer.

The Reynolds number for the flow of fluid in the boundary layer is expressed as

$$Re_x = \frac{Ux}{v}$$

where x = distance from the leading edge.

U = free-stream velocity of the fluid flow.

v = kinematic viscosity of the fluid.

Reynolds number for laminar boundary layer is

$$Re_x \leq 3 \times 10^5$$

$$\frac{Ux}{v} \leq 3 \times 10^5$$

It means, Reynolds number (Re_x) at point '*E*' is 3×10^5. If the Reynolds number (Re_x) > 3×10^5 then the next transition zone begins. So the laminar boundary layer is maintained upto point '*E*' at which $Re_x = 3 \times 10^5$.

Experiments have shown that for laminar boundary layer, the velocity profile is parabolic, and the velocity profiles at different locations along the plate are geometrically similar.

3.2.2 Transition Boundary Layer

The short length over which the boundary layer flow changes from laminar to turbulent is called transition zone. This is shown by distance *EF* in Fig. 3.4. The formation of boundary layer over transition zone is called transition boundary layer.

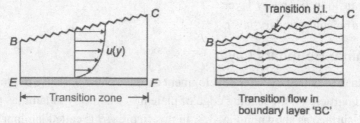

Fig. 3.4: Transition Boundary Layer.

The boundary layer thickness will go on increasing in the direction of flow. The range of Reynolds number for the zone is

$$Re_x > 3 \times 10^5 \text{ and } < 5 \times 10^5$$

i.e., $\qquad 3 \times 10^5 < Re_x < 5 \times 10^5$

3.2.3 Turbulent Boundary Layer

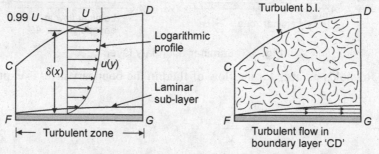

Fig. 3.5: Turbulent Boundary Layer.

The zone next to transition zone is called turbulent zone. This is shown by FG in Fig. 10.5. In this zone, the fluid flows in zig-zag manner.* If Reynold's number (Re_x) is $\geq 5 \times 10^5$ then formation of this zone begins. The boundary layer thickness will go on increasing in the direction of flow. The experiments have shown that for turbulent boundary layer, the velocity profile is logarithmic, and the velocity profile at different locations along the plate are geometrically similar.

---

*The formation of *b.l.* over this zone is called turbulent boundary layer.

3.2.4 Laminar Sub-layer

If the plate surface is very smooth, a thin layer develops very close to the plate surface (or boundary of solid body) due to viscous effects of the fluid, in turbulent zone. This thin layer is called laminar sub-layer (or viscous sub-layer). The nature of the flow in this layer is laminar.

As shown in Figs. 3.2 and 3.5, the turbulent boundary layer has a double layered structure, the large part of boundary layer due to turbulent fluctuation motion (*i.e.*, exchange of momentum in a transverse direction) and is unaffected by the viscosity. The thin laminar sub-layer close to the solid surface in turbulent zone is only affected by the viscosity. Due to thin laminar sub-layer, we assume that velocity variation is linear and so the velocity gradient can be considered constant. Therefore, the shear stress in the laminar sub-layer would be constant and equal to the boundary shear stress (τ_0). Thus the shear stress in the laminar sub-layer is

$$\tau_0 = \mu \left(\frac{\partial u}{\partial y} \right)_{y=0}$$

$$\tau_0 = \mu \frac{u}{y} \qquad\qquad \because \text{ for linear variation, } \frac{\partial u}{\partial y} = \frac{u}{y}$$

3 .3 BOUNDARY LAYER THICKNESS: δ

When real fluid flows over a solid body, the fluid particles contact to surface of solid body and attain zero velocity. These fluid particles, then retards the motion of fluid particles in the adjoining fluid layer, which retards the motion of fluid particles in next layer and so on, until a distance $y = \delta$ from the surface of solid body reaches where these effects become negligible. The distance δ normal to the surface of solid body is called boundary layer thickness. It is denoted by δ (greek letter 'delta').

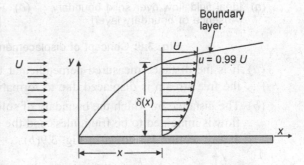

Fig. 10.6: Boundary layer thickness.

The boundary layer thickness (δ) is defined as the normal distance from the surface of solid body to the point where the velocity of flow (u) is 99% of the free stream velocity (U).

Mathematically, it is defined

$$y = \delta \quad \text{for } u = 99\% \ U = 0.99 \ U$$

Fig. 3.7: Boundary layer thickness increasing in the direction of flow.

As shown in Fig. 10.7, the boundary layer thickness is zero at leading edge *i.e.*, at point 1 on the flat plate and increasing in the direction of flow.

i.e., $\delta_4 > \delta_3 > \delta_2 > \delta_1$

3.4 DISPLACEMENT THICKNESS: δ^*

The velocity distribution caused by boundary layer formation displaces the flow rate slightly outward from the solid boundary. As in case of ideal flow, no boundary layer formation occurs hence no displacement of flow takes place. Figure 3.8(*b*) shows the streamlines in boundary layer. The deflecting streamlines and the widening gap in between indicate retardation of flow and a small vertical velocity component.

The displacement thickness (δ^*) may be defined in any one of the following ways:

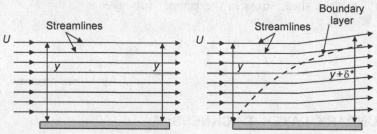

(a) Ideal fluid flow over solid boundary   (b) Real fluid flow over solid boundary
      [Absence of boundary layer]   [Formation of boundary layer]

Fig. 3.8: Concept of displacement thickness.

(*a*) It is the distance measured perpendicular to the solid boundary, by which the free stream is displaced due to formation of boundary layer.

(*b*) The distance by which the boundary of solid body is displaced if the entire flow is imagined to be frictionless and the same mass flow rate maintained at any section is shown in Fig. 3.9(*b*).

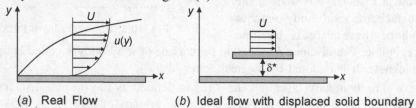

(a) Real Flow   (b) Ideal flow with displaced solid boundary

Fig. 3.9: Displacement thickness.

Now consider the flow of a fluid having free stream velocity (U) over a smooth flat plate at zero incidence as shown in Fig. 3.10. At a certain distance x from the leading edge consider a section 1–1.

The velocity at point B is zero and at point C is nearly U, where BC is equal to the thickness of boundary layer *i.e.*,

Distance $BC = \delta$

At the section 1–1, consider a small fluid element strip of thickness dy and distance y from the plate surface.

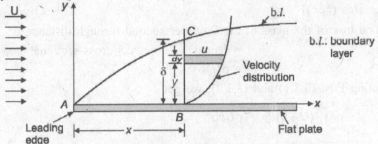

Fig. 3.10: Flow over a flat plate.

Let u = velocity of fluid at a small element strip.

b = width of a flat plate, not shown in Fig. 3.10.

∴ The cross-sectional area of small fluid element strip:

$$dA = b \times dy$$

Mass of fluid per second flowing through small element strip:

$\dot{m}$ = density × velocity

× cross-sectional area of small element strip

$$= \rho u dA = \rho u b dy \qquad \because dA = b dy$$

If there is no plate in the flow, then the fluid will flow with a constant velocity equal to free stream velocity (U) at the section 1–1.

Then mass of fluid per second flowing through small element strip:

$\dot{M}$ = density × velocity

× cross-sectional area of small element strip

$$\dot{M} = \rho U dA = \rho U b dy$$

We have $U > u$, the formation of the boundary layer due to the presence of the plate in the fluid will be a reduction in mass flowing per second through the small element strip.

∴ The reduction in mass per second flowing through small element strip

$$= \dot{M} - \dot{m} = \rho U b dy - \rho u b dy = \rho b (U - u) dy$$

∴ Total reduction in mass of fluid per second flowing through BC due to presence of plate in the fluid

$$= \int_0^\delta \rho b(U - u) dy \qquad \because BC = \delta$$

$$= \rho b \int_0^\delta (U - u) dy \mid \text{if fluid is incompressible } i.e., \rho = C$$

$$= \rho b \int_0^\delta (U - u) dy \qquad \qquad \dots (3.3.1)$$

Let the plate is displaced by a vertical distance δ^* and velocity of flow for the distance δ^* is equal to the free stream velocity U [*i.e.*, frictionless flow passing through an area $(b\delta^*)$]

∴ The loss of the mass of the fluid per second through distance δ^*

$$= \text{density} \times \text{velocity} \times \text{cross-sectional area of flow}$$
$$= \rho U b \delta^* \qquad \qquad \qquad \qquad ... (3.3.2)$$

Equating Eqs. (3.3.1) and (3.3.2), we get

$$\rho b \int_0^\delta (U - u)dy = \rho U b \delta^*$$

or
$$\delta^* U = \int_0^\delta (U - u)dy$$

$$\delta^* = \int_0^\delta \left(1 - \frac{u}{U}\right) dy$$

Displacement thickness (δ^*) tells us how far the streamlines of the flow are displaced outwards due to the decrease in velocity in the boundary layer.

3.5 MOMENTUM THICKNESS: θ

It is defined as the distance measured perpendicular to the solid boundary, by which the boundary is displaced to compensate the reduction in momentum of the flow fluid due to formation of boundary layer. It is denoted by θ.

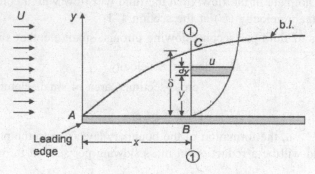

Fig. 3.11: Flow over a flat plate.

Now consider the flow over a plate as shown in Fig. 3.11. Let the section 1–1 is at a distance x from leading edge. Consider a small fluid element strip of thickness (dy) and distance y from the plate surface.

[**Note:** As derived in a similar manner for a previous article 3.4]

Mass of fluid per second flowing through small element strip: $\dot{m} = \rho u dA = \rho u b dy$

If there is no plate in the flow, then the fluid will be flowing with a constant velocity equal to free stream velocity (U) at the section 1–1.

Then

Mass of fluid per second flowing through small element strip:

$$\dot{M} = \rho U dA = \rho U b \cdot dy$$

∴ The reduction in mass per second flowing through small element strip

$$= \dot{M} - \dot{m} = \rho Ubdy - \rho ubdy = \rho b(U - u)dy$$

The reduction in momentum per second through small element strip

$$= \text{reduction in mass/s} \times \text{velocity of flow}$$

$$= (\dot{M} - \dot{m})u = \rho b(U - u)dy \times u = \rho bu(U - u)dy$$

Total reduction in momentum per second through BC (*i.e.*, δ)

$$= \int_0^\delta \rho bu(U - u)dy \qquad \qquad \dots (3.5.1)$$

Let θ = vertical distance by which plate is displaced when the fluid is flowing with a free stream velocity U [*i.e.*, frictionless flow passing through an area ($b\theta$).]

∴ Loss of momentum/s of fluid flowing through distance θ with a free stream velocity (U)

$$= \text{mass/s of fluid flow through distance } \theta \times \text{free stream velocity } U$$

$$= [\rho \times \text{Area} \times U] \times U$$

$$= \rho b\theta \ U^2 \qquad \qquad \qquad \qquad \because \text{Area} = b\theta$$

$$= \rho b\theta \ U^2 \qquad \qquad \dots (3.5.2)$$

Equating Eqs. (3.5.2) and (3.5.1), we get

$$\rho b\theta U^2 = \int_0^\delta \rho bu(U - u)dy$$

$$\rho b\theta U^2 = \rho b \int_0^\delta u(U - u)dy$$

$$[\text{Let fluid is incompressible } i.e., \ \rho = C]$$

$$\theta U^2 = \int_0^\delta u(U - u)dy$$

$$\theta = \int_0^\delta \frac{u}{U^2}(U - u)dy$$

$$\theta = \int_0^\delta \frac{u}{U}\left[1 - \frac{u}{U}\right]dy$$

For parallel flow past a flat plate held at zero incidence, the approximate relationship among the boundary layer, displacement and momentum thicknesses are:

$$\delta = 3\delta^* = 7.5\theta$$

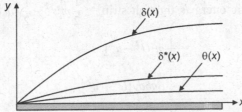

Fig. 3.12: Variation of boundary layer, displacement and momentum thickness over flat plane in the direction of flow.

The ratio of displacement thickness (δ^*) to momentum thickness (θ) is called the shape factor. It is denoted by H.

Mathematically,

$$\text{Shape factor: } H = \frac{\text{Displacement thickness: } \delta^*}{\text{Momentum thickness: } \theta}$$

$$H = \frac{\delta^*}{\theta}$$

3.6 ENERGY THICKNESS: δ^{**}

It is defined as the distance measured normal to the solid boundary, by which the boundary is displaced to compensate the reduction in kinetic energy of the flowing fluid due to formation of boundary layer. It is denoted by δ^{**}.

Now consider the flow over a plate as shown in Fig. 3.13. Let the sections 1–1 is at a distance x from leading edge.

Consider a small fluid element strip of thickness (dy) and distance y from the plate surface.

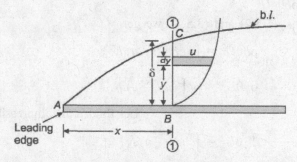

Fig. 3.13: Flow over a flat plate.

The mass of fluid per second flowing through small element strip:

$$\dot{m} = \rho u dA = \rho u b dy$$

Kinetic energy/s of fluid through strip $= \dfrac{1}{2}\dot{m} u^2$

Kinetic energy/s of fluid through strip in the absence of boundary layer $= \dfrac{1}{2}\dot{m} U^2$

$\therefore$ Loss of kinetic energy/s through strip $= \dfrac{1}{2}\dot{m}U^2 - \dfrac{1}{2}\dot{m}u^2$

$$= \frac{1}{2}\dot{m}[U^2 - u^2]$$

$$= \frac{1}{2}\rho u b dy\,[U^2 - u^2] \qquad\qquad \because \dot{m} = \rho u b dy$$

$$= \frac{1}{2}\rho b u [U^2 - u^2]dy$$

$\therefore$ Total loss of kinetic energy/s of fluid passing through BC (δ)

$$= \int_0^\delta \frac{1}{2} \rho b u [U^2 - u^2] dy$$

$$= \frac{1}{2} \rho b \int_0^\delta u [U^2 - u^2] dy \qquad \text{... (3.6.1)}$$

Let δ^{**} = vertical distance by which the plate is displaced to compensate for the reduction in kinetic energy per second.

∴ Loss of kinetic energy/s through distance δ^{**} of fluid flowing with a free stream velocity (U):

$$= \frac{1}{2} \times \text{mass} \times (\text{velocity})^2$$

$$= \frac{1}{2} \times \rho b \delta^{**} U \times U^2 \qquad\qquad [\dot{M} = \rho b \delta^* U]$$

$$= \frac{1}{2} \rho b \delta^{**} U^3 \qquad \text{... (3.6.2)}$$

Equating Eqs. (3.6.2) and (3.6.1), we get

$$\frac{1}{2} \rho b \delta^{**} U^3 = \frac{1}{2} \rho b \int_0^\delta u [U^2 - u^2] dy$$

$$\delta^{**} - \frac{1}{U^3} \int_0^\delta u [U^2 - u^2] dy$$

$$\delta^{**} = \int_0^\delta \frac{u}{U^3} [U^2 - u^2] dy$$

$$\delta^{**} = \int_0^\delta \frac{u}{U} \left[\frac{U^2}{U^2} - \frac{u^2}{U^2} \right] dy$$

$$\delta^{**} = \int_0^\delta \frac{u}{U} \left[1 - \frac{u^2}{U^2} \right] dy$$

With the energy thickness (δ^{**}) known, the loss of kinetic energy/s can be determined from the following relation:

$$E_L = \frac{1}{2} \rho b \delta^{**} U^3$$

Problem 3.1: A linear distribution of velocity in the boundary on a flat plate is given by

$$\frac{u}{U} = \frac{y}{\delta}$$

where u = velocity at a distance y from the flat plate and
= U at y = δ   where δ is boundary layer thickness.

Find: (i) $\dfrac{\delta^*}{\delta}$ (ii) $\dfrac{\delta^*}{\theta}$ (iii) $\dfrac{\delta^*}{\delta^{**}}$

Solution: Given:

Linear distribution of velocity in the boundary layer:

$$\frac{u}{U} = \frac{y}{\delta}$$

The displacement thickness:

$$\delta^* = \int_0^\delta \left(1 - \frac{u}{U}\right) dy$$

$$= \int_0^\delta \left(1 - \frac{y}{\delta}\right) dy$$

$$= \left[y - \frac{y^2}{2\delta}\right]_0^\delta \qquad \qquad \because \ \delta = C, \text{ at particular section}$$

$$= \delta - \frac{\delta^2}{2\delta} = \delta - \frac{\delta}{2} = \frac{\delta}{2}$$

The momentum thickness:

$$\theta = \int_0^\delta \left(1 - \frac{u}{U}\right)\frac{u}{U} dy = \int_0^\delta \left(1 - \frac{y}{\delta}\right)\frac{y}{\delta} dy$$

$$= \int_0^\delta \left(\frac{y}{\delta} - \frac{y^2}{\delta^2}\right) dy$$

$$= \left[\frac{y^2}{2\delta} - \frac{y^3}{3\delta^2}\right]_0^\delta = \frac{\delta^2}{2\delta} - \frac{\delta^3}{3\delta^2} = \frac{\delta}{2} - \frac{\delta}{3} = \frac{\delta}{6}$$

The energy thickness: $\delta^{**} = \int_0^\delta \left(1 - \frac{u^2}{U^2}\right)\frac{u}{U} dy = \int_0^\delta \left(\frac{u}{U} - \frac{u^3}{U^3}\right) dy$

$$= \int_0^\delta \left(\frac{y}{\delta} - \frac{y^3}{\delta^3}\right) dy = \left[\frac{y^2}{2\delta} - \frac{y^4}{4\delta^3}\right]_0^\delta$$

$$= \left[\frac{\delta^2}{2\delta} - \frac{\delta^4}{4\delta^3}\right]_0^\delta = \frac{\delta}{2} - \frac{\delta}{4} = \frac{\delta}{4}$$

(i) $\dfrac{\delta^*}{\delta} = \dfrac{\delta}{2 \times \delta} = \dfrac{1}{2} = \mathbf{0.5}$

(ii) $\dfrac{\delta^*}{\theta} = \dfrac{\delta/2}{\delta/6} = \dfrac{6}{2} = \mathbf{3}$

(iii) $\dfrac{\delta^*}{\delta^{**}} = \dfrac{\delta/2}{\delta/4} = \dfrac{4}{2} = \mathbf{2}$

Problem 3.2:   Determine the displacement thickness and momentum thickness in terms of boundary layer thickness for the given velocity profile.

$$\frac{u}{U} = 2\left(\frac{y}{\delta}\right) - \left(\frac{y}{\delta}\right)^2$$

where u is the velocity at a height y above the surface and U is the free-stream velocity.

(*GGSIP University Delhi, 2004, 2007, 2008*)

Solution: Given data:

Velocity profile: $\quad \dfrac{u}{U} = 2\left(\dfrac{y}{\delta}\right) - \left(\dfrac{y}{\delta}\right)^2$

We know, displacement thickness: $\delta^* = \displaystyle\int_0^\delta \left(1 - \dfrac{u}{U}\right) dy$ $\qquad$ ... (i)

Substituting the value of $\dfrac{u}{U} = 2\left(\dfrac{y}{\delta}\right) - \left(\dfrac{y}{\delta}\right)^2$ in Eq. (i), we get

$$\delta^* = \int_0^\delta \left[1 - 2\left(\dfrac{y}{\delta}\right) + \left(\dfrac{y}{\delta}\right)^2\right] dy = \left[y - \dfrac{2}{\delta}\cdot\dfrac{y^2}{2} + \dfrac{1}{\delta^2}\dfrac{y^3}{3}\right]_0^\delta$$

$$= \delta - \dfrac{1}{\delta}\times\delta^2 + \dfrac{1}{\delta^2}\times\dfrac{\delta^3}{3} = \delta - \delta + \dfrac{\delta}{3} = \dfrac{\delta}{3}$$

and momentum thickness: $\theta = \displaystyle\int_0^\delta \dfrac{u}{U}\left[1 - \dfrac{u}{U}\right] dy$ $\qquad$ ... (ii)

Substituting the value of $\dfrac{u}{U} = 2\left(\dfrac{y}{\delta}\right) - \left(\dfrac{y}{\delta}\right)^2$ in Eq. (ii), we get

$$\theta = \int_0^\delta \left[2\left(\dfrac{y}{\delta}\right) - \left(\dfrac{y}{\delta}\right)^2\right]\left[1 - 2\left(\dfrac{y}{\delta}\right) + \left(\dfrac{y}{\delta}\right)^2\right] dy$$

$$= \int_0^\delta \left[2\left(\dfrac{y}{\delta}\right) - 4\dfrac{y^2}{\delta^2} + 2\dfrac{y^3}{\delta^3} - \dfrac{y^2}{\delta^2} + \dfrac{2y^3}{\delta^3} - \dfrac{y^4}{\delta^4}\right] dy$$

$$= \int_0^\delta \left[2\dfrac{y}{\delta} - 5\dfrac{y^2}{\delta^2} + 4\dfrac{y^3}{\delta^3} - \dfrac{y^4}{\delta^4}\right] dy$$

$$= \left[\dfrac{2}{\delta}\times\dfrac{y^2}{2} - \dfrac{5}{\delta^2}\times\dfrac{y^3}{3} + \dfrac{4}{\delta^3}\times\dfrac{y^4}{4} - \dfrac{1}{\delta^4}\times\dfrac{y^5}{5}\right]_0^\delta$$

$$= \dfrac{\delta^2}{\delta} - \dfrac{5}{\delta^2}\times\dfrac{\delta^3}{3} + \dfrac{4}{\delta^3}\times\dfrac{\delta^4}{4} - \dfrac{1}{\delta^4}\times\dfrac{\delta^5}{5} = \delta - \dfrac{5}{3}\delta + \delta - \dfrac{\delta}{5}$$

$$= 2\delta - \dfrac{5}{3}\delta - \dfrac{\delta}{5} = \dfrac{30\delta - 25\delta - 3\delta}{15} = \dfrac{2\delta}{15}$$

Problem 3.3: The velocity distribution in the boundary layer over a high spillway face is to have the following form:

$$\left(\dfrac{u}{U}\right) = \left(\dfrac{y}{\delta}\right)^{0.22}$$

Prove that the displacement thickness, the momentum thickness and the energy thickness can be expressed as

$$\dfrac{\delta^*}{\delta} = 0.180, \dfrac{\theta}{\delta} = 0.125 \text{ and } \dfrac{\delta^{**}}{\delta} = 0.217 \text{ respectively.}$$

(GGSIP University, Delhi, Dec. 2005)

Solution: Given data:

Velocity distribution: $\dfrac{u}{U} = \left(\dfrac{y}{\delta}\right)^{0.22}$

We know,

Displacement thickness: $\delta^* = \displaystyle\int_0^\delta \left(1 - \dfrac{u}{U}\right) dy$

Momentum thickness: $\theta = \displaystyle\int_0^\delta \dfrac{u}{U}\left(1 - \dfrac{u}{U}\right) dy$

and energy thickness: $\delta^{**} = \displaystyle\int_0^\delta \dfrac{u}{U}\left[1 - \dfrac{u^2}{U^2}\right] dy$

Now displacement thickness: $\delta^* = \displaystyle\int_0^\delta \left(1 - \dfrac{u}{U}\right) dy$ ... (i)

Substituting the value of $\dfrac{u}{U} = \left(\dfrac{y}{\delta}\right)^{0.22}$ in Eq. (i), we get

$$\delta^* = \int_0^\delta \left(1 - \frac{y^{0.22}}{\delta^{0.22}}\right) dy$$

$$\delta^* = \left[y - \frac{y^{0.22+1}}{\delta^{0.22} \times 1.22}\right]_0^\delta = \left[\delta - \frac{\delta^{1.22}}{1.22\,\delta^{0.22}}\right]$$

$$= \left[\delta - \frac{\delta^{0.22} \cdot \delta}{1.22\,\delta^{0.22}}\right] = \left[\delta - \frac{\delta}{1.22}\right]$$

$$= \delta\left[1 - \frac{1}{1.22}\right] = \delta\left[\frac{1.22 - 1}{1.22}\right] = \delta \times \frac{0.22}{1.22}$$

$$\delta^* = 0.180\,\delta$$

or $\dfrac{\delta^*}{\delta} = \mathbf{0.180}$

Similarly for momentum thickness: $\theta = \displaystyle\int_0^\delta \dfrac{u}{U}\left[1 - \dfrac{u}{U}\right] dy$ ... (ii)

Substituting the value of $\dfrac{u}{U} = \left(\dfrac{y}{\delta}\right)^{0.22}$ in Eq. (ii), we get

$$= \int_0^\delta \left(\frac{y}{\delta}\right)^{0.22}\left[1 - \left(\frac{y}{\delta}\right)^{0.22}\right] dy$$

$$= \int_0^\delta \left[\left(\frac{y}{\delta}\right)^{0.22} - \left(\frac{y}{\delta}\right)^{0.44}\right] dy$$

$$= \left[\frac{1}{\delta^{0.22}} \cdot \frac{y^{1.22}}{1.22} - \frac{1}{\delta^{0.44}} \cdot \frac{y^{1.44}}{1.44}\right]_0^\delta$$

$$= \frac{1}{\delta^{0.22}} \times \frac{\delta^{1.22}}{1.22} - \frac{1}{\delta^{0.44}} \times \frac{\delta^{1.44}}{1.44}$$

$$= \frac{\delta}{1.22} - \frac{\delta}{1.44} = \delta\left[\frac{1}{1.22} - \frac{1}{1.44}\right] = \delta[0.819 - 0.694]$$

$$\theta = 0.125\delta$$

or $\qquad\qquad \dfrac{\theta}{\delta} = \mathbf{0.125}$

and energy thickness: $\delta^{**} = \displaystyle\int_0^\delta \frac{u}{U}\left[1 - \frac{u^2}{U^2}\right] dy$ $\qquad\qquad$ ... (iii)

Substituting the value of $\dfrac{u}{U} = \left(\dfrac{y}{\delta}\right)^{0.22}$ in Eq. (iii), we get

$$\delta^{**} = \int_0^\delta \left(\frac{y}{\delta}\right)^{0.22}\left[1 - \left(\frac{y}{\delta}\right)^{0.44}\right] dy$$

$$= \int_0^\delta \left[\left(\frac{y}{\delta}\right)^{0.22} - \left(\frac{y}{\delta}\right)^{0.66}\right] dy$$

$$= \left[\frac{1}{\delta^{0.22}}\frac{y^{1.22}}{1.22} - \frac{1}{\delta^{0.66}} \times \frac{y^{1.66}}{1.66}\right]_0^\delta$$

$$= \frac{1}{\delta^{0.22}} \times \frac{\delta^{1.22}}{1.22} - \frac{1}{\delta^{0.66}} \times \frac{\delta^{1.66}}{1.66}$$

$$= \frac{\delta}{1.22} - \frac{\delta}{1.66}$$

$$= \delta\left[\frac{1}{1.22} - \frac{1}{1.66}\right] = \delta[0.819 - 0.602]$$

or $\qquad\qquad \delta^{**} = \delta \times 0217$

or $\qquad\qquad \dfrac{\delta^{**}}{\delta} = \mathbf{0.217}$

Problem 3.4: Find the displacement thickness and momentum thickness in terms of the normal boundary layer thickness δ in respect to the following velocity profile in the boundary layer on a flat plate:

$\qquad$ (i) $\dfrac{u}{U} = \left(\dfrac{y}{\delta}\right)^{\frac{1}{m}}$ $\qquad\qquad\qquad$ (ii) $\dfrac{u}{U} = \left(\dfrac{y}{\delta}\right)^{\frac{1}{7}}$.

where $\qquad\qquad$ u = velocity at a height y above the surface of flat plate and $\qquad\qquad\qquad$ U = free stream velocity.

Solution:

(i) Given:

Velocity profile: $\dfrac{u}{U} = \left(\dfrac{y}{\delta}\right)^{\frac{1}{m}}$

The displacement thickness:

$$\delta^* = \int_0^\delta \left(1 - \frac{u}{U}\right) dy = \int_0^\delta \left[1 - \left(\frac{y}{\delta}\right)^{1/m}\right] dy$$

$$= \left[y - \frac{1}{\delta^{\frac{1}{m}}} \cdot \frac{y^{\frac{1}{m}+1}}{\left(\frac{1}{m}+1\right)}\right]_0^\delta = \left[y - \frac{1}{\delta^{\frac{1}{m}}}\left(\frac{m}{m+1}\right) \cdot y^{\frac{m+1}{m}}\right]_0^\delta$$

$$= \left[\delta - \frac{1}{\delta^{\frac{1}{m}}} \cdot \left(\frac{m}{m+1}\right)\delta^{\frac{m+1}{m}}\right] = \delta - \left(\frac{m}{m+1}\right) \cdot \delta^{\frac{m+1}{m}-\frac{1}{m}}$$

$$= \delta - \left(\frac{m}{m+1}\right)\delta$$

$$= \left(1 - \frac{m}{m+1}\right)\delta = \left(\frac{m+1-m}{m+1}\right)\delta = \left(\frac{1}{m+1}\right)\delta$$

The momentum thickness: $\theta = \int_0^\delta \left[1 - \frac{u}{U}\right]\frac{u}{U}\,dy = \int_0^\delta \left[\frac{u}{U} - \left(\frac{u}{U}\right)^2\right]dy$

$$= \int_0^\delta \left[\left(\frac{y}{\delta}\right)^{\frac{1}{m}} - \left(\frac{y}{\delta}\right)^{\frac{2}{m}}\right]dy = \int_0^\delta \left[\frac{1}{\delta^{\frac{1}{m}}}y^{\frac{1}{m}} - \frac{1}{\delta^{\frac{2}{m}}}y^{\frac{2}{m}}\right]dy$$

$$= \int_0^\delta \left[\delta^{\frac{-1}{m}}y^{\frac{1}{m}} - \delta^{\frac{-2}{m}}y^{\frac{2}{m}}\right]dy$$

$$= \left[\frac{\delta^{\frac{-1}{m}} \cdot y^{\frac{1}{m}+1}}{\left(\frac{1}{m}+1\right)} - \frac{\delta^{\frac{-2}{m}} \cdot y^{\frac{2}{m}+1}}{\left(\frac{2}{m}+1\right)}\right]_0^\delta$$

$$= \left(\frac{m}{m+1}\right)\delta^{\frac{-1}{m}}\delta^{\frac{m+1}{m}} - \left(\frac{m}{m+2}\right)\delta^{\frac{-2}{m}}\delta^{\frac{m+2}{m}}$$

$$= \frac{m}{m+1}\delta - \left(\frac{m}{m+2}\right)\delta = \left(\frac{m}{m+1} - \frac{m}{m+2}\right)\delta$$

$$= \left(\frac{m(m+2) - m(m+1)}{(m+1)(m+2)}\right)\delta = \left(\frac{m^2 + 2m - m^2 - m}{(m+1)(m+2)}\right)\delta$$

$$\theta = \frac{m}{(m+1)(m+2)}\delta$$

(*ii*)Given:

Velocity profile: $\dfrac{u}{U} = \left(\dfrac{y}{\delta}\right)^{\frac{1}{7}}$

The displacement thickness: $\delta^* = \displaystyle\int_0^\delta \left(1 - \dfrac{u}{U}\right) dy = \int_0^\delta \left[1 - \left(\dfrac{y}{\delta}\right)^{\frac{1}{7}}\right] dy$

$$= \int_0^\delta \left[1 - \dfrac{y^{\frac{1}{7}}}{\delta^{\frac{1}{7}}}\right] dy = \int_0^\delta \left(1 - \delta^{\frac{-1}{7}} y^{\frac{1}{7}}\right) dy$$

$$= \left[y - \dfrac{\delta^{\frac{-1}{7}} \cdot y^{\frac{1}{7}+1}}{\left(\dfrac{1}{7}+1\right)}\right]_0^\delta = \delta - \dfrac{\delta^{\frac{-1}{7}} \cdot \delta^{\frac{8}{7}}}{\dfrac{8}{7}} = \delta - \dfrac{7}{8}\delta$$

$$= \dfrac{\delta}{8}$$

The momentum thickness: $\theta = \displaystyle\int_0^\delta \left(1 - \dfrac{u}{U}\right)\dfrac{u}{U} dy = \int_0^\delta \left[1 - \left(\dfrac{y}{\delta}\right)^{\frac{1}{7}}\right]\left(\dfrac{y}{\delta}\right)^{\frac{1}{7}} dy$

$$= \int_0^\delta \left[\left(\dfrac{y}{\delta}\right)^{\frac{1}{7}} - \left(\dfrac{y}{\delta}\right)^{\frac{2}{7}}\right] dy$$

$$= \int_0^\delta \left(\delta^{\frac{-1}{7}} \cdot y^{\frac{1}{7}} - \delta^{\frac{-2}{7}} y^{\frac{2}{7}}\right) dy$$

$$= \left[\dfrac{\delta^{\frac{-1}{7}} y^{\frac{1}{7}+1}}{\left(\dfrac{1}{7}+1\right)} - \dfrac{\delta^{\frac{-2}{7}} y^{\frac{2}{7}+1}}{\left(\dfrac{2}{7}+1\right)}\right]_0^\delta$$

$$= \left[\dfrac{7}{8}\delta^{\frac{-1}{7}} \delta^{\frac{8}{7}} - \dfrac{7}{9}\delta^{\frac{-2}{7}} \cdot \delta^{\frac{9}{7}}\right]$$

$$\theta = \dfrac{7}{8}\delta - \dfrac{7}{9}\delta = \delta 7\left[\dfrac{1}{8} - \dfrac{1}{9}\right]$$

$$\theta = 7\delta\left[\dfrac{9-8}{72}\right] = \dfrac{7}{72}\delta$$

3.7 DRAG FORCE ON A FLAT PLATE DUE TO BOUNDARY LAYER

As we know, the boundary layer (b.l.) is developed when liquid flows over flat plate. Let a small length dx of a flat plate at a distance x from the leading edge O,

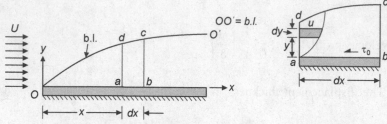

(a) Drag force on a flat plate   (b) Enlarged view of the
     due to boundary layer   small length of plate

Fig. 3.14: Drag Force

as shown in Fig. 3.14(a), and the enlarged view shown in Fig. 3.14(b).

The shear stress exerted by the fluid on the flat plate: τ_0

$$\tau_0 = \mu \left(\frac{du}{dy} \right)_{y=0}$$

where $\left(\dfrac{du}{dy} \right)_{y=0}$ is velocity gradient near the flat plate at $y = 0$.

∴ The shear force or drag force on a small length dx: dF_0

$$dF_0 = \text{shear stress} \times \text{surface area}$$
$$= \tau_0 dA$$
$$= \tau_0 \, dxb \qquad\qquad \text{... (3.7.1)} \because dA = dx \cdot b$$

where b = width of flat plate, not shown in Fig. 3.14

Consider $abcd$ is a control volume of the fluid over the distance dx as shown in Fig. 3.14(b). The edge dc represents the outer edge of the boundary layer.

Let u = velocity at any point within the b.l.

The mass flow rate entering through face ad: m_{ad}

$$m_{ad} = \int_0^\delta \rho \times \text{velocity} \times \text{cross-section area of}$$
$$\text{strip of thickness } dy$$
$$= \int_0^\delta \rho u \, b \, dy$$

The mass flow rate leaving through face bc : m_{bc}

$$m_{bc} = m_{ad} + \frac{\partial}{\partial x} m_{ad} \times dx$$
$$= \int_0^\delta \rho u \, b \, dy + \frac{\partial}{\partial x} \left[\int_0^\delta \rho u \, b \, dy \right] dx$$

According to continuity equation for a steady and incompressible fluid flow;

The mass flow rate entering through face ad + mass flow rate entering through face dc = mass flow rate leaving through face bc

$$m_{ad} + m_{dc} = m_{bc}$$
or
$$m_{dc} = m_{bc} - m_{ad}$$

$$= \int_0^\delta \rho u b \, dy + \frac{\partial}{\partial x}\left[\int_0^\delta \rho u b \, dy\right]dx - \int_0^\delta \rho u b \, dy$$

$$m_{dc} = \frac{\partial}{\partial x}\left[\int_0^\delta \rho u b \, dy\right]dx$$

Now find out momentum per second along x-axis:

The rate of change of momentum entering face ad:

$$= m_{ab} \times u = \int_0^\delta \rho u b \, dy \times u = \int_0^\delta \rho u^2 b \, dy$$

The rate of change of momentum leaving face bc:

$$= m_{bc} \times u = \int_0^\delta \rho u^2 b \, dy + \frac{\partial}{\partial x}\left[\int_0^\delta \rho u^2 b \, dy\right]dx$$

The rate of change of momentum entering face dc:

$= m_{dc} \times U[\because$ fluid entering through face dc with uniform velocity $U]$

$$= \frac{\partial}{\partial x}\left[\int_0^\delta \rho u b \, dy\right]dx \cdot U$$

$$= \frac{\partial}{\partial x}\left[\int_0^\delta \rho U u b \, dy\right]dx \cdot U \qquad [\because U = \text{constant, it can be taken inside}$$
the differential and integral for simplification]

Now the rate of change of momentum of the control volume – the rate of change of momentum entering face ad + entering face dc – leaving face bc

$$= \int_0^\delta \rho u^2 b \, dy + \frac{\partial}{\partial x}\left[\int_0^\delta \rho u b \, dy\right]dx - \int_0^\delta \rho u^2 b \, dy - \frac{\partial}{\partial x}\left[\int_0^\delta \rho u^2 b \, dy\right]dx$$

$$= \frac{\partial}{\partial x}\left[\int_0^\delta \rho U u b \, dy\right]dx - \frac{\partial}{\partial x}\left[\int_0^\delta \rho u^2 b \, dy\right]dx$$

$$= \frac{\partial}{\partial x}\left[\int_0^\delta (\rho U u b - \rho u^2 b) \, dy\right]dx$$

$$= \frac{\partial}{\partial x}\left[\rho b \int_0^\delta (U u - u^2) \, dy\right]dx \qquad\qquad \because \rho \text{ and } b \text{ are constant}$$

$$= \rho b \frac{\partial}{\partial x}\left[\int_0^\delta U^2\left(\frac{u}{U} - \frac{u^2}{U^2}\right) dy\right]dx$$

$$= \rho b U^2 \frac{\partial}{\partial x}\left[\int_0^\delta \left(\frac{u}{U} - \frac{u^2}{U^2}\right) dy\right]dx$$

$$= \rho b U^2 \frac{\partial}{\partial x}\left[\int_0^\delta \left(1 - \frac{u}{U}\right)\frac{u}{U} dy\right]dx$$

$$= \rho b U^2 \frac{\partial \theta}{\partial x} dx \qquad\qquad \ldots (3.7.2)$$

where θ = momentum thickness

$$= \int_0^\delta \left(1 - \frac{u}{U}\right)\frac{u}{U}\,dy$$

In the absence of any pressure and gravity forces the drag or shear force at the flat plate surface must be balanced by the net rate of change of momentum of the control volume.

i.e.,   Eq. (3.7.1) = Eq. (3.7.2)

$$\tau_0 dx \cdot b = \rho b U^2 \frac{\partial \theta}{\partial x} \cdot dx$$

$$\tau_0 = \rho U^2 \frac{\partial \theta}{\partial x}$$

or $\dfrac{\tau_0}{\rho U^2} = \dfrac{\partial \theta}{\partial x}$ ... (3.7.3)

Equation (3.7.3) is known as **Von-Karman momentum integral equation** for the hydrodynamic boundary layer over a flat plate.

From Eq. (10.7.3), the drag force on a small length dx: $dF_D = \tau_0\,dx\,b$

The total drag force on the plate of length l on one side: F_D

$$F_D = \int_0^l dF_0 = \int_0^l \tau_0 b\,dx$$

Local skin friction coefficient (C_f): It is defined as the ratio of the local wall shear stress (τ_0) to the dynamic pressure $\left(\dfrac{1}{2}\rho U^2\right)$ of the uniform flow stream-

Mathematically,

Local skin friction coefficient:

$$C_f = \frac{\text{Local wall shear stress}}{\text{Dynamic pressure of the free stream}}$$

$$C_f = \frac{\tau_0}{\dfrac{1}{2}\rho U^2} = \frac{\mu\left(\dfrac{du}{dy}\right)_{y=0^+}}{\dfrac{1}{2}\rho U^2}$$

The local skin friction coefficient is also local coefficient of drag.

Average skin friction coefficient ($\overline{C_f}$): It is defined as the ratio of the total drag force (F_D) to the dynamic force $\left(\dfrac{1}{2}\rho A U^2\right)$. It is also called average coefficient of drag.

Mathematically,

$$\overline{C_f} = \frac{F_D}{\dfrac{1}{2}\rho A U^2}$$

where ρ = density of fluid

                                        A = area of the plate

                                        U = free stream velocity.

Drag force: F_D

The force exerted by flowing fluid on a body in the direction of flow is called drag force or simply drag. The drag force can be measured directly by simply attaching the body subjected to fluid flow to a calibrated spring and measuring the displacement in the flow direction.

3.8 ESTIMATION OF THE LAMINAR BOUNDARY LAYER THICKNESS

For laminar boundary layer, the boundary layer thickness can easily be estimated as follow:

In the laminar boundary layer, the inertia force and the friction force are same:

The inertia for per unit volume = the friction force per unit volume

$$\frac{ma}{V} = \frac{\tau \cdot A}{V}$$

$$\frac{\rho V}{V}\frac{du}{dt} = \mu \frac{du}{dy}\frac{A}{V}$$

$$\rho \frac{du}{dt} = \mu \frac{du}{dy} \times \frac{1}{y} \qquad\qquad \left(\because\ y = \frac{V}{A}\right)$$

$$\rho \frac{du}{dx} = \mu \frac{du}{dy^2}$$

$$\frac{du}{}$$

$$\rho \frac{d^2 u}{dx} = \mu \frac{du}{dy^2} \qquad\qquad\qquad ...\ (10.8.1)$$

where x is length of the plate, $\dfrac{\partial u}{\partial x}$ is proportional to $\dfrac{U}{x}$

where U is the velocity of free stream. The velocity gradient perpendicular to the plate: $\dfrac{du}{dy}$ is proportional to $\dfrac{U}{\delta}$.

where δ is boundary layer thickness. Eq. (3.8.1) becomes

$$\rho \frac{U^2}{x} = k'\mu \frac{u}{\delta^2}$$

where k' is constant

$$\rho \frac{U}{x} = k' \frac{\mu}{\delta^2}$$

or $$\delta^2 = \frac{k'\mu x}{\rho U}$$

or
$$\delta = \sqrt{\frac{k' \mu x}{\rho U}}$$

$$\delta = \sqrt{k'} \sqrt{\frac{\mu x}{\rho U}}$$

$$\delta = k\sqrt{\frac{\mu x}{\rho U}} \qquad \qquad \text{... (3.8.2)}$$

where k is numerical constant factor. By exact analytical solution of boundary layer equation found by H. Blasius (1908), is 5 (*i.e.*, $k = 5$)

$$\delta = 5\sqrt{\frac{\mu x}{\rho U} \times \frac{x}{x}} = 5\sqrt{\frac{\mu x^2}{\rho U x}}$$

$$\delta = 5x\sqrt{\frac{\mu}{\rho U x}}$$

$$\delta = 5x\sqrt{\frac{v}{Ux}} = \frac{5x}{\sqrt{\frac{Ux}{v}}}$$

$$\delta = \frac{5x}{\sqrt{\text{Re}_x}} \qquad \qquad \because \ \text{Re}_x = \frac{\rho U x}{\mu} = \frac{Ux}{v}$$

$$\therefore \qquad \qquad \delta = \frac{5x}{\sqrt{\mathbf{Re}_x}} \qquad \qquad \text{... (3.8.3)}$$

From Eq. (3.8.3), the following important conclusion can be drawn:

(*i*) At a given section along the plate length (*i.e.*, x = constant), the boundary layer thickness (δ) decrease with the increasing Reynolds number (Re_x).

For a given fluid, Reynolds number will depend on the free stream velocity (U). Thus for a high velocity flow, the boundary layer will be very thin. For limiting case of frictionless flow (*i.e.*, ideal fluid flow), with $\text{Re}_x = \infty$, the boundary layer thickness vanishes

(*ii*) For a given fluid flowing at a certain free stream velocity (*i.e.*, v and U both are constant), the boundary layer thickness (δ) increases in proportion to $x^{1/2}$. (*i.e.*, $\delta \propto x^{1/2}$)

Local drag coefficient: $C_{fx} = \dfrac{\tau_0}{\dfrac{1}{2}\rho U^2}$

As we know that the shear stress on the plate: τ_0

$$\tau_0 = \mu\left(\frac{\partial u}{\partial y}\right)_{y=0} = C_1\mu\frac{U}{\delta}$$

$$\text{... (3.8.4)}$$

Substituting the value of δ from Eq. (3.8.2) in above Eq. (3.8.4), we get

$$\tau_0 = C_1 \mu \frac{U}{k\sqrt{\dfrac{\mu x}{\rho U}}} = \frac{C_1}{k} \mu U \sqrt{\frac{\rho U}{\mu x}}$$

$$\tau_0 = C_2 \sqrt{\frac{\rho \mu U^3}{x}}$$

where $\qquad C_2 = \dfrac{C'}{k}$, another constant

Local drag coefficient: $\quad C_f = \dfrac{C_2 \sqrt{\dfrac{\rho \mu U^3}{x}}}{\dfrac{1}{2} \rho U^2}$

$$C_f = 2 C_2 \sqrt{\frac{\mu}{\rho U x}}$$

$$C_f = \frac{C}{\sqrt{Re_x}} \qquad \because \text{ constant: } C = 2C_2 \quad \dots (3.8.5)$$

In Eq. (3.8.5) the value of the constant of proportionality has been obtained by H.Blasius by exact analytical solution for the laminar boundary layer equations as 0.66 u. Thus Eq. (3.8.5) becomes

$$C_f = \frac{0.664}{\sqrt{Re_x}}$$

$$C_f = \frac{\tau_0}{\dfrac{1}{2} \rho U^2} = \frac{0.664}{\sqrt{Re_x}} \qquad \qquad \dots (3.8.6)$$

$$\tau_0 = \frac{0.664}{2} \frac{\rho U^2}{\sqrt{Re_x}} = 0.332 \frac{\mu U}{x \sqrt{\dfrac{\rho U x}{\mu}}} \frac{\rho U x}{\mu}$$

$$\tau_0 = 0.332 \frac{\mu U}{x} \sqrt{Re_x}$$

The total horizontal drag force (F_D) on one side of the plate on which laminar boundary layer is developed can be obtained as:

$$F_D = \int_0^l \tau_0 b\, dx$$

where $\qquad b$ = width of the plate

$\qquad\qquad\quad l$ = length of the plate

Surface area of the plate: $A = b.l$.

The average drag coefficient $\left(\bar{C}_f\right)$ may be obtained as

$$\bar{C}_f = \frac{\dfrac{F_D}{bl}}{\dfrac{1}{2}\rho U^2} = \frac{F_D}{\dfrac{1}{2}bl\rho U^2}$$

$$\bar{C}_f = \frac{1.328}{\sqrt{\mathrm{Re}_l}}$$

$$\bar{C}_f = \frac{1.328}{\sqrt{\mathrm{Re}_l}} = \frac{F_D}{\dfrac{1}{2}bl\rho U^2}$$

or

$$\frac{F_D}{\dfrac{1}{2}bl\rho U^2} = \frac{1.328}{\sqrt{\mathrm{Re}_l}}$$

or

$$F_D = \frac{1.328}{2}\frac{bl\rho U^2}{\sqrt{\mathrm{Re}_l}}$$

$$= 0.664\frac{bl\rho U^2}{\sqrt{\mathrm{Re}_l}} = 0.664\frac{\mu bU}{\sqrt{\mathrm{Re}_l}} \times \frac{\rho Ul}{\mu}$$

$$F_D = 0.664\,\mu bU \times \sqrt{\frac{\rho Ul}{\mu}}$$

From the exact analytical solution of the boundary layer equations by H. Blasius, the following expression for displacement thickness (δ^*) and the momentum thickness has been obtained.

$$\frac{\delta^*}{x} = \frac{1.729}{\sqrt{\mathrm{Re}_x}} \quad \text{and} \quad \frac{\theta}{x} = \frac{0.664}{\sqrt{\mathrm{Re}_x}}$$

Problem 3.5: Find an expression for boundary layer thickness (δ), shear stress (τ_0), average coefficient of drag $\left(\bar{C}_f\right)$ in terms of Reynolds number for the following velocity profile for laminar boundary layer given as:

(i) $\dfrac{u}{U} = \dfrac{y}{\delta}$

(ii) $\dfrac{u}{U} = 2\left(\dfrac{y}{\delta}\right) - \left(\dfrac{y}{\delta}\right)^2$

(iii) $\dfrac{u}{U} = \dfrac{3}{2}\left(\dfrac{y}{\delta}\right) - \dfrac{1}{2}\left(\dfrac{y}{\delta}\right)^3$

(iv) $\dfrac{u}{U} = 2\left(\dfrac{y}{\delta}\right) - 2\left(\dfrac{y}{\delta}\right)^3 + \left(\dfrac{y}{\delta}\right)^4$

(v) $\dfrac{u}{U} = \sin\left(\dfrac{\pi}{2}\dfrac{y}{\delta}\right)$.

Solution: Given:

(i) The velocity profile: $\dfrac{u}{U} = \dfrac{y}{\delta}$ ... (i)

We know that the Von Karman momentum integral equation:

$$\frac{\tau_0}{\rho U^2} = \frac{\partial \theta}{\partial x}$$

where

$$\theta = \int_0^\delta \frac{u}{U}\left(1 - \frac{u}{U}\right) dy$$

$\therefore$

$$\frac{\tau_0}{\rho U^2} = \frac{\partial}{\partial x}\left[\int_0^\delta \frac{u}{U}\left(1 - \frac{u}{U}\right) dy\right]$$

$$\frac{\tau_0}{\rho U^2} = \frac{\partial}{\partial x}\left[\int_0^\delta \frac{y}{\delta}\left(1 - \frac{y}{\delta}\right) dy\right]$$

$$= \frac{\partial}{\partial x}\left[\int_0^\delta \left(\frac{y}{\delta} - \frac{y^2}{\delta^2}\right) dy\right]$$

$$= \frac{\partial}{\partial x}\left[\frac{y^2}{2\delta} - \frac{y^3}{3\delta^2}\right]_0^\delta = \frac{\partial}{\partial x}\left[\frac{\delta^2}{2\delta} - \frac{\delta^3}{3\delta^2}\right]$$

$$= \frac{\partial}{\partial x}\left[\frac{\delta}{2} - \frac{\delta}{3}\right] = \frac{\partial}{\partial x}\left(\frac{\delta}{6}\right)$$

$$\frac{\tau_0}{\rho U^2} = \frac{1}{6}\frac{\partial \delta}{\partial x}$$

or

$$\tau_0 = \frac{1}{6}\delta U^2 \frac{\partial \delta}{\partial x}$$ ... (ii)

Also the shear stress at the surface in laminar layer:

$$\tau_0 = \mu \left(\frac{du}{dy}\right)_{y=0}$$ ... (iii)

From Equation (i), we get

$$u = U\frac{y}{\delta}$$

$\therefore$

$$\frac{\partial u}{\partial y} = \frac{U}{\delta}$$

or

$$\left(\frac{\partial u}{\partial y}\right)_{y=0} = \frac{U}{\delta}$$

Substituting this value in Eq. (iii), we get

$$\tau_0 = \mu \frac{U}{\delta}$$ ... (iv)

Equating Eqs. (ii) and (iv), we get

$$\frac{1}{6}\rho U^2 \frac{\partial \delta}{\partial x} = \mu \frac{U}{\delta}$$

or

$$\delta \frac{\partial \delta}{\partial x} = 6\frac{\mu U}{\rho U^2} = \frac{6\mu}{\rho U}$$

or

$$\delta \partial \delta = \frac{6\mu}{\rho U} dx$$

As

$$\delta = f(x) \text{ only}$$

Hence partial derivative can be changed to total derivative.

$$\delta d\delta = \frac{6\mu}{\rho U} dx$$

By integrating the above expression, we get

$$\frac{\delta^2}{2} = \frac{6\mu}{\rho U}x + C \qquad \left(\because \frac{\mu}{\rho U} = \text{constant} \right)$$

where C is constant of integration and its value is determined by boundary condition at $x = 0$, $\delta = 0$ and hence $C = 0$.

$$\therefore \qquad \frac{\delta^2}{2} = \frac{6\mu x}{\rho U}$$

$$\delta^2 = \frac{12\mu x}{\rho U}$$

$$\delta = \sqrt{\frac{12\mu x}{\rho U}} = 3.464\sqrt{\frac{\mu x}{\rho U}} \qquad \qquad \dots (v)$$

$$\delta = 3.464\sqrt{\frac{\mu x}{\rho U} \times \frac{x}{x}} = 3.464\sqrt{\frac{\mu x^2}{\rho U x}}$$

$$\delta = 3.464 \sqrt{\frac{x^2}{\dfrac{\rho U x}{\mu}}}$$

$$\delta = 3.464\frac{x}{\sqrt{\dfrac{\rho U x}{\mu}}} = 3.464\frac{x}{\sqrt{Re_x}} \qquad \left(\because Re_x = \frac{\rho U x}{\mu} \right)$$

$$\boldsymbol{\delta = 3.464\frac{x}{\sqrt{Re_x}}} \qquad \qquad \dots (vi)$$

It is clear from Eq. (v), as μ, ρ and U are constants hence $d \propto \sqrt{x}$ i.e., thickness of laminar boundary layer is directly proportional to square root of the distance from the leading edge. Equation (vi) shows that the thickness of laminar boundary layer is inversely proportional to the square root of

Reynolds number.

Shear stress (τ_0) in terms of Reynolds number Re_x:
From Eq. (iv), the shear stress:

$$\tau_0 = \mu \frac{U}{\delta}$$

Substituting the value of δ from Eq. (vi) in above equation, we get

$$\tau_0 = \mu \frac{U}{3.464 \dfrac{x}{\sqrt{Re_x}}} = \frac{1}{3.464} \frac{\mu U \sqrt{Re_x}}{x}$$

$$\tau_0 = 0.288 \frac{\mu U \sqrt{Re_x}}{x}$$

Average coefficient of drag ($\bar{C}_f$) in terms of Reynolds number:

By definition of average coefficient of drag: $\bar{C}_f = \dfrac{F_D}{\dfrac{1}{2}\rho A U^2}$ $\qquad (A = bl)$

$$\bar{C}_f = \frac{2 F_D}{\rho b l U^2}$$

where

$$F_D = \int_0^l \tau_0 b\, dx$$

$$= \int_0^l 0.288 \frac{\mu U \sqrt{Re_x}}{x} \cdot b\, dx$$

$$= \int_0^l 0.288 \frac{\mu U}{x} \sqrt{\frac{\rho U x}{\mu}}\, b\, dx$$

$$= 0.288 \mu b U \sqrt{\frac{\rho U}{\mu}} \int_0^l \frac{\sqrt{x}}{x}\, dx$$

$$= 0.288 \mu b U \sqrt{\frac{\rho U}{\mu}} \int_0^l \frac{1}{\sqrt{x}}\, dx$$

$$= 0.288 \mu b U \sqrt{\frac{\rho U}{\mu}} \int_0^l x^{-1/2}\, dx$$

$$= 0.288 \mu b U \sqrt{\frac{\rho U}{\mu}} \left[\frac{x^{1/2}}{1/2}\right]_0^l$$

$$= 0.288\, \mu b U \sqrt{\frac{\rho U}{\mu}}\, 2\left[l^{1/2}\right]$$

$$F_D = 0.576 \mu b U \sqrt{\frac{\rho U l}{\mu}}$$

$$\therefore \qquad \bar{C}_f = \frac{2 \times 0.576 \mu b U}{\rho b l U^2} \sqrt{\frac{\rho U l}{\mu}}$$

$$= \frac{1.152 \mu}{\rho l U} \sqrt{\frac{\rho U l}{\mu}} = 1.152 \sqrt{\frac{\mu}{\rho l U}}$$

$$\bar{C}_f = \frac{1.152}{\sqrt{\mathbf{Re}_l}} \qquad\qquad \because Re_l = \frac{\rho l U}{\mu}$$

(ii) Given velocity profile: $\dfrac{u}{U} = 2\left(\dfrac{y}{\delta}\right) - \left(\dfrac{y}{\delta}\right)^2$ ... (vii)

We know, Von Karman momentum integral equation:

$$\frac{\tau_0}{\rho U^2} = \frac{\partial \theta}{\partial x}$$

where

$$\theta = \int_0^\delta \frac{u}{U}\left(1 - \frac{u}{U}\right) dy$$

$$\therefore \qquad \frac{\tau_0}{\rho U^2} = \frac{\partial}{\partial x}\left[\int_0^\delta \frac{u}{U}\left(1 - \frac{u}{U}\right) dy\right]$$

$$= \frac{\partial}{\partial x}\left[\int_0^\delta \left\{\frac{2y}{\delta} - \frac{y^2}{\delta^2}\right\}\left\{1 - \left(\frac{2y}{\delta} - \frac{y^2}{\delta^2}\right)\right\} dy\right]$$

$$= \frac{\partial}{\partial x}\left[\int_0^\delta \left(\frac{2y}{\delta} - \frac{y^2}{\delta^2}\right)\left(1 - \frac{2y}{\delta} + \frac{y^2}{\delta^2}\right) dy\right]$$

$$= \frac{\partial}{\partial x}\left[\int_0^\delta \left(\frac{2y}{\delta} - \frac{4y^2}{\delta^2} + \frac{2y^3}{\delta^3} - \frac{y^2}{\delta^2} + \frac{2y^3}{\delta^3} - \frac{y^4}{\delta^4}\right) dy\right]$$

$$= \frac{\partial}{\partial x}\left[\int_0^\delta \left(\frac{2y}{\delta} - \frac{5y^2}{\delta^2} + \frac{4y^3}{\delta^3} - \frac{y^4}{\delta^4}\right) dy\right]$$

$$= \frac{\partial}{\partial x}\left[\frac{2y^2}{2\delta} - \frac{5y^3}{3\delta^2} + \frac{4y^4}{4\delta^3} - \frac{y^5}{5\delta^4}\right]_0^\delta$$

$$= \frac{\partial}{\partial x}\left[\frac{\delta^2}{\delta} - \frac{5}{3}\frac{\delta^3}{\delta^2} + \frac{\delta^4}{\delta^3} - \frac{\delta^5}{5\delta^4}\right]$$

$$= \frac{\partial}{\partial x}[\delta - 1.6666\delta + \delta - 0.2\delta] = \frac{\partial}{\partial x}(0.1334\delta)$$

$$\frac{\tau_0}{\rho U^2} = 0.1334\frac{\partial \delta}{\partial x}$$

or
$$\tau_0 = 0.1334\rho U^2 \frac{\partial \delta}{\partial x} \qquad \qquad \text{... (viii)}$$

Also the shear stress at the surface in laminar layer:

$$\tau_0 = \mu \left(\frac{\partial u}{\partial y} \right)_{y=0} \qquad \qquad \text{... (ix)}$$

From Eq. (vii), we get

$$u = U \left[\frac{2y}{\delta} - \frac{y^2}{\delta^2} \right]$$

$\therefore$

$$\frac{\partial u}{\partial y} = U \left[\frac{2}{\delta} - \frac{2y}{\delta^2} \right]$$

$$\left(\frac{\partial u}{\partial y} \right)_{y=0} = U \left[\frac{2}{\delta} - \frac{2 \times 0}{\delta^2} \right] = \frac{2U}{\delta}$$

Substituting this value in Eq. (ix), we get

$$\tau_0 = \mu \frac{2U}{\delta} = \frac{2\mu U}{\delta} \qquad \qquad \text{... (x)}$$

Equating Eqs. (viii) and (x), we get

$$0.1334\rho U^2 \frac{\partial \delta}{\partial x} = 2\frac{\mu U}{\delta}$$

$$\delta \partial \delta = 14.99 \frac{\mu U}{\rho U^2} \partial x$$

$$\delta \partial \delta = 14.99 \frac{\mu U}{\rho U} \partial x$$

As
$$\delta = f(x) \text{ only}$$
Hence partial derivative can be changed to total derivative.

$$\delta d\delta = 14.99 \frac{\mu}{\rho U} dx$$

By integrating the above expression, we get

$$\frac{\delta^2}{2} = 14.99 \frac{\mu x}{\rho U} + C$$

$x = 0$, $\delta = 0$ and hence $C = 0$

$\therefore$

$$\delta_2 = 29.98 \frac{\mu x}{\rho U}$$

$$\delta = \sqrt{29.98 \frac{\mu x}{\rho U}} = 5.475 \sqrt{\frac{\mu x}{\rho U}}$$

$$\delta = 5.48 \sqrt{\frac{\mu x}{\rho U} \times \frac{x}{x}} = 5.48 \sqrt{\frac{\mu x^2}{\rho U x}}$$

$$\delta = 5.48 \frac{x}{\sqrt{\dfrac{\rho U x}{\mu}}} = \frac{5.48x}{\sqrt{Re_x}}$$

$$\delta = \frac{5.48x}{\sqrt{Re_x}} \qquad\qquad (xi)$$

Shear stress (τ_0) in terms of Reynolds number:

From Eq. (x), the shear stress:

$$\tau_0 = \frac{2\mu u}{\delta}$$

Substituting the value of δ from Eq. (xi) in above equation, we get

$$\tau_0 = \frac{2\mu u}{5.48 \dfrac{x}{\sqrt{Re_x}}} = 0.365 \frac{\mu U}{x} \sqrt{Re_x}$$

$$\tau_0 = 0.365 \frac{\mu U}{x} \sqrt{Re_x}$$

Average coefficient of drag ($\bar{C}_f$) in terms of Reynolds number:

By definition of average coefficient of drag:

$$\bar{C}_f = \frac{F_D}{\dfrac{1}{2}\rho A U^2} = \frac{2F_D}{\rho b l U^2} \qquad\qquad \because A = bl$$

where

$$F_D = \int_0^l \tau_0 \, b \, dx$$

$$= \int_0^l 0.365 \frac{\mu U}{x} \sqrt{Re_x} \, b dx$$

$$= 0.365 \int_0^l \frac{\mu U}{x} \sqrt{\frac{\rho U x}{\mu}} \, b dx$$

$$= 0.365 \int_0^l \mu U \sqrt{\frac{\rho U}{\mu}} \times \frac{1}{\sqrt{x}} \, b dx$$

$$= 0.365 \mu U \sqrt{\frac{\rho U}{\mu}} \times b \int_0^l x^{-1/2} \, dx$$

$$= 0.365 \mu U \sqrt{\frac{\rho U}{\mu}} \times b \left[\frac{x^{1/2}}{1/2} \right]_0^l$$

$$= 0.365 \mu U \sqrt{\frac{\rho U}{\mu}} \times 2b \, l^{1/2}$$

$$F_D = 0.73 \mu U b \sqrt{\frac{\rho U l}{\mu}}$$

$$\therefore \qquad \bar{C}_f = \frac{2 \times 0.73 \mu U b}{\rho b l U^2} \sqrt{\frac{\rho U l}{\mu}}$$

$$= 1.46 \frac{\mu}{\rho l U} \sqrt{\frac{\rho U l}{\mu}} = 1.46 \sqrt{\frac{\mu}{\rho l U}}$$

$$\bar{C}_f = \frac{1.46}{\sqrt{\mathbf{Re}_l}} \qquad\qquad \because \ \mathrm{Re}_l = \frac{\rho l U}{\mu}$$

(*iii*) Given velocity profile: $\dfrac{u}{U} = \dfrac{3}{2} \dfrac{y}{\delta} - \dfrac{1}{2} \left(\dfrac{y}{\delta}\right)^3$ ... (*xii*)

We know, Von-Karman momentum integral equation:

$$\frac{\tau_0}{\rho U^2} = \frac{\partial \theta}{\partial x}$$

where $\theta = \displaystyle\int_0^\delta \frac{u}{U}\left(1 - \frac{u}{U}\right) dy$

$$\therefore \quad \frac{\tau_0}{\rho U^2} = \frac{\partial}{\partial x}\left[\int_0^\delta \frac{u}{U}\left(1 - \frac{u}{U}\right)\right] dy$$

$$= \frac{\partial}{\partial x}\left[\int_0^\delta \left[\frac{3}{2}\left(\frac{y}{\delta}\right) - \frac{1}{2}\left(\frac{y}{\delta}\right)^3\right]\left[1 - \left\{\frac{3}{2}\left(\frac{y}{\delta}\right) - \frac{1}{2}\left(\frac{y}{\delta}\right)^3\right\}\right] dy\right]$$

$$= \frac{\partial}{\partial x}\left[\int_0^\delta \left(\frac{3y}{2\delta} - \frac{y^3}{2\delta^3}\right)\left(1 - \frac{3y}{2\delta} + \frac{y^3}{2\delta^3}\right) dy\right]$$

$$= \frac{\partial}{\partial x}\left[\int_0^\delta \left(\frac{3y}{2\delta} - \frac{9y^2}{4\delta^2} + \frac{3y^4}{2\delta^4} - \frac{y^3}{2\delta^3} + \frac{3y^4}{4\delta^4} - \frac{y^6}{4\delta^6}\right) dy\right]$$

$$= \frac{\partial}{\partial x}\left[\frac{3y^2}{2\times2\delta} - \frac{9y^3}{3\times4\delta^2} + \frac{3y^5}{5\times4\delta^4} - \frac{y^4}{4\times2\delta^3} + \frac{3y^5}{5\times4\delta^4} - \frac{y^2}{7\times4\delta^6}\right]_0^\delta$$

$$= \frac{\partial}{\partial x}\left[\frac{3\delta^2}{4\delta} - \frac{3\delta^3}{4\delta^2} + \frac{3}{20}\frac{\delta^5}{\delta^4} - \frac{1}{8}\frac{\delta^4}{\delta^3} + \frac{3}{20}\frac{\delta^5}{\delta^4} - \frac{1}{28}\frac{\delta^7}{\delta^6}\right]$$

$$= \frac{\partial}{\partial x}\left[\frac{3}{4}\delta - \frac{3}{4}\delta + \frac{3}{20}\delta - \frac{1}{8}\delta + \frac{3}{20}\delta - \frac{1}{28}\delta\right]$$

$$\frac{\tau_0}{\rho U^2} = \frac{\partial}{\partial x}\left[\frac{6}{20}\delta - \frac{1}{8}\delta - \frac{1}{28}\delta\right] = \frac{\partial\delta}{\partial x}\left[\frac{84 - 35 - 10}{280}\right] = \frac{39}{280}\frac{\partial\delta}{\partial x}$$

or $\tau_0 = \rho U^2 \times \dfrac{39}{280}\dfrac{\partial\delta}{\partial x} = \dfrac{39}{280}\rho U^2 \dfrac{\partial\delta}{\partial x}$ ... (*xiii*)

Also the shear stress at the surface in laminar layer:

$$\tau_0 = \mu\left(\frac{du}{dy}\right)_{y=0} \qquad\qquad \dots (xiv)$$

From Eq. (xii), we get

$$u = U\left[\frac{3}{2}\frac{y}{8} - \frac{y^3}{2\delta^3}\right]$$

$$\therefore \qquad\qquad \frac{du}{dy} = U\left[\frac{3}{2\delta} - \frac{3y^2}{2\delta^3}\right]$$

Hence

$$\left(\frac{du}{dy}\right)_{y=0} = U\left[\frac{3}{2\delta} - \frac{3}{2\delta^3} \times 0\right] = \frac{3U}{2\delta}$$

$$\therefore \qquad\qquad \tau_0 = \mu\left(\frac{du}{dy}\right)_{y=0} = \mu\frac{3U}{2\delta} = \frac{3}{2}\frac{\mu U}{\delta} \qquad\qquad \dots (xv)$$

Equating Eqs. (xiii) and (xv), we get

$$\frac{39}{280}\rho U^2 \frac{\partial\delta}{\partial x} = \frac{3}{2}\frac{\mu U}{\delta}$$

$$\therefore \qquad\qquad \delta\partial\delta = \frac{3}{2}\mu U \times \frac{280}{39} \times \frac{1}{\rho U^2}\partial x = \frac{420}{39}\frac{\mu}{\rho U}\partial x$$

On integrating above equation, we get

$$\frac{\delta^2}{2} = \frac{420}{39}\frac{\mu}{\rho U}x + C$$

where $x = 0$, $\delta = 0$, $\therefore C = 0$

$$\therefore \qquad\qquad \frac{\delta^2}{2} = \frac{420}{39} \cdot \frac{\mu}{\rho U}x$$

or

$$\delta = \sqrt{\frac{420\times 2}{39}\frac{\mu x}{\rho U}}$$

$$= 4.64\sqrt{\frac{\mu}{\rho U x}}x = 4.64\sqrt{\frac{\mu x \times x}{\rho U x}}$$

$$\delta = 4.64\sqrt{\frac{\mu}{\rho U x}}x = \frac{4.64x}{\sqrt{Re_x}}$$

$$\delta = \frac{4.64x}{\sqrt{Re_x}} \qquad\qquad \dots (xvi)$$

Shear stress (τ_0) in terms of Reynolds number:

From equation (xv), the shear stress:

$$\tau_0 = \frac{3}{2}\frac{\mu U}{\delta}$$

Substituting the value of δ from Eq. (*xvi*) in above equation, we get

$$\tau_0 = \frac{3}{2}\frac{\mu U}{\dfrac{4.64x}{\sqrt{\text{Re}_x}}} = \frac{3}{9.28}\frac{\mu U\sqrt{\text{Re}_x}}{x}$$

$$\tau_0 = 0.323\frac{\mu U}{x}\sqrt{\text{Re}_x}$$

Average coefficient of drag $(\bar{C}_f)$ in terms of Reynolds number:

By definition of average coefficient of drag:

$$\bar{C}_f = \frac{F_D}{\dfrac{1}{2}\rho A U^2} = \frac{2F_D}{\rho b l U^2} \qquad\qquad \because A = bl$$

where

$$F_D = \int_0^l \tau_0 b\,dx$$

$$= \int_0^l 0.323\frac{\mu U}{x}\sqrt{\text{Re}_x}\,b\,dx$$

$$= 0.323\int_0^l \frac{\mu U}{x}\sqrt{\frac{\rho U x}{\mu}}\times b \times dx$$

$$= 0.323\mu U\sqrt{\frac{\rho U}{\mu}}\times b\int_0^l \frac{1}{\sqrt{x}}\,dx$$

$$= 0.323\mu U\sqrt{\frac{\rho U}{\mu}}\times b\int_0^l x^{-1/2}\,dx$$

$$= 0.323\mu U\sqrt{\frac{\rho U}{\mu}}\times b\left[\frac{x^{1/2}}{\dfrac{1}{2}}\right]_0^l$$

$$= 0.323\times 2\mu U\sqrt{\frac{\rho U}{\mu}}\times b[\sqrt{l}]$$

$$F_D = 0.646\mu U b\sqrt{\frac{\rho U l}{\mu}}$$

$$\therefore \qquad \bar{C}_f = \frac{2\times 0.646\mu U b}{\rho b l U^2}\sqrt{\frac{\rho U l}{\mu}}$$

$$= 1.292\frac{\mu}{\rho l U}\sqrt{\frac{\rho U l}{\mu}} = 1.292\sqrt{\frac{\mu}{\rho l U}}$$

$$\bar{C}_f = \frac{\mathbf{1.292}}{\sqrt{\mathbf{Re}_l}}$$

(*iv*) Given velocity profile: $\dfrac{u}{U} = \dfrac{2y}{\delta} - \dfrac{2y^3}{\delta^3} + \dfrac{y^4}{\delta^4}$ ... (*xvii*)

We know, Von-Karman momentum integral equation:

$$\frac{\tau_0}{\rho U^2} = \frac{\partial \theta}{\partial x}$$

where

$$\theta = \int_0^\delta \frac{u}{U}\left(1 - \frac{u}{U}\right) dy$$

$$\therefore \quad \frac{\tau_0}{\rho U^2} = \frac{\partial}{\partial x}\left[\int_0^\delta \frac{u}{U}\left(1 - \frac{u}{U}\right) dy\right]$$

$$= \frac{\partial}{\partial x}\left[\int_0^\delta \left(\frac{2y}{\delta} - \frac{2y^3}{\delta^3} + \frac{y^4}{\delta^4}\right)\left(1 - \left\{\frac{2y}{\delta} - \frac{2y^3}{\delta^3} + \frac{y^4}{\delta^4}\right\}\right) dy\right]$$

$$= \frac{\partial}{\partial x}\left[\int_0^\delta \left(\frac{2y}{\delta} - \frac{2y^3}{\delta^3} + \frac{y^4}{\delta^4}\right)\left(1 - \frac{2y}{\delta} + \frac{2y^3}{\delta^3} - \frac{y^4}{\delta^4}\right) dy\right]$$

$$= \frac{\partial}{\partial x}\left[\int_0^\delta \left(\frac{2y}{\delta} - \frac{4y^2}{\delta^2} + \frac{4y^4}{\delta^4} - \frac{2y^5}{\delta^5} - \frac{2y^3}{\delta^3} + \frac{4y^4}{\delta^4} - \frac{4y^6}{\delta^6} + \right.\right.$$

$$\left.\left. \frac{2y^7}{\delta^7} + \frac{y^4}{\delta^4} - \frac{2y^5}{\delta^5} + \frac{2y^7}{\delta^7} - \frac{y^8}{\delta^8}\right) dy\right]$$

$$= \frac{\partial}{\partial x}\left[\int_0^\delta \left(\frac{2y}{\delta} - \frac{4y^2}{\delta^2} - \frac{2y^3}{\delta^3} + \frac{9y^4}{\delta^4} - \frac{4y^5}{\delta^5} - \frac{4y^6}{\delta^6} + \frac{4y^7}{\delta^7} - \frac{y^8}{\delta^8}\right) dy\right]$$

$$= \frac{\partial}{\partial x}\left[\frac{2y^2}{2\delta} - \frac{4y^3}{3\delta^2} - \frac{2y^4}{4\delta^3} + \frac{9y^5}{5\delta^4} - \frac{4y^6}{6\delta^5} - \frac{4y^7}{7\delta^6} + \frac{4y^8}{8\delta^7} - \frac{y^9}{9\delta^8}\right]_0^\delta$$

$$= \frac{\partial}{\partial x}\left[\delta - \frac{4}{3}\delta + \frac{9}{5}\delta - \frac{2}{3}\delta - \frac{4}{7}\delta - \frac{1}{9}\delta\right]$$

$$= \frac{\partial}{\partial x}\left[\frac{315 - 420 + 63 \times 9 - 210 - 45 \times 4 - 35}{315}\right]\delta$$

$$= \frac{\partial}{\partial x}\left[\frac{315 - 420 + 567 - 210 - 180 - 35}{315}\right]\delta$$

$$\frac{\tau_0}{\rho U^2} = \frac{\partial}{\partial x}\left[\frac{882 - 845}{315}\right]\delta = \frac{\partial}{\partial x}\left[\frac{37}{315}\right]\delta = \frac{37}{315}\frac{\partial \delta}{\partial x}$$

or $\qquad \tau_0 = \frac{37}{315}\rho U^2 \frac{\partial \delta}{\partial x}$ $\qquad \qquad$ ... (xviii)

Also the shear stress at the surface in laminar layer:

$$\tau_0 = \mu\left(\frac{\partial u}{\partial y}\right)_{y=0} \qquad \qquad \text{... (xix)}$$

From Eq. (*xvii*), we get

$$u = U\left(\frac{2y}{\delta} - \frac{2y^3}{\delta^3} + \frac{y^4}{\delta^4}\right)$$

$$\therefore \quad \frac{\partial u}{\partial y} = U\left[\frac{2}{\delta} - \frac{4y}{\delta^2} - \frac{4y^3}{\delta^4}\right]$$

$$\left(\frac{\partial u}{\partial y}\right)_{y=0} = U\left[\frac{2}{\delta} - \frac{4}{\delta^2}(0) - \frac{4}{\delta^4}(0)\right] = \frac{2U}{\delta}$$

$$\therefore \quad \tau_0 = \mu\left(\frac{\partial u}{\partial y}\right)_{y=0} = \mu \times \frac{2U}{\delta} = \frac{2U\mu}{\delta}$$

$$\tau_0 = \frac{2U\mu}{\delta} \qquad \qquad \text{... (xx)}$$

Equating Eqs. (*xviii*) and (*xx*), we get

$$\frac{37}{315}\rho U^2 \frac{\partial \delta}{\partial x} = \frac{2U\mu}{\delta}$$

or

$$\delta \partial \delta = \frac{315}{37} \times \frac{2U\mu}{\rho U^2} dx = \frac{630}{37}\frac{\mu}{\rho U} dx$$

As $\qquad\qquad \delta = f(x)$ only

Hence partial derivative can be changed to total derivative

$$\delta d\delta = \frac{630}{37}\frac{\mu}{\rho U} dx$$

By integrating the above expression, we get

$$\frac{\delta^2}{2} = \frac{630}{37}\frac{\mu}{\rho U}x + C$$

where C = constant of integration As $x = 0$, $\delta = 0$ and hence $C = 0$

$$\frac{\delta^2}{2} = \frac{630}{37}\frac{\mu}{\rho U}x$$

$$\delta = \sqrt{\frac{630 \times 2}{37}\frac{\mu}{\rho U}}x = 5.84\sqrt{\frac{\mu x}{\rho U}}$$

$$= 5.84\sqrt{\frac{\mu x \times x}{\rho U x}} = 5.84\sqrt{\frac{\mu}{\rho U x}} \times x$$

$$\delta = \frac{5.84x}{\sqrt{Re_x}} \qquad\qquad \text{... (xxi)}$$

Shear stress (τ_0) in terms of Reynolds number:
From Eq. (*xx*), the shear stress:

$$\tau_0 = \frac{2U\mu}{\delta}$$

Substituting the value of δ from Eq. (*xxi*) in above equation, we get

$$\tau_0 = \frac{2U\mu}{\frac{5.84x}{\sqrt{Re_x}}} = \frac{2U\mu}{5.84x\sqrt{Re_x}}$$

$$\tau_0 = 0.34\frac{\mu U}{x}\sqrt{Re_x}$$

Average coefficient of drag $(\bar{C}_f)$ in terms of Reynolds number:

By definition of average coefficient of drag:

$$\bar{C}_f = \frac{F_D}{\frac{1}{2}\rho A U^2} = \frac{2F_D}{\rho b l U^2} \qquad\qquad \because A = bl$$

where

$$F_D = \int_0^l \tau_0 b dx$$

$$= \int_0^l 0.34\frac{U\mu}{x}\sqrt{Re_x}\, bdx$$

$$= \int_0^l 0.34\frac{U\mu}{x}\sqrt{\frac{\rho Ux}{\mu}}\, bdx$$

$$= 0.34\mu U\sqrt{\frac{\rho U}{\mu}} \times b\int_0^l \frac{1}{\sqrt{x}}\, dx$$

$$= 0.34\mu Ub\sqrt{\frac{\rho U}{\mu}}\int_0^l x^{-1/2}\, dx$$

$$= 0.34\mu Ub\sqrt{\frac{\rho U}{\mu}}\left[\frac{x^{1/2}}{\frac{1}{2}}\right]_0^l$$

$$= 0.34\times 2\mu Ub\sqrt{\frac{\rho U}{\mu}}\sqrt{l}$$

$$F_D = 0.68\mu Ub\sqrt{\frac{\rho Ul}{\mu}}$$

$$\therefore \qquad \bar{C}_f = \frac{2\times 0.68\mu Ub}{\rho b l U^2}\sqrt{\frac{\rho Ul}{\mu}}$$

$$= 1.36\times\frac{\mu}{\rho Ul}\times\sqrt{\frac{\rho Ul}{\mu}} = \frac{1.36\times 1}{\sqrt{\frac{\rho Ul}{x}}}$$

$$\therefore \qquad \bar{C}_f = \frac{1.36}{\sqrt{\text{Re}_x}}$$

(v) Given velocity profile: $\dfrac{u}{U} = \sin\left(\dfrac{\pi}{2}\dfrac{y}{\delta}\right)$ ... (xxii)

We know, Von-Karman momentum integral equation:

$$\frac{\tau_0}{\rho U^2} = \frac{\partial \theta}{\partial x}$$

where $\qquad \theta = \displaystyle\int_0^\delta \frac{u}{U}\left(1 - \frac{u}{U}\right) dy$

$$\therefore \qquad \frac{\tau_0}{\rho U^2} = \frac{\partial}{\partial x}\left[\int_0^\delta \frac{u}{U}\left(1 - \frac{u}{U}\right) dy\right]$$

$$= \frac{\partial}{\partial x}\left[\int_0^\delta \frac{u}{U}\left(1 - \frac{u}{U}\right) dy\right]$$

$$= \frac{\partial}{\partial x}\left[\int_0^\delta \sin\left(\frac{\pi}{2}\frac{y}{\delta}\right)\left[1 - \sin\left(\frac{\pi}{2}\cdot\frac{y}{\delta}\right)\right] dy\right]$$

$$= \frac{\partial}{\partial x}\left[\int_0^\delta \left[\sin\left(\frac{\pi}{2}\frac{y}{\delta}\right) - \sin^2\left(\frac{\pi}{2}\frac{y}{\delta}\right)\right] dy\right]$$

$$= \frac{\partial}{\partial x}\left[\left[\frac{-\cos\dfrac{\pi y}{2\delta}}{\dfrac{\pi}{2\delta}}\right] - \left[\frac{\dfrac{\pi y}{2\delta} \times \dfrac{1}{2}}{\dfrac{\pi}{2\delta}} - \frac{\sin 2\left(\dfrac{\pi}{2}\dfrac{y}{\delta}\right)}{4\times\dfrac{\pi}{2\delta}}\right]\right]_0^\delta$$

$$\left[\because \int \sin^2\left(\frac{\pi}{2}\frac{y}{\delta}\right) dy = \frac{\dfrac{\pi y}{2\delta} \times \dfrac{1}{2}}{\dfrac{\pi}{2\delta}} - \frac{\sin 2\left(\dfrac{\pi}{2}\dfrac{y}{\delta}\right)}{4\times\dfrac{\pi}{2\delta}}\right]$$

$$\frac{\tau_0}{\rho U^2} = \frac{\partial}{\partial x}\left[\left[\frac{-\cos\dfrac{\pi}{2}\dfrac{\delta}{\delta}}{\dfrac{\pi}{2\delta}} + \frac{\cos\dfrac{\pi}{2}\times\dfrac{0}{\delta}}{\dfrac{\pi}{2\delta}}\right] - \left(\frac{\dfrac{\pi}{2}\dfrac{\delta}{\delta}\times\dfrac{1}{2}}{\dfrac{\pi}{2\delta}} - 0\right)\right]$$

$$\frac{\tau_0}{\rho U^2} = \frac{\partial}{\partial x}\left[\left[0 + \frac{1}{\dfrac{\pi}{2\delta}}\right] - \left(\frac{\left(\dfrac{\pi}{4}\right)}{\dfrac{\pi}{2\delta}}\right)\right] = \frac{\partial}{\partial x}\left[\frac{2\delta}{\pi} - \frac{\pi}{4}\times\frac{2\delta}{\pi}\right]$$

$$\frac{\tau_0}{\rho U^2} = \frac{\partial}{\partial x}\left[\frac{2\delta}{\pi} - \frac{\delta}{2}\right] = \frac{\partial}{\partial x}\left[\frac{4-\pi}{2\pi}\right]\delta = \left(\frac{4-\pi}{2\pi}\right)\frac{\partial\delta}{\partial x}$$

or
$$\tau_0 = \left(\frac{4-\pi}{2\pi}\right)\rho U^2 \frac{\partial \delta}{\partial x} \qquad \qquad ... (xxiii)$$

Also the shear stress at the surface in laminar layer:

$$\tau_0 = \mu\left(\frac{\partial u}{\partial y}\right)_{y=0} \qquad \qquad ... (xxiv)$$

From Eq. (xxii), we get

$$u = U\sin\left(\frac{\pi}{2}\frac{y}{\delta}\right)$$

$$\frac{\partial u}{\partial y} = U\cos\left(\frac{\pi}{2}\frac{y}{\delta}\right)\times\frac{\pi}{2\delta}$$

$$\left(\frac{\partial u}{\partial y}\right)_{y=0} = U\frac{\pi}{2\delta}\cos\left(\frac{\pi}{2}\times\frac{0}{\delta}\right)$$

$$= U\frac{\pi}{2\delta}\cos 0 \qquad \qquad \because \cos 0° = 1$$

$$= U\frac{\pi}{2\delta}$$

$$\therefore \qquad \tau_0 = \mu\left(\frac{\partial u}{\partial y}\right)_{y=0} = \mu\frac{U\pi}{2\delta}$$

$$\tau_0 = \frac{\mu U\pi}{2\delta} \qquad \qquad ... (xxv)$$

Equating Eqs. (xxiii) and (xxv), we get

$$\left(\frac{4-\pi}{2\pi}\right)\rho U^2 \frac{\partial \delta}{\partial x} = \mu U\frac{\pi}{2\delta}$$

$$\delta\partial\delta = \frac{\mu U\pi}{3}\times\frac{2\pi}{(4-\pi)}\times\frac{1}{\rho U^2}\partial x$$

$$\therefore \qquad \delta\partial\delta = \frac{\pi^2}{(4-\pi)}\frac{\mu U}{\rho U^2}\partial x$$

$$\delta\partial\delta = 11.4975\frac{\mu}{\rho U}\partial x$$

As
$$\delta = f(x) \text{ only}$$

Hence partial derivative can be changed to total derivative.

$$\delta d\delta = 11.4975\frac{\mu}{\rho U}dx$$

By integrating the above expression, we get

$$\frac{\delta^2}{2} = 11.4975\frac{\mu}{\rho U}x + C$$

where C = constant of integration

At $x = 0$, $\delta = 0$ and hence $C = 0$

$\therefore$

$$\frac{\delta^2}{2} = 11.4975 \frac{\mu}{\rho U} x$$

$$\delta = \sqrt{2 \times 11.4975 \frac{\mu}{\rho U} x} = 4.795 \sqrt{\frac{\mu}{\rho U} x}$$

$$= 4.795 \sqrt{\frac{\mu}{\rho U} x \times \frac{x}{x}} = 4.795 \sqrt{\frac{\mu}{\rho U x} \times x}$$

$$= \frac{4.795 x}{\sqrt{\frac{\rho U x}{\mu}}}$$

$$\delta = \frac{4.795 x}{\sqrt{Re_x}} \qquad \qquad ... (xxvi)$$

Shear stress (τ_0) in terms of Reynolds number:

From Eq. (xxv), the shear stress:

$$\tau_0 = \frac{\mu U \pi}{2\delta}$$

Substituting the value of δ from Eq. (xxvi) in above equation, we get

$$\tau_0 = \frac{\mu U \pi}{2 \times \dfrac{4.795 x}{\sqrt{Re_x}}} = \frac{\mu U \pi \sqrt{Re_x}}{2 \times 4.795 x}$$

$$= \frac{3.14}{2 \times 4.795} \times \frac{\mu U}{x} \sqrt{Re_x}$$

$$\tau_0 = 0.327 \frac{\mu U}{x} \sqrt{Re_x}$$

Average coefficient of drag ($\bar{C}_f$) in terms of Reynolds number:

By definition of average coefficient of drag:

$$\bar{C}_f = \frac{F_D}{\dfrac{1}{2} \rho A U^2} = \frac{2 F_D}{\rho b l U^2}$$

where

$$F_D = \int_0^l \tau_0 b \, dx$$

$$= \int_0^l 0.327 \frac{\mu U}{x} \sqrt{Re_x} \times b \, dx$$

$$= 0.327 \mu b U \int_0^l \frac{1}{x} \times \sqrt{\frac{\rho U x}{\mu}} \, dx$$

$$= 0.327 \mu U b \sqrt{\frac{\rho U}{\mu}} \int_0^l x^{-1/2} \, dx$$

$$= 0.327 \mu b U \sqrt{\frac{\rho U}{\mu}} \left[\frac{x^{1/2}}{\frac{1}{2}} \right]_0^l$$

$$= 0.327 \times 2 \, \mu b U \sqrt{\frac{\rho U}{\mu}} \sqrt{l}$$

$$F_D = 0.655 \, \mu b U \sqrt{\frac{\rho U l}{\mu}}$$

$$\therefore \qquad \bar{C}_f = \frac{2 \times 0.655 \, \mu b U}{\rho b l U^2} \times \sqrt{\frac{\rho U l}{\mu}}$$

$$= 1.31 \frac{\mu}{\rho U l} \times \sqrt{\frac{\rho U l}{\mu}} = 1.31 \times \sqrt{\frac{\mu}{\rho U l}} = \frac{1.31}{\sqrt{\frac{\rho U l}{\mu}}}$$

$$\bar{C}_f = \frac{1.31}{\sqrt{\mathrm{Re}_l}}$$

Table 3.1 Shows the values of boundary layer thickness (δ), shear stress (τ_0) and coefficient of drag ($\bar{C}_f$) in terms of Reynolds number for various velocity profiles:

S. No.	Velocity Profile	Boundary Layer Thickness: δ	Shear Stress: τ_0	Coeff-icient of Drag: $\bar{C}_f$	Drag Force: F_D
1.	$\dfrac{u}{U} = \dfrac{y}{\delta}$	$\dfrac{3.464x}{\sqrt{\mathrm{Re}_x}}$	$0.288\dfrac{\mu U}{x}\sqrt{\mathrm{Re}_x}$	$\dfrac{1.152}{\sqrt{\mathrm{Re}_l}}$	$0.576\mu b U\sqrt{\dfrac{\rho U l}{\mu}}$
2.	$\dfrac{u}{U} = 2\left(\dfrac{y}{\delta}\right) - \left(\dfrac{y}{\delta}\right)^2$	$\dfrac{5.48x}{\sqrt{\mathrm{Re}_x}}$	$0.365\dfrac{\mu U}{x}\sqrt{\mathrm{Re}_x}$	$\dfrac{1.46}{\sqrt{\mathrm{Re}_l}}$	$0.73\mu b U\sqrt{\dfrac{\rho U l}{\mu}}$
3.	$\dfrac{u}{U} = \dfrac{3}{2}\left(\dfrac{y}{\delta}\right) - \dfrac{1}{2}\left(\dfrac{y}{\delta}\right)^3$	$\dfrac{4.64x}{\sqrt{\mathrm{Re}_x}}$	$0.323\dfrac{\mu U}{x}\sqrt{\mathrm{Re}_x}$	$\dfrac{1.292}{\sqrt{\mathrm{Re}_l}}$	$0.646\,\mu b U\sqrt{\dfrac{\rho U l}{\mu}}$
4.	$\dfrac{u}{U} = 2\left(\dfrac{y}{\delta}\right) - 2\left(\dfrac{y}{\delta}\right)^3 + \left(\dfrac{y}{\delta}\right)^4$	$\dfrac{5.84x}{\sqrt{\mathrm{Re}_x}}$	$0.34\dfrac{\mu U}{x}\sqrt{\mathrm{Re}_x}$	$\dfrac{1.36}{\sqrt{\mathrm{Re}_l}}$	$0.68\,\mu b U\sqrt{\dfrac{\rho U l}{\mu}}$
5.	$\dfrac{u}{U} = \sin\left(\dfrac{\pi}{2}\dfrac{y}{\delta}\right)$	$\dfrac{4.795x}{\sqrt{\mathrm{Re}_x}}$	$0.327\dfrac{\mu U}{x}\sqrt{\mathrm{Re}_x}$	$\dfrac{1.31}{\sqrt{\mathrm{Re}_l}}$	$0.655\,\mu b U\sqrt{\dfrac{\rho U l}{\mu}}$
6.	Blasius's solution	$\dfrac{5x}{\sqrt{\mathrm{Re}_x}}$	$0.332\dfrac{\mu U}{x}\sqrt{\mathrm{Re}_x}$	$\dfrac{1.328}{\sqrt{\mathrm{Re}_l}}$	$0.664\,\mu b U\sqrt{\dfrac{\rho U l}{\mu}}$

Problem 3.6: Find the thickness of the boundary layer at the end of the flat plate and the drag force on one side of a plate 0.9 m long and 0.6 m wide when placed in water flowing with a velocity of 0.12 m/s. Find also the value of coefficient of drag. For the following velocity profiles:

$$(i) \quad \frac{u}{U} = \frac{y}{\delta} \qquad\qquad (ii) \quad \frac{u}{U} = 2\left(\frac{y}{\delta}\right) - \left(\frac{y}{\delta}\right)^2$$

Take dynamic viscosity for water = 0.011 poise.

Solution: Given data:

Length of plate: $l = 0.9$ m
Width of plate: $b = 0.6$ m
Velocity of water: $U = 0.12$ m/s

Dynamics of viscosity: $\mu = 0.011$ poise $= \dfrac{0.011}{10}$ N.s/m^2 = 0.0011 Ns/m^2

Reynolds number at the end of the plate *i.e.*, at a distance of 0.9 m from the leading edge is given by,

$$Re_l = \frac{\rho U l}{\mu} = \frac{1000 \times 0.12 \times 0.9}{0.0011} = 98181.81$$

As we know that the laminar boundary layer exists up to Reynolds number = 3 × 10^5. Hence given data for laminar boundary layer:

(*i*) Given velocity profile: $\dfrac{u}{U} = \dfrac{y}{\delta}$

We know that the thickness of boundary layer for given velocity profile at $x = 0.9$ m:

$$\delta = \frac{3.464x}{\sqrt{Re_x}} = \frac{3.464 \times 0.9}{\sqrt{98181.81}}$$

$$= 9.95 \times 10^{-3} \text{ m} = \textbf{9.95 mm}$$

The drag force on one side of the plate: F_D

$$F_D = 0.576 \ \mu b U \sqrt{\frac{\rho U l}{\mu}}$$

$$= 0.576 \times 0.0011 \times 0.6 \times 0.12 \times \sqrt{98181.81}$$

$$= 0.01429 \text{ N} = \textbf{14.29 kN}$$

Coefficient of drag:

$$\bar{C}_f = \frac{1.152}{\sqrt{Re_l}} = \frac{1.152}{\sqrt{98181.81}} = 3.676 \times 10^{-3} = \textbf{0.00367}$$

(*ii*) Given velocity profile: $\dfrac{u}{U} = 2\left(\dfrac{y}{\delta}\right) - \left(\dfrac{y}{\delta}\right)^2$

We know that the thickness of boundary layer for given velocity profile:

$$\delta = \frac{5.48x}{\sqrt{Re_x}}$$

$$= \frac{5.48 \times 0.9}{\sqrt{98181.81}} = 0.01574 \text{ m} = \mathbf{15.74 \text{ mm}}$$

The drag force on one side of the plate: F_D

$$F_D = 0.73\mu b U \sqrt{\frac{\rho U l}{\mu}}$$

$$= 0.73 \times 0.0011 \times 0.6 \times 0.12 \times \sqrt{98181.81}$$
$$= 0.018116 \text{ N} = \mathbf{18.116 \text{ kN}}$$

Coefficient of drag:

$$\bar{C}_f = \frac{1.46}{\sqrt{Re_l}} = \frac{1.46}{\sqrt{98181.81}}$$

$$= 3.1914 \times 10^{-3} = \mathbf{0.00319}$$

Problem 3.7: Air is flowing over a smooth flat plate with a velocity of 8 m/s. The length of the plate is 1.5 m and with 0.7 m. If laminar boundary layer exists upto the value of Re = 3 × 10⁵, find the maximum distance from the leading edge upto which laminar boundary layer exists. Find also the maximum thickness of laminar boundary layer if the velocity profile is given as

$$\frac{u}{U} = 2\left(\frac{y}{\delta}\right) - 2\left(\frac{y}{\delta}\right)^2 + \left(\frac{y}{\delta}\right)^4$$

Take kinematic viscosity for air = 0.14 stokes.

Solution: Given data:

Velocity of air: $U = 9$ m/s
Length of plate: $l = 1.5$ m
Width of plate: $b = 0.7$ m
Reynolds number: Re = 3 × 10⁵ upto which laminar b.l. exists
Kinematic viscosity: $v = 0.14$ stokes or cm²/s
                     $= 0.14 \times 10^{-4}$ m²/s

Now

We know, Reynolds number: $Re_x = \frac{\rho U x}{\mu} = \frac{U x}{v}$

where x = distance from the leading edge at which Reynolds number is 3 × 10⁵.

$$\therefore \qquad 3 \times 10^5 = \frac{9 \times x}{0.14 \times 10^{-4}}$$

or $x = 0.46666$ m = **466.66 mm**

The maximum thickness of the laminar boundary for given velocity profile:

$$\delta = \frac{5.84x}{\sqrt{Re_x}}$$

$$= \frac{5.84 \times 0.466}{\sqrt{3 \times 10^5}} = 0.004968 \text{ m} = \textbf{4.96 mm}$$

Problem 3.8: Air is flowing over a flat plate 600 mm long and 500 mm wide with a velocity of 4 m/s. The kinematic viscosity of air is given as 0.15 stokes. Find (*i*) the boundary layer thickness at the end of the plate, (*ii*) shear stress at 150 mm from the leading edge, and (*iii*) drag force one side of the plate if the velocity profile is given by

$$\frac{u}{U} = \sin\left(\frac{\pi}{2} \frac{y}{\delta}\right)$$

Take density of air is 1.23 kg/m³.

Solution: Given data:

Length of plate: $l = 600$ mm $= 0.6$ m

Width of plate: $b = 500$ mm $= 0.5$ m

Velocity of air: $U = 4$ m/s

Kinematic viscosity: $v = 0.15$ stokes or cm²/s $= 0.15 \times 10^{-4}$ m²/s

Density of air: $\rho = 1.23$ kg/m³

Reynolds number at the end of the plate *i.e.*, at a distance of 0.6 m from the leading edge is given by

$$\text{Re}_l = \frac{\rho U l}{\mu} = \frac{U l}{v} = \frac{4 \times 0.6}{0.15 \times 10^{-4}} = 16 \times 10^4$$

As we know that the laminar boundary layer exists upto Reynolds number is 3 × 10⁵. Hence given data shows that the laminar boundary layer is maintained on the whole length of flat plate.

For given velocity profile: $\dfrac{u}{U} = \sin\left(\dfrac{\pi}{2} \dfrac{y}{\delta}\right)$

We know that the thickness of boundary layer for a given velocity profile at $x = l = 0.6$ m

$$\delta = \frac{4.795 x}{\sqrt{\text{Re}_x}} = \frac{4.795 \times 0.6}{\sqrt{160000}} = 0.00719 \text{ m} = \textbf{7.19 mm}$$

The shear stress at any distance from leading edge for given profile: τ_0

$$\tau_0 = 0.327 \frac{\mu U}{x} \sqrt{\text{Re}_x}$$

At $x = 150$ mm $= 0.15$ m

$$\text{Re}_x = \frac{U x}{v} = \frac{4 \times 0.15}{0.15 \times 10^{-4}} = 40000$$

$$\therefore \qquad \tau_0 = \frac{0.327 \times v \rho \times 4}{0.15} \times \sqrt{40000} \qquad \qquad \because \mu = v\rho$$

$$= \frac{0.327 \times 0.15 \times 10^{-4} \times 1.23 \times 4}{0.15} \times \sqrt{40000}$$

$$= 0.03217 \text{ N/m}^2 = \mathbf{31.17 \ kN/m^2}$$

Drag force on one side of the plate for given velocity profile: F_D

$$F_D = 0.655\mu bU \times \sqrt{\frac{\rho Ul}{\mu}} = 0.655 v\rho bU \times \sqrt{\frac{Ul}{v}}$$

$$= 0.655 \times 0.15 \times 10^{-4} \times 1.23 \times 0.5 \times 4 \ \sqrt{Re_l}$$

$$= 2.416 \times 10^{-5} \times \sqrt{160000} = 0.00966 \text{ N} = \mathbf{9.66 \ kN}$$

Problem 3.9: A thin plate is moving in still atmospheric air at a velocity of 4.5 m/s. The length of the plate is 0.7 m and width 0.6 m. Find:

(*i*) Thickness of boundary layer at the end of the plate, and

(*ii*) Drag force on both sides of the plate.

Take density and kinematic viscosity of air as 1.24 kg/m³ and 0.15 stokes respectively.

Solution: Given data:

Velocity of plate: $\quad U = 4.5$ m/s

Length of plate: $\quad l = 0.7$ m

Width of plate: $\quad b = 0.6$ m

Density of air: $\quad \rho = 1.24$ kg/m³

Kinematic viscosity: $\quad v = 0.15$ stokes or cm²/s $= 0.15 \times 10^{-4}$ m²/s

Reynolds number: $\quad Re = \dfrac{Ul}{v} = \dfrac{4.5 \times 0.7}{0.15 \times 10^{-4}} = 210000$

It is clear, Re $< 3 \times 10^5$, hence boundary layer is laminar over the whole length of the plate.

(*i*) Thickness of boundary layer at the end of the plate by Blasius's solution is

$$\delta = \frac{5x}{\sqrt{Re_x}} = \frac{5 \times 0.7}{\sqrt{210000}} = 0.00763 \text{ m} = \mathbf{7.63 \ mm}$$

(*ii*) Drag force on both sides of the plate: $2F_D$

Drag force on one side of the plate: F_D

$$F_D = 0.664\,\mu bU \sqrt{\frac{\rho Ul}{\mu}} \quad \text{from Blasius's solution}$$

$$= 0.664\,\mu bU \sqrt{Re_l} = 0.664 \times \rho vbU \sqrt{Re_l}$$

$$= 0.664 \times 1.28 \times 0.15 \times 10^{-4} \times 0.6 \times 4.5 \times \sqrt{210000}$$

$$\because \ Re_x = Re_l$$

$$= 0.01577 \text{ N} = 15.77 \text{ kN}$$

∴ Drag force on both sides of the plate:

$$= 2F_D = 2 \times 15.77 \text{ kN} = \mathbf{31.54 \ kN}$$

Problem 3.10: A viscous fluid flows over a flat plate such that the boundary layer thickness at a diameter 1.3 m from the leading edge is 12 mm. Assuming the flow to be laminar, determine the boundary layer thickness at a distance of (*i*) 0.2 m, (*ii*) 2 m, and (*iii*) 10 m from the leading edge.

Solution: Given data:

$$x = 1.3 \text{ m}$$

$$\delta = 12 \text{ mm} = 0.012 \text{ m}$$

We know that the boundary layer thickness for laminar boundary layer is

$$\delta = \frac{5x}{\sqrt{Re_x}}$$

where x is the distance from the leading edge.

$$\therefore \qquad 0.012 = \frac{5 \times 1.3}{\sqrt{Re_x}} = \frac{5 \times 1.3}{Re_x^{1/2}}$$

or $\qquad\qquad Re_x^{1/2} = \frac{5 \times 1.3}{0.012} = 541.66$

or $\qquad\qquad Re_x = 293402.77$

also $\qquad\qquad Re_x = \frac{\rho U x}{\mu}$

$$\therefore \qquad 293402.77 = \frac{\rho U}{\mu} \times 1.3$$

or $\qquad\qquad \dfrac{\rho U}{\mu} = 225694.43$

(i) $\qquad\qquad x_1 = 0.2 \text{ m}$

$$Re_1 = \frac{\rho U x_1}{\mu} \quad \text{assuming } \rho, U, \text{ and } \mu \text{ are constant}$$

$$= 225694.43 \times 0.2 = \mathbf{45138.88}$$

Therefore, the boundary layer thickness at $x_1 = 0.2$ m becomes

$$\delta_1 = \frac{5x_1}{\sqrt{Re_1}} = \frac{5 \times 0.2}{\sqrt{45138.88}}$$

$$= 4.706 \times 10^{-3} \text{ m} = \mathbf{4.706 \text{ mm}}$$

(ii) $\qquad\qquad x_2 = 2 \text{ m}$

$$Re_2 = \frac{\rho U x_2}{\mu} = 225694.43 \times 2 = 451388.86$$

Therefore, the boundary layer thickness at $x_2 = 2$ m becomes

$$\delta_2 = \frac{5\,x_2}{\sqrt{Re_2}} = \frac{5\times2}{\sqrt{451388.86}}$$

$$= 0.01488 \text{ m} = \textbf{14.88 mm}$$

(iii) $x_3 = 10$ m

$$Re_3 = \frac{\rho U x_3}{\mu} = 225694.43 \times 10 = 2256944.3$$

Therefore, the boundary layer thickness at $x_2 = 10$ m becomes

$$\delta_3 = \frac{5\,x_3}{\sqrt{Re_3}} = \frac{5\times10}{\sqrt{2256944.3}}$$

$$= 0.03328 \text{ m} = \textbf{33.28 mm}$$

3.9 TURBULENT BOUNDARY LAYER ON A FLAT PLATE

As we know turbulent boundary layer exists next to transition boundary layer at which the Reynolds number (Re) $> 5 \times 10^5$. In this boundary layer, the fluid flows in zig-zag manner. The turbulent boundary layer is thicker than transition boundary layer and transition boundary layer is thicker that laminar boundary layer thickness (*i.e.*, the boundary layer thickness increases from leading edge to the direction of flow, hence turbulent boundary layer thickness > transition b.l. thickness > laminar b.l. thickness). The velocity distribution in a turbulent boundary layer follows a logarithmic law (*i.e.*, $u \propto \log_e y$), which can also be represented by a power law as:

$$\frac{u}{U} = \left(\frac{y}{\delta}\right)^n$$

where the value of the exponent $n = \dfrac{1}{7}$ for Reynolds number $\left(Re_l = \dfrac{Ul}{\nu}\right)$ ranging from 5×10^5 to 10^7 over a flat plate.

$\therefore$ $$\frac{u}{U} = \left(\frac{y}{\delta}\right)^{1/7} \qquad\qquad\qquad \ldots (3.9.1)$$

$$u = \frac{U}{\delta^{1/7}} y^{1/7}$$

Differentiating above equation w.r.t. y, we get

$$\frac{du}{dy} = \frac{U}{\delta^{1/7}} \frac{1}{7} y^{1/7-1}$$

$$\frac{du}{dy} = \frac{1}{7} \frac{U}{\delta^{1/7}} y^{-6/7}$$

$$\frac{du}{dy} = \frac{1}{7} \frac{U}{\delta^{1/7} y^{6/7}}$$

$$\left(\frac{du}{dy}\right)_{y=0} = \frac{1}{0} = \infty$$

Hence Eq. (3.9.1) cannot be applied at the surface of plate because $\left(\dfrac{du}{dy}\right)_{y=0} = \infty$.

However, immediately adjacent to the surface of plate, there is laminar sublayer, which is so thin that its velocity profile may be taken as linear and tangential to the 'seventh root' profile at the point where the laminar sublayer merges with the turbulent part of the boundary layer.

For turbulent boundary layer:

Boundary layer thickness:δ $\boxed{\dfrac{0.376\,x}{\text{Re}_x^{1/5}}}$ ... (3.9.2)

$$\delta = \frac{0.376x}{\left(\dfrac{Ux}{v}\right)^{1/5}} = \frac{0.376x}{(U/v)^{1/5}} \cdot x^{-1/5} = \frac{0.376x^{4/5}}{(U/v)^{1/5}}$$

At constant values of U and v.

$$\delta \propto x^{4/5}$$

As we know that in laminar boundary layer, $\delta \propto x^{1/2}$, but in case of turbulent boundary layer, $\delta \propto x^{4/5}$. Hence the turbulent boundary layer thickness increases faster than that of laminar boundary layer thickness.

Remember some other expressions for turbulent boundary layer:

Average coefficient of drag: $\overline{C}_f = \dfrac{0.074}{\text{Re}_{e_x}^{1/5}}$

Local coefficient of drag: $C_f = \dfrac{0.059}{\text{Re}_x^{1/5}}$

Average coefficient proposed by Prandtl as

$$\overline{C}_f = \frac{0.074}{\text{Re}_l^{1/5}} - \frac{C}{\text{Re}_l} \qquad\qquad ... (3.9.3)$$

where constant C in Eq. (3.9.3) depends on the value of the Reynolds number (Re_x) at which the laminar boundary layer becomes turbulent. The values of C for various values of critical Reynold's number are as given below:

S. No.	Critical Reynold's Number: Re_x	Constant: C
1	3×10^5	1050
2	5×10^5	1700
3	10^6	3300
4	3×10^6	8700

In maximum cases the value of critical Reynolds number (Re_x) may be taken as 5×10^5 and hence $C = 1700$, Eq. (3.9.3) then becomes

$$\bar{C}_f = \frac{0.074}{Re_l^{1/5}} - \frac{1700}{Re_l}$$

Problem 3.11: Find the thickness of the boundary layer at the trailing edge of a smooth plate of length 5 m and a width 1.2 m, when the plate is moving with a moving of 5 m/s in stationary air. Take kinematic viscosity of air as 0.11 stokes.

Solution: Given data:

Length of plate:	$l = 5$ m	
Width of plate:	$b = 1.2$ m	
Velocity of plate:	$U = 5$ m/s	
Kinematic viscosity:	$v = 0.11$ stokes $= 0.11 \times 10^{-4}$ m²/s	

Reynolds number: $Re_l = \dfrac{Ul}{v} = \dfrac{5 \times 5}{0.11 \times 10^{-4}} = 227.27 \times 10^4$

As the Reynolds number is more than 5×10^5 and hence the boundary layer at the trailing edge is turbulent.

The boundary layer thickness for turbulent b.l.: δ

$$\delta = \frac{0.376x}{Re_x^{1/5}} = \frac{0.376 \times 5}{(227.27 \times 10^4)^{1/5}} \qquad \because \ x = l, \ Re_x = Re_l$$

$$= 0.10065 \text{ m} = \textbf{100.65 mm}$$

Problem 3.12: It is required to determine the frictional drag of a submarine. The length of the hull is 75 m and its surface area is 3000 m². The submarine is travelling at a constant speed of 5 m/s. Critical Reynolds number at which the flow in the boundary layer changes from laminar to turbulent is 5×10^5. Assuming that the boundary layer at the leading edge is laminar, obtain the frictional drag and the power required to propel the submarine at 5 m/s. Take kinematic viscosity $= 0.01$ stokes and density $= 1000$ kg/m³.

Solution: Given data:

Length of the hull:	$l = 75$ m
Surface area:	$A = 3000$ m²
Speed of submarine:	$U = 5$ m/s
Critical Reynolds number:	$Re_c = 3 \times 10^5$
Kinematic viscosity:	$v = 0.01$ stokes or cm²/s $= 0.01 \times 10^{-4}$ m²/s
Density:	$\rho = 1000$ kg/m³

Reynolds number: $Re_l = \dfrac{Ul}{v} = \dfrac{5 \times 75}{0.01 \times 10^{-4}} = 3.75 \times 10^8$

Since at leading edge b.l. is laminar, it changes from laminar to turbulent on the surface of the submarine

At critical Reynolds number: $Re_c = 5 \times 10^5$

Average drag coefficient: $\quad \bar{C}_f = \dfrac{0.074}{(Re_l)^{1/5}} - \dfrac{1700}{Re_l}$

$$= \dfrac{0.074}{(3.75 \times 10^8)^{1/5}} - \dfrac{1700}{3.75 \times 10^8} = 1.422 \times 10^{-3}$$

The frictional drag: $\quad F_D = \bar{C}_f \rho A \dfrac{U^2}{2}$

$$= 1.422 \times 10^{-3} \times 1000 \times 3000 \times \dfrac{(5)^2}{2}$$

$$= 53325 \text{ N} = \textbf{53.325 kN}$$

Power required: $\quad P = F_D \cdot U = 53.325 \times 5 = \textbf{266.625 kW}$

Problem 3.13: Oil with a free stream velocity of 2 m/s flows over a thin plate 2 m wide and 2 m long. Calculate the boundary layer thickness and the shear stress at the trailing end point and determine the total surface resistance of the plate. Take specific gravity as 0.86 and kinematic viscosity as 10^{-5} m²/s.

Solution: Given data:

Free stream velocity of oil: $\qquad U = 2$ m/s
Width of plate: $\qquad b = 2$ m
Length of plate: $\qquad l = 2$ m
∴ Area of plate: $\qquad A = b \times l = 2 \times 2 = 4$ m²
Specific gravity of oil: $\qquad S = 0.86$
∴ Density of oil: $\qquad \rho = 0.86 \times 1000 = 860$ kg/m³
Kinematic viscosity: $\qquad v = 10^{-5}$ m²/s

Now the Reynolds number at the trailing end:

$$Re_l = \dfrac{Ul}{v} = \dfrac{2 \times 2}{10^{-5}} = 4 \times 10^5.$$

Since Re_l is less than 5×10^5, the boundary layer is laminar over the entire length of the plate.

∴ Thickness of boundary layer at the end of the plate from Blasius's solution is:

$$\delta = \dfrac{5x}{\sqrt{Re_x}} \qquad\qquad \text{Here } x = l, \ Re_x = Re_l$$

$$= \dfrac{5 \times 2}{\sqrt{4 \times 10^5}} = .01581 \text{ m} = \textbf{15.81 mm}$$

Shear stress at the end of the plate: τ_0

$$\tau_0 = 0.332 \dfrac{\mu U}{x} \sqrt{Re_x} = 0.332 \times \dfrac{v\rho U}{x} \sqrt{Re_x}$$

$$= 0.332 \times \dfrac{10^{-5} \times 860 \times 2}{2} \times \sqrt{4 \times 10^5} = \textbf{1.805 N/m}^2$$

Surface resistance on one side of the plate: F_D

$$F_D = \dfrac{1}{2}\bar{C}_f \rho A U^2$$

where
$$\bar{C}_f = \frac{1.328}{\sqrt{Re_l}} = \frac{1.328}{\sqrt{4\times10^5}} = 0.0021$$

∴
$$F_D = \frac{1}{2}\times0.0021\times860\times4\times(2)^2 = 14.448 \text{ N}$$

∴ Total resistance $= 2F_D = 2 \times 14.448 = \mathbf{28.896 \text{ N}}$

Problem 3.14: Calculate the friction drag on a plate 15 cm wide and 45 cm long placed longitudinal in a stream of oil specific gravity 0.90 m and kinematic viscosity = 0.90 stokes, flowing with a free stream velocity of 6 m/s. Also, find out of the thickness of the boundary layer and shear stress at the trailing edge.

Solution: Given Data:

$$\text{Width} : b = 15 \text{ cm} = 0.15 \text{ m}$$
$$\text{Length} : l = 45 \text{ cm} = 0.45 \text{ m}$$

∴
$$\text{Area of plate} : A = b \times l$$
$$= 0.15 \times 0.45$$
$$= 0.0675 \text{ m}^2$$

$$\text{Specific gravity} : S = 0.90$$

∴
$$\text{Density of oil} : \rho = 1000 \text{ S}$$
$$= 1000 \times 0.90 \text{ kg/m}^3$$
$$= 900 \text{ kg/m}^3$$

$$\text{Kinematic viscosity:} \ v = 0.90 \text{ stokes}$$
$$= 0.90 \times 10^{-4} \text{ m}^2/\text{s}$$

$$\text{Free stream velocity:} \ U = 6 \text{ m/s}$$

$$\text{Reynolds number at end of the place:} \ Re_l = \frac{\rho U l}{\mu} = \frac{Ul}{v}$$

$$= \frac{6\times0.45}{0.90\times10^{-4}} = 3 \times 10^4$$

As we know that the laminar boundary layer exist upto Reynolds number is 3×10^5. Hence, given data shows that the laminar boundary layer is maintained on the whole length of flat plate. Therefore, the skin friction coefficient is

$$\bar{C}_f = \frac{1.328}{\sqrt{Re_l}} = \frac{1.325}{\sqrt{3\times10^4}} = 0.00764$$

Friction drag on a plate :
$$F_D = \frac{1}{2}\bar{C}_f\rho A U^2$$

$$= \frac{1}{2} \times 0.00764 \times 900 \times 0.0675 \times (6)^2$$

$$= 8.35 \text{ N}$$

∴   Total friction drag $= 2 F_D = 2 \times 8.35 = 16.7 \text{ N}$

. The boundary layer thickness at the trailing edge of the plate is

$$\delta = \frac{5l}{\sqrt{Re_l}} = \frac{5 \times 0.45}{\sqrt{3 \times 10^4}}$$

$$= 0.01299 \text{ mm} = 12.99 \text{ mm}$$

$$\approx \textbf{13 mm}$$

Shear stress at the starling edge,

$$\tau_o = 0.332 \frac{\mu U}{l} \sqrt{Re_l}$$

$$= 0.332 \frac{v\rho U}{l} \sqrt{Re_l}$$

$$= 0.332 \times \frac{0.90 \times 10^{-4} \times 900}{0.45} \sqrt{3 \times 10^4} = \textbf{10.35 N/m}^2$$

3.10 BOUNDARY LAYER ON ROUGH SURFACES

In previous sections the development of the boundary layer along smooth plates has been considered. However, in most practical applications related with boundary layer development on solid boundary such as airplane wings, turbine blades etc., the surface cannot be considered smooth. For a rough plate if k is the average height of roughness projections on the surface of the plate and δ is the boundary layer thickness, then the relative roughness $\left(\dfrac{k}{\delta}\right)$ is a significant parameter showing the behaviour of the boundary surface. For $k = C$

Then $\dfrac{k}{\delta}$ decreases along the plate in the direction of flow because δ increases in the direction of flow.

Nikuradse was introducing a dimensionless roughness parameter $\left(\dfrac{V_* k_s}{v}\right) \cdot$ on which smooth and rough surfaces are differentiates:

$\dfrac{V_* k_s}{v} < 5$ for hydrodynamical smooth boundary

$5 < \dfrac{V_* k_s}{v} < 70$ for transition

$\dfrac{V_* k_s}{v} > 70$ for hydrodynamically rough (or completely rough)

where $V_* = \sqrt{\dfrac{\tau_0}{\rho}}$, shear or friction velocity

k_s = equivalent sand grain roughness defined as that value of the roughness which would offer the same resistance to the flow past the plate as that due to the actual roughness on the surface of the plate.

v = kinematic viscosity

For completely rough zone, the local drag coefficient (C_f) and the average drag coefficient $(\bar{C}_f)$ are given as:

$$C_f = \left[2.87 + 1.58 \log_{10} \left(\frac{l}{k_s} \right) \right]^{-2.5}$$

$$\bar{C}_f = \left[1.89 + 1.62 \log_{10} \left(\frac{l}{k_s} \right) \right]^{-2.5}$$

which are valid for $10^2 < \dfrac{l}{k_s} < 10^6$

3.11 SEPARATION OF BOUNDARY LAYER

As we know if a solid body is immersed in a flow of real fluid, a thin layer of fluid developed close to the body surface in which velocity gradient $\left(\dfrac{du}{dy} \right)$ exists normal to the surface of the body. This thin layer across the narrow region of the solid body is called boundary layer. The fluid particles moving on the wall (or the body in the boundary layer are decelerated because of surface friction due to viscous effect, and thus their kinetic energy is decreased. If there is a pressure drop in the direction of flow ($dp/dx < 0$, favourable or desirable pressure gradient), the pressure can overcome the deceleration of the particles so that they continue to move along the surface of the body. Somewhere the pressure of fluid particles increases in the direction of flow ($dp/dx > 0$, unfavourable or adverse pressure gradient) due to passage of flow (i.e., diverging passage acts diffuser). In this way, some kinetic energy of fluid particles is utilized to increase the pressure and the remaining part of kinetic energy is not enough to overcome the surface friction of the solid body. Due to above reason the thickness of the boundary layer increases, the layer of fluid particles stop and then separate from the wall (or boundary of solid body). This phenomenon which separates the boundary layer from the wall is called separation or separation of boundary layer. The point on the surface of the solid body at which the separation takes place is called separation point.

At separation point:

(i) Adverse pressure gradient i.e., $\dfrac{dp}{dx} > 0$.

(ii) $\left(\dfrac{\partial u}{\partial y} \right)_{y=0} = 0$

(iii) The wall shear stress is zero i.e., $\tau_0 = 0$.

Beyond the separation point, the adverse pressure gradient is large enough $\left(\dfrac{\partial p}{\partial x} \gg 0\right)$, the wall shear stress is negative due to the negative value of $\left(\dfrac{\partial u}{\partial y}\right)_{y=0}$. That region is called reverse flow (or back flow).

Effect of pressure gradient along the direction of flow on boundary layer separation:

Consider a fluid flow on a convex shape solid body (*i.e.*, fluid flow on a converging-diverging of solid body) as shown in Fig. 10.15. The boundary of solid body is converging from A to C and diverging from C to F.

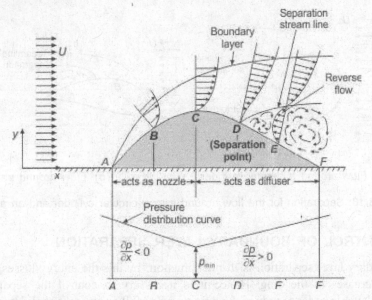

Fig. 3.15: Separation of boundary layer: Fluid flow on convex shape of solid body.

The converging zone (A to C) acts nozzle in which the velocity of fluid particles increase and the pressure decrease in the direction of the flow and hence the pressure gradient is negative $\left(\dfrac{\partial p}{\partial x} < 0\right)$. This negative pressure gradient is favourable or desirable because the loss of kinetic energy of the fluid layer moving on the boundary of solid body against surface friction is recovered by decrease in pressure. As long as negative pressure gradient is maintained, the layer of fluid particles is easily moving on the boundary of solid body.

The diverging zone (C to F) acts diffuser in which the velocity of fluid particles decrease and the pressure increases in the direction of the flow and hence the pressure gradient is positive $\left(\dfrac{\partial p}{\partial x} > 0\right)$. This positive pressure is unfavourable or not desirable because some kinetic energy of fluid particles is utilized to increase the pressure and the remaining part of kinetic energy is not enough to overcome the surface friction of the solid body. Beyond point C, say point D as shown in Fig. 10.15, the layer of fluid particles stop and then separate from the boundary of solid body. Beyond point D, the adverse pressure gradient is large enough $\left(\dfrac{\partial p}{\partial x} >> 0\right)$, the wall shear stress (τ_0) is negative due to negative value of velocity gradient. $\left[\left(\dfrac{\partial u}{\partial y}\right)_{y=0} < 0\right]$. That region is called reverse flow (or back flow).

Similar explanation of the boundary layer separation for the flow around a long circular cylinder and an airfoil are shown in Fig. 3.16 (a) and (b) respectively.

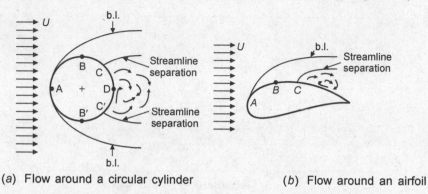

(a) Flow around a circular cylinder	(b) Flow around an airfoil

Fig. 3.16: Separation for the flow around a long circular cylinder and an airfoil

3.12 CONTROL OF BOUNDARY LAYER SEPARATION

The boundary layer separation is ill phenomenon, by this the energy losses increase due to increases in the drag. Hence it is necessary to control the separation of boundary layer in order to reduce the drag on an airfoil or any other solid body. Some methods for the control of boundary layer separation are discussed below:

3.12.1 Suction Method

By this method the slow moving fluid particles layer on the boundary of solid body is removed by suction through slots as shown in Fig. 3.17, so that on the downstream of the point of suction a new boundary layer starts developing which is able to withstand an adverse pressure gradient and hence separation is prevented:

Fig. 3.17: Suction of fluid from boundary layer.

Moreover, the suction of the fluid from the boundary layer also delays its transition from laminar to turbulent flow due to which skin friction drag is reduced.

This method is successfully used in the design of aircraft wings.

3.12.2 By Pass Method

By this method, additional energy is supplied to the fluid particles layer being moved on the surface of solid body. This additional supplied energy accelerates the fluid particles and hence prevents separation of boundary layer.

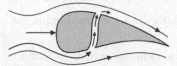

Fig. 3.18: By-pass slotted wing.

This can be achieved by diverting a portion of the fluid of the main stream from the region of high pressure to the retarded region of boundary layer through a slot provided in body as in the case of the slotted wing as shown in Fig. 3.18. This method is also known as acceleration of boundary layer. Disadvantage of this method is that if the fluid is injected into a laminar b.l., it undergoes a transition from laminar to turbulent layer which results in an increased skin friction drag.

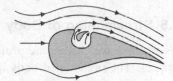

Fig. 3.19: Control separation by blowing of fluid.

3.12.3 Injection Method

The injection of fluid through porous wall controls the boundary layer separation. This may be achieved by blowing high energy fluid particles tangentially from the region where the chance of separation takes place. This is shown in Fig. 3.19. The injection of fluid promotes turbulence and thereby increases skin friction drag. But the form drag is reduced considerably due to suppression of flow separation and this reduction can be of significant magnitude so as to ignore the enhanced skin friction drag.

3.12.4 Rotating of Cylinder Method

As we know, the formation of the boundary layer is due to the difference between the velocity of the flowing fluid and the solid boundary. According to this method to eliminate the formation of a boundary layer, the solid body is rotated in the flowing fluid.*

---

* The drag on the solid body is basically caused by pressure difference. This force brought about by the pressure difference is known as form drag whereas the shear stress at the wall gives rise to skin friction drag. Generally, these two drag forces together are responsible for resultant drag on a body.

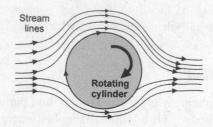

Fig. 3.20: Flow Past a Rotating Cylinder.

In this method, a rotating cylinder is provided in the stream of fluid as shown in Fig. 3.20.

On the upper side of the cylinder, where the fluid as well as the cylinder move in the same direction, the boundary layer does not form and hence the separation is completely eliminated. However, on the lower side of the rotating cylinder, where the fluid motion is opposite to cylinder motion, separation would occur.

3.12.5 Streamlining of Body Shape

In this method, we change the profile of the body to a streamlined shape as shown in Fig.3.21. The streamlined body has rounded nose and long tapered tail, to reduce the magnitude of the positive (or adverse) pressure gradient. As positive pressure gradient can be reduced to the minimum, separation can be delayed or eliminated.

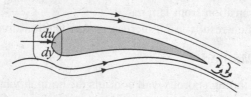

Fig. 3.21: Delay of separation by use of the profile a streamlined shape.

SUMMARY

1. A thin layer of fluid developed close to the body surface and in which velocity gradient   exists normal to the surface of the body. This thin layer across the narrow region of the solid body is called boundary layer or friction layer.
   $$\frac{Ux}{v}$$

2. Boundary layer formation over a flat plate is classified into three stages as:
   (*i*) Laminar boundary layer.
   (*ii*) Transition boundary layer.
   (*iii*) Turbulent boundary layer.

Contd...

3. **Laminar boundary layer:** The initial stage of boundary layer development exhibits characteristics of laminar flow *i.e.*, the fluid particles at the leading edge of plate move orderly in laminas parallel to the flat plate surface. The Reynolds number for the flow of fluid in the boundary layer is expressed as:

$$Re_x =$$

where x = distance from the leading edge,
   U = free-stream velocity of the fluid flow
   v = kinematic viscosity of the fluid.

Reynolds number for laminar boundary layer is;

$$Re_x \leq 3 \times 10^5$$

4. **Transition boundary layer:** The length of plate over which the b.l. flow changes from laminar to turbulent is called transition zone and the formation of b.l. over this zone is called transition boundary layer.

The range of Reynolds number for this zone is;

$$3 \times 10^5 < Re_x < 5 \times 10^5$$

5. **Turbulent boundary layer:** The zone next to transition zone is called turbulent zone. In this zone, the fluid flows in zig-zag manner. The formation of b.*l.* over this zone is called turbulent boundary layer. If Reynolds number (Re_x) is $\geq 5 \times 10^5$ then formation of this zone begins.

6. **Laminar sub-layer:** If the plate is very smooth, a thin layer develops very close to the plate surface due to only by viscous effects of the fluid, in turbulent zone. This layer is called laminar sub-layer. The nature of the flow in this layer is laminar.

7. **Boundary layer thickness:** δ

It is defined as the normal distance from the surface of solid body to the point where the velocity of flow (u) is 99% of the free stream velocity (U)

Mathematically, it is defined as

$y = \delta$ for $u = 99\% \ U = 0.99 \ U$

8. **Displacement thickness:** $\delta^* = \int_0^\delta \left(1 - \frac{u}{U}\right) dy$

Displacement thickness (δ^*) tells us how far the streamlines of the flow are displaced outward due to the decrease in velocity in the boundary layer.

9. **Momentum thickness:** θ $\quad = \int_0^\delta \frac{u}{U}\left(1 - \frac{u}{U}\right) dy$

10. **Energy thickness:** δ^{**} $\quad = \int_0^\delta \frac{u}{U}\left(1 - \frac{u^2}{U^2}\right) dy$

Contd...

11. Von-Karman momentum integral equation for the hydrodynamic b.l. over a flat plate:

$$\frac{\tau_0}{\rho U^2} = \frac{\partial \theta}{\partial x}$$

where θ = momentum thickness

$$= \int_0^\delta \frac{u}{U}\left(1 - \frac{u}{U}\right) dy$$

12. Local skin friction coefficient: C_f

$$C_f = \frac{\text{Local wall shear stress: } \tau_0}{\text{Dynamic pressure of the free stream: } \frac{1}{2}\rho U^2}$$

$$C_f = \frac{\tau_0}{\frac{1}{2}\rho U^2}$$

13. Average coefficient of drag: $\overline{C}_f$

$$\overline{C}_f = \frac{F_D}{\frac{1}{2}\rho A U^2}$$

where F_D = drag force

$$= \int_0^l \tau_0 b\, dx$$

where b = width of the plate
          l = length of the plate

14. In Laminar boundary layer:
Velocity distribution is parabolic boundary layer thickness: $\delta \propto x^{1/2}$

15. In turbulent boundary layer: Velocity distribution is logarithmic, boundary layer thickness: $\delta \propto x^{4/5}$

16. Boundary layer separation: The layer of fluid particles which separates from the boundary (surface) of solid body is called separation of boundary layer. The point on the surface of the solid body at which the separation takes place is called separation point.

At separation point:

(*i*) Adverse pressure gradient *i.e.*, $\dfrac{\partial p}{\partial x} > 0$

(*ii*) $\left(\dfrac{\partial u}{\partial y}\right)_{y=0} = 0$

(*iii*) The wall shear stress is zero *i.e.*, $\tau_0 = 0$

Beyond the separation point, the adverse pressure gradient is large enough $\left(\dfrac{\partial p}{\partial x} >> 0\right)$, the wall shear stress is negative due to the negative value of $\left(\dfrac{\partial u}{\partial y}\right)_{y=0}$. That region is called reverse flow (or back flow)

17. **Methods for the control of boundary layer separation:**

    (*i*) Suction method

    (*ii*) By-pass method

    (*iii*) Injection method

    (*iv*) Rotating of cylinder method

    (*v*) Streamlining of body shape

ASSIGNMENT - 1

1. What is a boundary layer?
2. Explain briefly boundary layer and its significances.

                                                (GGSIP Uniersity of Delhi Dec. 2008)

3. Define:

    (*i*) Laminar boundary layer

    (*ii*) Turbulent boundary layer

    (*iii*) Laminar sub-layer and

    (*iv*) Boundary layer thickness.

4. Define and explain any two of the following:

    (*i*) Displacement thickness,

    (*ii*) Momentum thickness,

    (*iii*) Energy thickness as applied to boundary layer flow.

5. Prove that the momentum thickness (θ) and energy thickness for boundary layer flows are given by:

$$\theta = \int_0^\delta \frac{u}{U}\left(1 - \frac{u}{U}\right) dy \quad \text{and}$$

$$\delta^{**} = \int_0^\delta \frac{u}{U}\left(1 - \frac{u^2}{U^2}\right) dy$$

6. Define and explain the terms: displacement thickness and momentum thickness as applied to boundary layer flows. Derive also the expression for the above two quantities.

7. Define and explain the terms: Laminar boundary layer, turbulent boundary layer and laminar sub-layer.   (*GGSIP University Delhi, Dec. 2007*)

8. Derive expression for the momentum thickness and the energy thickness for boundary layer flows.   (*GGSIP University Delhi, Dec. 2006*)

9. Derive an expression for Von-Karman momentum integral equation.

10. Discuss:

    (*i*) the concept of the boundary with reference to fluid motion over a flat plate;

    (*ii*) phenomenon of separation for flow over curved surfaces;

    (*iii*) the prevention of separation.

11. Define boundary layer and explain the fundamental causes of its existence. Also discuss the various methods of controlling the boundary layer.

ASSIGNMENT - 2

1. Show that for linear distribution of velocity in the boundary layer; $\dfrac{\delta}{\theta} = 6$.

2. Show that the ratio of the boundary layer thickness to displacement thickness $\left(\dfrac{\delta}{\delta^*}\right)$ for the velocity profile given by $\dfrac{u}{U} = \sin\left(\dfrac{\pi\delta}{2}\right)$ is 2.75.

3. Show that for the velocity profile,

$$\frac{u}{U} = 2\eta - \eta^2$$

where $\eta = \dfrac{y}{\delta}$, that ratio $\dfrac{\delta}{\delta^*} = 3$.

4. The velocity distribution in the boundary layer was found to fit the equation

$$\frac{u}{U} = \left(\frac{y}{\delta}\right)^n$$

Show that the displacement and the energy thicknesses can be expressed as

$$\delta^* = \delta - \frac{\delta}{n+1}$$

$$\delta^{**} = \frac{\delta}{n+1} - \frac{\delta}{3n+1}$$

5. For a velocity profile within the boundary layer:

$$\frac{u}{U} = \frac{3}{2}\left(\frac{y}{\delta}\right) - \frac{1}{2}\left(\frac{y}{\delta}\right)^3$$

Find the ratio of displacement thickness to normal boundary layer thickness.

$$\left[\text{Ans.} \quad \frac{\delta^*}{\delta} = \frac{3}{8}\right]$$

6. For the velocity profile in laminar boundary layer as,

$$\frac{u}{U} = \frac{3}{2}\left(\frac{y}{\delta}\right) - \frac{1}{2}\left(\frac{y}{\delta}\right)^3$$

   find the thickness of the boundary layer and the shear stress 1.5 m from the leading edge of a plate. The plate is 2 m long and 1.4 m wide and is placed in water which is moving with a velocity of 200 mm per second. Find the total drag force on the plate if m for water = 0.01 poise.

   Ans. 12.7 mm, 0.2276 N

7. Air is flowing over a smooth plate with a velocity of 10 m/s. The length of the plate is 1.2 m and width 0.8 m. If laminar boundary layer exists up to a value of $Re = 2 \times 10^5$, find the maximum distance from the leading edge up to which laminar boundary layer exists. Find the maximum thickness of laminar boundary layer if the velocity profile is given by

$$\frac{u}{U} = 2\left(\frac{y}{\delta}\right) - \left(\frac{y}{\delta}\right)^2$$

   Take kinematic viscosity for air = 0.15 stokes.   **Ans.** 300 mm, 3.67 mm

8. Air is flowing over a flat plate 500 mm long and 600 mm wide with a velocity of 4 m/s. The kinematic viscosity of air is given as 0.15×10^{-4} m^2/s. Find (i) the boundary layer thickness at the end of the plate, (ii) shear stress at 200 mm from the leading edge and (iii) drag force on one side of the plate.

   Take the velocity profile over the plate as $\dfrac{u}{U} = \sin\left(\dfrac{\pi}{2} \cdot \dfrac{y}{\delta}\right)$ and density of air

   1.24 kg/m^3.   **Ans.** 6.56 mm, 0.02805 N/m^2, 0.01086 N

9. A plate of 600 mm length and 400 mm wide is immersed in a fluid of specific gravity 0.9 and kinematic viscosity 10^{-4} m^2/s. The fluid is moving with a velocity of 6 m/s. Determine (i) boundary layer thickness, (ii) shear stress at the end of the plate, and (iii) drag force on one side of the plate.

   Ans. (i) 15.81 mm, (ii) 56.6 N/m^2, (iii) 26.78 N

□□□

4

Flow Through Pipe

4.1 INTRODUCTION

A closed conduit, carrying fluid under pressure is called pipe. The terms pipe and duct are usually used interchangeably for flow sections. In general, flow sections of circular cross section are referred to as pipes (especially when the fluid is a liquid), and flow sections of non-circular cross section as ducts (especially when the fluid is a gas). Small diameter pipes are usually referred as tubes.

You have probably noticed that most fluids, especially liquid, are transported in circular pipes. This is because pipes can withstand large pressure differences between the inside and the outside without undergoing significant distortion. Ducts are usually used in the air-conditioning such as the heating and cooling system of buildings where the pressure difference is relatively small, the manufacturing and installation costs are lower, and the available space is limited for ductwork.

The pressure in a pipe may be above or below atmosphere pressure as desired. This chapter deals with the study of energy (or head) losses when a fluid is flowing through a long pipe of constant cross-sectional area, various fittings, valves, bends, tees, inlets, exists, sudden enlargement, and sudden contraction in addition to the pipes.

4.2 ENERGY LOSSES IN PIPES

When a real fluid is flowing through a pipe, energy (or head) losses through pipes are classified in two categories:

    (*i*) Major losses   (*ii*) Minor losses

4.2.1 Major Losses

Energy (or head) losses due to friction among fluid particles or between fluid particles and surface area of a pipe at constant cross-sectional area of fluid flow are known as **major losses**. It is calculated by the following formulae:

    (*a*) Darcy-Weisbach formula and   (*b*) Chezy's formula

© The Author(s) 2023
S. Kumar, *Fluid Mechanics (Vol. 2)*,
https://doi.org/10.1007/978-3-030-99754-0_4

(a) Darcy-Weisbach Formula:

Let V = mean velocity of fluid flow through pipe.

   d = diameter of pipe.

   l = length of pipe between sections (1) and (2) as shown in Fig. 11.1.

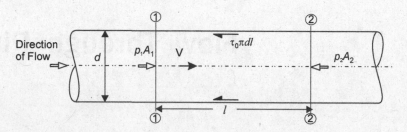

Fig. 4.1: Pressure and shear forces in pipe flow.

p_1, p_2 = pressure at sections (1) and (2) respectively.

A_1, A_2 = cross-sectional areas at sections (1) and (2) respectively.

$\therefore$ $A_1 = A_2 = A = \dfrac{\pi}{4} d^2$

τ_0 = shear stress at pipe surface which offers the resistance to motion

According to momentum equation, which states that the net force exerted in the direction of flow is equal to the rate of change of momentum in the direction of flow.

$$p_1 A_1 - p_2 A_2 - \tau_0 \pi dl = m(V_2 - V_1)$$

$$p_1 A - p_2 A - \tau_0 \pi dl = 0 \qquad\qquad \because V_1 = V_2 = V$$

$$(p_1 - p_2)A = \tau_0 \pi dl \qquad\qquad \because \tau_0 = \frac{f \rho V^2}{2},$$

$$(p_1 - p_2)\frac{\pi}{4} d^2 = \frac{f \rho V^2}{2} \pi dl \qquad \text{it can be shown from dimensional analysis.}$$

or $$\frac{p_1 - p_2}{\rho} = \frac{4 f \, l V^2}{2 d}$$

Dividing by 'g' on both sides, we get

$$\frac{p_1 - p_2}{\rho g} = \frac{4 f \, l V^2}{2 g d} \qquad\qquad\qquad \text{... (4.2.1)}$$

where f = coefficient of friction.

   ρ = density of fluid.

Applying modified Bernoulli's equation at sections (1) and (2), we get

$$\frac{P_1}{\rho g} + \frac{V_1^2}{2g} + z_1 = \frac{P_2}{\rho g} + \frac{V_2^2}{2g} + z_2 + h_f$$

where h_f = loss of head due to friction.

$V_1 = V_2 = V$ i.e., velocity of fluid flow is uniform.

and $z_1 = z_2$, pipe is horizontal.

$\therefore$

$$\frac{P_1}{\rho g} - \frac{P_2}{\rho g} = h_f$$

$$\frac{P_1 - P_2}{\rho g} = h_f \qquad\qquad ... (4.2.2)$$

Equating Eqs. (4.2.2) and (4.2.1), we get

$$h_f = \frac{4f\ lV^2}{2gd} \qquad\qquad ... (4.2.3)$$

$$= \frac{f'lV^2}{2gd} \qquad\qquad ... (4.2.4)$$

where f' = friction factor.

$f' = 4f$

= 4 times coefficient of friction.

Equations (4.2.3) and (4.2.4) are called **Darcy-Weisbach formula.**

f, f' = both coefficient of friction and friction factor are function of Reynold's number.

$$\left.\begin{array}{l} f = \dfrac{16}{Re} \\[2mm] f' = 4f \\[2mm] f' = \dfrac{64}{Re} \end{array}\right\} \quad \text{for } Re < 2000, \text{ laminar flow}$$

$$\left.\begin{array}{l} f = \dfrac{0.079}{Re^{1/4}} \\[4mm] \\ f' = \dfrac{0.316}{Re^{1/4}} \end{array}\right\} \quad \text{for } 4000 < Re < 10^6, \text{ turbulent}$$

flow

Decrease in pressure from Eq. (4.2.2), we get

$$P_1 - P_2 = \rho g h_f \qquad\qquad\qquad \text{For horizontal pipe}$$

Loss of fluid power: $P = mgh_f$ watt $\quad Q = mv = \dfrac{m}{\rho}$ or $m = \rho Q$

$$= \rho Q g h_f \text{ watt}$$

where m = mass flow rate, kg/s

g = 9.81 m/s^2

Q = discharge, m^3/s

ρ = density of fluid, kg/m^3

h_f = head loss due to friction, m

(b) **Chezy's formula:**

Starting from Darcy-Weisbach formula, loss of head due to friction:

$$h_f = \frac{4flV^2}{2gd}$$

$$\frac{h_f}{l} = \frac{4fV^2}{2gd}$$

$$i = \frac{4fV^2}{2gd}$$

where $i = \dfrac{h_f}{l}$,   loss of head due to friction per unit length of pipe is called **slope of hydraulic gradient.**

$$i = \frac{4fV^2}{2gd} \times \frac{\pi d}{\pi d}$$

$$= \frac{fV^2}{2g} \times \frac{\pi d}{\dfrac{\pi d^2}{4}}$$

$$= \frac{fV^2}{2g} \times \frac{P}{A}$$

$$= \frac{fV^2}{2g} \times \frac{1}{m}$$

where $m = \dfrac{A}{P}$, **hydraulic mean depth** OR **hydraulic radius.**

Cross-sectional area: $A = \dfrac{\pi}{4}d^2$

Perimeter: $P = \pi d$

$\therefore$ $$i = \frac{\rho}{\rho} \frac{fV^2}{2gm}$$

$$i = \frac{f_1}{\rho g} \frac{V^2}{m}$$

where $f_1 = \dfrac{\rho f}{2}$, coefficient depending on the roughness of the pipe.

Let $$C^2 = \frac{\rho g}{f_1}, \quad \text{where } C \text{ is called Chezy's constant.}$$

$$i = \frac{1}{C^2}\frac{V^2}{m}$$

or $$V^2 = C^2 im$$

or $$V = C\sqrt{im} \qquad \qquad \ldots (4.2.5)$$

Equation (4.2.5) is known as **Chezy's formula.** Thus, the loss of head due to friction in pipe from Chezy's formula can be obtained if the velocity of flow through pipe and the value of C is known. The value of m for pipe is always equal to $\dfrac{d}{4}$ $\left(i.e., m = \dfrac{A}{P} = \dfrac{\pi\, d^2}{4\, \pi d} = \dfrac{d}{4} \right)$.

Determination of Friction Factor Using Moody's Diagram

The diagram developed by Prof. Lewis F. Moody for commercial pipes has become a convenient and reliable tool for solving practical problems in pipe flow. Moody's diagram gives the value of friction factor f' of any pipe provided its relative roughness ϵ/d and Reynolds number of flow Re are known. ϵ is absolute (or average) roughness, or equivalent sand roughness and d is diameter of pipe.

We make the following observations from the Moody's diagram:

(*i*) For laminar flow, the friction factor f' decreases with increasing Reynold's number, and it is independent of surface roughness. The friction factor f' is also found by the following equation for laminar flow.

Friction factor: $f' = \dfrac{64}{Re}$ for $Re < 2000$

(*ii*) The critical (or transition) region from the laminar to turbulent regime $(2000 < Re < 4000)$ is indicated by the shaded area in the Moody's diagram. The flow in this region may be laminar or turbulent, depending on flow disturbances, or it may alternate between laminar and turbulent, and thus the friction factor may also alternate between the values for laminar and turbulent flow. The data in this range are the least reliable and not of much practical significance.

(*iii*) At very large Reynold's number (*i.e.*, $Re \geq 10^6$) the friction factor curves corresponding to specified relative roughness curve are nearly horizontal, and thus the friction factor is independent of the Reynold's number and dependent upon only relative roughness $\dfrac{\epsilon}{d}$ as shown in Fig. 4.2. The flow in that region is called fully rough turbulent flow or just fully rough flow because the thickness of the viscous sublayer decreases with increasing Reynold's number.

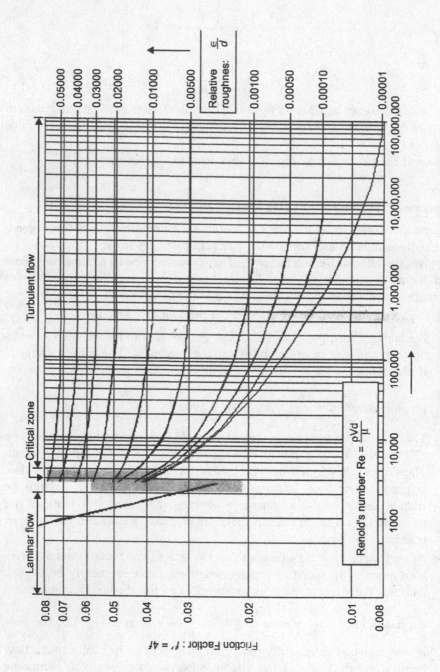

Fig. 4.2: Moody's diagram.

The friction factor f' is also found by the following equation for turbulent flow:

Friction factor: $f' = \dfrac{0.316}{Re^{1/4}}$ only for $4000 < Re < 10^6$

Table 4.1: Equivalent Roughness Values for New Commercial Pipes

S. No.	Material	Absolute Roughness: ∈ mm
1.	Glass, plastic	0 (smooth)
2.	Concrete	0.3 – 3
3.	Riveted steel	0.9 – 9
4.	Wood slave	0.18 – 0.9
5.	Rubber smoothed	0.01
6.	Copper or brass tubing	0.0015
7.	Cast iron	0.26
8.	Galvanized iron	0.15
9.	Wrought iron	0.046
10.	Stainless steel	0.002
11.	Commercial steel	0.045

The procedure for determination of the friction factor of any pipe is demonstrated through the following example. Let the velocity of water through a new galvanized iron pipe of 200 mm diameter be 5 m/s.

Let $\rho = 1000$ kg/m^3, $\mu = 0.002$ Ns/m^2

From table 11.1, absolute roughness:

$$\in = 0.15 \text{ mm}$$

Hence relative roughness:

$$\frac{\in}{d} = \frac{0.15}{200} = 0.00075$$

Reynold's number: $Re = \dfrac{\rho V d}{\mu} = \dfrac{1000 \times 5 \times 0.2}{0.002} = 500000$

For these values of $\dfrac{\in}{d} = 0.00075$ and $Re = 500000$

Moody's diagram gives friction factor: $f' = 0.018$

Problem 4.1: Determine the head lost due to friction in a pipe of diameter 400 mm and length 100 m, through which water is flowing at a velocity 2.5 m/s by using (a) Darcy-Weisbach formula (b) Chezy's formula for which $C = 60$.

Take kinematic viscosity (v) for water = 0.01 stoke.

Solution: Given data:

Diameter of pipe:	$d = 400$ mm $= 0.4$ m
Length of pipe:	$l = 100$ m
Velocity of flow:	$V = 2.5$ m/s
Chezy's constant:	$C = 60$
Kinematic viscosity:	$v = 0.01$ stoke $= 0.01$ cm²/s $\qquad \because$ 1 stoke $= 1$ cm²/s
	$= 0.01 \times 10^{-4}$ m²/s

(*a*) **By using Darcy-Weisbach formula**

Head lost due to friction: $h_f = \dfrac{4 f l V^2}{2gd}$

where f = coefficient of friction is a function of Reynold's number, Re

Reynold's number: $Re = \dfrac{Vd}{v} = \dfrac{2.5 \times 0.4}{0.01 \times 10^{-4}} = 100 \times 10^4 = 10 \times 10^5$

Hence, flow is turbulent because of $Re > 4000$

$\therefore$ Coefficient of friction: $f = \dfrac{0.079}{Re^{0.25}} = \dfrac{0.079}{(10 \times 10^5)^{0.25}} = 0.00249$

$\therefore$ Head lost: $h_f = \dfrac{4 \times 0.00249 \times 100 \times (2.5)^2}{2 \times 9.81 \times 0.4}$

$= \textbf{0.7931 m of water}$

(*b*) **By Chezy's formula**

Velocity: $V = C\sqrt{im}$

where $C = 60$

$$m = \dfrac{\text{Area}}{\text{Perimeter}} = \dfrac{A}{P} = \dfrac{\frac{\pi}{4} d^2}{\pi d} = \dfrac{d}{4}$$

$$m = \dfrac{0.4}{4} \quad 0.1 \text{ m}$$

$\therefore$ $2.5 = 60\sqrt{i \times 0.1}$

or $i = 0.01736$

also $i = \dfrac{h_f}{l}$

$$0.01736 = \dfrac{h_f}{100}$$

or Head lost: $h_f = \textbf{1.73 m of water}$

Head lost: $\boldsymbol{h_f = 0.7931}$ **m of water by using Darcy-Weisbach formula.**
$= 1.73$ m of water by using Chezy's formula.

Problem 4.2: Water is flowing through a pipe 1.5 km long with a velocity of 1.5 m/s. What should be the diameter of the pipe, if the loss of head due to friction 10 m. Take coefficient of friction $f = 0.01$.

Solution: Given data:

Length of pipe: $l = 1.5$ km $= 1500$ m

Velocity of water: $V = 1.5$ m/s

Loss of head due to friction: $h_f = 10$ m

Coefficient of friction: $f = 0.01$

We know that the loss of head due to friction:

$$h_f = \frac{4flV^2}{2gd} \qquad \text{(By Darcy's formula)}$$

$$10 = \frac{4 \times 0.01 \times 1500 \times (1.5)^2}{2 \times 9.81 \times d}$$

or $d = 0.68807$ m $= \mathbf{688.07}$ **mm**

Problem 4.3: A reservoir has been built 3 km away from a college compus having 4000 inhabitants. Water is to be supplied from the reservoir to the campus. It is estimated that each inhabitant will consume 200 litres of water per day, and that half of the daily supply is pumped within 10 hours. Find the size of the supply main, if the loss of head due to friction in pipeline is 15 m. Take coefficient of friction for the pipe is 0.007.

Solution: Given data:

Length of pipe: $l = 3$ km $= 3000$ m

Number of inhabitants: $n = 4000$

Consumption of water by each inhabitant $= 200$ litre/day $= 0.2$ m³/day

∴   Total supply $=$ number of inhabitants $\times$ consumption by each
                            inhabitant
                    $= 4000 \times 0.2$ m³/day $= 800$ m³/day

Since half of this supply is to be pumped in 10 hours,

∴ Maximum flow through pipe:

$$Q = \frac{1}{2} \times \frac{800}{10 \times 60 \times 60} \text{ m}^3/\text{s} = 0.0111 \text{ m}^3/\text{s}.$$

Loss of head due to friction: $h_f = 15$ m

Coefficient of friction: $f = 0.007$

We know that the loss of head due to friction:

$$h_f = \frac{4flV^2}{2gd} \qquad \text{(By Darcy's formula)}$$

$$h_f = \frac{4fl}{2gd} \times \frac{16Q^2}{\pi^2 d^4} \qquad \left| \quad Q = AV = \frac{\pi}{4}d^2V \text{ or } V = \frac{4Q}{\pi d^2} \right.$$

$$h_f = \frac{32flQ^2}{g\pi^2 d^5}$$

$$15 = \frac{32 \times 0.007 \times 3000 \times (0.0111)^2}{9.81 \times (3.14)^2 \times d^5}$$

or

$$d^5 = 5.7068 \times 10^{-5}$$

$$d = 0.14167 \text{ m} = \textbf{141.67 mm}$$

Problem 4.4: In a refinery, crude oil of specific gravity 0.85 and viscosity 0.1 poise, flow through a pipe of 300 mm diameter with an average velocity of 3 m/s. Find the pumping power required to maintain the flow per km length of the pipe.

Solution: Given data:

Specific gravity: $S = 0.85$

∴ Density: $\rho = S \times 1000 \text{ kg/m}^3 = 0.85 \times 1000 = 850 \text{ kg/m}^3$

Viscosity: $\mu = 0.1 \text{ poise} = \dfrac{0.1}{10} \text{ Ns/m}^2 = 0.01 \text{ Ns/m}^2$

Diameter of pipe: $d = 300 \text{ mm} = 0.3 \text{ m}$

∴ Cross-sectional area: $A = \dfrac{\pi}{4}d^2 = \dfrac{3.14}{4} \times (0.3)^2 = 0.067065 \text{ m}^2$

Average velocity: $V = 3 \text{ m/s}$

Reynold's number: $Re = \dfrac{\rho V d}{\mu} = \dfrac{850 \times 3 \times 0.3}{0.01} = 76500$

The nature of flow is turbulent, because of Reynold's number $Re > 4000$.

Assuming that the pipe surface is smooth, the coefficient of friction can be obtained as:

$$f = \frac{0.079}{Re^{0.25}}$$

$$f = \frac{0.079}{(76500)^{0.25}} = 0.00475$$

Loss of head due to friction per km length (*i.e.*, $l = 1$ km $= 1000$ m) of pipe is

$$h_f = \frac{4flV^2}{2gd} = \frac{4 \times 0.00475 \times 1000 \times (3)^2}{2 \times 9.81 \times 0.3} = 29.05 \text{ m}$$

Discharge: $Q = AV = 0.7065 \times 3 = 0.2119 \text{ m}^2$

Power required to maintain the flow: P

$$P = \rho Q g h_f = 850 \times 0.2119 \times 9.81 \times 29.05$$
$$= 51329.26 \text{ W} \approx \textbf{51.33 kW}$$

Problem 4.5: A town having a population of 100000 is to be supplied with water from a reservoir at 6 km distance. It is estimated that one-half of the daily supply of 125 litres per head should be delivered within eight hours. Find the size of the pipe through which water is supplied, if the head available is 16 m. Take $C = 45$ in Chezy's formula.

Solution: Give data:

Population of town: $n = 100000$

Length of pipe: $l = 6$ km $= 6000$ m

$$\text{Daily supply} = \frac{1}{2} \times 125 \text{ litre/head} = 62.5 \text{ litre/head} = 0.0625 \text{ m}^3/\text{head}$$

∴   Total supply = population of town × daily supply

$$= 100000 \times 0.0625 = 6250 \text{ m}^3/\text{day}$$

Heat available = loss of head due to friction

∴ $h_f = 16$ m

Chezy's constant: $C = 45$

Since supply of water in 8 hours;

$$Q = \frac{6250}{8 \times 60 \times 60} = 0.217 \text{ m}^3/\text{s}$$

also $Q = AV = \dfrac{\pi}{4} d^2 V$

$$0.217 = \frac{3.14}{4} d^2 V$$

or $d^2 V = 0.27643$

or $V = \dfrac{0.27643}{d^2}$

We know that the hydraulic mean depth for a circular pipe: $m = \dfrac{d}{4}$

and $i = \dfrac{h_f}{l} = \dfrac{16}{6000} = 0.00266$

According to Chezy's formula,

Velocity of water: $V = C\sqrt{im}$

$$\frac{0.27643}{d^2} = 45\sqrt{0.00266 \times \frac{d}{4}}$$

or $0.27643 = \dfrac{45d^2}{\sqrt{4}} \sqrt{0.00266} \times \sqrt{d}$

$$0.27643 = \frac{45}{2} \times d^{5/2} \times \sqrt{0.00266}$$

or $d^{5/2} = 0.23821$

or $d = 0.56335$ m $= \mathbf{563.35}$ **mm**

Problem 4.6:   An oil of specific gravity 0.85 and viscosity 0.05 poise is flowing through a pipe of diameter 250 mm at the rate of 75 litre/s. Find the head lost due to friction for a 600 m length of pipe. Find also the power required to maintain the flow.

Solution: Given data:

Specific gravity of oil: $S = 0.85$

∴ Density of oil: $\rho = S \times 1000 = 0.85 \times 1000 = 850$ kg/m³

Viscosity of oil: $\mu = 0.05$ poise $= \dfrac{0.05}{10}$ Ns/m² $= 0.005$ Ns/m²

Diameter of pipe: $d = 250$ mm $= 0.25$ m
∴ Cross-sectional area of pipe:

$$A = \frac{\pi}{4}d^2 = \frac{3.14}{4} \times (0.25)^2 = 0.04906 \text{ m}^2$$

Discharge: $Q = 75$ litre/s $= \dfrac{75}{1000}$ m³/s $= 0.075$ m³/s

also $Q = AV$
∴ $0.075 = 0.04906 \times V$
or $V = 1.528$ m/s
Length of pipe: $l = 600$ m

Now Reynold's number: $Re = \dfrac{\rho V d}{\mu} = \dfrac{850 \times 1.528 \times 0.25}{0.005} = 64940$

Hence, flow is turbulent because of $Re > 4000$

∴ Coefficient of friction: $f = \dfrac{0.079}{Re^{0.25}} = \dfrac{0.079}{(64940)^{0.25}} = 0.00494$

∴ Head lost due to friction:

$$h_f = \frac{4flV^2}{2gd} = \frac{4 \times 0.00494 \times 600 \times (1.528)^2}{2 \times 9.81 \times 0.25}$$

$$= \textbf{5.64 m of water}$$

∴ Power required to maintain the flow: P

$$P = mgh_f = \rho Q g h_f \qquad\qquad \because m = \rho Q$$
$$= 850 \times 0.075 \times 9.81 \times 5.64$$
$$= 3527.18 \text{ W} = \textbf{3.527 kW}$$

Problem 4.7:   The friction factor for turbulent flow through rough pipes can be determined by Karman-Prandtl equation,

$$\frac{1}{\sqrt{f'}} = \log_{10}\left(\frac{d}{2\epsilon}\right)^2 + 1.74$$

where $f' = $ friction factor,
            $d = $ diameter of pipe,
            $\epsilon = $ average roughness.

Two reservoirs with a surface level difference of 20 m are to be connected by 1 m diameter pipe 6 km long. What will be the discharge when a cast iron pipe of roughness $\epsilon = 0.26$ mm is used? What will be the percentage increase in the discharge if the cast iron pipe is replaced by a galvanized iron pipe of roughness $\epsilon = 0.15$ mm?

Neglect all minor losses.

Solution: Given data:

Difference in surface level between two reservoirs: $h = 20$ m

Diameter of pipe: $\qquad\qquad d = 1$ m

Length of pipe: $\qquad\qquad l = 6$ km $= 6000$ m

Roughness of cast iron pipe: $\qquad \epsilon = 0.26$ mm $= 0.00026$ m

Roughness of galvanized iron pipe: $\epsilon = 0.15$ mm $= 0.00015$ m

Case I: Cast iron pipe:

Given equation: $\qquad \dfrac{1}{\sqrt{f'}} = \log_{10}\left(\dfrac{d}{2\epsilon}\right)^2 + 1.74$

$$\dfrac{1}{\sqrt{f'}} = \log_{10}\left(\dfrac{1}{2\times 0.00026}\right)^2 + 1.74$$

$$\dfrac{1}{\sqrt{f'}} = 6.568 + 1.74 = 8.308$$

Squaring both sides, we get

$$\dfrac{1}{f'} = (8.308)^2 = 69.02 = \dfrac{1}{69.02} = 0.01448$$

also $\qquad\qquad f' - 4f$

or Coefficient of friction:

$$f = \dfrac{f'}{4} = \dfrac{0.01448}{4} = 0.00362$$

Now loss of head due to friction:

$$h_f = h = \dfrac{4flV^2}{2gd} \qquad\qquad \because \text{ Minor losses are neglected}$$

$$\therefore \qquad\qquad 20 = \dfrac{4\times 0.00362\times 6000\times V^2}{2\times 9.81\times 1}$$

or $\qquad\qquad V^2 = 4.5165$

or $\qquad\qquad V = 2.125$ m/s

$\therefore$ Discharge flow through pipe:

$$Q = AV = \dfrac{\pi}{4}d^2\times V = \dfrac{3.14}{4}\times (1)^2\times 2.125 \text{ m}^3/\text{s}$$

$$= \mathbf{1.668} \text{ m}^3/\text{s}$$

Case II: Galvanized iron pipe:

Given equation: $\qquad \dfrac{1}{\sqrt{f'}} = \log_{10}\left(\dfrac{d}{2\epsilon}\right)^2 + 1.74$

$$\dfrac{1}{\sqrt{f'}} = \log_{10}\left(\dfrac{1}{2\times 0.00015}\right)^2 + 1.74$$

$$\dfrac{1}{\sqrt{f'}} = 7.045 + 1.74 = 8.785$$

Squaring both sides, we get

$$\frac{1}{f'} = (8.785)^2 = 77.176$$

or $$f' = \frac{1}{77.176} = 0.01295$$

also $$f' = 4f$$

or Coefficient of friction:

$$f = \frac{f'}{4} = \frac{0.01295}{4} = 0.00323$$

Now loss of head due to friction: $h_f = h = 20$ m

$$h_f = \frac{4flV^2}{2gd} \qquad\qquad \because \text{ Minor losses are neglected}$$

$$20 = \frac{4 \times 0.00323 \times 6000 V^2}{2 \times 9.81 \times 1}$$

or $$V^2 = 5.062$$

or $$V = 2.249 \text{ m/s}$$

$\therefore$ Discharge flow through pipe:

$$Q' = AV = \frac{\pi}{4}d^2 \times V = \frac{3.14}{4} \times (1)^2 \times 2.249 = 1.765 \text{ m}^3/\text{s}$$

% Increase in the discharge $= \left[\dfrac{Q'-Q}{Q}\right] \times 100 = \left[\dfrac{1.765-1.668}{1.668}\right] \times 100 = \textbf{5.81 \%}$

Problem 4.8:   Glycerine (specific gravity 1.26, viscosity 0.9 Pa.s) is pumped at the rate of 20 litre/s through a straight pipe, diameter 100 mm, 45 m long and inclined upward at 15° to the horizontal. The gauge pressure at inlet is 590 kPa. Find the gauge pressure at the outlet end and the average shear stress at the wall.

Solution: Given data:

Specific gravity: $S = 1.26$

$\therefore$ Density: $\rho = S \times 1000 \text{ kg/m}^3 = 1.26 \times 1000 = 1260 \text{ kg/m}^3$

Viscosity: $\mu = 0.9 \text{ Pa.s} = 0.9 \text{ Ns/m}^2$

Discharge: $Q = 20 \text{ litre/s} = \dfrac{20}{1000} \text{ m}^3/\text{s} = 0.02 \text{ m}^3/\text{s}$

Diameter of pipe: $d = 100 \text{ mm} = 0.1 \text{ m}$

$\therefore$ Cross-sectional area: $A = \dfrac{\pi}{4}d^2 = \dfrac{3.14}{4} \times (0.1)^2 = 7.85 \times 10^{-3} \text{ m}^3$

Length of pipe: $l = 45 \text{ m}$

Angle: $\theta = 15°$

Pressure at inlet: $p_1 = 590 \text{ kPa} = 590 \times 10^3 \text{ N/m}^2$

Mean velocity of flow: $V = \dfrac{\text{Discharge: } Q}{\text{Cross-sectional area: } A} = \dfrac{0.02}{7.85 \times 10^{-3}} = 2.547 \text{ m/s}$

Reynolds number: $Re = \dfrac{\rho V d}{\mu} = \dfrac{1260 \times 2.547 \times 0.1}{0.9} = 355.6$

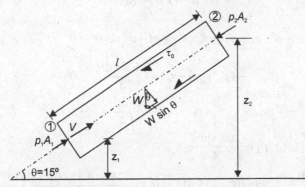

Fig. 4.3: Schematic for Problem 4.8

The nature of given flow is laminar, because of Reynolds number $Re < 2000$

$\therefore$ Coefficient of friction: $f = \dfrac{16}{Re} = \dfrac{16}{355.6} = 0.045$

and head lost due to friction:

$$h_f = \dfrac{4flV^2}{2gd} = \dfrac{4 \times 0.045 \times 45 \times (2.547)^2}{2 \times 9.81 \times 0.1} - 26.78 \text{ m of glycerine}$$

Now applying modified Bernoulli's equation between two ends, we get

$$\dfrac{p_1}{\rho g} + \dfrac{V_1^2}{2g} + z_1 = \dfrac{p}{\rho g} + \dfrac{V_2^2}{2g} + z_2 + h_f$$

$$\dfrac{p_1}{\rho g} + z_1 = \dfrac{p_2}{\rho g} + z_2 + h_f \qquad\qquad \because V_1 = V_2$$

$$\dfrac{p_2}{\rho g} = \dfrac{p_1}{\rho g} + z_1 - z_2 - h_f$$

$$= \dfrac{p_1}{\rho g} - (z_2 - z_1) - h_f$$

$$\therefore \qquad \dfrac{p_2}{\rho g} = \dfrac{p_1}{\rho g} - l\sin\theta - h_f$$

$$\dfrac{p_2}{1260 \times 9.81} = \dfrac{590 \times 10^3}{1260 \times 9.81} - 45\sin 15° - 26.78$$

$$\dfrac{p_2}{1260 \times 9.81} = 9.305 \qquad\qquad \sin\theta = \dfrac{z_2 - z_1}{l}$$

or $\qquad p_2 = 9.305 \times 1260 \times 9.81 \qquad$ or $(z_2 - z_1) = l\sin\theta$

$$= 115015.38 \text{ N/m}^2 = 115.015 \times 10^3 \text{ Pa} = \textbf{115.015 kPa}$$

According to momentum equation which states that the net force exerted in the direction of flow is equal to the rate of change of momentum in the direction of flow.

$$p_1 A_1 - p_2 A_2 - \tau_0 \pi dl - W\sin\theta = m(V_2 - V_1)$$

where $\qquad\qquad V_1 = V_2 \ $ and $ \ A_1 = A_2 = \dfrac{\pi}{4}d^2$

$$W = \text{Mg} = \rho \times \text{volume} \times g$$

$$= \rho \frac{\pi}{4} d^2 \cdot l \times g = \rho g l \frac{\pi}{4} d^2$$

$\therefore$
$$\frac{\pi}{4} d^2 (p_1 - p_2) - \tau_0 \pi dl - \rho g l \frac{\pi}{4} d^2 \sin\theta = 0$$

$$\frac{d}{4}(p_1 - p_2) - \tau_0 l - \frac{\rho g \, d \sin\theta}{4} = 0$$

$$\frac{d}{4}(p_1 - p_2) - \frac{\rho g \, d \sin\theta}{4} = \tau_0 l$$

or $$\tau_0 = \frac{d}{4l}(p_1 - p_2) - \frac{\rho g d \sin\theta}{4}$$

$$= \frac{0.1}{4 \times 45}(590 \times 10^3 - 115.015 \times 10^3) - \frac{1260 \times 9.81 \times 0.1 \times \sin 15°}{4}$$

$$= 263.88 - 79.97 = \mathbf{183.91 \ N/m^2}$$

OR

Head lost is due to the viscous forces.

i.e.,
$$\rho g h_f \cdot \frac{\pi}{4} d^2 = \tau_0 \times \pi dl$$

$$\rho g h_f \frac{d}{4} = \tau_0 l$$

$$1260 \times 9.81 \times 26.78 \times \frac{0.1}{4} = \tau_0 \times 45$$

or $$\tau_0 = \mathbf{183.89 \ N/m^2}$$

4.2.2 Minor Losses

The loss of energy (or head) due to local disturbances in pipelines such as sudden enlargements, sudden contractions, pipe bends, valves and other fittings is called local or secondary or minor losses. They are so called because the friction loss is the major loss in long pipe. If $\frac{l}{d} \geq 1000$, minor losses may be neglected as they are relatively small in magnitude. In short pipes, however, minor losses assume considerable significance. In some cases, the minor losses may be greater than the major losses. This is the case, for example, in system with several turns and valve in a short distance.

The minor losses in pipeline problem are discussed below:

  (*i*) Loss of head due to sudden enlargement.

 (*ii*) Loss of head due to sudden contraction.

(*iii*) Loss of head at the entrance to a pipe.

(*iv*) Loss of head at the exit of a pipe.

 (*v*) Loss of head due to an obstruction in pipe.

(*vi*) Loss of head in pipe fittings.

Loss of head due to sudden enlargement

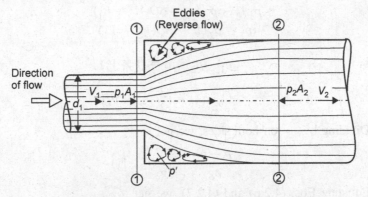

Fig. 4.4: Flow through a sudden enlargement.

The loss of energy or head in sudden enlargement or expansion is mainly due to turbulent eddies that are formed in the corner as shown in Fig. 4.4.

Consider two sections (1) – (1) and (2) – (2) before and after the enlargement.

Let p_1 = pressure intensity at section (1) – (1).

V_1 = velocity of flow at section (1) – (1).

A_1 = cross-sectional area of pipe at section (1) – (1).

p_2, V_2 and A_2 = pressure, velocity and cross-sectional area at section (2) – (2) respectively.

p' = pressure intensity of the liquid eddies on the area $(A_2 - A_1)$.

h_e = loss of head due to sudden enlargement.

Applying modified Bernoulli's equation to sections (1) – (1) and (2) – (2), we get

$$\frac{p_1}{\rho g} + \frac{V_1^2}{2g} + z_1 = \frac{p_2}{\rho g} + \frac{V_2^2}{2g} + z_2 + \text{loss of head due to sudden enlargement.}$$

$$\frac{p_1}{\rho g} + \frac{V_1^2}{2g} = \frac{p_2}{\rho g} + \frac{V_2^2}{2g} + h_e \qquad \because z_1 = z_2, \text{ pipe is horizontal}$$

or $$\frac{p_1}{\rho g} - \frac{p_2}{\rho g} = \frac{V_2^2}{2g} - \frac{V_1^2}{2g} + h_e \qquad \qquad \qquad ... (4.2.6)$$

According to momentum equation which states that the net force exerted in the direction of flow is equal to the rate of change of momentum in the direction of flow.

$$p_1 A_1 + p'(A_2 - A_1) - p_2 A_2 = m(V_2 - V_1)$$

Actual measurements have shown that the pressure p' in the eddies is equal to p_1.

i.e., $$p' = p_1$$

$$\therefore \quad p_1 A_1 + p_1(A_2 - A_1) - p_2 A_2 = \rho Q(V_2 - V_1)$$

$$\left[\begin{array}{l} Q = mv \\ Q = \dfrac{m}{\rho} \\ \text{or} \therefore m = \rho Q \end{array}\right]$$

$$p_1A_1 + p_1A_2 - p_1A_1 - p_2A_2 = \rho Q(V_2 - V_1)$$
$$p_1A_2 - p_2A_2 = \rho Q(V_2 - V_1)$$
$$(p_1 - p_2)A_2 = \rho Q(V_2 - V_1)$$

or
$$p_1 - p_2 = \rho \frac{Q}{A_2}(V_2 - V_1) \qquad\qquad \left| V_2 = \frac{Q}{A_2} \right.$$

$$\frac{p_1 - p_2}{\rho} = V_2(V_2 - V_1)$$

Dividing by 'g' on both sides, we get

$$\frac{p_1 - p_2}{\rho g} = \frac{V_2}{g}(V_2 - V_1) \qquad\qquad \text{... (4.2.7)}$$

Equating Eqs. (4.2.6) and (4.2.7), we get

$$\frac{V_2^2}{2g} - \frac{V_1^2}{2g} + h_e = \frac{V_2}{g}(V_2 - V_1)$$

or
$$h_e = \frac{V_2}{g}(V_2 - V_1) - \frac{V_2^2}{2g} + \frac{V_1^2}{2g}$$

$$= \frac{V_2^2}{g} - \frac{V_1 V_2}{g} - \frac{V_2^2}{2g} + \frac{V_1^2}{2g}$$

$$= \frac{2V_2^2 - 2V_1V_2 - V_2^2 + V_1^2}{2g}$$

$$h_e = \frac{V_1^2 + V_2^2 - 2V_1V_2}{2g} = \frac{(V_1 - V_2)^2}{2g}$$

$$h_e = \frac{(V_1 - V_2)^2}{2g} \qquad\qquad \text{... (4.2.8)}$$

Change in pressure:
$$\Delta p = p_2 - p_1$$

$$= \frac{\rho}{2}(V_1^2 - V_2^2) - \rho g h_e \qquad \text{for horizontal pipe}$$

$$= \frac{\rho}{2}(V_1^2 - V_2^2) + \rho g(z_1 - z_2) - \rho g h_e$$
$$\text{for inclined pipe}$$

Loss of fluid power:
$$P = mgh_e \quad \text{watt}$$
$$= \rho Q g h_e \quad \text{watt}$$

where
m = mass flow rate, kg/s

g = 9.81 m/s^2

h_e = loss of head due to sudden enlargement, m

ρ = density of fluid, kg/m^3

Q = discharge, m^3/s

Total energy line (TEL) and hydraulic grade line (HGL) for flow through pipe with sudden enlargement:

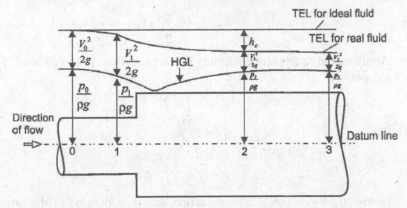

Fig. 4.5: Total energy and hydraulic grade lines for flow through a sudden enlargement.

Hydraulic grade line (HGL):

Hydraulic grade line is obtained by joining piezometric head (*i.e.*, sum of pressure and datum heads) in the direction of flow.

Total energy line (TEL):

Total energy line is obtained by joining total head (*i.e.*, sum of pressure, datum and velocity heads) in the direction of flow. In present case, datum line is passing through the centre of pipe line so, datum head is zero at all points at the centre of pipe.

Loss of head due to sudden contraction

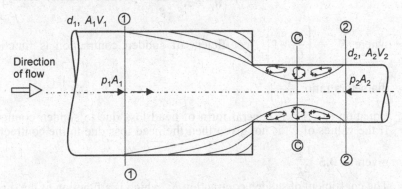

Fig. 4.6: Flow through a sudden contraction.

Consider a liquid flowing in a pipe which has a sudden contraction in area as shown in Fig. 4.6. Consider two sections (1) – (1) and (2) – (2) before and after the contraction. As the liquid flows from large pipe to small pipe, the area of flow goes on decreasing and becomes minimum at a section (C) – (C) as shown in Fig. 4.6. This section at which the area of flow is minimum is called vena-contracta. [*i.e.*, the minimum cross-sectional area of flow at which the area of flow changes from a contraction to expansion is called the vena-contracta].

After section $(C) - (C)$, a sudden enlargement of the area of flow takes place. Based on this assumption, application of Eq. (4.5.3) gives

$$h_c = \frac{(V_c - V_2)^2}{2g} = \frac{V_2}{2g}\left[\frac{V_c}{V_2} - 1\right]^2 \qquad \dots (4.2.9)$$

Applying the equation of continuity between sections $(C) - (C)$ and $(2) - (2)$, we get

$$A_c V_c = A_2 V_2$$

or

$$\frac{V_c}{V_2} = \frac{A_2}{A_c}$$

$$\frac{V_c}{V_2} = \frac{1}{A_c/A_2} = \frac{1}{C_c}$$

where C_c = coefficient of contraction and it is defined as the ratio of a cross-sectional area at vena-contracta to cross-sectional area of section $(2) - (2)$.

i.e.,

$$C_c = \frac{A_c}{A_2}$$

Substituting the value of $\dfrac{V_c}{V_2}$ in Eq. (4.5.4), we get

$$h_c = \frac{V_2^2}{2g}\left[\frac{1}{C_c} - 1\right]^2 \qquad \dots (4.2.10)$$

$$h_c = K_c \frac{V_2^2}{2g} \qquad \dots (4.2.11)$$

where $K_c = \left[\dfrac{1}{C_c} - 1\right]^2$, coefficient of sudden contraction is function of

diameters ratio $\left(\dfrac{d_2}{d_1}\right)$.

Equation (4.2.11) is general form of head loss due to sudden contraction. If the values of C_c is not given then the head loss due to the contraction is given as $0.5 \dfrac{V_2^2}{2g}$.

The coefficient of sudden contraction K_c which is a function of the diameter ratio $\dfrac{d_2}{d_1}$. Its value is given in Table 4.2 for different values of $\dfrac{d_2}{d_1}$.

Table 4.2

$\dfrac{d_2}{d_1}$	0	0.2	0.4	0.6	0.8	1.0
K_c	0.5	0.42	0.36	0.28	0.15	0

Change in pressure: $\quad\quad\quad \Delta p = p_2 - p_1$

$$= \frac{\rho}{2}(V_1^2 - V_2^2) - \rho g h_c \text{ for horizontal pipe}$$

$$= \frac{\rho}{2}(V_1^2 - V_2^2) + \rho g(z_1 - z_2) - \rho g h_c$$

for inclined pipe.

Loss of fluid power: $\quad\quad\quad \boldsymbol{P = mgh_c \text{ watt}}$

$$\boldsymbol{= \rho Q g h_c \text{ watt}}$$

where ρ = density of fluid, kg/m^3

$\quad\quad g = 9.81$ m/s^2

$\quad\quad h_c$ = loss of head due to sudden contraction, m

$\quad\quad m$ = mass flow rate, kg/s

$\quad\quad Q$ = discharge, m^3/s

Total energy line (TEL) and hydraulic grade line (HGL) for flow through pipe with sudden contraction:

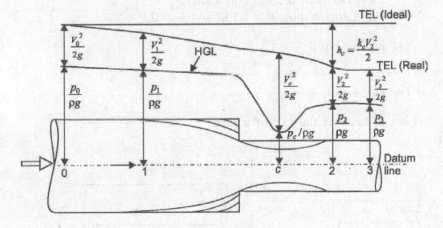

Fig. 4.7: Total energy and hydraulic grade lines for flow through a sudden contraction.

Loss of head at the entrance to a pipe

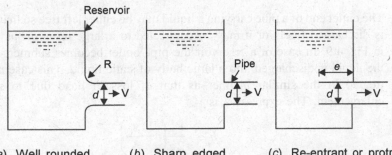

(a) Well rounded   (b) Sharp edged   (c) Re-entrant or protruding
     entrance   entrance   entrance
     (R > 0.14 d)   (e > 0.5 d)

Fig. 4.8: Various pipe entrance.

When liquid enters in a pipe from a large vessel *e.g.*, tank or reservoir, some loss of head or energy occurs at the entrance to the pipe which is known as inlet loss of energy or head. While the liquid enters the pipe, it gets contracted to a narrow neck and again expands to the entire cross-section of the pipe. The expression for loss of head at entrance 'h_i' will be similar to that the loss of head due to sudden contraction 'h_c', starting the expression from Eq. (4.5.5), in loss of head due to sudden contraction:

$$h_i = \frac{V^2}{2g}\left(\frac{1}{C_c}-1\right)^2 \qquad \because V = V_2$$

Such type of flow will give rise to a condition in which cross sectional area of tank $A \to \infty$ and diameter ratio $\dfrac{D}{d}$ tends to zero. Thus, from Table 4.2, loss coefficient K_L corresponding to this condition is 0.5. The various kinds of pipe entrances are shown in Fig. 4.8.

Let d = diameter of pipe

R = radius of bend in Fig. 4.8(*a*)

For Fig. 4.8(*a*), radius: $R > 0.14d$ and $h_i = 0$

For Fig. 4.8(*b*), $h_i = 0.5\dfrac{V^2}{2g}$

For Fig. 4.8(*c*), $h_i = \dfrac{V^2}{2g}$

Loss of head at the exit of a pipe

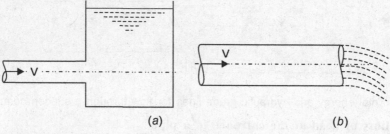

(a)   (b)

Fig. 4.9: Exit losses.

The outlet end of a pipe carrying a liquid may be either left free so that liquid is discharged freely or it may be connected to a large reservoir as shown in Fig. 4.9. In case of a reservoir the pipe outlet becomes submerged and the liquid is discharged into a large body of static liquid. This case may be treated in the similar manner as that of loss of head due to sudden enlargement. The expression is

$$h_o = \frac{V^2}{2g}\left(1-\frac{A_1}{A_2}\right)^2$$

For $A_2 \to \infty$

$\therefore$

$$h_o = \frac{V^2}{2g}$$

The velocity head in the pipe $\left(\dfrac{V^2}{2g}\right)$ which corresponds to the kinetic energy

per unit weight is lost in turbulence of eddies in the reservoir. This loss is usually termed as the exit loss for the pipe.

Loss of fluid power: $P = mgh_o$ watt

$\qquad\qquad\qquad\qquad\qquad\qquad = \rho Q g h_o$ watt

where ρ = density of fluid, kg/m^3

$\qquad g = 9.81$ m/s^2

$\qquad Q$ = discharge, m^3/s

$\qquad m$ = mass flow rate, kg/s

$\qquad h_o$ = loss of head at the exit pipe, m

Loss of head due to an obstraction in pipe

Consider a pipe of cross-sectional area 'A' having an obstruction as shown in Fig. 4.10.

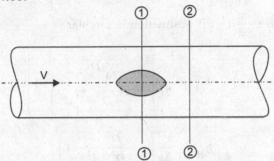

Fig. 4.10: An obstruction in a pipe.

Let a = maximum cross-sectional area of obstruction.

$\qquad\quad V$ = velocity of liquid in pipe.

Then $(A - a)$ = cross-section of liquid at section (1) – (1)

As the liquid flows and passes through section (1) – (1), a vena-contracta is formed beyond section (1) – (1), after which the stream of liquid widens again and velocity of flow at section (2) – (2) becomes uniform and equal to velocity V in the pipe. This situation is similar to the flow of liquid through sudden enlargement.

Let V_c = velocity of liquid at vena-contracta. Then, loss of head due to obstruction is equal to the head due to enlargement from vena-contracta to section (2) – (2),

$$h_{ob} = \frac{(V_c - V)^2}{2g} \qquad\qquad ... \,(4.2.12)$$

From continuity equation, we have

$$A_c V_c = AV \qquad\qquad\qquad ... \,(4.2.13)$$

where a_c = cross-sectional area at vena-contracta.

if C_c = coefficient of contraction

Then, $C_c = \dfrac{\text{cross-sectional area at vena-contracta}}{\text{cross-sectional area of flow at section (1) – (1)}} = \dfrac{a_c}{A - a}$

$\therefore$ $\qquad\qquad a_c = C_c(A - a)$

Substituting the value of a_c in Eq. (4.2.13), we get

$\qquad C_c(A - a) V_c = AV$

or $\qquad\qquad V_c = \dfrac{AV}{C_c(A-a)}$

Substituting the value of V_c in Eq. (4.2.12), we get

$$h_{ob} = \frac{\left(\dfrac{AV}{C_c(A-a)} - V\right)^2}{2g}$$

$$h_{ob} = \frac{V^2}{2g}\left(\frac{A}{C_c(A-a)} - 1\right)^2 \qquad\qquad \dots (4.2.14)$$

where $\quad A = \dfrac{\pi}{4}D^2$

$\qquad\quad a = \dfrac{\pi}{4}d^2 \quad$ if obstruction is circular

$\therefore \qquad\qquad h_{ob} = \dfrac{V^2}{2g}\left(\dfrac{\dfrac{\pi}{4}D^2}{C_c\left(\dfrac{\pi}{4}D^2 - \dfrac{\pi}{4}d^2\right)} - 1\right)^2$

$$h_{ob} = \frac{V^2}{2g}\left(\frac{D^2}{C_c(D^2 - d^2)} - 1\right)^2 \qquad\qquad \dots (4.2.15)$$

Loss of fluid power: $\qquad P = mgh_{ob}$ watt

$\qquad\qquad\qquad\qquad\qquad = \rho Q g h_{ob}$ watt

where

$\qquad\qquad\qquad \rho$ = density of fluid, kg/m^3

$\qquad\qquad\qquad Q$ = discharge, m^3/s

$\qquad\qquad\qquad g = 9.81$ m/s^2

$\qquad\qquad\qquad h_{ob}$ = loss of head due to obstruction, m

$\qquad\qquad\qquad m$ = mass flow rate, kg/s

Loss of head in pipe fitting

The loss of head in the various pipe fittings, such as valves, elbows and bends *etc.,* occurs because of their irregular interior surfaces which produce large scale turbulence.

The loss of head or energy in various pipe fittings is expressed as:

$$h_{\text{fitting}} = K_L \frac{V^2}{2g}$$

where V is the mean velocity in the pipe and K_L is the head loss coefficient which is dependent on Reynolds number and the magnitude of which depends upon the shape of the fitting (*i.e.*, angle of bend, radius of bend, opening of the valve etc.). The value of K_L is found by experiment. Table 4.3 gives the values of K_L for various common pipe fitting.

Loss of fluid power:

$$P = mgh_{\text{fitting}} \text{ watt}$$
$$= \rho Q g h_{\text{fitting}} \text{ watt}$$

where

ρ = density of fluid, kg/m^3

$g = 9.81 \text{ m/s}^2$

m = mass of flow rate, kg/s

Q = discharge, m^3/s

h_{fitting} = loss of head in pipe fitting, m

Table 4.3: Head Loss Coefficient for Pipe fitting

S. No.	Name of Pipe Fitting		Loss Coefficient: K_L
1.	Globe valve:	Fully open	10
		Half open	20
2.	Gate valve:	Fully open	0.2
		Three-fourth open	1.15
		Half open	5.6
		One-fourth open	24.0
3.	Pump's foot valve		1.5
4.	45° elbow		0.4
	90° elbow:	Short radius	0.9
		Medium radius	0.75
		Large radius	0.60
5.	90° bend		0.6
	180° bend		2.2
6.	Check valve:	Fully open	3

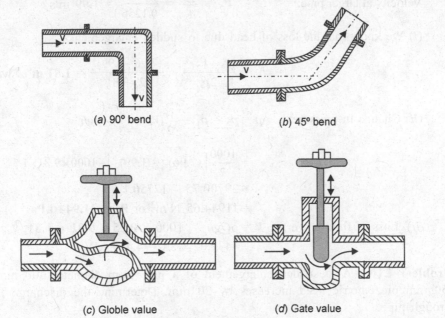

(a) 90° bend   (b) 45° bend

(c) Globle value   (d) Gate value

Fig. 4.11: Pipe fittings.

Problem 4.9: The rate of flow of water through a horizontal pipe is 250 litre/s. The pipe diameter 200 mm is suddenly enlarged to 400 mm. Find

(*i*) loss of head

(*ii*) change in pressure

(*iii*) loss of fluid power.

Solution: Given data:

Discharge: $\qquad Q = 250 \text{ litre/s} = \dfrac{250}{1000} = 0.25 \text{ m}^3/\text{s}$

Diameter of smaller pipe: $\qquad d_1 = 200 \text{ mm} = 0.2 \text{ m}$

∴ Cross-sectional area of smaller pipe:

$$A_1 = \frac{\pi}{4} d_1^2 = \frac{3.14}{4} \times (0.2)^2 = 0.0314 \text{ m}^2$$

Diameter of larger pipe: $\qquad d_2 = 400 \text{ mm} = 0.4 \text{ m}$

∴ Cross-sectional area of larger pipe:

$$A_2 = \frac{\pi}{4} d_2^2 = \frac{3.14}{4} \times (0.4)^2 = 0.1256 \text{ m}^2$$

We know that the discharge: $\qquad Q = A_1 V_1 = A_2 V_2$

Velocity at smaller pipe: $\qquad V_1 = \dfrac{Q}{A_1} = \dfrac{0.25}{0.0314} = 7.96 \text{ m/s}$

Velocity at larger pipe: $\qquad V_2 = \dfrac{Q}{A_2} = \dfrac{0.25}{0.1256} = 1.99 \text{ m/s}$

(*i*) We know that the loss of head due to sudden enlargement:

$$h_e = \frac{(V_1 - V_2)^2}{2g} = \frac{(7.96 - 1.99)^2}{2 \times 9.81} = \textbf{1.81 m of water}$$

(*ii*) Change in pressure: $\quad \Delta p = p_2 - p_1 = \dfrac{\rho}{2}(V_1^2 - V_2^2) - \rho g h_e$

$$= \frac{1000}{2}\left[(7.96)^2 - (1.99)^2\right] - 1000 \times 9.81 \times 1.81$$

$$= 29700.75 - 17756.1$$

$$= 11944.65 \text{ N/m}^2 \text{ or Pa} = \textbf{11.944 kPa}$$

(*iii*) Loss of fluid power: $\quad P = \rho Q g h_e = 1000 \times 0.25 \times 9.81 \times 1.81$

$$= 4439.02 \text{ watt} = \textbf{4.439 kW}$$

Problem 4.10: At a sudden enlargement of a pipe from diameter 300 mm to 450 mm piezometric head increases by 20 mm. Determine the discharge flow through pipe.

Solution: Given data:

Diameter of smaller pipe: $\quad\quad\quad d_1 = 300$ mm $= 0.3$ m

$\therefore$ Cross-sectional area: $\quad\quad A_1 = \frac{\pi}{4}d_1^2 = \frac{3.14}{4}\times(0.3)^2 = 0.0706$ m^2

Diameter of larger pipe: $\quad\quad\quad d_2 = 450$ mm $= 0.45$ m

$\therefore$ Cross-sectional area: $\quad\quad A_2 = \frac{\pi}{4}d_2^2 = \frac{3.14}{4}\times(0.45)^2 = 0.15896$ m^2

Increases of piezometric head $= 20$ mm $= 0.02$ m

i.e., $\quad\quad\quad \left(\frac{p_2}{\rho g}+z_2\right)-\left(\frac{p_1}{\rho g}+z_1\right) = 0.02$ m

Loss of head due to sudden enlargement:

$$h_e = \frac{(V_1-V_2)^2}{2g}$$

$$A_1V_1 = A_2V_2 \quad \text{by continuity equation}$$

or $\quad\quad\quad V_1 = \frac{A_2}{A_1}V_2 = \frac{0.15896}{0.0706}V_2 = 2.25\ V_2$

$\therefore \quad\quad\quad h_e = \frac{(2.25V_2-V_2)^2}{2g} = \frac{1.562V_2^2}{2g}$

Now applying modified Bernoulli's equation,

$$\frac{p_1}{\rho g}+\frac{V_1^2}{2g}+z_1 = \frac{p_2}{\rho g}+\frac{V_2^2}{2g}+z_2+h_e$$

$$\frac{V_1^2}{2g} = \left(\frac{p_2}{\rho g}+z_2\right)-\left(\frac{p_1}{\rho g}+z_1\right)+\frac{V_2^2}{2g}+h_e$$

$$\frac{(2.25V_2)^2}{2g} = 0.02+\frac{V_2^2}{2g}+\frac{1.562V_2^2}{2g}$$

$$\frac{5.062V_2^2}{2g} = 0.02+\frac{V_2^2}{2g}+\frac{1.562V_2^2}{2g}$$

or $\quad\quad \frac{5.062V_2^2}{2g}-\frac{V_2^2}{2g}-\frac{1.562V_2^2}{2g} = 0.02$

$$(5.062-1-1.562)\frac{V_2^2}{2g} = 0.02$$

$$2.5\frac{V_2^2}{2g} = 0.02$$

or $\quad\quad\quad V_2^2 = \frac{0.02\times2\times9.81}{2.5} = 0.15696$

or $\qquad\qquad\qquad\qquad\qquad\qquad V_2 = 0.396$ m/s

∴ Discharge: $\qquad\qquad\qquad Q = A_2 V_2 = 0.15896 \times 0.396 = 0.06295$ m³/s

$\qquad\qquad\qquad\qquad\qquad\qquad\qquad = $ **62.95 litre/s**

Problem 4.11: The rate of flow through a horizontal pipe is 0.3 m³/s. The diameter of the pipe is suddenly enlarged from 250 mm to 500 mm. The pressure in the smaller pipe is 13.734 N/cm². Determine:

(*i*) The loss of head due to sudden enlargement.

(*ii*) The pressure in the larger pipe.

Solution: Given data:

Discharge: $\qquad\qquad\qquad Q = 0.3$ m³/s

Diameter of smaller pipe: $d_1 = 250$ mm $= 0.25$ m

∴ Cross-sectional area: $A_1 = \dfrac{\pi}{4} d_1^2 = \dfrac{3.14}{4} \times (0.25)^2 = 0.049$ m²

Diameter of large pipe: $d_2 = 500$ mm $= 0.5$ m

∴ Cross-sectional area: $A_2 = \dfrac{\pi}{4} d_2^2 = \dfrac{3.14}{4} \times (0.5)^2 = 0.19625$ m²

Pressure intensity in smaller pipe:

$\qquad\qquad\qquad p_1 = 13.734$ N/cm² $= 13.734 \times 10^4$ N/m²

We know that $\qquad\qquad Q = A_1 V_1 = A_2 V_2$

∴ $\qquad\qquad\qquad V_1 = \dfrac{Q}{A_1} = \dfrac{0.3}{0.049} = 6.122$ m/s

and $\qquad\qquad\qquad V_2 = \dfrac{Q}{A_2} = \dfrac{0.3}{0.19625} = 1.528$ m/s

(*i*) Loss of head due to sudden enlargement:

$$h_e = \frac{(V_1 - V_2)^2}{2g} = \frac{(6.122 - 1.528)^2}{2 \times 9.81} = \textbf{1.075 m of oil}$$

(*ii*) Pressure in the large pipe: p_2

According to modified Bernoulli's equation

$$\frac{p_1}{\rho g} + \frac{V_1^2}{2g} + z_1 = \frac{p_2}{\rho g} + \frac{V_2^2}{2g} + z_2 + h_e$$

or $\qquad\qquad \dfrac{p_1}{\rho g} + \dfrac{V_1^2}{2g} = \dfrac{p_2}{\rho g} + \dfrac{V_2^2}{2g} + h_e \qquad\qquad |\because z_1 = z_2$ for horizontal pipe

or $\qquad\qquad \dfrac{V_1^2 - V_2^2}{2g} - h_e = \dfrac{p_2 - p_1}{\rho g}$

or $\qquad\qquad p_2 - p_1 = \dfrac{\rho}{2}\left(V_1^2 - V_2^2\right) - \rho g \, h_e$

$$p_2 - 13.734 \times 10^4 = \frac{1000}{2}((6.122)^2 - (1.528)^2 - 1000 \times 9.81 \times 1.075$$

$$p_2 - 13.734 \times 10^4 = 17572.05 - 10545.75$$

$$p_2 - 13.734 \times 10^4 = 7026.3$$

or
$$p_2 = 14.436 \times 10^4 \text{ N/m}^2 = \mathbf{14.436 \text{ N/cm}^2}$$

Problem 4.12: The rate of flow of water through a horizontal pipe is 250 litre/s. The diameter of the pipe which is 200 mm is suddenly enlarged to 400 mm. The pressure intensity in the smaller pipe is 150 kPa. Find

(i) loss of head

(ii) pressure intensity in the larger pipe

(iii) loss of power.

Solution: Given data:

Discharge: $\qquad$ $Q = 250 \text{ litre/s} = \dfrac{250}{1000} \text{ m}^3/\text{s} = 0.25 \text{ m}^3/\text{s}$

Diameter of smaller pipe: $\qquad$ $d_1 = 200 \text{ mm} = 0.2 \text{ m}$

∴ Cross-sectional area: $\qquad$ $A_1 = \dfrac{\pi}{4} d_1^2 = \dfrac{3.14}{4} \times (0.2)^2 = 0.0314 \text{ m}^2$

Diameter of large pipe: $\qquad$ $d_2 = 400 \text{ mm} = 0.4 \text{ m}$

∴ Cross-sectional area: $\qquad$ $A_2 = \dfrac{\pi}{4} d_2^2 = \dfrac{3.14}{4} \times (0.4)^2 = 0.1256 \text{ m}^2$

Pressure intensity in smaller pipe:

$$p_1 = 150 \text{ kPa} = 150 \times 10^3 \text{ N/m}^2$$

We know that $\qquad$ $Q = A_1 V_1 = A_2 V_2$

∴ $\qquad$ $V_1 = \dfrac{Q}{A_1} = \dfrac{0.25}{0.314} = 7.96 \text{ m/s}$

and $\qquad$ $V_2 = \dfrac{Q}{A_2} = \dfrac{0.25}{0.1256} = 1.99 \text{ m/s}$

(i) Loss of head due to sudden enlargement:

$$h_e = \frac{(V_1 - V_2)^2}{2g} = \frac{(7.96 - 1.99)^2}{2 \times 9.81}$$

$$= \mathbf{1.816 \text{ m of water}}$$

(ii) Pressure intensity in the large pipe: p_2

According to modified Bernoulli's equation,

$$\frac{p_1}{\rho g} + \frac{V_1^2}{2g} + z_1 = \frac{p_2}{\rho g} + \frac{V_2^2}{2g} + z_2 + h_e$$

$$\frac{p_1}{\rho g} + \frac{V_1^2}{2g} = \frac{p_2}{\rho g} + \frac{V_2^2}{2g} + h_e \qquad \because z_1 = z_2 \text{ for horizontal pipe}$$

$$\frac{V_1^2 - V_2^2}{2g} - h_e = \frac{p_2 - p_1}{\rho g}$$

or $$p_2 - p_1 = \frac{\rho}{2}(V_1^2 - V_2^2) - \rho g h_e$$

or $$p_2 - 150 \times 10^3 = \frac{1000}{2}\left[(7.96)^2 - (1.99)^2\right] - 1000 \times 9.81 \times 1.816$$

$$p_2 - 150 \times 10^3 = 29.7 \times 10^3 - 17.815$$

or $$p_2 = 161.885 \times 10^3 \text{ N/m}^2 = \textbf{161.885 kPa}$$

(iii) Loss of power: $$P = \rho Q g h_e = 1000 \times 0.25 \times 9.81 \times 1.816 \text{ watt}$$
$$= 4453.74 \text{ W} = \textbf{4.453 kW}$$

Problem 4.13: A horizontal pipe carries water at the rate of 0.04 m³/s. Its diameter, which is 300 mm reduces abruptly to 150 mm. Calculate the pressure loss across the contraction. Take the coefficient of contraction $C_c = 0.62$.

(GGSIP Univeristy, Delhi. Dec. 2005)

Solution: Given data:

Discharge: $Q = 0.04 \text{ m}^3/\text{s}$
Diameter of pipe at section (1) – (1): $d_1 = 300 \text{ mm} = 0.3 \text{ m}$
∴ Cross-sectional area at section (1) – (1):

$$A_1 = \frac{\pi}{4}d_1^2 = \frac{3.14}{4} \times (0.3)^2 = 0.07065 \text{ m}^2$$

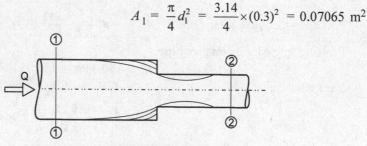

Fig. 4.12: Schematic for Problem 4.13

Diameter of pipe at section (2) – (2):

$$d_2 = 150 \text{ mm} = 0.15 \text{ m}$$

∴ Cross-sectional area at section (2) – (2):

$$A_2 = \frac{\pi}{4}d_2^2 = \frac{3.14}{4} \times (0.15)^2 = 0.01766 \text{ m}^2$$

Coefficient of contraction: $C_c = 0.62$
According to continuity equation:

$$Q = A_1 V_1$$

or Velocity at section (1) – (1): $$V_1 = \frac{Q}{A_1} = \frac{0.04}{0.07065} = 0.5661 \text{ m/s}$$

and similarly the velocity at section (2) – (2):

$$V_2 = \frac{Q}{A_2} = \frac{0.04}{0.01766} = 2.26 \text{ m/s}$$

We know that the head loss due to sudden contraction: $h_c = K_c \dfrac{V_2^2}{2g}$

where K_c = coefficient of sudden contraction.

$$= \left[\frac{1}{C_c} - 1\right]^2$$

$$\therefore \qquad h_c = \left[\frac{1}{C_c} - 1\right]^2 \cdot \frac{V_2^2}{2g} = \left[\frac{1}{0.62} - 1\right]^2 \times \frac{(2.26)^2}{2 \times 9.81}$$

$$= [1.612 - 1]^2 \times 0.2603 = 0.0974 \text{ m of water.}$$

Now applying Bernoulli's equation at section (1) – (1) and at section (2) – (2), we get

$$\frac{p_1}{\rho g} + \frac{V_1^2}{2g} + z_1 = \frac{p_2}{\rho g} + \frac{V_2^2}{2g} + z_2 + h_c$$

$$\frac{p_1}{\rho g} - \frac{p_2}{\rho g} = \frac{V_2^2}{2g} - \frac{V_1^2}{2g} + h_c \qquad\qquad \because z_1 = z_2$$

$$= \frac{(2.26)^2}{2 \times 9.81} - \frac{(0.5661)^2}{2 \times 9.81} + 0.0974$$

$$= 0.34139 \text{ m of water}$$

∴ Pressure loss across the contraction: $p_1 - p_2 = 0.34139 \times \rho g$

$$= 0.34139 \times 1000 \times 9.81 = 3349.03 \text{ N/m}^2$$

$$= \textbf{3.349 kPa}$$

Problem 4.14: A horizontal pipe of diameter 500 mm is suddenly contracted to a diameter of 250 mm. The rate of flow of water is 300 litre/s. Find the pressure loss across the contraction. Take coefficient of contraction as 0.6.

Solution: Given data:

Diameter of large pipe: $d_1 = 500$ mm $= 0.5$m

∴ Cross-sectional area: $A_1 = \dfrac{\pi}{4}d_1^2 = \dfrac{3.14}{4} \times (0.5)^2 = 0.19625 \text{ m}^2$

Diameter of smaller pipe: $d_2 = 250$ mm $= 0.25$ m

∴ Cross-sectional area: $A_2 = \dfrac{\pi}{4}d_2^2 = \dfrac{3.14}{4} \times (0.25)^2 = 0.049 \text{ m}^2$

Rate of flow: $Q = 300$ litre/s $= \dfrac{300}{1000}$ m^3/s $= 0.3$ m^3/s

Coefficient of contraction: $C_c = 0.6$

We know that the rate of flow:

$$Q = A_1V_1 = A_2V_2$$

$$\therefore \qquad V_1 = \frac{Q}{A_1} = \frac{0.3}{0.19625} \text{ m/s} = 1.528 \text{ m/s}$$

and

$$V_2 = \frac{Q}{A_2} = \frac{0.3}{0.049} \text{ m/s} = 6.122 \text{ m/s}$$

We know that the loss of head due to contraction:

$$h_c = \frac{V_2^2}{2g}\left(\frac{1}{C_c} - 1\right)^2 = \frac{(6.122)^2}{2 \times 9.81}\left(\frac{1}{0.6} - 1\right)^2$$

$$= 1.91 \times (0.6666)^2 = 0.8487 \text{ m}$$

Now applying modified Bernoulli's equation before and after contraction, we get

$$\frac{p_1}{\rho g} + \frac{V_1^2}{2g} + z_1 = \frac{p_2}{\rho g} + \frac{V_2^2}{2g} + z_2 + h_c$$

$$\frac{p_1}{\rho g} - \frac{p_2}{\rho g} = \frac{V_2^2}{2g} - \frac{V_1^2}{2g} + h_c \quad \because z_1 = z_2 \text{ for horizontal pipe}$$

$$\frac{p_1 - p_2}{\rho g} = \frac{V_2^2}{2g} - \frac{V_1^2}{2g} + h_c$$

$$p_1 - p_2 = \frac{\rho}{2}(V_2^2 - V_1^2) + \rho g h_c$$

$$p_1 - p_2 = \frac{1000}{2}[(6.122)^2 - (1.528)^2] + 1000$$
$$\times 9.81 \times 0.8487$$

$$p_1 - p_2 = 17572.05 + 8325.74 = 25897.79 \text{ N/m}^2 \text{ or Pa}$$
$$= \textbf{25.897 kPa}$$

Problem 4.15: In a city water supply system, water is following through a pipe line 30 cm in diameter. The pipe diameter is suddenly reduced to 20 cm. Estimate the discharge through the pipe if the difference in pressure across the sudden is 5 kPa.

Solution: Given data:

Diameter of pipe at section (1)–(1): $d_1 = 30$ cm $= 0.3$ m

$\therefore$ Cross-sectional area : $\qquad A_1 = \frac{\pi}{4}d_1^2 = \frac{3.14}{4} \times (0.3)^2 = 0.07065 \text{ m}^2$

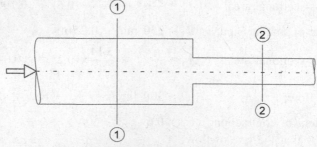

Fig. 4.13: Schematic for Problem 4.15

Diameter of pipe at section (2)–(2): $d_2 = 20$ cm $= 0.20$ m

$\therefore$ Cross-sectional area: $A_2 = \dfrac{\pi}{4} d_2^2 = \dfrac{3.14}{4} \times (0.20)^2 = 0.0314$ m^2

Pressure difference across a sudden contraction,
$$p_1 - p_2 = 5 \text{ kPa} = 5 \times 10^3 \text{ Pa}$$

According to continuity equation,
$$Q = A_1 V_1 = A_2 V_2$$

or
$$A_1 V_1 = A_2 V_2$$
$$0.07065 \times V_1 = 0.0314 \times V_2$$

or
$$V_1 = 0.444 V_2$$

We know that the head loss to sudden contraction: $h_c = K_c \dfrac{V_2^2}{2g}$

where
$$K_c = \text{coefficient sudden contraction.}$$

$$= \left(\frac{1}{C_c} - 1 \right)^2$$

Assume coefficient of contraction: $C_c = 0.62$

$\therefore$
$$K_c = \left(\frac{1}{0.62} - 1 \right)^2 = 0.375$$

Therefore,
$$h_c = \frac{0.375 V_2^2}{2g}$$

$$\boxed{h_c = \frac{0.1875 V_2^2}{g}}$$

Now applying Bernoulli's equation at sections (1)–(1) and (2)–(2), we get

$$\frac{p_1}{\rho g} + \frac{V_1^2}{2g} + z_1 = \frac{p_2}{\rho g} + \frac{V_2^2}{2g} + z_2 + h_c$$

$$\frac{p_1}{\rho g} - \frac{p_2}{\rho g} + \frac{V_1^2}{2g} = \frac{V_2^2}{2g} + h_c \qquad\qquad \therefore z_1 = z_2$$

$$\frac{p_1 - p_2}{\rho g} + \frac{(0.444 V_2)^2}{2g} = \frac{V_2^2}{2g} + \frac{0.1875 V_2^2}{g}$$

or
$$\frac{p_1 - p_2}{\rho} + \frac{(0.444 V_2)^2}{2} = \frac{V_2^2}{2} + 0.1875\ V_2^2$$

$$\frac{5\times10^3}{1000} + 0.0985 \; V_2^2 = 0.5\,V_2^2 + 0.1875V_2^2$$

$$5 + 0.0985 \; V_2^2 = 0.6875 \; V_2^2$$

or $$V_2^2 = 8.4889$$

or $$V_2 = 2.91 \; \text{m/s}$$

∴ $$Q = A_2 V_2 = 0.0314 \times 2.91 = 0.09137 \; \text{m}^3/\text{s}$$

$$= \textbf{91.37 liter/s}$$

Problem 4.16: Oil of specific gravity 0.85 is flowing through a horizontal pipe of diameter 250 mm at a velocity of 2 m/s. A circular solid plate of diameter 150 mm is placed in the pipe to obstruct the flow. Find the loss of head due to obstruction in the pipe. Take coefficient of contraction as 0.7. Find also the loss of fluid power.

Solution: Given data:

Specific gravity of oil: $S = 0.85$

∴ Density of oil: $\rho = S \times \rho_{water} = 0.85 \times 1000 \; \text{kg/m}^3 = 850 \; \text{kg/m}^3$

Diameter of pipe: $D = 250 \; \text{mm} = 0.25 \; \text{m}$

Velocity of oil through pipe: $V = 2 \; \text{m/s}$

Diameter of obstruction: $d = 150 \; \text{mm} = 0.15 \; \text{m}$

Coefficient of contraction: $C_c = 0.7$

Loss of head due to obstruction in pipe is given by Eq. (11.5.10):

$$h_{ob} = \frac{V^2}{2g}\left(\frac{D^2}{C_c(D^2-d^2)}-1\right)^2$$

$$= \frac{2^2}{2\times9.81}\times\left(\frac{0.25^2}{0.7(0.25^2-0.15^2)}-1\right)^2$$

$$= \textbf{0.3095 m of oil.}$$

Loss of fluid power: $P = \rho Q g h_{ob}$

where $Q = AV$

$$= \frac{\pi}{4}D^2\times V = \frac{3.14}{4}\times(0.25)^2\times2 = 0.09812 \; \text{m}^3/\text{s}$$

∴ $$P = 850 \times 0.09812 \times 9.81 \times 0.3095 \; \text{W} = \textbf{253.22 W}$$

Problem 4.17: Find the rate of flow of water through pipe of diameter 300 mm and length 100 m when one end of the pipe is connected to a tank and other end of the pipe is open to the atmosphere. The pipe is horizontal and the height of water in the tank is 6 m above the centre of the pipe. Consider all minor losses and take coefficient of friction as 0.004.

Solution: Given data:

Diameter of pipe: $\qquad\qquad\qquad\qquad\quad$ $d = 300$ mm $= 0.3$ m
Length of pipe: $\qquad\qquad\qquad\qquad\qquad$ $l = 100$ m
Head of water available at inlet of pipe: $H = 6$ m
Coefficient of friction: $\qquad\qquad\qquad$ $f = 0.004$

Now net available head at inlet of pipe = minor losses + major loss

$$H = (h_i + h_o) + h_f$$

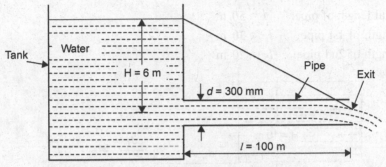

Fig. 4.14: Schematic for Problem 4.17

where $h_i = 0.5 \dfrac{V^2}{2g}$, loss of head at entrance of pipe.

$\qquad$ $h_o = \dfrac{V^2}{2g}$, loss of head at exit of pipe.

$\qquad$ $h_f = \dfrac{4flV^2}{2gd}$, loss of head due to friction.

$\therefore \qquad\qquad\qquad\qquad H = 0.5 \dfrac{V^2}{2g} + \dfrac{V^2}{2g} + \dfrac{4flV^2}{2gd}$

$$6 = \left(0.5 + 1 + \frac{4fl}{d}\right)\frac{V^2}{2g}$$

$$6 = \left(0.5 + 1 + \frac{4 \times 0.004 \times 100}{0.3}\right)\frac{V^2}{2g}$$

$$6 = (0.5 + 1 + 5.33)\frac{V^2}{2g}$$

$$6 = 6.83 \frac{V^2}{2g}$$

or $\qquad\qquad\qquad V^2 = \dfrac{6}{6.83} \times 2 \times 9.81 = 17.23$ $\quad$ or $\quad V = 4.15$ m/s

We know that the rate of flow:

$$Q = AV = \frac{\pi}{4}d^2 V$$

$$= \frac{3.14}{4} \times (0.3)^2 \times 4.15 \text{ m}^3/\text{s} = 0.29319 \text{ m}^3/\text{s}$$

$$= \mathbf{293.19} \text{ litre/s}$$

Problem 4.18: Find the rate of flow through a horizontal pipe line 50 m long which is connected to a water tank at one end and discharged freely into the atmosphere at the other end. For the first 30 m of its length from the tank, the pipe is 200 mm diameter and its diameter is suddenly enlarged to 300 mm. The height of water level in the tank is 10 m above the centre line of the pipe. Considering all minor losses. Take $f = 0.02$ for both sections of the pipe.

Solution: Given data:

Total length of pipe: $l = 50$ m
Length of 1st pipe: $l_1 = 30$ m
Length of 2nd pipe: $l_2 = 20$ m

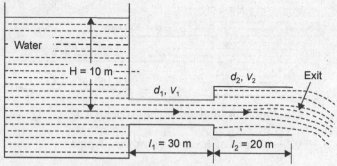

Fig. 4.15: Schematic for Problem 4.18

Diameter of 1st pipe: $d_1 = 200$ mm $= 0.2$ m
Diameter of 2nd pipe: $d_2 = 300$ mm $= 0.3$ m
Head of water available at inlet of pipe: $H = 10$ m
Coefficient of friction: $f = 0.02$
Now net available head at inlet of pipe = minor losses + major losses

$$H = h_i + h_{f_1} + h_e + h_{f_2} + h_o$$

where $h_i = 0.5\dfrac{V_1^2}{2g}$, loss of head at entrance of pipe (*i.e.*, minor loss)

$h_{f_1} = \dfrac{4f l_1 V_1^2}{2gd_1}$, loss of head due to friction in 1st pipe (*i.e.*, major loss)

$h_e = \dfrac{(V_1 - V_2)^2}{2g}$, loss of head due to sudden enlargement (*i.e.*, minor loss)

$h_{f_2} = \dfrac{4f l_2 V_2^2}{2gd_2}$, loss of head due to friction in 2nd pipe (*i.e.*, major loss)

$h_o = \dfrac{V_2^2}{2g}$, loss of head at exit of pipe (*i.e.*, minor loss)

$\therefore$ $H = 0.5\dfrac{V_1^2}{2g} + \dfrac{4f l_1 V_1^2}{2gd_1} + \dfrac{(V_1 - V_2)^2}{2g} + \dfrac{4f l_2 V_2^2}{2gd_2} + \dfrac{V_2^2}{2g}$ ...(i)

According to continuity equation $Q = A_1V_1 = A_2V_2$

or
$$A_1V_1 = A_2V_2$$

$$\frac{\pi}{4}d_1^2V_1 = \frac{\pi}{4}d_2^2V_2$$

$$d_1^2V_1 = d_2^2V_2$$

or
$$V_1 = \left(\frac{d_2}{d_1}\right)^2 V_2 = \left(\frac{0.3}{0.2}\right)^2 V_2 = 2.25V_2$$

Substituting the value of $V_1 = 2.25V_2$ in above Eq. (i), we get

$$H = 0.5\frac{(2.25)^2V_2^2}{2g} + \frac{4fl_1}{2gd_1}(2.25)^2V_2^2 + \frac{(2.25V_2 - V_2)^2}{2g} + \frac{4fl_2V_2^2}{2gd_2} + \frac{V_2^2}{2g}$$

$$10 = 2.53\frac{V_2^2}{2g} + \frac{4\times0.02\times30\times(2.25)^2}{0.2}\frac{V_2^2}{2g} + 1.56\frac{V_2^2}{2g} + \frac{4\times0.02\times20}{0.3}\frac{V_2^2}{2g} + \frac{V_2^2}{2g}$$

$$10 = [2.53 + 60.75 + 1.56 + 5.33 + 1]\frac{V_2^2}{2g}$$

$$10 = 71.17\frac{V_2^2}{2g}$$

or $V_2^2 = \dfrac{10}{71.17}\times2g = \dfrac{10\times2\times9.81}{71.17} = 2.756$

or $V_2 = 1.66$ m/s

$\therefore$ Rate of flow: $Q = A_2V_2 = \dfrac{\pi}{4}d_2^2V_2 = \dfrac{3.14}{4}\times(0.3)^2\times1.66$ m³/s

$$= 0.11728 \text{ m}^3/\text{s} = \textbf{117.28 litre/s}$$

Problem 4.19: Water at 10°C flows from a large reservoir to a smaller one through a 50 mm diameter cast iron piping system, as shown in Fig. 4.16. Determine the elevation z_1 for a flow rate of 6 litre/s.

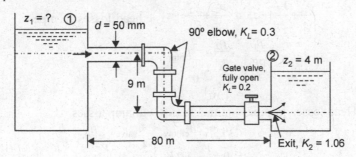

Fig. 4.16: Schematic for Problem 4.19

Properties of water at 10°C,

$$\rho = 999.7 \text{ kg/m}^3, \ \mu = 1.307 \times 10^{-3} \text{ Ns/m}^2$$

Solution: Given data:

Diameter of pipe: $\quad d = 50$ mm $= 0.05$ m

∴ Cross-sectional area of pipe:

$$A = \frac{\pi}{4}d^2 = \frac{3.14}{4} \times (0.05)^2 = 1.962 \times 10^{-3} \text{ m}^2$$

Discharge: $\quad\quad Q = 6$ litre/s $= 0.006$ m³/s

also $\quad\quad\quad\quad Q = AV$

∴ Average velocity in pipe:

$$V = \frac{Q}{A} = \frac{0.006}{1.962 \times 10^{-3}} = 3.058 \text{ m/s}$$

Total length of pipe: $l = 80 + 9 = 89$ m

The roughness of cast iron pipe:

$$\epsilon = 0.26 \text{ mm from Table 4.1.}$$
$$= 0.00026 \text{ m}$$

Relative roughness: $\quad \dfrac{\epsilon}{d} = \dfrac{0.00026}{0.05} = 0.0052$

Reynold's number: $\quad Re = \dfrac{\rho V d}{\mu} = \dfrac{999.7 \times 3.058 \times 0.05}{1.307 \times 10^{-3}} = 116950.36$

The flow is turbulent since $Re > 4000$.

The friction factor f' can be determined from the Moody chart on the basis of

$Re = 116950.36$ and relative roughness: $\dfrac{\epsilon}{d} = 0.0052$

i.e., Moody's chart gives:

$$f' = 0.0315$$

also $\quad\quad\quad\quad f' = 4f$

or   Coefficient of friction: $f = \dfrac{f'}{4} = \dfrac{0.0315}{4} = 0.00787$

Applying the energy equation points (1) and (2) on the free surfaces of the two reservoirs.

$$\frac{p_1}{\rho g} + \frac{V_1^2}{2g} + z_1 = \frac{p_2}{\rho g} + \frac{V_2^2}{2g} + z_2 + \text{head losses in pipe}$$

where

$$\frac{p_1}{\rho g} = \frac{p_2}{\rho g} = p_{atm}$$
$$V_1 = V_2 = 0$$
$$z_2 = 4 \text{ m}$$

Head losses in pipe $=$ minor losses $+$ major losses

$$= h_i + 2h_{elbow} + h_{valve} + h_o + h_f$$

$$= 0.5\frac{V^2}{2g} + 2K_L\frac{V^2}{2g} + K_L\frac{V^2}{2g} + K_L\frac{V^2}{2g} + \frac{4flV^2}{2gd}$$

$$= \left[0.5 + 2 \times 0.3 + 0.2 + 1.06 + \frac{4fl}{d}\right]\frac{V^2}{2g}$$

$$= \left[0.5 + 0.6 + 0.2 + 1.06 + \frac{4 \times 0.00787 \times 89}{0.05}\right] \times \frac{(3.058)^2}{2 \times 9.81}$$

$$= [58.394] \times 0.4766 = 27.83 \text{ m}$$

$$\therefore \qquad p_{atm} + 0 + z_1 = p_{atm} + 0 + 4 + 27.83$$

or $\qquad\qquad z_1 = \mathbf{31.83 \text{ m}}$

4.3 SIPHON

A long bent pipe which rises above its hydraulic grade line and has negative pressure, is called **siphon**. In the other words, siphon is a long bent pipe which carries liquid from a reservoir at a higher level to another reservoir at a lower level when the two reservoirs are separated by a hill or high obstruction as shown in Fig. 4.17. The portion of the pipe which lies above the hydraulic gradient line, marked by M, C, N, has negative pressures, is called a siphon. The highest point of the siphon has the largest negative pressure, is called the summit, marked by C. The pressure at C can be reduced theoretically to -10.3 m of water but in actual practice this pressure is only -7.6 m of water (or $10.3 - 7.6 = 2.7$ m of water absolute). The absolute pressure head at the point C should not be less than 2.7 m of water. If it is less than this, the flow would stop due to separation, water will start vapourising and the dissolved gases will be given off from the water. Thus, **the maximum height of the hill should not be more than 7.6 m above the hydraulic gradient line.**

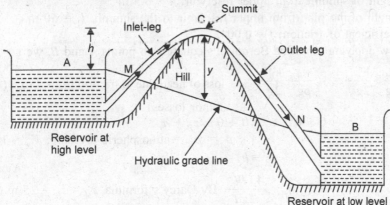

Fig. 4.17: Flow through a siphon.

The siphon can be made to work by exhausting air, thus creating vacuum in it, or by filling it completely with water.

Problem 4.20: A siphon of diameter 150 mm connects two reservoirs having a difference in elevation of 25 m. The length of the siphon is 400 m and the summit is 3.5 m above the water level in the upper reservoir. The length of the pipe from upper reservoir to the summit is 50 m. Determine:

(*i*) Discharge through the siphon and

(*ii*) Pressure at the summit.

Neglect minor losses. The coefficient of friction, $f = 0.004$.

Solution: Given data:

Diameter of siphon: $d = 150$ mm $= 0.15$ m

Length of siphon: $l = 400$ m

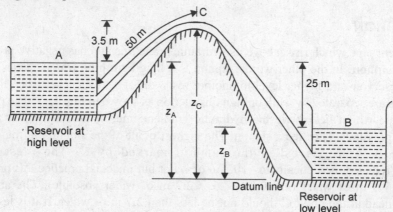

Fig. 4.18: Schematic for Problem 4.20

Difference between level of two reservoirs: $z_A - z_B = 25$ m.

Height of summit from upper reservoir: $h = 3.5$ m

Length of the pipe from upper reservoir to the summit: $l_1 = 50$ m

Coefficient of friction: $f = 0.004$.

Now applying modified Bernoulli's equation to points A and B, we get

$$\frac{p_A}{\rho g} + \frac{V_A^2}{2g} + z_A = \frac{p_B}{\rho g} + \frac{V_B^2}{2g} + z_B + \text{loss of head due to friction from } A \text{ to } B \text{ (neglect}$$

$$\text{minor losses)}$$

or $\qquad 0 + 0 + z_A = 0 + 0 + z_B + h_f$

$$[\because p_A = p_B = \text{atmospheric pressure}, V_A = V_B = 0]$$

$\therefore \qquad\qquad z_A - z_B = h_f$

$$\because h_f = \frac{4flV^2}{2gd} \quad \text{By Darcy's formula, } z_A - z_B = 25 \text{ m (given)}$$

$$25 = \frac{4flV^2}{2gd}$$

$$25 = \frac{4 \times 0.004 \times 400 \times V^2}{2 \times 9.81 \times 0.15}$$

or $\qquad 25 = 2.174V^2$

or $\qquad V^2 = 11.494$

or $\qquad V = 3.39$ m/s

(*i*) Discharge through the siphon: Q

Now discharge flow through the siphon:

$$Q = \text{velocity} \times \text{cross-sectional area of the siphon}$$

$$= VA = 3.39 \times \frac{\pi}{4} d^2 = 3.39 \frac{3.14}{4} \times (0.15)^2$$

$$= 0.05987 \text{ m}^3/\text{s} = \textbf{59.87 litre/s}$$

(*ii*) Pressure at the summit: p_c

Now applying modified Bernoulli's equation to points A and C, we get

$$\frac{p_A}{\rho g}+\frac{V_A^2}{2g}+z_A = \frac{p_C}{\rho g}+\frac{V_C^2}{2g}+z_C + \text{loss of head due to friction}$$

$$\text{between } A \text{ and } C$$

$$0 + 0 + z_A = \frac{p_C}{\rho g}+\frac{V_c^2}{2g}+z_C+h_{f_1}$$

or

$$\frac{p_C}{\rho g} = z_A-z_C-\frac{V_c^2}{2g}-h_{f_1} = -(z_C-z_A)-\frac{V_c^2}{2g}-h_{f_1}$$

$$= -3.5-\frac{(3.39)^2}{2\times9.81}-\frac{4fl_1V^2}{2gd} \qquad \because V_c = V$$

$$= -3.5 - 0.585 - \frac{4\times0.004\times50\times(3.39)^2}{2\times9.81\times0.15}$$

$$= -3.5 - 0.586 - 3.123$$

$$\frac{p_C}{\rho g} = 7.208 \text{ m of water}$$

$$\frac{p_C}{\rho g} = 10.3 - 7.208 \text{ m of water absolute}$$

$$\frac{p_C}{\rho g} = 3.092 \text{ m of water absolute}$$

$$p_C = 3.092 \times \rho g = 3.092 \times 1000 \times 9.81 \text{ N/m}^2$$
$$= 30332.52 \text{ N/m}^2 = \mathbf{30.33 \text{ kN/m}^2 \text{ absolute.}}$$

Problem 4.21: Two reservoirs are connected by a pipeline which rises above the level of the higher reservoir. What will be the highest point of the siphon above the water surface level in the higher tank if the length of the pipe leading upto this point is 450 m, the total length of the pipeline is 1000 m and the diameter of the pipe is 300 mm? The difference in the levels of the two reservoirs is 12.5 m. Take coefficient of friction as 0.01. The siphon must run full. Assume that the separation will occur if the absolute pressure in the pipe falls below 2.44 m of water.

Solution: Given data:

Length of pipe from upper reservoir to the summit:

$$l_1 = 450 \text{ m}$$

Length of siphon:

$$l = 1000 \text{ m}$$

Diameter of siphon:

$$d = 300 \text{ m} = 0.3 \text{ m}$$

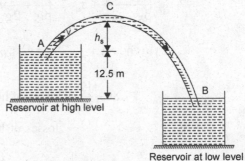

Fig. 4.19: Schematic for Problem 4.21

Difference between level of two reservoirs: $z_A - z_B = 12.5$ m

Coefficient of friction: $f = 0.01$

Absolute pressure head at summit $= 2.44$ m of water

∴ Vacuum pressure head at summit:

$$\frac{p_C}{\rho g} = 10.3 - 2.44 = 7.86 \text{ m of water}$$

i.e., Pressure head at summit: $\dfrac{p_C}{\rho g} = -7.86$ m of water

Now applying modified Bernoulli's equation at points A and B, we get

$$\frac{p_A}{\rho g} + \frac{V_A^2}{2g} + z_A = \frac{p_B}{\rho g} + \frac{V_B^2}{2g} + z_B + \text{losses of head}$$

$$0 + 0 + z_A = 0 + 0 + z_B + \text{losses of head}$$

or
$$z_A - z_B = \text{losses of head}$$

where
$$\text{losses of head} = h_i + h_f + h_o$$

$$= 0.5\frac{V^2}{2g} + \frac{4flV^2}{2gd} + \frac{V^2}{2g}$$

∴
$$z_A - z_B = \frac{0.5V^2}{2g} + \frac{4flV^2}{2gd} + \frac{V^2}{2g}$$

$$z_A - z_B = \left[0.5 + \frac{4fl}{d} + 1\right]\frac{V^2}{2g}$$

$$12.5 = \left[0.5 + \frac{4 \times 0.01 \times 1000}{0.3} + 1\right]\frac{V^2}{2g}$$

$$12.5 = 134.83 \times \frac{V^2}{2 \times 9.81}$$

or
$$V^2 = 1.8189$$

or
$$V = 1.348 \text{ m/s}$$

Now applying modified Bernoulli's equation at points A and C, we get

$$\frac{p_A}{\rho g} + \frac{V_A^2}{2g} + z_A = \frac{p_C}{\rho g} + \frac{V_C^2}{2g} + z_C + \text{losses of head}$$

$$0 + 0 + z_A = -7.86 + \frac{V_C^2}{2g} + z_C + \text{losses of head}$$

or
$$7.86 - \frac{V_C^2}{2g} - \text{losses of head} = z_C - z_A$$

where
$$\text{losses of head} = h_i + h_{f1} + h_o$$

$$= 0.5\frac{V^2}{2g} + \frac{4fl_1V^2}{2gd} + \frac{V^2}{2g}$$

and
$$V_c = V$$
$$z_C - z_A = h_s$$

$\therefore$ $$7.86 - \frac{V^2}{2g} - \left(0.5\frac{V^2}{2g} + \frac{4fl_1V^2}{2gd} + \frac{V^2}{2g}\right) = h_s$$

$$7.86 - \frac{V^2}{2g} - 0.5\frac{V^2}{2g} - \frac{4fl_1V^2}{2gd} - \frac{V^2}{2g} = h_s$$

$$7.86 - \left(1 + 0.5 + \frac{4fl_1}{d} + 1\right)\frac{V^2}{2g} = h_s$$

$$7.86 - \left(1 + 0.5 + \frac{4 \times 0.01 \times 450}{0.3} + 1\right)\frac{(1.348)^2}{2 \times 9.81} = h_s$$

$$7.86 - (1 + 0.5 + 60 + 1) \times 0.0926 = h_s$$
$$7.86 - 5.787 = h_s$$

or $$h_s = \textbf{2.073 m}$$

4.4 PIPES IN SERIES: COMPOUND PIPES

In piping systems, two or more pipes of different diameters are connected end to end, is called pipes in series or compound pipes. Let three pipes in series as shown in Fig. 4.20.

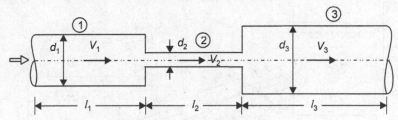

Fig. 4.20: Pipe in series.

l_1, l_2, l_3 = length of pipes 1, 2 and 3 respectively
d_1, d_2, d_3 = diameter of pipes 1, 2 and 3 respectively
V_1, V_2, V_3 = velocity of flow through pipes 1, 2 and 3 respectively
f_1, f_2, f_3 = coefficient of frictions for pipes 1, 2 and 3 respectively.

The rate of flow through the entire system remains constant regardless of the diameters of the individual pipes in the system. This is a natural consequence of the conservation of mass principle for steady incompressible flow.

Mathematically,
$$Q_1 = Q_2 = Q_3$$
$$A_1V_1 = A_2V_2 = A_3V_3$$
$$\frac{\pi}{4}d_1^2V_1 = \frac{\pi}{4}d_2^2V_2 = \frac{\pi}{4}d_3^2V_3$$

or $$d_1^2V_1 = d_2^2V_2 = d_3^2V_3$$

The total head loss in this case is equal to the sum of the head losses in individual pipes in the system, including the minor losses.

Mathematically,

$$\text{Total head loss} = h_{f_1} + h_c + h_{f_2} + h_e + h_{f_3}$$

where

$$h_{f_1} = \frac{4f_1 l_1 V_1^2}{2gd_1}, \text{ loss of head due to friction in pipe-1.}$$

$$h_c = 0.5\frac{V_2^2}{2g}, \text{loss of head due to sudden contraction from pipe-1 to pipe-2.}$$

$$h_{f_2} = \frac{4f_2 l_2 V_2^2}{2gd_2}, \text{ loss of head due to friction in pipe-2.}$$

$$h_e = \frac{(V_2 - V_3)^2}{2g}, \text{loss of head due to sudden enlargement from pipe-2 to pipe-3.}$$

$$h_{f_3} = \frac{4f_3 l_3 V_3^2}{2gd_3}, \text{ loss of head due to friction in pipe-3.}$$

$\therefore$ **Total head loss** $= \dfrac{4f_1 l_1 V_1^2}{2gd_1} + 0.5\dfrac{V_2^2}{2g} + \dfrac{4f_2 l_2 V_2^2}{2gd_2} + \dfrac{(V_2 - V_3)^2}{2g} + \dfrac{4f_3 l_3 V_3^2}{2gd_3}$

Two reservoirs are connected through pipes in series:

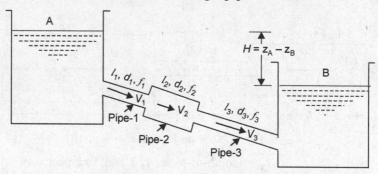

Fig. 4.21: Two reservoirs are connected through pipes in series

$$Q_1 = Q_2 = Q_3$$
$$A_1 V_1 = A_2 V_2 = A_3 V_3$$
$$\frac{\pi}{4}d_1^2 V_1 = \frac{\pi}{4}d_2^2 V_2 = \frac{\pi}{4}d_3^2 V_3$$

or

$$d_1^2 V_1 = d_2^2 V_2 = d_3^2 V_3$$

$$\text{Total head loss} = h_i + h_{f_1} + h_e + h_{f_2} + h_c + h_{f_3} + h_o$$

$$h_i = 0.5\frac{V_1^2}{2g}, \text{ loss of head at the entrance of pipe-1.}$$

$$h_{f_1} = \frac{4f_1 l_1 V_1^2}{2gd_1}, \text{loss of head due friction in pipe-1}$$

$$h_e = \frac{(V_1 - V_2)^2}{2g}, \text{ loss of head due to sudden enlargement from pipe-1}$$

$$h_{f_2} = \frac{4 f_2 l_2 V_2^2}{2 g d_2}, \text{ loss of head due to friction in pipe-2}$$

$$h_c = \frac{0.5 V_3^2}{2g}, \text{loss of head due to sudden contraction from pipe-2 to pipe-3}$$

$$h_{f_3} = \frac{4 f_3 l_3 V_3^2}{2 g d_3}, \text{ loss of head due to friction in pipe-3.}$$

$$h_o = \frac{V_3^3}{2g}, \text{ loss of head at the exit of pipe-3.}$$

$$\therefore \quad \text{Total head loss} = \frac{0.5 V_1^2}{2g} + \frac{4 f_1 l_1 V_1^2}{2 g d_1} + \frac{(V_1 - V_2)^2}{2g} + \frac{4 f_2 l_2 V_2^2}{2 g d_2} + \frac{0.5 V_3^2}{2g} + \frac{4 f_3 l_3 V_3^2}{2 g d_3} + \frac{V_3^2}{2g}$$

$$\text{...(4.4.1)}$$

Now applying modified Bernoulli's equation at points A and B (*i.e.*, points on the free surfaces of liquid in reservoirs), we get

$$\frac{p_A}{\rho g} + \frac{V_A^2}{2g} + z_A = \frac{p_B}{\rho g} + \frac{V_B^2}{2g} + z_B + \text{total head loss}$$

$$0 + 0 + z_A = 0 + 0 + z_B + \text{total head loss} \qquad \text{...(4.4.2)}$$

$$z_A - z_B = \text{total head loss}$$

or $\qquad\qquad H = \text{total head loss}$

From Eqs. (4.4.1) and (4.4.2), we get

$$H = \frac{0.5 V_1^2}{2g} + \frac{4 f_1 l_1 V_1^2}{2 g d_1} + \frac{(V_1 - V_2)^2}{2g} + \frac{4 f_2 l_2 V_2^2}{2 g d_2} + \frac{0.5 V_3^2}{2g} + \frac{4 f_3 l_3 V_3^2}{2 g d_3} + \frac{V_3^2}{2g}$$

4.5 CONCEPT OF EQUIVALENT LENGTH AND EQUIVALENT PIPE

4.5.1 Equivalent Length

The loss of head in pipe fittings are expressed in terms of an equivalent length which is the length of uniform diameter pipe in which an equal loss of head (*i.e.*, due to friction) would occur for the same discharge.

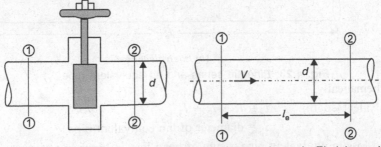

Fig. 4.22: Head loss caused by a component (valve as shown in Fig.) is equivalent to the head loss caused by a same cross-sectional area of the pipe whose length is the equivalent length.

The equivalent length l_e is obtained by equating minor loss equal to major loss.

$$K_L \frac{V^2}{2g} = \frac{4f \, l_e V^2}{2gd}$$

or

$$K_L = \frac{4f \, l_e}{d}$$

$$K_L = \frac{f' l_e}{d} \qquad\qquad \because 4f = f'$$

or Equivalent length: $\quad l_e = K_L \dfrac{d}{f}$

where $\qquad\qquad\qquad f$ = coefficient of friction

$\qquad\qquad\qquad\qquad f'$ = friction factor

For a known value of f' which depends upon the Reynold's number and the pipe roughness, the equivalent length for a given pipe fitting can be expressed in terms of pipe diameter.

4.5.2 Equivalent Pipe

It is defined as the pipe of uniform diameter having loss of head equal to the total loss of head as the compound pipe (*i.e.,* pipes in series) at same discharge (Q). The uniform diameter of the equivalent pipe is called equivalent size of the pipe. The length of equivalent pipe is equal to sum of lengths of the compound pipe.

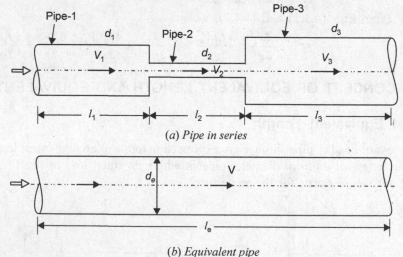

(a) *Pipe in series*

(b) *Equivalent pipe*

Fig. 4.23: Pipes in series and its equivalent pipe

Mathematically,

Equivalent length: $\qquad l_e = l_1 + l_2 + l_3$

Let $\qquad\qquad\qquad d_e$ = diameter of the equivalent pipe.

Head loss in equivalent pipe = sum of head loss in pipes in series

$$h_{f_e} = h_{f_1} + h_c + h_{f_2} + h_e + h_{f_3}$$

where $h_{f_e} = \dfrac{4 f_e \, l_e V^2}{2g \, d_e}$, loss of head due to friction in equivalent pipe.

$h_{f_1} = \dfrac{4 f_1 \, l_1 V_1^2}{2g \, d_1}$, loss of head due to friction in pipe-1.

$h_c = \dfrac{0.5 V_2^2}{2g}$, loss of head due to sudden contraction from pipe-I to pipe-2

$h_{f_2} = \dfrac{4 f_2 \, l_2 V_2^2}{2g \, d_2}$, loss of head due to friction in pipe-2.

$h_e = \dfrac{(V_2 - V_3)^2}{2g}$, loss of head due to sudden enlargement from pipe-2-

to pipe-3

$h_{f_3} = \dfrac{4 f_3 \, l_3 V_3^2}{2g \, d_3}$, loss of head due to friction in pipe-3.

$\therefore \quad \dfrac{4 f_e \, l_e V^2}{2g d_e} = \dfrac{4 f_1 \, l_1 V_1^2}{2g d_1} + 0.5 \dfrac{V^2}{2g} + \dfrac{4 f_2 \, l_2 V_2^2}{2g d_2} + \dfrac{(V_2 - V_3)^2}{2g} + \dfrac{4 f_3 \, l_3 V_3^2}{2g d_3}$

$\dfrac{4 f_e \, l_e V^2}{2g d_e} = \dfrac{4 f_1 \, l_1 V_1^2}{2g d_1} + \dfrac{4 f_2 \, l_2 V_2^2}{2g d_2} + \dfrac{4 f_3 \, l_3 V_3^2}{2g d_3}$ ...(11.5.1)

[∵ neglecting minor losses]

Assuming $\qquad f_e = f_1 = f_2 = f_3 = f$

Equation (4.5.1) becomes

$$\frac{l_e V^2}{d_e} = \frac{l_1 V_1^2}{d_1} + \frac{l_2 V_2^2}{d_2} + \frac{l_3 V_3^2}{d_3} \qquad ...(4.5.2)$$

Discharge: $\qquad Q = A_e V = A_1 V_1 = A_2 V_2 = A_3 V_3$

or $\qquad V = \dfrac{Q}{A_e} = \dfrac{Q}{\dfrac{\pi}{4} d_e^2} = \dfrac{4Q}{\pi d_e^2}$

$V_1 = \dfrac{Q}{A_1} = \dfrac{Q}{\dfrac{\pi}{4} d_1^2} = \dfrac{4Q}{\pi d_1^2}$

$V_2 = \dfrac{Q}{A_2} = \dfrac{4Q}{\dfrac{\pi}{4} d_2^2} = \dfrac{4Q}{\pi d_2^2}$

and $\qquad V_3 = \dfrac{Q}{A_3} = \dfrac{4Q}{\dfrac{\pi}{4} d_3^2} = \dfrac{4Q}{\pi d_3^2}$

Substituting the values of V_1, V_2, and V_3 in above equation (4.5.2), we get

$$\frac{l_e}{d_e} \times \frac{16Q^2}{\pi^2 d_e^4} = \frac{l_1}{d_1} \times \frac{16Q^2}{\pi^2 d_1^4} + \frac{l_2}{d_2} \times \frac{16Q^2}{\pi^2 d_2^4} + \frac{l_3}{d_3} \times \frac{16Q^2}{\pi^2 d_3^4}$$

or
$$\frac{l_e}{d_e^5} = \frac{l_1}{d_1^5} + \frac{l_2}{d_2^5} + \frac{l_3}{d_3^5} \qquad (4.5.3)$$

The above Eq. (4.5.3) is known as **Dupuit's equation**. In this equation l_e = $l_1 + l_2 + l_3$ and d_1, d_2 and d_3 are known. Hence diameter of the equivalent pipe d_e can be determined.

If the diameter of the equivalent pipe d_e is fixed up, its length l_e can be determined from Eq. (4.5.3). In that case, $l_e \neq l_1 + l_2 + l_3$

If coefficient of friction is not same, the above Eq. (4.5.3) is written as:

$$\frac{f_e l_e}{d_e^5} = \frac{f_1 l_1}{d_1^5} + \frac{f_2 l_2}{d_2^5} + \frac{f_3 l_3}{d_3^5}$$

4.6 PIPES IN PARALLEL

In piping system, two or more pipes are connected between the main pipe, is called pipes in parallel as shown in Fig. 4.24. The rate of flow through main pipe is equal to the sum of rate of flow through branch pipes. Hence from Fig. 4.24, we get

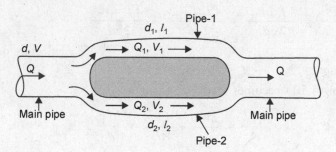

Fig. 4.24: Two pipes in parallel.

$$Q = Q_1 + Q_2$$

$$\frac{\pi}{4} d^2 V = \frac{\pi}{4} d_1^2 V_1 + \frac{\pi}{4} d_2^2 V_2$$

or
$$d^2 V = d_1^2 V_1 + d_2^2 V_2$$

The same head loss occurs in each branch between the main pipe.

i.e.,
$$h_{f1} = h_{f2}$$

$$\frac{4 f_1 \, l_1 V_1^2}{2 g d_1} = \frac{4 f_2 \, l_2 V_2^2}{2 g d_2}$$

$$\frac{f_1 \, l_1 V_1^2}{d_1} = \frac{f_2 \, l_2 V_2^2}{d_2}$$

where f_1, f_2 are coefficients of friction in pipes 1 and 2 respectively.

4.6.1 Three Pipes in Parallel

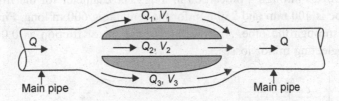

Fig. 4.25: Three pipes in parallel.

Discharge through main pipe:

$$Q = Q_1 + Q_2 + Q_3$$

and head loss: $h_{f1} = h_{f2} = h_{f3}$

$$\frac{4f_1\, l_1 V_1^2}{2gd_1} = \frac{4f_2\, l_2 V_2^2}{2gd_2} = \frac{4f_3\, l_3 V_3^2}{2gd_3}$$

or

$$\frac{f_1 l_1 V_1^2}{d_1} = \frac{f_2 l_2 V_2^2}{d_2} = \frac{f_3 l_3 V_3^2}{d_3}$$

where f_1, f_2, f_3 are coefficients of friction in pipes 1, 2 and 3 respectively.

4.6.2 Four Pipes in Parallel

Discharge through main pipe:

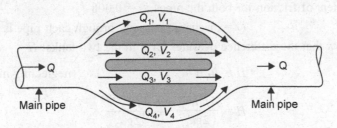

Fig. 4.26: Four pipes in parallel.

$$Q = Q_1 + Q_2 + Q_3 + Q_4$$

and head loss: $h_f = h_{f2} = h_{f3} = h_{f4}$

$$\frac{4f_1\, l_1 V_1^2}{2gd_1} = \frac{4f_2\, l_2 V_2^2}{2gd_2} = \frac{4f_3\, l_3 V_3^2}{2gd_3} = \frac{4f_4\, l_4 V_4^2}{2gd_4}$$

or

$$\frac{f_1 l_1 V_1^2}{d_1} = \frac{f_2 l_2 V_2^2}{d_2} = \frac{f_3 l_3 V_3^2}{d_3} = \frac{f_4 l_4 V_4^2}{d_4}$$

where f_1, f_2, f_3, f_4 are coefficients of friction in pipes 1, 2, 3 and 4 respectively.

Problem 4.22: What is discharged from one tank to another with 30 m difference of water levels through a pipe 1200 m long. The diameter for the first 600 m length of the pipe is 400 mm and 250 mm for the remaining 600 m long. Find the discharge in litre/s through the pipe. Assume the coefficient of friction as 0.008 for both the pipes, neglecting minor losses.

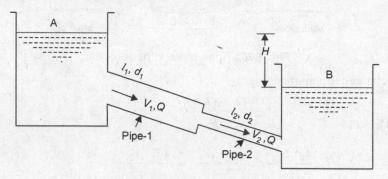

Fig. 4.27: Schematic for Problem 4.22

Solution: Given data:

Difference of water level in the two tanks: $H = 30$ m

Length of pipe-1: $l_1 = 600$ m

Diameter of pipe-1: $d_1 = 400$ mm $= 0.4$ m

Length of pipe-2: $l_2 = 600$ m

Diameter of pipe-2: $d_2 = 250$ mm $= 0.250$ m

Coefficient of friction for both the pipes : $f = 0.008$

Let $Q =$ discharge passing through each pipe is same.

We know that the difference of water level in the two tanks: H

$$H = h_{f_1} + h_{f_2} \qquad\qquad \text{(neglecting minor losses)}$$

$$H = \frac{4f\, l_1 V_1^2}{2gd_1} + \frac{4f\, l_2 V_2^2}{2gd_2}$$

where $V_1 = \dfrac{Q}{A_1} = \dfrac{Q}{\dfrac{\pi}{4}d_1^2} = \dfrac{4Q}{\pi d_1^2}$

and $V_2 = \dfrac{4Q}{\pi d_2^2}$

$\therefore$ $H = \dfrac{4f\, l_1}{2gd_1} \times \dfrac{16Q^2}{\pi^2 d_1^4} + \dfrac{4f l_2}{2gd_2} \times \dfrac{16Q^2}{\pi^2 d_2^4}$

$$H = \frac{32f\, l_1 Q^2}{g\pi^2 d_1^5} + \frac{32f\, l_2 Q^2}{g\pi^2 d_2^5}$$

$$H = \frac{32f Q^2}{g\pi^2}\left[\frac{l_1}{d_1^5} + \frac{l_2}{d_2^5}\right]$$

$$30 = \frac{32 \times 0.008 \times Q^2}{9.81 \times (3.14)^2} \left[\frac{600}{(0.4)^5} + \frac{600}{(0.25)^5} \right]$$

$$11334.688 = Q^2 \, [58593.75 + 614400]$$

or
$$Q^2 = 0.01684$$

$$Q = 0.12976 \text{ m}^3/\text{s} = 0.12976 \times 1000 \text{ litres/s} = \textbf{129.76 litre/s}$$

Problem 4.23: Two reservoirs are connected by a 1500 m pipeline. The first 700 m of the pipe-line is 300 mm diameter and has a frictional of 0.020. The remaining pipe is 600 mm in diameter and its frictional coefficient is 0.018. Estimate the difference of water level in the two reservoirs when the flow is 9000 liter/minute and the ends of the pipe line at the reservoirs are sharp.

Solution: Given data:

Length of pipe : $l = 1500$ m

Length of pipe–1 : $l_1 = 700$ m

Diameter of pipe–1 : $d_1 = 300$ mm $= 0.30$ m

Frictional coefficient or friction factor,

$$f_1' = 0.020$$

Length of pipe–2 : $l_2 = l - l_1 = 1500 - 700 = 800$ m

Diameter of pipe–2 : $d_2 = 600$ mm $= 0.60$ m

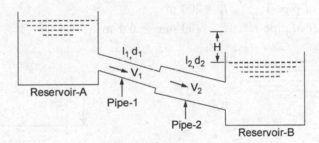

Fig. 4.28: Schematic for Problem 4.23

Frictional coefficient or friction factor,

$$f'_2 = 0.018$$

Discharge: $Q = 9000$ litre/minute

$$= \frac{9000}{1000 \times 600} \text{ m}^3/\text{s} = 0.15 \text{ m}^3/\text{s}$$

We known that the difference of water level in the two tanks;

$$H = \text{head loss in pipe–1} + \text{head loss in pipe–2}$$

$$H = h_{f1} + h_{f2}$$

$$= \frac{f_1' \, l_1 \, v_1^2}{2 \, g \, d_1} + \frac{f_2' \, l_2 \, V_2^2}{2 \, g \, d_2}$$

where $\qquad V_1 = \dfrac{Q}{A_1} = \dfrac{Q}{A_1} = \dfrac{4Q}{\pi d_1^2}$

and $\qquad V_2 = \dfrac{4Q}{\pi d_2^2}$

$\therefore \qquad H = \dfrac{f_1' \, l_1}{2gd_1}\left(\dfrac{4Q}{\pi d_1^2}\right)^2 + \dfrac{f_2' \, l_2}{2gd_2}\left(\dfrac{4Q}{\pi d_2^2}\right)^2$

$\qquad H = \dfrac{8 f_1' \, l_1 Q^2}{\pi^2 g d_1^5} + \dfrac{8 f_2' \, l_2 Q^2}{\pi^2 g d_2^5}$

$\qquad = \dfrac{8 \times 0.020 \times 700 \times (0.15)^2}{(3.14)^2 \times 9.81 \times (0.30)^5} + \dfrac{8 \times 0.018 \times 800 \times (0.15)^2}{(3.14)^2 \times 9.81 \times (0.60)^5}$

$\qquad = 10.72 + 0.34 = \mathbf{11.06 \ m}$

Problem 4.24: The difference in water surface levels in two tanks, which are connected by three pipes in series of length 200 m, 300 m and 150 m and of diameters 400 mm, 200 mm and 300 mm respectively is 20 m. Find the rate of water if coefficients of friction are 0.004, 0.0045, 0.005 respectively, considering: (*i*) minor losses (*ii*) neglecting minor losses.

Solution: Given data:

Length of pipe-1: $\qquad l_1 = 200$ m

Diameter of pipe-1: $\quad d_1 = 400$ mm = 0.4 m

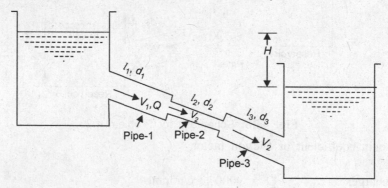

Fig. 4.29: Schematic for Problem 4.24

Coefficients of friction of pipe – 1:

$\qquad\qquad f_1 = 0.004$

Length of pipe-2: $\qquad l_2 = 300$ m

Diameter of pipe-2: $\quad d_2 = 200$ mm = 0.2 m

Coefficient of friction of pipe : 2:

$\qquad\qquad f_2 = 0.0045$

Length of pipe-3: $l_3 = 150$ m

Diameter of pipe-3: $d_3 = 300$ m $= 0.3$ m

Coefficient of friction of pipe-3:

$$f_3 = 0.005$$

Difference of water level in the two tank: $H = 20$ m

Let Q = discharge passing through each pipe is same

We know that the difference of water level in the two tank: H

$$H = h_{f_1} + h_{f_2} + h_{f_3} \qquad \text{(neglecting minor losses)}$$

$$H = \frac{4 f_1 \, l_1 V_1^2}{2 g d_1} + \frac{4 f_2 \, l_2 V_2^2}{2 g d_2} + \frac{4 f_3 \, l_3 V_3^2}{2 g d_3}$$

where $$V_1 = \frac{Q}{A_1} = \frac{Q}{\frac{\pi}{4} d_1^2} = \frac{Q}{\pi d_1^2}$$

$$V_2 = \frac{4Q}{\pi d_2^2}$$

and $$V_3 = \frac{4Q}{\pi d_3^2}$$

$$\therefore \quad H = \frac{4 f_1 l_1}{2 g d_1} \times \frac{16 Q^2}{\pi^2 d_1^4} + \frac{4 f_2 l_2}{2 g d_2} \times \frac{16 Q^2}{\pi^2 d_2^4} + \frac{4 f_3 l_3}{2 g d_3} \times \frac{16 Q^2}{\pi^2 d_3^4}$$

$$H = \frac{32 Q^2}{g \pi^2} \left[\frac{f_1 f_1}{d_1^5} + \frac{f_2 f_2}{d_2^5} + \frac{f_3 f_3}{d_3^5} \right]$$

$$20 = \frac{32 \times Q^2}{9.81 \times (3.14)^2} \left[\frac{0.004 \times 200}{(0.4)^5} + \frac{0.0045 \times 300}{(0.25)^5} + \frac{0.005 \times 150}{(0.3)^5} \right]$$

$$20 = 0.3308 \; Q^2 \; [78.125 + 4218.75 + 308.64]$$

$$20 = 0.3308 \; Q^2 \times 4605.515$$

or $$Q^2 = 0.013127$$

$$Q = 0.114572 \text{ m}^3/\text{s} = \mathbf{114.572 \text{ litre/s}}$$

Problem 4.25: A 30 m long pipe line connects two reservoirs, both of which are open to the atmosphere. The difference in their water level is 12 m. The pipe has three equal sections of 10 m each. The first and last sections are 60 mm in diameter and the intermediate section is 40 mm in diameter. The value of f for the pipes is 0.0054. Calculate the flow rate and draw the total energy and hydraulic grade lines.

Solution: Given data:

Length of pipe: $l = 30$ m

Difference of water level between two reservoirs: $H = 12$ m

$$l_1 = l_2 = l_3 = 10 \text{ m}$$

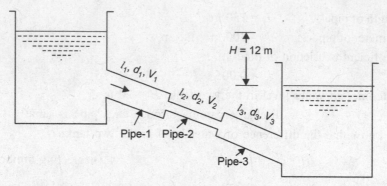

Fig. 4.30: Schematic for Problem 4.25

Diameter of pipe-1: $d_1 = 60$ mm $= 0.06$ m

Diameter of pipe-2: $d_2 = 40$ mm $= 0.04$ m

Diameter of pipe-3: $d_3 = d_1 = 0.06$ m

Coefficient of friction: $f = 0.0054$

We know that the difference of water level between two reservoirs: H

$$H = h_{f_1} + h_{f_2} + h_{f_3} \qquad \text{(neglecting minor losses)}$$

where $h_{f_1} = \dfrac{4f\, l_1 V_1^2}{2gd}$, loss of head due to friction in pipe-1

$$h_{f_2} = \dfrac{4f\, l_2 V_2^2}{2gd_2}, \text{ loss of head due to friction in pipe-2}$$

and $h_{f_3} = \dfrac{4f\, l_3 V_3^2}{2gd_3}$, loss of head due to friction in pipe-3

$$\therefore \qquad H = \dfrac{4f\, l_1 V_1^2}{2gd_1} + \dfrac{4f\, l_2 V_2^2}{2gd_2} + \dfrac{4f\, l_3 V_3^2}{2gd_3}$$

$$H = \dfrac{4f l_1}{2g} \left[\dfrac{V_1^2}{d_1} + \dfrac{V_2^2}{d_2} + \dfrac{V_3^2}{d_3} \right] \qquad \because l_1 = l_2 = l_3$$

$$H = \dfrac{4f l_1}{2g} \left[\dfrac{V_1^2}{d_1} + \dfrac{V_2^2}{d_2} + \dfrac{V_3^2}{d_3} \right]$$

We know that the discharge flow pipes in series are constant.

i.e., $Q = A_1 V_1 = A_2 V_2 = A_3 V_3$

or $V_1 = \dfrac{Q}{A_1} = \dfrac{Q}{\dfrac{\pi}{4} d_1^2} = \dfrac{4Q}{\pi d_1^2}$

Similarly $V_2 = \dfrac{4Q}{\pi d_2^2}$

and $V_3 = \dfrac{4Q}{\pi d_3^2}$

$$\therefore \quad H = \frac{4f\,l_1}{2g}\left[\frac{16Q^2}{\pi^2 d_1^5} + \frac{16Q^2}{\pi^2 d_2^5} + \frac{16Q^2}{\pi^2 d_3^5}\right]$$

$$H = \frac{4f\,l_1}{2g} \times \frac{16Q^2}{\pi^2}\left[\frac{1}{d_1^5} + \frac{1}{d_2^5} + \frac{1}{d_3^5}\right]$$

$$12 = \frac{4\times0.0054\times10\times16Q^2}{2\times9.81\times(3.14)^2}\left[\frac{1}{(0.06)^5} + \frac{1}{(0.04)^5} + \frac{1}{(0.06)^5}\right]$$

$$12 = 0.01786\ Q^2\ [0.1286 \times 10^7 + 0.976 \times 10^7 + 0.1286 \times 10^7]$$

$$12 = 0.01786\ Q^2 \times 1.2337\times 10^7$$

or
$$Q^2 = 0.00005446$$

$$Q = 0.00737\ \text{m}^3/\text{s} = \textbf{7.379 litre/s}$$

Total Energy Line (TEL) and Hydraulic Grade Line (HGL) for neglecting minor losses.

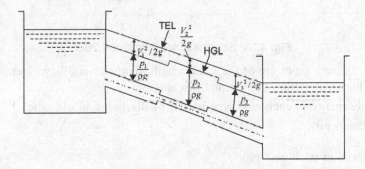

Fig. 4.31: Total energy and hydraulic grade lines for neglecting minor losses.

Total Energy Line (TEL) and Hydraulic Grade Line (HGL) for considering minor losses.

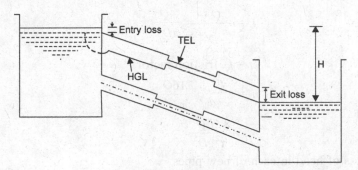

Fig. 4.32: Total energy and hydraulic grade lines for considering minor losses.

Problem 4.26: An old water supply distribution pipe of 250 mm diameter of a city is to be replaced by two parallel pipes of smaller equal diameter having equal lengths and identical friction factor values. Find out the new diameter required.

(GGSIP University Delhi, Dec. 2001)

Solution: Given data:

Diameter of old pipe: $D = 250$ mm $= 0.25$ m

Let Q = discharge in old pipe

d = diameter of each of the new pipes

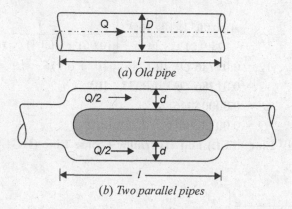

(a) Old pipe

(b) Two parallel pipes

Fig. 4.33: Schematic for Problem 4.26

Since the new pipes are of the same length, diameter and have identical value of friction factor, thus they have equal discharges.

(*i.e.,* discharge in each new pipe is half of discharge in old pipe)

Mathematically,

Discharge in each new pipe $= \dfrac{Q}{2}$

We know that the loss of head in the old pipe:

$$h_f = \frac{4fl\,V^2}{2gd}$$

where $V = \dfrac{Q}{\dfrac{\pi}{4}D^2} = \dfrac{4Q}{\pi D^2}$

$4f = f'$, friction factor

$$h_f = \frac{f'l}{2gd} \times \frac{16Q}{\pi D^4}$$

$$h_f = \frac{8f'l\,Q^2}{\pi^2 gD^5} \qquad\qquad \ldots(i)$$

and loss of head in each of new pipes

$$h_f = \frac{8f'l\,(Q/2)^2}{\pi^2 gd^5} = \frac{2f'l\,Q^2}{\pi^2 gd^5} \qquad\qquad \ldots(ii)$$

Equating Eqs. (*i*) and (*ii*), we get

$$\frac{8f'l\,Q^2}{\pi^2 gD^5} = \frac{2f'l\,Q^2}{\pi^2 gd^5}$$

or
$$\frac{4}{D^5} = \frac{1}{d^5}$$

or
$$d^5 = \frac{D^5}{4} = \frac{(0.25)^5}{4}$$
$$d^5 = 0.000244$$

or
$$d = 0.18944 \text{ m} = \mathbf{189.44 \text{ mm}}$$

Problem 4.27: A straight 25 cm diameter pipeline 5 km long is laid between two reservoirs having a difference of levels of 40 m. To increase the capacity of the system, an additional 2.5 km long 25 cm diameter pipe is laid parallel from the reservoir to the mid-point of the original pipe. Find the increase in discharge due to installation of a new pipe. Assume $4f = 0.025$ in the Darcy-Weisbach equation for both the pipes.

(*GGSIP University, Delhi, Dec. 2008*)

Solution: Given data:

Case-1

Diameter of pipeline: $d = 25$ cm $= 0.25$ m
Length of pipeline: $l = 5$ km $= 5000$ m
Difference of levels between two reservoirs: $H = 40$ m
$$4f = 0.025$$

or   Coefficient of friction:

$$f = \frac{0.025}{4} = 0.00625$$

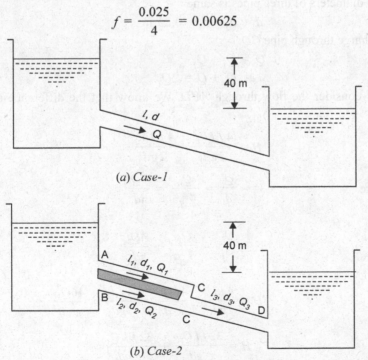

(*a*) *Case-1*

(*b*) *Case-2*

Fig. 4.34: Schematic for Problem 4.27

We know that the difference of water level in the two reservoirs: H

$$H = h_f$$

$$H = \frac{4f\,l\,V^2}{2gd}$$

$$H = \frac{4f\,l}{2gd} \times \frac{16Q^2}{\pi^2 d^4} \qquad \left| \quad \because V = \frac{Q}{A} = \frac{Q}{\dfrac{\pi}{4}d^2} = \frac{4Q}{\pi d^2} \right.$$

$$H = \frac{32f\,lQ^2}{g\pi^2 d^5}$$

$$40 = \frac{32 \times 0.00625 \times 5000 \times Q^2}{9.81 \times (3.14)^2 \times (0.25)^5}$$

or
$$Q^2 = 0.003778$$
$$Q = 0.06146 \text{ m}^3/\text{s}$$

Case-2: The old pipe 5000 m length and diameter 25 cm (0.25 m) with $f = 0.00625$. In addition to this, a new pipe 2500 m length and 25 cm (0.25 m) diameter with same value of $f = 0.00625$ is installed as AC as shown in Fig. 4.32 (b).

Length of pipe $\qquad AC$ = length of pipe BC = length of pipe CD

i.e., $\qquad\qquad l_1 = l_2 = l_3 = 2500$ m

coefficients of friction for pipes is same

i.e., $\qquad\qquad f_1 = f_2 = f_3 = 0.00625$

Also diameters of three pipes is same

i.e., $\qquad\qquad d_1 = d_2 = d_3 = 0.25$ m

Discharge through pipe CD:

$$Q_3 = Q_1 + Q_2$$
$$= Q_1 + Q_1 = 2Q_1 \qquad\qquad \because Q_2 = Q_1$$

Now consider the flow through ACD. We know that the difference of water levels (H),

$$H = \frac{4f_1 l_1 V_1^2}{2gd_1} + \frac{4f_3 l_3 V_3^2}{2gd_3}$$

where
$$V_1 = \frac{Q_1}{A_1} = \frac{Q_1}{\dfrac{\pi}{4}d_1^2} = \frac{4Q_1}{\pi d_1^2}$$

and
$$V_3 = \frac{Q_3}{A_3} = \frac{Q_3}{\dfrac{\pi}{4}d_3^2} = \frac{4Q_3}{\pi d_3^2}$$

$$H = \frac{4f_1 l_1}{2gd_1} \times \frac{16Q_1^2}{\pi^2 d_1^4} + \frac{4f_3 l_3}{2gd_3} \times \frac{16Q_3^2}{\pi^2 d_3^4}$$

$$H = \frac{32f_1 l_1 Q_1^2}{\pi^2 gd_1^5} + \frac{32f_3 l_3 Q_3^2}{\pi^2 gd_3^5}$$

$$H = \frac{32 f_1 l_1 Q_1^2}{\pi^2 g d_1^5} + \frac{32 f_1 l_1 (2Q_1)^2}{\pi^2 g d_1^5}$$

$$[\because f_3 = f_1, l_3 = l_1, Q_3 = 2Q_1, d_3 = d_1]$$

$$H = \frac{160 f_1 l_1 Q_1^2}{\pi^2 g d_1^5}$$

$$40 = \frac{160 \times 0.00625 \times 2500 \times Q_1^2}{(3.14)^2 \times 9.81 \times (0.25)^5}$$

or $\qquad Q_1^2 = 0.001511$

or $\qquad Q_1 = 0.03887$ m³/s

$\therefore \qquad Q_3 = 2Q_1 = 2 \times 0.03887 = 0.07774$ m³/s

$\therefore$ Increase in discharge $= Q_3 - Q = 0.07774 - 0.06146 = \textbf{0.01628 m}^3\textbf{/s}$

OR

Increase in discharge $= \dfrac{Q_3 - Q}{Q} = \dfrac{0.0774 - 0.06146}{0.06146}$

$$= 0.2648 = \textbf{26.48\%}$$

Problem 4.28: Find the capacity of a pump that is required in a 75 mm line so that 20 litre/s flow through each pipe as shown in Fig. 4.33. Neglect minor losses. Assume that water is flowing at 20°C in smooth pipes. At 20°C, for water; viscosity $= 10^{-3}$ Ns/m² and density $= 10^3$ kg/m³.

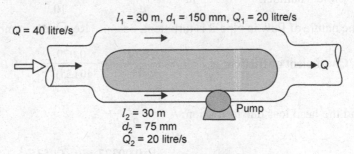

Fig. 4.35: Pipes in parallel with a pump.

Solution: Given data:

Pipe-1:

Length of pipe: $\qquad l_1 = 30$ m

Diameter of pipe: $\qquad d_1 = 150$ mm $= 0.15$ m

$\therefore$ Cross-sectional area: $\quad A_1 = \dfrac{\pi}{4} d_1^2 = \dfrac{3.14}{4} \times (0.15)^2 = 0.01766$ m²

Discharge: $\qquad Q_1 = 20$ litre/s $= \dfrac{20}{1000}$ m³/s $= 0.02$ m³/s

$\therefore$ Velocity of water: $\qquad V_1 = \dfrac{Q_1}{A_1} = \dfrac{0.02}{0.01766} = 1.132$ m/s

Reynolds number: $\qquad Re_1 = \dfrac{\rho V_1 d_1}{\mu} = \dfrac{10^3 \times 1.132 \times 0.15}{10^{-3}} = 169800$

The nature of flow is turbulent, because Reynolds number $Re_1 > 4000$.

$\therefore$ Coefficient of friction: $\quad f_1 = \dfrac{0.079}{Re_1^{0.25}} = \dfrac{0.079}{(169800)^{0.25}} = 0.00389$

and the head loss due to friction:

$$h_{f_1} = \frac{4 f_1 l_1 V_1^2}{2 g d_1} = \frac{4 \times 0.00386 \times 30 \times (1.132)^2}{2 \times 9.81 \times (0.15)}$$

$$= 0.2016 \text{ m}$$

For pipe-2:
Length of pipe: $\qquad\qquad l_2 = 30$ m
Diameter: $\qquad\qquad\qquad d_2 = 75$ mm $= 0.075$ m

$\therefore$ Cross-sectional area: $\quad A_2 = \dfrac{\pi}{4} d_2^2 = \dfrac{3.14}{4} \times (0.075)^2 = 0.00441$ m²

$\qquad$ Discharge: $\qquad\qquad Q_2 = 20$ litre/s $= 0.02$ m³/s

$\therefore$ Velocity: $\qquad\qquad V_2 = \dfrac{Q_2}{A_2} = \dfrac{0.02}{0.00441} = 4.535$ m/s

Reynold's number: $\qquad Re_2 = \dfrac{\rho V_2 d_2}{\mu} = \dfrac{10^3 \times 4.535 \times 0.075}{10^{-3}} = 340125$

The nature of flow in pipe-2 is turbulent, because Reynold's number $Re_2 > 4000$.

$\therefore$ Coefficient of friction: $\quad f_2 = \dfrac{0.079}{Re_2^{0.25}} = \dfrac{0.079}{(340125)^{0.25}} = 0.00327$

and the head loss due to friction: $h_{f_2} = \dfrac{4 f_2 l_2 V_2^2}{2 g d_2}$

$$= \frac{4 \times 0.00327 \times 30 \times (4.535)^2}{2 \times 9.81 \times 0.075} = 5.484 \text{ m}$$

As we know that the head loss due to friction in each pipe in parallel is same. But in present case, head loss due to friction in pipe-2 is more than head loss due to friction in pipe-1. So, we condude that the pump placed in pipe-2 must provide the power required to overcome this extra loss of head due to friction.

$$h_f = h_{f_2} - h_{f_1} = 5.484 - 0.2016 = 5.282 \text{ m}$$

$\therefore$ Power of pump: $\qquad P = \rho\, Q_2\, g\, h_f$

$$= 1000 \times 0.02 \times 9.81 \times 5.282 \text{ W}$$

$$= 1036.32 \text{ W} = \textbf{1.036 kW}$$

Problem 4.29: Calculate the quantity of water from the reservoir through the pipe system shown in Fig. 11.36. The water drives a water turbine which develops 100 kW.

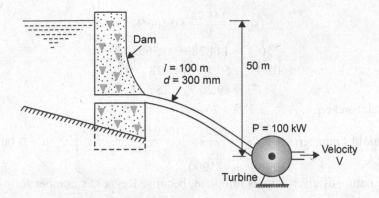

Fig. 4.36: Power produced by a turbine.

Solution: Given data:

Power produced by the turbine: $P = 100$ kW $= 100 \times 10^3$ W

Gross head: $H_G = 50$ m

Length of pipe: $l = 100$ m

Diameter of pipe: $d = 300$ mm $= 0.3$ m

$\therefore$ Cross-sectional area: $A = \dfrac{\pi}{4}d^2 = \dfrac{3.14}{4} \times (0.3)^2 = 0.07065$ m^2

Power produced by the turbine: $P = \rho Q g H$

or $H = \dfrac{P}{\rho Q g}$

where $H =$ head utilized by the turbine.

$$H = \frac{100 \times 10^3}{1000 \times AV \times 9.81}$$

$$= \frac{100 \times 10^3}{1000 \times 0.07065 \times V \times 9.81} = \frac{144.28}{V}$$

By energy balance, we get

Gross head = head utilized + loss of head due to friction + head at outlet of turbine

$$H_G = H + h_f + \frac{V^2}{2g}$$

where $$h_f = \frac{4flV^2}{2gd}$$

Let $f = 0.0025$

$\therefore$

$$50 = \frac{144.28}{V} + \frac{4 \times 0.0025 \times 100 \times V^2}{2 \times 9.81 \times 0.1} + \frac{V^2}{2 \times 9.81}$$

$$50 = \frac{144.28}{V} + 0.5096V^2 + 0.0509V^2$$

$$50 = \frac{144.28}{V} + 0.5005V^2$$

or $\qquad\qquad 50\ V = 144.28 + 0.5605\ V^3$

or $\qquad\qquad 0.5605\ V^3 - 50\ V + 144.28 = 0$

or $\qquad\qquad V^3 - 89.20\ V + 257.41 = 0$

By trial and error: $\qquad V = 3.25$ m/s

Reynold's number: $\qquad Re = \dfrac{\rho dV}{\mu} = \dfrac{1000 \times 0.3 \times 3.25}{0.001}$ $\qquad \left| \quad \because\ \mu = 0.001\ \text{Ns/m}^2 \right.$

$\qquad\qquad\qquad\qquad\quad = 975000 \qquad\qquad\qquad\qquad\qquad\qquad\qquad\quad$ for water

The nature of given flow is turbulent, because Reynold's number $Re > 4000$.

Coefficient of friction: $\qquad f = \dfrac{0.079}{Re^{0.25}} = \dfrac{0.079}{(975000)^{0.25}} = 0.0025$

We get the coefficient of friction (f) almost equal to 0.0025 and therefore, we can accept the value of V as 3.25 m/s

$\therefore$ Discharge: $\qquad\qquad Q = AV = 0.07065 \times 3.25 = 0.22961$ m³/s

$\qquad\qquad\qquad\qquad\qquad\qquad = \mathbf{229.61\ litre/s}$

4.7 TRANSMISSION OF HYDRAULIC POWER THROUGH PIPELINES

The transmission of hydraulic power through pipe lines is commonly used for working of several hydraulic machines. The hydraulic power transmitted depends upon:

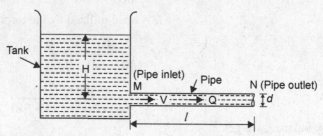

Fig. 4.37: Transmission of hydraulic power through pipe line.

(*a*) The discharge of liquid flowing through the pipe and

(*b*) The total head available at the end of the pipe (at point N in present case.)

Consider a pipe MN connected to a high level storage tank as shown in Fig. 4.37.

Let $\qquad\qquad\qquad\qquad d =$ diameter of the pipe

$\qquad\qquad\qquad\qquad\quad l =$ length of the pipe

$\qquad\qquad\qquad\qquad\quad Q =$ discharge flow through pipe

$\qquad\qquad\qquad\qquad\quad V =$ velocity of flow in pipe

$\qquad\qquad\qquad\qquad\quad h_f =$ loss of head in pipe MN, due to friction

$\qquad\qquad\qquad\qquad\quad H =$ total head available at the inlet of pipe or total head supplied.

The head available at the outlet of the pipe N (neglecting minor losses) = total head at inlet of pipe at point M – loss of head due to friction in pipe MN

$$= H - h_f$$

$$= H - \frac{4flV^2}{2gd} \quad \text{[by Darcy's formula, } h_f = \frac{4flV^2}{2gd}\text{]}$$

Mass of water flowing through the pipe per second,

$$m = \rho Q$$

Weight of water flowing through the pipe per second,

$$\dot{W} = mg = \rho g Q$$

Power available at inlet of the pipe at point M,

= weight of water per second × head available at inlet of the pipe

$$= \rho g Q H$$

and power available at outlet of the pipe at point N,

P = weight of water per second × head available at the outlet of the pipe at point N

$$= \rho g Q [H - h_f]$$

where $H - h_f$ – head available at outlet of the pipe.

$$P = \rho g Q \left[H - \frac{4flV^2}{2gd} \right]$$

Efficiency of power transmission:

$$\eta = \frac{\text{Power available at outlet of the pipe}}{\text{Power available at the inlet of the pipe}}$$

$$= \frac{\rho g Q [H - h_f]}{\rho g Q H} = \frac{H - h_f}{H} \qquad \qquad ...(i)$$

4.7.1 Condition for Maximum Transmission Power

The power available at the pipe at point N,

$$P = \rho g Q \left[H - \frac{4f\,lV^2}{2\,gd} \right]$$

$$= \rho g A V \left[H - \frac{4f\,lV^2}{2\,gd} \right] \qquad (\because Q = AV)$$

$$= \rho g A \left[HV - \frac{4f\,lV^3}{2\,gd} \right] \qquad \qquad ...(ii)$$

It is evident from Eq. (ii) that power transmitted depends upon the velocity of water through pipe (V), as the other things are constant.

∴ Power transmitted will be maximum, when

$$\frac{dP}{dV} = 0$$

$$\frac{d}{dV}\left[\rho g A\left(HV - \frac{4flV^3}{2gd}\right)\right] = 0$$

or

$$\rho g A\left(H.1 - \frac{4fl}{2gd} \times 3V^2\right) = 0$$

or

$$H - \frac{4f\ lV^2}{2g\ d} \times 3 = 0$$

or

$$H - h_f \times 3 = 0 \qquad\qquad \left(h_f - \frac{4flV^2}{2g\ d}\right)$$

or

$$3\ h_f = H$$

$$h_f = \frac{H}{3}$$

It means that power transmitted through the pipe is maximum, when the loss of head due to friction is one-third of the total head at inlet (or total head supplied).

4.7.2 Maximum Efficiency of Transmission of Power

Efficiency of power transmission though pipe is given by Eq. (*i*) as:

$$\eta = \frac{H - h_f}{H}$$

The maximum efficiency would correspond to the maximum power transmitted,

$$\left[h_f = \frac{H}{3}\right]$$

∴   Maximum efficiency:

$$\eta_{max} = \frac{H - \dfrac{H}{3}}{H}$$

$$= 1 - \frac{1}{3} = \frac{2}{3} = 0.66666 = 66.666\%$$

$$= 66.67\%$$

Problem 4.30:   A pipe of diameter 300 mm and length 3000 m is used for transmission of power by water. The total head at the inlet of pipe is 400 m. Find the maximum power available at the outlet of pipe, assume *f* = 0.005.

(GGSIP University, Delhi, May-June, 2007).

Solution: Given data:

Diameter of the pipe: *d* = 300 mm = 0.3 m

Length of the pipe: $l = 3000$ m
Total head at inlet: $H = 400$ m
                            $f = 0.005$
Condition for maximum power transmission

$$h_f = \frac{H}{3}$$

$\therefore$ $h_f = \frac{400}{3} = 133.33$ m

also, $h_f = \frac{4flV^2}{2gd}$

$$133.33 = \frac{40 \times 0.005 \times 3000 \times V^2}{2 \times 9.81 \times 0.3}$$

or $V^2 = 13.079$
                            $V = 3.616$ m/s

$\therefore$ Discharge: $Q = \frac{\pi}{4}d^2 \times V = \frac{3.14}{4} \times (0.3)^2 \times 3.616 = 0.2554$ m³/s

Maximum head available at the outlet of pipe $= H - h_f$
Maximum power available at the outlet of pipe $= \rho g Q [H - h_f]$
                            $= 1000 \times 9.81 \times 0.2554 \times (400 - 133.33)$
                            $= 668134.75$ W $= \mathbf{668.134\ kW}$

Problem 4.31: The maximum power is to be transmitted through a pipeline. Workout the conditions for maximum transmission of power. It is desired to develop 1000 kW of power at 85% efficiency by supplying water to a hydraulic turbine through a horizontal pipe 500 m long. Determine the necessary flow rate and minimum diameter of pipe to carry that discharge. Water is available at a head of 150 m. Take $f = 0.006$ in the formula.

$$h_f = 4f \times l \times \frac{V^2}{d \times 2g}$$

(GGSIP University, Delhi, May 2003)

Solution: Given data:

Power transmission: $P = 1000$ kW $= 1000 \times 10^3$ W
Efficiency: $\eta = 85\% = 0.85$
Length of pipe: $l = 500$ m
Water is available at a head:
                        $H = 150$ m
                        $f = 0.006$
Given head loss due to friction:

$$h_f = \frac{4flV^2}{2gd}$$

Now, condition for maximum transmission of power:

Head loss due to friction is one-third of available head and

$$\eta_{max} = 66.67\%$$

i.e., $h_f = \dfrac{H}{3}$ and $\eta_{max} = 66.67\%$

But given maximum efficiency (85%) is greater than theoretical maximum efficiency (66.67%)

So, $h_f = \dfrac{H}{3}$ cannot be applied in given case

We know, efficiency: $\eta = \dfrac{\text{Head available at outlet of the pipe}}{\text{Head available at inlet of the pipe}} = \dfrac{H - h_f}{H}$

$$0.85 = \frac{150 - h_f}{150}$$

or $127.5 = 150 - h_f$

or $h_f = 22.5$ m

Power transmission: $P = \rho \, Q \, g \, (H - h_f)$

$$1000 \times 10^3 = 1000 \times Q \times 9.81 \times (150 - 22.5)$$

or   Discharge: $Q = \mathbf{0.7995\ m^3/s}$

We know, discharge: $Q = AV = \dfrac{\pi}{4} d^2 \times V$

or $V = \dfrac{4Q}{\pi d^2} = \dfrac{4Q}{\pi d^2}$

Head loss due to friction:

$$h_f = \frac{4 f \, l \, V^2}{2gd} = \frac{4 f \, l}{2gd} \left(\frac{4Q}{\pi d^2} \right)^2 = \frac{64 f \, l \, Q^2}{2g \, \pi^2 d^5}$$

$$22.5 = \frac{64 \times 0.006 \times 500 \times (0.7995)^2}{2. \times 9.81 \times (3.14)^2 \times d^5}$$

or $d^5 = 0.028196$

Minimum diameter of the pipe: $d = \mathbf{0.4898\ m}$.

Problem 4.32: Power is to be transmitted hydraulically along a distance of 7500 m through a number of 125 mm diameter pipes, laid in parallel. The pressure at the discharge end is maintained constant at 6000 kPa. Determine the minimum number of pipes required to ensure an efficiency of at least 85% when the power delivered is 156 kW. Take coefficient of friction, $f = 0.006$ for all pipes and neglect losses other than pipe friction.

Solution: Given data:

Length of each pipe: $l = 7500$ m

Diameter of each pipe: $\qquad d = 125$ mm $= 0.125$ m

The pipes are used in parallel *i.e.*, head loss due to friction is same in each pipe

Pressure at the discharge end $\qquad = 6000$ kPa $= 6000 \times 10^3$ Pa or N/m^2

$\quad$ Head available at the end of pipe = head available at inlet of the pipe

$$- \text{head loss due to friction}$$

$$= H - h_f$$

∴ Pressure at the discharge end $\quad = \rho Q\,(H - h_f)$

$$6000 \times 10^3 = 1000 \times 9.81\,(H - h_f)$$

or $\qquad\qquad\qquad\qquad (H - h_f) = 611.62$ m

We know, Efficiency: $\qquad\qquad \eta = \dfrac{H - h_f}{H}$

$$0.85 = \dfrac{611.62}{H}$$

or $\qquad\qquad\qquad\qquad H = 719.55$ m

∴ $\quad$ Head loss due to friction: $\quad h_f = 719 - 611.62 = 107.38$ m

Power delivered: $\qquad\qquad\qquad P = 156$ kW $= 156 \times 10^3$ W

Also power: $\qquad\qquad\qquad\qquad P = \rho Q g\,(H - h_f)$

$$156 \times 10^3 = 1000 \times Q \times 9.81 \times 611.62$$

or $\;$ Total discharge: $\qquad\qquad Q = 0.0260$ m^3/s

Head loss due to friction in each pipe is same.

$$h_f = \frac{4\,flV^2}{2\,gd}$$

Let q is discharge passing through to each pipe.

Discharge: $\qquad\qquad\qquad\qquad q = \frac{\pi}{4}d^2V$

$$V = \frac{4q}{\pi d^2}$$

∴ $\qquad\qquad\qquad\qquad h_f = \frac{4\,fl}{2gd}\cdot\left(\frac{4q}{\pi d^2}\right)^2$

$$107.38 = \frac{4 \times 0.006 \times 7500}{2 \times 9.81 \times 0.125} \times \frac{16q^2}{(3.14)^2 \times (0.125)^4}$$

or $\qquad\qquad\qquad\qquad q^2 = 2.201 \times 10^{-4}$

or $\qquad\qquad\qquad\qquad q = 0.0483$ m^3/s

∴ $\quad$ Number of pipes required $\quad = \dfrac{Q}{q} = \dfrac{0.0260}{0.1483} = 1.75 \approx 2$

Hence number of pipes required $\quad = \mathbf{2.}$

4.8 WATER HAMMER

When flow of a liquid in a long pipe is decreased or stopped by closing the valve rapidly, velocity of liquid decreases. This decrease in velocity always gives increase in pressure and increase in pressure causes the formation of pressure wave which propagates throughout the liquid. A sudden rise in pressure has the effect of hammering action on the walls of the pipe. This phenomenon of sudden increase of pressure caused by a rapid closing of valve is known as water hammer or hammer blow. The water hammer is a very serious problem in case of pipes and penstock (hydroelectric power plant component), in supply lines for drinking water, in discharge lines for sewage water and in oil transmission pipe lines. A common example of water hammer is the knocking often heard in domestic water pipes when the tap is closed quickly.

The rise in pressure in some cases may be so large that pipes may even burst and therefore it is essential to take into account this pressure rise in the design of the pipes and penstocks. The magnitude of pressure rise due to water hammer depends on:

(*i*) Velocity of flow of liquid in the pipe

(*ii*) Length of the pipe

(*iii*) Elastic properties of the pipe material, whether rigid or elastic

(*iv*) Time taken to close the valve; gradual or quick closure of the valve.

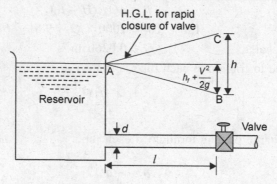

Fig. 4.38: Pressure rise due to closure of valve.

4.8.1 Pressure Rise due to Gradual Closure of Valve

Considering liquid to be incompressible and pipe walls to be rigid and inelastic, gradual closure of valve will cause uniform retardation and less increase in pressure.

The closure of valve is said to be gradual when $t > \dfrac{2l}{a}$

where

t = time taken to close the valve in seconds

l = length of pipe in metre

a = velocity of pressure wave in m/s

$= \sqrt{K/\rho}$, where K is bulk modulus of water, ρ is density of water.

The time $\left(\dfrac{2l}{a}\right)$ is called critical time (t_c)

i.e., $t_c = \dfrac{2l}{a}$

Mass of water in the pipe = $\rho \times$ volume of water = ρAl

where A = cross-sectional area of pipe

Let t be time required for gradual closure of the valve, i.e., the time required to bring the water to zero velocity from initial velocity, V.

$$\text{Retardation} = \frac{\text{Change of velocity}}{\text{Time}}$$

$$= \frac{\text{Initial velocity} - \text{Final velocity}}{t}$$

$$= \frac{V - 0}{t} = \frac{V}{t}$$

Retarding force = mass $\times$ retardation

$$= \rho Al \times V/t$$

If p is the rise in pressure due to closure of the valve, then force due to pressure wave

$$= pA$$

If friction is neglected, then the equilibrium between pressure and retarding force gives:

$$pA = \rho\, Al\frac{V}{t}$$

or $p = \dfrac{\rho l V}{t}$ ...(i)

$\therefore$ The pressure head due to rise in pressure: h

$$h = \frac{p}{\rho g} = \frac{\dfrac{\rho l V}{t}}{\rho g} = \frac{\rho V}{gt} \qquad \qquad ...(ii)$$

As shown in Fig. 4.38, the sudden rise in pressure head is due to gradual closure of the valve. The pressure head is maximum at the valve and linearly decreases towards the reservoir end. The line AB is the hydraulic gradient line just after the closure of the valve.

4.8.2 Pressure Rise due to Instantaneous Closure of Valve

In this case the valve is instantaneously closed, i.e., when the time to closure of valve is zero, the pressure increases to attain an infinite valve.

From Eq. (i), the relation for the sudden rise in pressure, $p = \dfrac{\rho l V}{t}$

As $t = 0$, then $p = \infty$ for instantaneous closure of valve. But instantaneous closure of the valve is not possible and such concept of infinite pressure rise is only hypothetical. In reality at such higher pressure the liquid actually gets compressed. Further it is not possible to close the valve in a zero interval. Moreover the pipe material being elastic, it will also expand. The relation above therefore does not hold good for the condition $t = 0$.

So, practically instantaneous closure of valve means that time taken to close the valve is less than the critical time.

i.e., $t < \dfrac{2l}{a}$

or $t < t_c$

where $t_c = \text{critical time} = \dfrac{2l}{a}$

Let the flow of liquid in the pipe line be brought to rest instantaneously by the closure of a valve. As the flow of liquid is brought to rest, its kinetic energy is transferred into strain energy of the liquid, then strain energy in the liquid is absorbed by pipe material.

Kinetic energy of the flow of liquid before the valve closure $= \dfrac{1}{2} m V^2$

where $m = \text{mass of liquid in whole pipe}$
                                        $= \text{density of liquid} \times \text{volume}$
                                        $= \rho. \, Al$

$\therefore$ Kinetic energy $= \dfrac{1}{2} (\rho.Al) V^2$

Let the pressure intensity caused due to valve closure is p

$\therefore$ Strain energy of the liquid in pipe $= \dfrac{p^2}{2K} \times \text{volume of liquid in pipe}$

                                        $= \dfrac{p^2}{2K} \times Al$

where $K = \text{bulk modulus of liquid.}$

Equating the loss of kinetic energy to the gain of strain energy.

$\therefore$ $\dfrac{1}{2} (\rho.Al) V^2 = \dfrac{p^2}{2K} Al$

or $p^2 = \rho.K V^2$

or $p = \sqrt{\rho K} \, . V$

As we know, the velocity of pressure wave:

$$a = \sqrt{\dfrac{K}{\rho}}$$

or
$$a^2 = \frac{K}{\rho}$$

or
$$K = \rho a^2$$

$\therefore$
$$p = \sqrt{\rho.\rho a^2}.V = \rho Va \qquad \qquad ...(iii)$$

Equation (iii) is for rigid pipe material and compressible liquid flow through pipe. The pressure rise due to instantaneously closure of valve is equal to the product of the density of liquid, velocity of liquid flow pipe before closure of valve and velocity of pressure wave.

4.8.3 Pressure Rise due to Instantaneous Closure of Valve in an Elastic Pipe

The pressure rise due to rapid closure of the valve causes radial expansion of the walls of an elastic pipe. The hoop or circumferential and longitudinal stresses are produced and consequently some of the kinetic energy of the liquid is absorbed by the pipe as strain energy.

As we know, hoop or circumferential stress,

$$f_1 = \frac{pd}{2t}$$

and longitudinal stress: $\quad f_2 = \frac{pd}{4t}$

where $\qquad\qquad\qquad t$ = thickness of the pipe wall.

$\therefore$ Strain energy per unit volume stored in the pipe material

$$= \frac{1}{2E}[f_1^2 + f_2^2 - 2\mu f_1 f_2]$$

where $\qquad\qquad\qquad E$ = modulus of elasticity of the pipe material.

and $\qquad\qquad\qquad \mu$ = poisson's ratio of the pipe material,

Volume of the pipe material $= \pi\, d\, t.\, l$

$\therefore\quad$ Strain energy of the material

$$= \frac{\pi dt\, l}{2E}[f_1^2 + f_2^2 - 2\mu f_1 f_2]$$

$$= \frac{\pi dt\, l}{2E}\left[\left(\frac{pd}{2t}\right)^2 + \left(\frac{pd}{4t}\right)^2 - 2\mu \times \left(\frac{pd}{2t}\right)\left(\frac{pd}{4t}\right)\right]$$

$$= \frac{\pi dt\, l}{2E}\left[\frac{p^2 d^2}{4t^2} + \frac{p^2 d^2}{16t^2} - \frac{\mu p^2 d^2}{4t^2}\right]$$

Taking Poisson's ratio: $\quad \mu = \frac{1}{4}$

$$= \frac{\pi dt\, l}{2E}\left[\frac{p^2 d^2}{4t^2} + \frac{p^2 d^2}{16t^2} - \frac{p^2 d^2}{16t^2}\right] = \frac{\pi dt\, l}{2E}\left[\frac{p^2 d^2}{4t^2}\right]$$

$$= \frac{\pi p^2 d^3 l}{8t\, E}$$

$$= \frac{p^2 d\, l}{2\, E\, t} \times \frac{\pi d^2}{4}$$

$$\left[\because \text{ cross-sectional area of the pipe: } A = \frac{\pi}{4} d^2 \right]$$

$$= \frac{p^2 d\, l}{2\, E\, t} . A$$

Applying Energy Balance equation

Kinetic energy of liquid = strain energy of liquid + strain energy of pipe material

$$\frac{1}{2} \rho A l V^2 = \frac{p^2}{2K} Al + \frac{p^2 l d}{2Et} . A$$

$$\rho V^2 = \frac{p^2}{K} + \frac{p^2 d}{Et}$$

or

$$\rho V^2 = p^2 \left[\frac{1}{K} + \frac{d}{Et} \right]$$

$$p^2 = \frac{\rho V^2}{\left(\frac{1}{K} + \frac{d}{Et} \right)}$$

or

$$p = \sqrt{\frac{\rho V^2}{\left(\frac{1}{K} + \frac{d}{Et} \right)}} = V \sqrt{\frac{\rho}{\left(\frac{1}{K} + \frac{d}{Et} \right)}}$$

$\therefore$ Pressure rise due to instantaneous closure valve in an elastic pipe:

$$p = V \sqrt{\frac{\rho}{\left(\frac{1}{K} + \frac{d}{Et} \right)}}$$

For a rigid pipe, modulus of elasticity: $E = \infty$

or

$$\frac{d}{Et} = 0$$

Then above equation is reduced to

$$p = V \sqrt{\frac{\rho}{\frac{1}{K}}} = V \sqrt{\rho K}$$

which is same as equation derived for pressure rise due to instantaneous closure of valve in rigid pipe.

Problem 4.33: Water is flowing in a pipe line of 200 mm diameter at the rate of the 40 litre/s. A valve is introduced in the pipe line at a distance of 600 m. The valve is gradually closed in a time of 1.5 seconds. Calculate the increase in pressure intensity.

Solution: Given data:

Diameter of pipe: $d = 200$ mm $= 0.2$ m

Discharge: $Q = 40$ litre/s $= \dfrac{40}{1000}$ m³/s $= 0.04$ m³/s

Length of pipe: $l = 600$ m

Time taken to gradually close the valve: $t = 1.5$ s

Now, average velocity of flow:

$$V = \dfrac{\text{Discharge}}{\text{Cross-sectional area of the pipe}} = \dfrac{Q}{A}$$

$$= \dfrac{0.04}{\dfrac{\pi}{4}d^2} = \dfrac{0.04}{\dfrac{3.14}{4}(0.2)^2} = 1.27 \text{ m/s}$$

Pressure rise due to gradual closure of valve:

$$p = \dfrac{\rho l V}{t} = \dfrac{1000 \times 600 \times 1.27}{1.5} - 508000 \text{ N/m}^2 \text{ ro Pa.}$$

$$= \mathbf{508\ kPa.}$$

Problem 4.34: A rigid pipe conveying water is 2500 m long. The velocity of flow is 1.2 m/s. Calculate the rise of pressure caused within the pipe due to valve closure in (i) 10 seconds (ii) 2.5 seconds.

Take bulk modulus of water equal to 20×10^8 N/m²

Solution. Given data:

Length of pipe: $l = 2500$ m

Velocity of water flowing through pipe:

$$V = 1.2 \text{ m/s}$$

Now, velocity of the pressure wave:

$$a = \sqrt{\dfrac{K}{\rho}} = \sqrt{\dfrac{20 \times 10^8}{1000}} = 1414.21 \text{ m/s}$$

Critical time: $t_c = \dfrac{2l}{a} = \dfrac{2 \times 2500}{1414.21} = 3.53$ s

(i) When the valve is closed in 10 seconds

i.e., $t = 10$ s and $t_c = 3.53$ s

$$t < t_c$$

∴ This is a case of gradual valve closure

∴ Pressure rise: $\qquad p = \dfrac{\rho l V}{t} = \dfrac{1000 \times 2500 \times 12}{10}$

$\qquad\qquad\qquad = 300000 \text{ N/m}^2 = \textbf{300 kPa} \quad (\because 1 \text{ N/m}^2 = 1 \text{ Pa})$

(*ii*) When the valve is closed in 2.5 seconds

i.e., $\qquad\qquad\qquad\qquad t = 2.5 \text{ s and } t_c = 3.53 \text{ s}$

$\qquad\qquad\qquad t < t_c$

∴ This will be regarded as instantaneous valve closure.

Since the pipe is rigid, pressure rise: $p = \rho V a$

$\qquad\qquad p = 1000 \times 1.2 \times 1414.21 = 1697052 \text{ N/m}^2 \text{ or Pa}$

$\qquad\qquad = \textbf{1697.05 kPa.}$

Problem 4.35: A 800 mm diameter steel pipe carries water at the rate of 0.75 m³/s. The pipe wall has a thickness of 10 mm. The elastic modulus of steel is 20 × 10¹⁰ N/m² and the bulk modulus of water is 2 × 10⁹ N/m². Determine the increase in pressure if the valve at the end of 3500 m long pipeline is closed in 5 seconds.

Solution: Given data:

Diameter of pipe:	$d = 800 \text{ mm} = 0.8 \text{ m}$
Discharge:	$Q = 0.75 \text{ m}^3/\text{s}$
Thickness of pipe:	$t = 10 \text{ mm} = 0.01 \text{ m}$
Elastic modulus of steel:	$E_s = 20 \times 10^{10} \text{ N/m}^2$
Elastic modulus of water:	$E = 2 \times 10^9 \text{ N/m}^2$
Length of pipe:	$l = 3500 \text{ m}$
Time of valve closure:	$t = 5 \text{ s}$

Now, the combined modulus of elasticity,

$$K = \dfrac{E}{1 + \dfrac{d}{t} \cdot \dfrac{E}{E_s}}$$

$$= \dfrac{2 \times 10^9}{1 + \dfrac{0.8}{0.01} \times \dfrac{2 \times 10^9}{20 \times 10^{10}}} = \dfrac{2 \times 10^9}{1 + 0.8}$$

$$= 1.11 \times 10^9 \text{ N/m}^2$$

Velocity of pressure wave: $\qquad a = \sqrt{\dfrac{K}{\rho}} = \sqrt{\dfrac{1.11 \times 10^9}{1000}} = 1053.56 \text{ m/s}$

The critical time for valve closure is given by

$$t_c = \dfrac{2l}{a} = \dfrac{2 \times 3500}{1053.58} = 6.64$$

since $t < t_c$ the valve closure is rapid.

The average velocity of flow: $\qquad V = \dfrac{Q}{A} = \dfrac{Q}{\dfrac{\pi}{4} d^2} = \dfrac{4Q}{\pi d^2}$

$$= \frac{4 \times 0.75}{3.14 \times (0.8)^2} = 1.49 \text{ m/s}$$

∴   The rise of pressure: $p = \rho V a$

$$= 1000 \times 1.49 \times 1053.56$$

$$= 1569904.4 \text{ N/m}^2 \text{ or Pa} = \textbf{1569.90 kPa} .$$

4.9 PIPE NETWORKS

Pipe networks consist of multiple pipes interconnected in series, parallel and forming several loops or circuits. A typical example of such system is the municipal water distribution system in our cities. Multiple pipe systems reach a limit of complexity in the problems of distribution of flow in pipe-networks. A sample pipe networks is shown in Fig. 4.39.

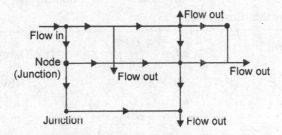

Fig. 4.39: A simple pipe network.

The most important factor is that the flow rate through the entire system (*i.e.,* in each pipe) should remain constants regardless of the diameters of the individual pipe in the system. The fluid flow in a pipe network must satisfy the following conditions:

1. The total flow into each junction must be equal to the total flow out of the junction (*i.e.,* to satisfy the continuity equation).

2. The head loss between two junctions must be the same for all points between the two junctions.

3. The algebraic sum of head losses in each loop must be zero. A head loss is taken to be positive for flow in the clockwise direction and negative for flow in anti-clockwise direction.

4. The head loss in each pipe is expressed by:

$$h_f = KQ^n \qquad\qquad (i)$$

   where K is a constant for each pipe which depends upon the fluid viscosity and the length of pipe, diameter of pipe and coefficient of friction of pipe.

   For turbulent flow,

$$n = 2 \qquad \text{for rough pipe}$$

$$n = 1.85 \qquad \text{for smooth pipe}$$

4.9.1 Hardy Cross Method (HCM)

The pipe network problems are generally solved by trail and error procedure. The most practical and widely used method of flow analysis is that of successive approximations developed by Hardy Cross.

The procedure of Hardy cross method is as follows:

(a) Assume the reasonable discharge in each pipe which satisfied the continuity equation at each junction.

(b) The head loss in each pipe is calculated according to Eq. (i),

$$h_f = KQ^n$$

where K is constant for each pipe which depends upon μ, d, l, and f and for turbulent flow:

$$n = 2 \qquad \text{for rough pipe}$$
$$n = 1.85 \qquad \text{for smooth pipe}$$

We know that head loss:

$$h_f = \frac{4 flV^2}{2gd} \qquad \text{(by Darcy-Weisbach formula)}$$

Discharge:
$$Q = AV \qquad \text{(by continuity equation)}$$

$$Q = \frac{\pi}{4} d^2 V$$

or
$$V = \frac{4Q}{\pi d^2}$$

$$\therefore \qquad h_f = \frac{4 fl}{2gd} \times \frac{16Q^2}{\pi^2 d^4} = \frac{32 flQ^2}{g\,\pi^2 d^5}$$

$$\boldsymbol{h_f = K\,Q^2}$$

where
$$K = \frac{32 fl}{g\,\pi^2 d^5}$$

$n = 2$, for turbulent flow in rough pipe

(c) The net head in each loop, i.e., $\Sigma h_f = \Sigma KQ^n$. A head loss is taken to be positive for flow in the clockwise direction and negative for flow in anti-clockwise direction.

(d) In order to satisfy condition (3), $\Sigma h_f = \Sigma KQ^n = 0$, if the net head loss due to assumed value of discharge Q_0 of the loop is zero, then the assumed value of Q_0 in that loop is correct. But if the net head loss due to assumed value of Q_0 is not zero, then the assumed value Q_0 is corrected by introducing a correction ΔQ for the flow till a correction is balanced.

For any pipe we may write

$$Q = Q_0 - \Delta Q$$

where
$$Q_0 = \text{assumed discharge}$$
$$Q = \text{correct discharge}$$

Then for each pipe

$$h_f = KQ^n = K(Q_0 - \Delta Q)^n$$
$$= K(Q_0^n - nQ_0^{n-1}\Delta Q \dots)$$

where the remaining terms of the series may be ignored if ΔQ is small as compared with Q_0.

$$h_f = K(Q_0^n - nQ_0^{n-1}\Delta Q)$$

For a single loop or circuit, with ΔQ same for all pipes.

$$\Sigma h_f = \Sigma KQ^n$$
$$= \Sigma KQ_0^n - \Delta Q \Sigma nKQ_0^{n-1} = 0$$

or $$\Sigma KQ_0^n - \Delta Q \Sigma nKQ_0^{n-1} = 0$$

or $$\Delta Q \Sigma nKQ_0^{n-1} = \Sigma KQ^n$$

or $$\Delta Q = \frac{\Sigma KQ_0^n}{\Sigma nKQ_0^{n-1}} \qquad \dots(ii)$$

It may be emphasised that numerator of Eq. (ii) is to be summed algebraically with due consideration of sign, while the denominator is summed arithmetically without any consideration of direction of flow. It may be noted that the assumption of considering only the first two terms of the series is not accurate if ΔQ is of the same order of magnitude as Q_0. The direction of ΔQ will be anti-clockwise if it is positive and the direction of ΔQ will be clockwise if it is negative.

(e) Revise the assumed discharge and repeat the process until the desired accuracy is obtained. Usually not more than three trials are necessary except for very complex system.

(f) After the corrections have been applied to each pipe in a loop and to all loops, a second trial calculation is made for all loops. The procedure is repeated till correction ΔQ becomes negligible.

Problem 4.36: A pipe network shown in Fig. 4.38 in where Q and h_f refer to discharges and head losses respectively. Find the values of discharges Q_B, Q_2, Q_4 and Q_5, and head losses h_{f_4} and h_{f_5} and gives these computed values at their respective places on a neat sketch of the network along with flow direction.

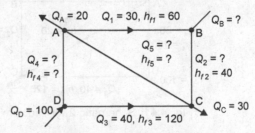

Fig. 4.40: Schematic 1 for Problem 4.36

Solution: At each junction, $\Sigma Q = 0$

 i.e., Discharge incoming = discharge leaving the junction

At junction D: $Q_o = Q_3 + Q_4$

                                                                   $100 = 40 + Q_4$

or $Q_4 = 100 - 40 = \mathbf{60}(Q_4$ leaves the junction $D)$

At junction A:

                                                                   $Q_4 = Q_A + Q_1 + Q_5$

                                                                   $60 = 20 + 30 + Q_5$

or $Q_5 = \mathbf{10}$ (Q_5 leaves the junction A)

At junction C:

                                                                   $Q_3 + Q_5 + Q_2 = Q_c$

                                                                   $40 + 10 + Q_2 = 30$

or $Q_2 = \mathbf{-20}$ (–ve sign shows that Q_2 leaves
                                                                                                        the junction C)

At junction B:

                                                                   $Q_1 + Q_2 = Q_B$

                                                                   $30 + 20 = Q_B$

or $Q_B = \mathbf{50}$ (Q_B leaves the junction B)

For each loop or circuit, $\Sigma h_f = 0$

For loop ACB:

$$h_{f_1} - h_{f_2} - h_{f_5} = 0$$

$$60 - 40 - h_{f_5} = 0$$

or $h_{f_5} = \mathbf{20}$

For loop ACD:

$$h_{f_5} - h_{f_3} - h_{f_4} = 0$$

$$20 - 120 - h_{f_4} = 0$$

or $h_{f_4} = \mathbf{100}$

The computed values and the flow directions are shown in the Fig. 4.39

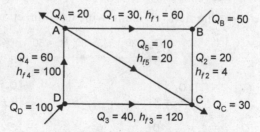

Fig. 4.41: Schematic 2 for Problem 4.36

Problem 4.37:   Find the distribution of discharge in pipe network as shown in Fig. 11.42. The head loss $h_f = KQ^2$. The value of K for each pipe is indicated in the figure.

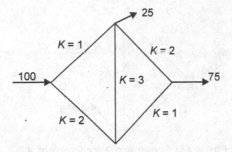

Fig. 4.42: Schematic 1 for Problem 4.37

Solution: By Hardy cross method:

Assume a suitable distribution of flow as shown in Fig. 4.41. So that the continuity equation is satisfied at each node. For this distribution of discharge, the corrections DQ for the loops ABC and BCD are calculated.

Given the head loss: $h_f = K Q^2$

where $n = 2$.

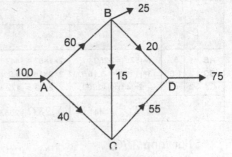

Fig. 4.43: Schematic 2 for Problem 4.37

Correction discharge: $\Delta Q = \dfrac{\Sigma KQ_0^2}{\Sigma 2KQ_0}$

			Loop ABC				Loop BCD	
Pipe	K	Q_0	$h_f = KQ_0^2$	$2KQ_0$	Pipe	K	Q_0	$h_f = KQ_0^2$	$2KQ_0$
AB	1	60	$1\times(60)^2 = 3600$	$2\times1\times60 = 120$	BD	2	20	$2\times(20)^2 = 800$	$2\times2\times20 = 80$
BC	3	15	$3\times(15)^2 = 675$	$2\times3\times15 = 90$	DC	1	55	$-1\times(55)^2 = -3025$	$2\times1\times55 = 110$
CA	2	40	$-2\times(40)^2 = -3200$	$2\times2\times40 = 160$	CB	3	15	$-3\times(15)^2 = 675$	$2\times3\times15 = 90$
			$\Sigma KQ_0^2 = 1075$	$\Sigma 2KQ_0 = 370$				$\Sigma KQ_0^2 = -2900$	$\Sigma 2KQ = 280$

For loop *ABC*:

$$\Delta Q = \frac{\Sigma KQ_0^2}{\Sigma 2KQ_0} = \frac{1075}{370} = 2.9 \approx 3$$

For loop *BCD*:

$$\Delta Q = \frac{\Sigma KQ_0^2}{\Sigma 2KQ_0} = \frac{-2900}{280} = -10.35$$

$+ ve\ \Delta Q$ is to be added to anti-clockwise flow and subtracted from clockwise and the $- ve\ \Delta Q$ added to clockwise and subtracted from anti-clockwise flow. The revised distribution of discharge is shown in Fig. 4.44

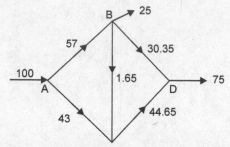

Fig. 4.44: Schematic 3 for Problem 4.37

Repeating the same procedure as above for the two circuits.

Loop *ABC*					Loop *BCD*			
Pipe	K	Q_0	$h_f = KQ_0^2$	$2KQ_0$	Pipe K Q_0	$h_f = KQ_0^2$	$2\,KQ_0$
AB	1	57	$1 \times (57)^2 = 3250$	$2 \times 1 \times 57 = 144$	BD 2 30.35	$2 \times (30.35)^2 = 1842.24$	$2 \times 2 \times 30.35 = 121.4$
BC	3	1.65	$3 \times (1.65)^2 = 8.16$	$2 \times 3 \times 16.5 = 9.9$	DC 1 44.65	$-1 \times (44.65)^2 = -1993.62$	$2 \times 1 \times 44.65 = 89.3$
CA	2	43	$-2 \times (43)^2 = -3700$	$2 \times 2 \times 43 = 172$	CB 3 1.65	$-3 \times (1.65)^2 = -8.16$	$2 \times 3 \times 16.5 = 9.9$
			$\sum KQ_0^2 = -441.84$	$\sum 2KQ_0 = 295.9$		$\sum KQ_0^2 = -159.54$	$\sum 2KQ = 220.6$

For loop *ABC*:

$$\Delta Q = \frac{\sum KQ_0^2}{\sum 2KQ_0} = \frac{-441.84}{295.9} = -1.49$$

For loop *BCD*:

$$\Delta Q = \frac{\sum KQ_0^2}{\sum 2KQ_0} = \frac{-159.54}{220.6} = -0.72$$

To apply the correction $\Delta Q = -1.49$ for loop *ABC* and $\Delta Q = -0.72$ for loop *BCD*, we obtain the distribution of flow as shown in Fig. 4.43 which may be taken as almost correct.

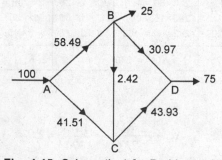

Fig. 4.45: Schematic 4 for Problem 4.37

4.10 SURGE TANK

Surge tank is main component of hydro-electric power plant. It is located on the penstock (penstock is a pipeline which carries the water from reservoir and supplies to the turbine inlet) and near to the turbine as far as possible. The height of surge tank is generally kept above the maximum water level in the reservoir. It is very necessary when the length of penstock is long. It serves the following two functions:

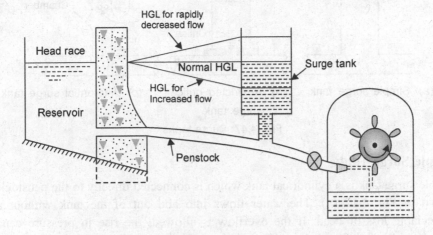

Fig. 4.46: Surge tank in hydro-electric power plant.

(1) It acts as a temporary storage device to store the water when the governor reduces the supply of water to the turbine. On the other hand, if the turbine needs more water, then load on turbine increases, the excess water made available to the turbine from storage tank.

(2) It protects the penstock against water hammer.

Working

When the flow of water in a penstock is decreased or stopped by closing the valve or turbine gate, the velocity of water decreases and consequently pressure increases. This increase in pressure causes the formation of pressure wave which raises the water into surge tank (if surge tank is not used, this pressure wave causes water hammer). On the other hand, when turbine needs more water, then excess water is supplied to the turbine by surge tank.

4.10.1 Types of Surge Tanks

Surge tank may be of the following types:

(1) Simple surge tank

(2) Restricted orifice type surge tank, and

(3) Differential surge tank.

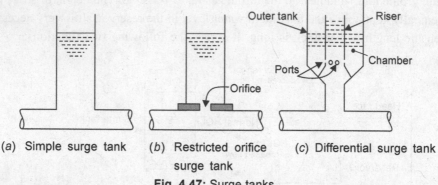

(a) Simple surge tank   (b) Restricted orifice   (c) Differential surge tank
                             surge tank
Fig. 4.47: Surge tanks.

Simple Surge Tank

Simple surge tank is a cylindrical tank which is connected directly to the penstock as shown in Fig. 4.47 (a). The water flows into and out of the tank without any appreciable loss of head. If the overflow is allowed, the rise in pressure can be eliminated but overflow surge tank is seldom satisfactory and usually uneconomical. It is built high enough so that water cannot overflow even with a full load removed on the turbine. But the use of this surge tank is restricted.

Restricted Orifice Type Surge Tank

The surge tank is connected to the penstock through an orifice provided at the tank as shown in Fig. 4.47 (b). The main objective of providing a throttle or restricted orifice is to create an appreciable friction loss when the water is flowing to or from the tank. When the load on the turbine is reduced, the surplus water passes through the orifice and a retarding head equal to the loss due to throttle is built up in the penstock. The size of the throttle can be designed according to designed retarding head. The size of the throttle adopted is usually such as the initial retarding head is equal to the rise of water surface in the tank when the full load is rejected by the turbine (*i.e.,* water supply to the turbine is stopped by closure of the gate valve). At small and full load changes, such surge tank is not very effective in speed regulation. The design is also complicated and hence this type of surge tank is not much in use.

Differential Surge Tank

It is much more efficient than simple and restricted orifice surge tanks. In this surge tank, the tank is connected at its bottom to the penstock through a small vertical pipe called riser as shown in figure 4.47 (c). The riser is provided with a number of ports at its bottom to connect with the tank. When there is decrease in load on the turbine, the water rises fast in the riser and provides a quick retarding effect. Similarly, when the load increases, the water falls first in the riser quickly, thus creating a large accelerating head on the penstock in a short time (*i.e.,* enough water supply to the turbine). The water falls slowly in the outer tank being a larger area and water flows through the ports into the riser pipe. In differential surge tank, the head building function is achieved by the riser (high height, small cross-sectional area vertical pipe), while the storage function is achieved through the outer tank.

4.11 THREE RESERVOIR PROBLEM

The three reservoir problem has three interconnected pipelines meeting at a common junction and originating from three reservoirs having different water surface levels. The three reservoir problem is about finding the direction and magnitude of the discharges in the three pipes when the geometric characteristics of the pipes and water surface elevations of the three reservoirs are known. Referring to Fig. 4.48, three reservoir A, B and C are connected by three pipes 1, 2 and 3 meet at the common junction J. The friction loss in the three pipe lines is sufficient large to permit the neglecting of minor losses (entry and exit losses). For solving any problem related to three reservoirs, the following basic principles apply:

(1) Continuity equation must be obeyed at junction J i.e., total flow into the junction must be equal total flow out.

(2) Darcy's equation must be satisfied for each pipe.

(3) There can only be one value of head at any point in the system.

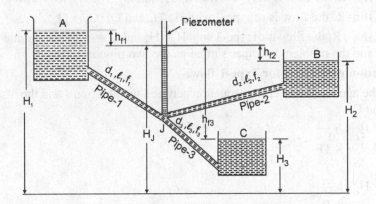

Fig. 4.48: Three reservoir problem.

Consider the reservoir A is the highest, B the interpediate and C the lowermost.

H_J = piezometric head at the junction J.

H_1 = head at the reservoir A

H_2 = head at the reservoir B

H_3 = head at the reservoir C

Q_1 = rate of flow in pipe-1

Q_2 = rate of flow in pipe-2

Q_3 = rate of flow in pipe-3

Following two methods are used to solve the three reservoir problem:

(1) Exact method, and

(2) Trial and error method.

In this type of problem the calculations are mostly simplified by putting

$$h_f = KQ^2 \qquad \text{where} \quad K = \frac{32 f \ell}{\pi^2 g d^5}$$

where h_f = head loss

K = resistance parameter which is constant for given pipe and its value can be calculated if pipe length l, diameter d and co-efficient of friction f and minor losses can be neglected.

4.11.1 Exact Method

The procedure described below is based on this method.

(*i*) To determine the direction of flow:

Let $\dfrac{H_1 - H_2}{H_2 - H_3} = h$ and $\dfrac{K_1}{K_3} = R$

(*a*) If h>R, the flow is of type-1 with h_J>H_2 and $Q_1 = Q_2 + Q_3$ as shown in Fig.4.46

(*b*) If h<R, the flow is of type-2 with H_J<H_2 and $Q_1 + Q_2 = Q_3$

(*c*) If h = R, the flow is of type-3 with $H_J = H_2$ and $Q_1 = Q_3$. This is a rare situation and the problem becomes similar as the two pipe in series.

(*ii*) **Solution procedure for Type-1 flow:**

By the application of energy equation between the junction J and the reservoir A, B and C.

$H_1 = H_J + h_{f1}$

$H_1 = H_J + K_1 Q^2$

$$\hspace{10cm} (1)$$

$H_2 = H_J - h_{f2}$

$H_2 = H_J - K_2 Q^2$

$$\hspace{10cm} (2)$$

and $H_3 = H_J - h_{f2}$

$H_3 = H_J - K_3 Q_3^2$

$$\hspace{10cm} (3)$$

We know that the continuity equation for type-1 flow,

$Q_1 = Q_2 + Q_3$

Let $Q_3 = nQ_2$

∴ $Q_1 = Q_2 + nQ_3$

$Q_1 = (1+n)\, Q_2$

$$\hspace{10cm} (4)$$

Eq (1) – Eq (2), we get

$H_1 - H_2 = K_1\, Q_1^2 + K_2\, Q_2^2$

Substituting the value of Q_1 from Eq. (4) in above equation, we get

$$H_1 - H_2 = K_1 (1+n)^2 Q_2^2 + K_2 Q_2^2$$

$$H_1 - H_2 = [K_1 (1+n)^2 + K_2] Q_2^2 \tag{5}$$

Eq (2) – Eq (3), we get

$$H_2 - H_3 = K_3 Q_3^2 - K_2 Q_2^2$$

$$H_2 - H_3 = K_3 n^2 Q_2^2 - K_2 Q_2^2 \qquad \because Q_3 = n Q_2$$

$$H_2 - H_3 = (K_3 n^2 - K_2) Q_2^2 \tag{6}$$

Dividing Eq (5) by Eq (6), we get

$$\frac{H_1 - H_2}{H_2 - H_3} = \frac{K_1 (1+n)^2 + K_2}{K_3\ n^2 - K_2} = h \tag{7}$$

Eq. (7) is a quadratic equation in n. Considering the positive root as the relative value, Q_2 is found by applying the value of n in Fig. (5) or Eq (6) and hence value of Q_3 is found by relation $Q_3 = n\ Q_2$ and Q_1 is found by continuity equation

$$Q_1 = Q_2 + Q_3$$

The head at the junction H_J is found by energy relationship between a reservoir and the junction, e.g., $H_1 = H_J + K_1 Q_1^2$.

(*iii*) **Solution procedure for Type-2 flow:**

By the application of energy equation between the junction J and the reservoir A, B and C.

$$H_1 = H_J + h_{f1}$$

$$H_1 = H_J + K_1 Q^2 \tag{8}$$

$$H_2 = H_J + h_{f2}$$

$$H_2 = H_J + K_2 Q_2^2 \tag{9}$$

and $H_3 = H_J - h_{f2}$

$$H_3 = H_J - K_3 Q_3^2 \tag{10}$$

We know that the continuity equation type-2 flow,

$$Q_1 + Q_2 = Q_3$$

Let $Q_3 = n\ Q_2$

$\therefore \quad Q_1 + Q_2 = n\ Q_2$

or $\quad Q_1 = (n-1)\ Q_2$

Eq (8) – Eq (9), we get

$$H_1 - H_2 = K_1 Q_1^2 - K_2 Q_2^2$$

$$H_1 - H_2 = K_1 (n-1)^2 Q_2^2 - K_2 Q_2^2$$

$$H_1 - H_2 = [K_1 (n-1)^2 - K_2] Q_2^2 \tag{11}$$

Eq (9) – Eq (10) , we get

$$H_2-H_3 = K_2 Q_2^2 + K_3 Q_3^2$$
$$H_2-H_3 = K_2 Q_2^2 + K_3 n^2 Q_2^2$$
$$H_2-H_3 = (K_2 + K_3 n^2) Q_2^2 \tag{12}$$

Dividing Eq (11) by Eq (12), we get

$$\frac{H_1 - H_2}{H_2 - H_3} = \frac{K_1 (n-1)^2 - K_2}{K_3 n^2 + K_2} = h \tag{13}$$

Eq (13) is a quadratic equation in n which two positive roots. Selecting the root where $n > 1$ as relevent (as $n < 1$ gives –ve Q_1 values) the discharge Q_2 is found from Eq (11), hence value of Q_3 is found by relation $Q_3 = n Q_2$ and Q_1 is found by continuity equation $Q_1 + Q_2 = Q_3$

The head at the junction H_J is found by relationship between a reservoir and the junction, e.g., $H_1 = H_J + K_1 Q_1^2$

4.11.2 Trial and Error Method

The procedure described below is based on this method.

(i) Assume a trial value of H_J. The first trial H_J may be taken around the average value of the lowest and highest reservoir levels.

(ii) For each H_J calculate Q_i in each pipeline with positive sign if it is towards the junction and negative sign if it is away from the junction. Find $\Delta Q = \Sigma Q_i$ and also find $\Sigma \left| \dfrac{Q}{h_f} \right|$.

(iii) The additive correction, to be added to the assumed value of H_J for purposes of next trial, is

$$\Delta H_J = \frac{2\Delta Q}{\Sigma \left| \dfrac{Q}{h_f} \right|}$$

(iv) For next trial, H_J = previous H_J + ΔH_J

(v) Continue till ΔQ is very small.

Notice: In this book, friction factor is denoted by f' and co-efficient of friction is denoted by f.

we know that

$$f' = 4f$$

Friction factor = 4 times co-efficient of friction.

Problem 4.38: Determine the rate of flow through the pipes shown in Fig 11.49. Take the friction factor: $f' = 0.02$ for all pipes.

Pipe	Diameter (mm)	Length (m)	Connectivity
1	150	350	AJ
2	100	200	BJ
3	100	250	CJ

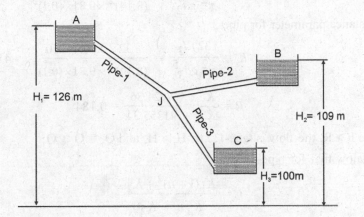

Fig. 4.49. : Schematic for Problem 4.38

Solution: Given data.

For pipe-1, $H_1 = 126$ m

$d_1 = 150$ mm $= 0.15$ m

$l_1 = 350$ m

For pipe -2,

$H_2 = 109$ m

$d_2 = 100$ mm $= 0.1$ m

$l_2 = 200$ m

For pipe-3,

$H_3 = 100$ m

$d_3 = 100$ mm $= 0.1$ m

$l_3 = 250$ m

Friction factor for all pipe: $f' = 0.02$

∴ Coefficient of friction: $f = \dfrac{f'}{4} = \dfrac{0.02}{4} = 0.005$

We know that

$$h = \frac{H_1 - H_2}{H_2 - H_3} = \frac{126 - 109}{109 - 100} = \frac{17}{9} = 1.88$$

Resistance parameter for pipe-1,

$$K_1 = \frac{32 f l_1}{\pi^2 g d_1^5} = \frac{32 \times 0.005 \times 350}{(3.14)^2 \times 9.81 \times (0.15)^5} = 7624.36$$

Resistance parameter for pipe-2,

$$K_2 = \frac{32 f l_2}{\pi^2 g d_2^5} = \frac{32 \times 0.005 \times 200}{(3.14)^2 \times 9.81 \times (0.1)^5} = 33084.27$$

Resistance parameter for pipe-3

$$K_3 = \frac{32 f l_3}{\pi^2 g d_3^5} = \frac{32 \times 0.005 \times 250}{(3.14)^2 \times 9.81 \times (0.1)^5} = 41355.34$$

$$R = \frac{K_1}{K_3} = \frac{7624.36}{41355.34} = 0.184$$

Since h > R, the flow is type-1, *i.e.*, $H_J > H_2$ and $Q_1 = Q_2 + Q_3$

We know that for type-1 flow,

$$h = \frac{K_1 (1+n)^2 + K_2}{K_3 n^2 - K_2}$$

$$1.88 = \frac{7624.36 (1+n)^2 + 33084.27}{41355.34 \, n^2 - 33084.27}$$

$$1.88 = \frac{(1+n)^2 + 4.34}{5.42 n^2 - 4.34}$$

$$10.18 \, n^2 - 8.15 = (1+n)^2 + 4.34$$

or $10.18 \, n^2 - 12.49 = 1 + n^2 + 2n$

or $9.18 \, n^2 - 2n - 13.49 = 0$

$$n = \frac{-b \pm \sqrt{b^2 - 4ac}}{2a}$$

$$= \frac{2 \pm \sqrt{4 - 4 \times 9.18 \times (-13.49)}}{2 \times 9.18}$$

$$n = \frac{2 \pm 23.09}{18.36} = 1.336$$

For type-1 flow,

$$H_1 - H_2 = [K_1 (1+n)^2 + K_2] \, Q_2^2$$

$$126 - 109 = [7624.36 \, (1+1.366)^2 + 33084.27] \, Q_2^2$$

$$17 = 75765.10 \, Q_2^2$$

or $Q_2^2 = 2.2437 \times 10^{-4}$

or $\mathbf{Q_2 = 0.01497 \ m^3/s}$

Discharge through pipe -3,

$$Q_3 = nQ_2 = 1.366 \times 0.01497 = \mathbf{0.02044 \ m^3/s}$$

Discharge through pipe-1,

$$Q_1 = Q_2 + Q_3 = 0.01497 + 0.02044 = \mathbf{0.03541\ m^3/s}$$

Problem 4.39:   The water levels in two reservoirs A and B are 104.5 m and 100 m respectively above the datum. A pipe joins each to a common point J, where pressure is 9.81 kn/m² gauge and height is 83.5 m above datum. Another pipe connects J to another tank C. What will be the height of water level in C assuming the same value of coefficient of friction f for all pipes.

Take $f = 0.0075$.

Pipe	Length m	Diameter mm
AJ	240	300
BJ	270	450
CJ	300	600

Solution: Given data:

$$H_1 = 104.5\ m$$
$$H_2 = 100\ m$$

Gauge pressure at junction J: $p_g = 98.1\ kN/m^2 = 98.1 \times 10^3\ N/m^2$

also $p_g = \rho gh$

∴ $98.1 \times 10^3 = 1000 \times 9.81 \times h$

Pressure head at junction J: $h = 10\ m$

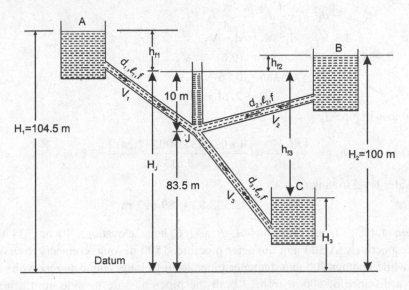

Fig. 4.50: Schematic for Problem 4.39

Position of junction J form datum= 8.35 m

∴ Piezometric head at J: $\quad H_J = 83.5 + 10 = 93.5$ m

Head loss between A and J: $h_{f1} = H_1 - H_J = 104.5 - 93.5 = 11$ m

Head loss between B and J: $h_{f2} = H_2 - H_J = 100 - 93.5 = 6.5$ m

Applying the Darcy's equation for pipe-1,

$$h_{f1} = \frac{4 f l_1 V_1^2}{2 g d_1}$$

$$11 = \frac{4 \times 0.0075 \times 240 \times V_1^2}{2 \times 9.81 \times 0.3}$$

or $\qquad\qquad V_1^2 = 8.992$

or $\qquad\qquad V_1 = 2.998$ m/s

Similarly applying the Darcy's equation for pipe-2,

$$h_{f2} = \frac{4 f l_2 V_2^2}{2 g d_2}$$

$$6.5 = \frac{4 \times 0.0075 \times 270 \times V_2^2}{2 \times 9.81 \times 0.45}$$

or $\qquad\qquad V_2^2 = 7.085$

or $\qquad\qquad V_2 = 2.66$ m/s

From the continuity equation,

$$Q_1 + Q_2 = Q_3$$

$$\frac{\pi}{4} d_1^2 V_1 + \frac{\pi}{4} d_2^2 V_2 = \frac{\pi}{4} d_3^2 V_3$$

or $\qquad d_1^2 V_1 + d_2^2 V_2 = d_3^2 V_3$

$$(0.3)^2 \times 2.998 + (0.45)^2 \times 2.66 = (0.6)^2 \times V_3$$

$$0.2698 + 0.5386 = 0.36 V_3$$

or $\qquad\qquad V_3 = 2.24$ m/s

Head loss in pipe-3,

$$h_{f3} = \frac{4 f l_3 V_3^2}{2 g d_3} = \frac{4 \times 0.0075 \times 300 \times (2.24)^2}{2 \times 9.81 \times 0.6} = 3.83\, m$$

∴ Water level in tank C,

$$H_3 = H_J - h_{f3} = 93.5 - 3.83 = \mathbf{89.697\ m}$$

Problem 4.40: Three reservoirs A, B and C have elevations 40 m , 34 m and 20 m respectively. A 300 mm diameter pipeline, 1500 m long, connects reservoirs A and B while another 300 mm diameter pipe, 2400 m long, connects reservoirs A and C. For a distance of 600 m from A both the pipes lie side by side upto a junction point J, where they are inter connected by a short branch pipe. Assuming that the pipe friction factor is 0.02, and neglecting losses at entry, exit and junction, find

(a) the direction of flow in reservoir B, and

(b) the flows leaving or entering the reservoir in litre/s.

Solution: Given data:
$$H_1 = 40 \text{ m}$$
$$H_2 = 34 \text{ m}$$
$$H_3 = 20 \text{ m}$$

For pipe-1,

Diameter: $\qquad d_1 = 300 \text{ mm} = 0.3 \text{ m}$

Length: $\qquad l_1 = 600 \text{ m}$

For pipe-2,

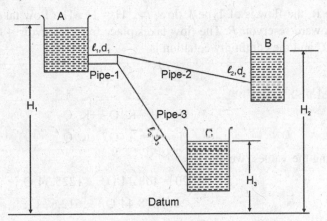

Fig. 4.51: Schematic for Problem 4.40

Diameter: $\qquad\qquad d_2 = 300 \text{ mm} = 0.3 \text{ m}$

Length: $\qquad\qquad l_2 = 900 \text{ m}$

For pipe-3:

Diameter: $\qquad\qquad d_3 = 300 \text{ mm} = 0.3 \text{ m}$

Length: $\qquad\qquad l_3 = 1800 \text{ m}$

Friction factor: $\qquad\qquad f' = 0.02$

Coefficient of friction : $\qquad f = \dfrac{f'}{4} = \dfrac{0.02}{4} = 0.005 \qquad\qquad \because f' = 4f$

Heads loss in the pipe line: $\quad h_f = KQ^2$

where $K = \dfrac{32fl}{\pi^2 gd^2}$, resistance parameter

Resistance parameter for pipe-1: $K_1 = \dfrac{32fl_1}{\pi^2 gd_1^5} = \dfrac{32 \times 0.005 \times 600}{(3.14)^2 \times 9.81 \times (0.3)^5} = 408.44$

Resistance parameter for pipe-2,

$$K_2 = \frac{32 f l_2}{\pi^2 g d_2{}^5} = \frac{32 \times 0.005 \times 900}{(3.14)^2 \times 9.81 \times (0.3)^2} = 612.67$$

Resistance parameter for pipe-3,

$$K_3 = \frac{32 f l_3}{\pi^2 g d_3{}^5} = \frac{22 \times 0.005 \times 1800}{(3.14)^2 \times 9.81 \times (0.3)^5} = 1225.34$$

$$R = \frac{K_1}{K_3} = \frac{408.44}{1225.34} = 0.333$$

and

$$h = \frac{H_1 - H_2}{H_2 - H_3} = \frac{40 - 34}{34 - 20} = 0.428$$

Since h > R, the flow is of Type-1 flow, *i.e.*, $H_J > H_2$. The flow takes place from the junction toward reservoir B. The flow takes place from reservoir A to reservoir B and reservoir C and the continuity equation is

$$2 Q_1 = Q_2 + Q_3 \qquad (1)$$

Using the Darcy's equation

$$H_1 - H_3 = K_1 Q_1{}^2 + K_3 Q_3{}^2$$
$$H_1 - H_2 = K_1 Q_1{}^2 + K_2 Q_2{}^2$$

Substituting the values, we get

$$20 = 408.44 \, Q_1{}^2 + 1225.34 \, Q_3{}^2 \qquad (2)$$

and

$$6 = 408.44 \, Q_1{}^2 + 612.67 \, Q_2{}^2 \qquad (3)$$

Let

$$\frac{Q_2}{Q_1} = n$$

$$Q_2 = n \, Q_1$$

Substituting $\quad Q_2 = nQ_1$ in Eq (2), we get

$$2Q_1 = nQ_1 + Q_3$$

or $\quad Q_3 = (2\text{-}n) \, Q_1$

Substituting the value of $\quad Q_3 = (2\text{-}n) \, Q_1$ in Eq. (2), we get

$$20 = 408.44 \, Q_1{}^2 + 1225.34 \, (2\text{-}n)^2 \, Q_1{}^2$$
$$20 = [408.44 + 1225.34 \, (2\text{-}n)^2] \, Q_1{}^2 \qquad (4)$$

Substituting the value of $\quad Q_2 = n \, Q_1$ in Eq. (3), we get

$$6 = 408.44 \, Q_1{}^2 + 612.67 \, n^2 \, Q_1{}^2$$
$$6 = [408.44 + 612.67 \, n^2] \, Q_1{}^2 \qquad (5)$$

Dividing Eq (4) by Eq (5), we get

$$\frac{20}{6} = \frac{408.44 + 1225.34(2-n)^2}{408.44 + 612.67\, n^2}$$

$$\frac{10}{3} = \frac{408.44\,[1+3(2-n)^2]}{408.44[1+1.5\,n^2]}$$

$$\frac{10}{3} = \frac{1+3(2-n)^2}{1+1.5\,n^2}$$

$$10 + 15\,n^2 = 3 + 9(4 + n^2 - 4n)$$

$$10 + 15\,n^2 = 3 + 36 + 9n^2 - 36n$$

$$6n^2 + 36n = 29 = 0$$

$$n = \frac{-b \pm \sqrt{b^2 - 4ac}}{2a} = \frac{-36 \pm \sqrt{(36)^2 - 4 \times 6 \times (-29)}}{2 \times 6}$$

$$= \frac{-36 \pm 44.63}{12} = 0.719$$

Substituting   n = 0.719 in Eq. (5) , we get

$$6 = [408.44 + 612.67 \times (0.719)^2]\ Q_1^2$$

$$6 = 725.16\ Q_1^2$$

or

$$Q_1^2 = \frac{6}{725.16} = 0.00827$$

$$Q_1 = 0.0909\,\text{m}^3\!/\text{s} = 90.9\,\text{litre/s}$$

$$Q_2 = nQ_1$$

$$= 0.719 \times 0.0909 = 0.06535\,\text{m}^3\!/\text{s} = 65.35\,\text{litre/s}$$

$$Q_3 = (2-n)Q_1 = (2-0.719) \times 0.0909 = 0.11644\,\text{m}^3\!/\text{s}$$

Flow leaving reservoir $A = 2Q_1 = 2 \times 0.0909 = 0.1818$ m³/s = **181.8 litre/s**

Flow entering reservoir B : $Q_2 = 0.06535$ m³/s = **65.35 litre/s**

Flow entering reservoir C : $Q_3 = 0.11644$ m³/s = **116.44 litre/s**

Problem 4.41: Determine the rate of flow through the pipes shown in Fig. 4.52. The free surface levels in the reservoir A, B and C are given, $l_1 = 700$ m, $l_2 = 1200$ m, $l_3 = 1000$ m. $d_1 = 400$ mm, $d_2 = 300$ mm, $d_3 = 200$ mm. Determine also the piezometic head at junction J. Take coefficient of friction: $f = 0.005$.

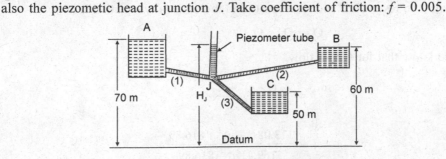

Fig. 4.42: Schematic for Problem 4.41

Solution: Given data:

$$H_1 = 70 \text{ m}$$
$$H_2 = 60 \text{ m}$$
$$H_3 = 50 \text{ m}$$
$$l_1 = 700 \text{ m}$$
$$l_2 = 1200 \text{ m}$$
$$l_3 = 1000 \text{ m}$$
$$d_1 = 400 \text{ mm} = 0.4 \text{ m}$$
$$d_2 = 300 \text{ mm} = 0.3 \text{ m}$$
$$d_3 = 300 \text{ mm} = 0.2 \text{ m}$$
$$f = 0.005$$

We know that
$$h = \frac{H_1 - H_2}{H_2 - H_3} = \frac{70 - 60}{60 - 50} = \frac{10}{10} = 1$$

Resistance parameter for pipe-1,

$$K_1 = \frac{32 f l_1}{\pi^2 g d_1^5} = \frac{32 \times 0.005 \times 700}{(3.14)^2 \times 9.81 \times (0.4)^5} = 113.08$$

Resistance parameter for pipe-2,

$$K_2 = \frac{32 f l_2}{\pi^2 g d_2^5} = \frac{32 \times 0.005 \times 1200}{(3.14)^2 \times 9.81 \times (0.3)^5} = 816.89$$

Resistance parameter for pipe-3,

$$K_3 = \frac{32 f l_3}{\pi^2 g d_3^5} = \frac{32 \times 0.005 \times 1000}{(3.14)^2 \times 9.81 \times (0.2)^5} = 5169.41$$

$$R = \frac{K_1}{K_2} = \frac{113.08}{5169.41} = 0.0218$$

Since h > R, the flow is of type-1 flow, *i.e.*, $H_J > H_2$. The flow takes place from reservoir *A* to reservoir *B* and reservoir *C* and the continuity equation is

$$Q_1 = Q_2 + Q_3$$

We know that for type-1 flow,

$$h = \frac{K_1 (1 + n)^2 + K_2}{K_3 n^2 - K_2}$$

$$1 = \frac{113.08 (1 + n)^2 + 816.89}{5169.41 n^2 - 816.89}$$

or $5169.41n^2 - 816.89 = 113.08(1+n)^2 + 816.89$

or $5169.41n^2 - 1633.76 = 113.08(1+n)^2$

$54.71n^2 - 14.44 = (1+n)^2$

$54.71n^2 - 14.44 = 1 + n^2 + 2n$

or $53.71n^2 - 2n - 15.44 = 0$

$$n = \frac{-b \pm \sqrt{b^2 - 2ac}}{2a} = \frac{2 \pm \sqrt{(2)^2 - 4 \times 53.71 \times (-15.44)}}{2 \times 53.71}$$

$$= \frac{2 \pm 57.62}{107.42} = 0.555$$

For type-1 flow,

$$H_1 - H_2 = [K_1 (1+n)^2 + K_2] Q_2^2$$

$$70-60 = [113.08 (1+0.555)^2 + 816.89] Q_2^2$$

$$10 = 1090.32 \ Q_2^2$$

or $Q_2^2 = 0.00917$

$Q_2 = 0.0957 \ m^3/s$

Discharge through pipe-3,

$$Q_3 = n Q_2 = 0.555 \times 0.0957 = 0.0531 \ m^3/s$$

Discharge through pipe-1,

$$Q_1 = Q_2 + Q_3 = 0.0957 + 0.0531 = 0.1488 \ m^3/s$$

By the application of energy equation between the junction J and the reservoir A.

$$H_1 = H_J + h_{f1}$$
$$H_1 = H_J + K_1 Q_1^2$$
$$70 = H_J + 113.08 \times (0.1488)^2$$
$$70 = H_J + 2.50$$

or $H_J = 70 - 2.50 = \textbf{67.5 m}$

Piezometric heat at junction J, $H_J = 67.5 \ m$

SUMMARY

1. When real fluid is flowing through a pipe, energy (or head) losses through pipe are classified in two categories:

   (i) Major losses

   (ii) Minor losses

2. **Major losses:** Energy (or head) losses due to friction among fluid particles or between fluid particles and surface area of a pipe at constant cross-sectional area of fluid flow is known as major losses. It is calculated by the following formulae:

Contd...

(*i*) **Darcy-Weisbach formula:**

Head loss due to friction: $h_f = \dfrac{4\,flV^2}{2gd}$

where f = coefficient of friction

$4f = f'$, friction factor

$h_f = \dfrac{f'lV^2}{2gd}$

$f = \dfrac{16}{Re}$ for $Re < 2000$, laminar flow

$f' = \dfrac{64}{Re}$

$f = \dfrac{0.079}{Re^{1/4}}$ for $4000 < Re < 10^6$, turbulent
flow

and $f' = \dfrac{0.316}{Re^{1/4}}$

(*ii*) **Chezy's formula:**

Velocity of flow: $V = C\sqrt{im}$

where C = Chezy's constant

m = hydraulic mean depth

$= \dfrac{A}{P}, \quad A = \dfrac{\pi}{4}d^2, \quad P = \pi d$

$m = \dfrac{d}{4}$ for pipe

$i = \dfrac{h_f}{l}$, loss of head due to friction per
unit length.

3. **Moody's diagram:** Moody's diagram gives the value of friction factor f' of any pipe provided its relative roughness $\in/d$ and Reynolds number of flow Re are known where $\in$ is absolute (or average) roughness, or equivalent sand roughness and d is diameter of pipe.

4. **Minor losses:** The loss of energy (or head) due to local disturbance in pipelines such as sudden enlargement, sudden contractions, pipe bends, valves and other fittings is called local or secondary or minor losses.

(*i*) Loss of head due to sudden enlargement: h_e

$$h_e = \dfrac{(V_1 - V_2)^2}{2g}$$

(*ii*) Loss of head due to sudden contraction: h_c

$$h_c = K_c \frac{V_2}{2g}$$

where $\qquad K_c = \left[\frac{1}{C_c}-1\right]^2$, coefficient of sudden

contraction

C_c = coefficient of contraction.

(iii) Loss of head at the entrance to a pipe: h_i

$$h_i = 0.5 \frac{V^2}{2g}$$

(iv) Loss of head at the exit of a pipe: h_o

$$h_o = \frac{V^2}{2g}$$

(v) Loss of head in pipe fitting: h_{fitting}

$$h_{\text{fitting}} = K_L \frac{V^2}{2g}$$

where $\qquad K_L$ = head loss coefficient

5. **Siphon:** A long bend pipe which rises above its hydraulic grade line has negative pressure, is called siphon.

6. **Pipes in series:** In piping systems, two or more pipes of different diameters are connected end to end, are called pipes in series or compound pipes. The rate of flow through the entire system remains constant regardless of the diameters of the individual pipes in the system.

Mathematically,

$$Q_1 = Q_2 = Q_3$$

7. **Equivalent length:** The loss of head in pipe fitting is expressed in terms of an equivalent length which is the length of uniform diameter pipe in which an equal loss of head would occur for same discharge.

8. **Equivalent pipe:** It is defined as the pipe of uniform diameter having loss of head equal to the total loss of head as the compound pipe (i.e., pipes in series) at same discharge. The uniform diameter of the equivalent pipe is called equivalent size of the pipe. The length of equivalent pipe is equal to sum of lengths of the compound pipe.

9. **Dupuit's equations:**

$$\frac{l_e}{d_e^5} = \frac{l_1}{d_1^5}+\frac{l_2}{d_2^5}+\frac{l_3}{d_3^5}+ ... \qquad \text{(for constant coefficient of friction)}$$

$$\frac{f_e l_e}{d_e^5} = \frac{f_1 l_1}{d_1^5}+\frac{f_2 l_2}{d_2^5}+\frac{f_3 l_3}{d_3^5}+ ... \quad \text{(for not constant coefficient of friction)}$$

10. Pipes in parallel: In piping systems, two or more pipes are connected between a main pipe, is called pipes in parallel. The rate of flow through main pipe is equal to the sum of rate of flow through branch pipes.

Mathematically,

$$Q = Q_1 + Q_2 + Q_3 + \dots\dots$$

11. Transmission of hydraulic power through pipe lines:

Power available at inlet of the pipe $= \rho QgH$

Power available at outlet of the pipe $= \rho Qg(H - h_f)$

where
H = total head available at the inlet of pipe

h_f = loss of head in pipe due to friction.

$$= \frac{4 flV^2}{2gd} \quad \text{(by Darcy's formula)}$$

Efficiency of power transmission: $\eta = \dfrac{H - h_f}{H}$

Power transmitted will be maximum when,

$$h_f = \frac{H}{3}$$

Then, maximum efficiency: $\eta_{max} = 66.67\%$

12. Water hammer: The phenomenon of sudden rise in pressure in a pipe when water flowing in it is suddenly brought to rest by closing the valve is known as water hammer or hammer blow.

(i) Pressure rise due to gradual closure of valve:

In this case, $t > t_c$

where
t = time taken to close the valve

$$t_c = \frac{2l}{a}, \text{ critical time}$$

l = length of pipe
a = velocity of pressure wave

Pressure rise:
$$p = \frac{\rho lV}{t}$$

where
V = velocity of liquid in the pipe.

(ii) Pressure rise due to instantaneous closure of valve in a rigid pipe.

In this case, $t < t_c$

Pressure rise: $p = \rho Va$

(iii) Pressure rise due to instantaneous closure of valve in an elastic pipe,

$$p = V\sqrt{\frac{\rho}{\left(\frac{1}{K} + \frac{d}{Et}\right)}}$$

where K = bulk modulus of liquid

E = modulus of elasticity of pipe

13. **Pipe networks:** Pipe networks consist of multiple pipes interconnected in series, parallel and forming several loops or circuits.

14. **Surge tank:** Surge tank is main component of hydro electric power plant. It is located on the penstock and near to the turbine as far as possible. It serves the following two functions:

   (*i*) It acts as a temporary storage device to store the water when the governor reduces the supply of water to the turbine. On the other hand, if the turbine needs more water, then load on turbine increases, the excess water made available to the turbine from surge tank.

   (*ii*) It protects the penstock against water hammer.

ASSIGNMENT - 1

1. What do you mean by the terms:
   (*i*) Major energy loss and
   (*ii*) Minor energy losses in pipes.

2. Explain briefly: Major and Minor losses in pipe lines.
                                          (*GGSIP University, Delhi, Dec. 2008*)

3. Derive an expression for the loss of head due to sudden enlargement in pipe diameter.   (*GGSIP University, Delhi, Dec. 2005*)

4. Derive an expression for the loss of head due to sudden contraction of a pipe.
                                          (*GGSIP University, Delhi, Dec. 2001*)

5. Derive an expression for the head loss due to sudden enlargement in pipe flow and therefrom deduce the head loss due to sudden contraction.
                                          (*GGSIP University, Delhi, Dec. 2008*)

6. Define and explain the terms: Pipes in series and pipes in parallel, equivalent pipe, minor energy losses and major energy loss.
                                          (*GGSIP University, Delhi, Dec. 2004*)

7. Define and explain: Equivalent pipe, minor and major losses in pipe, hydraulic gradient line and total energy line.   (*GGSIP University, Delhi, Dec. 2002*)

8. What is a syphon? On what principle it works?

9. Explain the terms:
   (*i*) Pipes in parallel
   (*ii*) Equivalent pipe and
   (*iii*) Equivalent size of the pipe.

10. What do you mean by the term pipe networks?

11. Show that for maximum transmission of power by means of water under pressure, the frictional loss of head in the pipe equals one-third of the total head supplied.

12. Explain the phenomenon of water hammer. Obtain an expression for the rise of pressure, when the flowing water in a pipe is brought to rest by closing the valve gradually.

13. What do you understand by water hammer? What provision is made in hydro electric power plant to minimize the effects of water hammer?

14. Water flowing in a pipe is brought to rest by instantaneous closure of valve in an elastic pipe. Show that the pressure rise due to valve closure is given by

$$p = V\sqrt{\frac{\rho}{\left(\frac{1}{K} + \frac{d}{Et}\right)}}$$

where
K = bulk modulus of liquid
E = modulus of elasticity of pipe
t = thickness of pipe

15. What is the function of surge tank in hydro-electric power plant?

ASSIGNMENT - 2

1. Determine the head loss due to friction in a pipe of diameter 300 mm and length 75 m, through which water is flowing at a velocity 3 m/s by using
   (a) Darcy-Weisbach formula
   (b) Chezy's formula for which $C = 6$. Take kinematic viscosity (v) for water = 0.01 stoke. **Ans.** (a) $h_f = 1.174$ m of water (b) $h_f = 2.5$ m of water

2. Water is flowing through a pipe 1.4 km long with a velocity of 0.9 m/s. What should be the diameter of the pipe, if the loss of head due to friction is 7 m. Take coefficient of friction: $f = 0.005$. **Ans.** 399.59 mm

3. A reservoir has been built 3.5 km away from a college campus having 5000 inhabitants. Water is to be supplied from the reservoir to the campus. It is estimated that each inhabitant will consume 250 litre of water per day, and that half of the daily supplied water is pumped within 10 hours. Find the size of the supply main, if the loss of head due to friction in pipeline is 20 m. Take coefficient of friction for the pipeline as 0.006. **Ans.** 159.95 mm

4. An oil of specific gravity 0.9 and viscosity 0.06 poise is flowing through a pipe of diameter 200 mm at the rate of 60 litre/s. Find the head lost due to friction for a 500 m length of pipe. Find also the power required to maintain the flow. **Ans.** 9.49 m of water, 5.027 kW

5. The rate of flow of water through a horizontal pipe is 300 litre/s. The pipe of diameter 200 mm is suddenly enlarged to 400 mm. Find:
   (i) Loss of head
   (ii) Change in pressure
   (iii) Loss of fluid power
   Ans. (i) 2.617 m of water (ii) 17.07 kPa (iii) 7.70 kW

6. At a sudden enlargement of a pipe from 240 mm to 480 mm and piezometric head increases by 10 mm. Find the rate of flow.   **Ans.** 32.70 litre/s

7. A horizontal pipe of diameter 150 mm is suddenly contracted to a diameter of 100 mm. The rate of flow of water is 30 litre/s. Find the pressure loss across the contraction. Take C_c= 0.6**Ans.** 9.094 kPa

8. Water is following through a horizontal pipe of diameter 200 m at a velocity of 3 m/s. A circular solid plate of diameter 150 mm is placed in the pipe due to obstruction in the pipe. Take coefficient of contraction in the pipe as 0.65. Find also loss of fluid power.   **Ans.** 2.905 m of water, 2.684 kW

9. Find the rate of flow of water through pipe of diameter 200 mm and length 50 m when one end of the pipe is connected to a tank and other end of the pipe is open to the atmosphere. The pipe is horizontal and the height of water in the tank is 4 m above the centre of the pipe. Consider all minor losses and take coefficient of friction as 0.009.**Ans.** 85.72 litre/s

10. Find the rate of flow through a horizontal pipe line 40 m long connected to a water tank at one end and discharges freely into the atmosphere at the other end. For the first 25 m of its length from the tank, the pipe is 150 mm diameter and its diameter is suddenly enlarged to 300 mm. The height of water level in the tank is 8 m above the centre of the pipe. Considering all minor losses take coefficient of friction as 0.01.   **Ans.** 78.63 litre/s

11. A pipe line AB of diameter 300 mm and length 400 mm carries water at the rate of 50 litre/s. The flow takes place from A to B where point B is 30 m above A. Find the pressure at point A if the pressure at point B is 19.62 N/cm². Take coefficient of friction as 0.008.   **Ans.** 50.12 N/cm²

12. What is syphon? On what principle it works? A syphon of diameter 200 mm connects two reservoirs having a difference of water level as 20 m. The total length of syphon is 500 m and the summit is 3 m above the water level in the upper reservoir. The length of the pipe from upper reservoir to the summit is 100 m. Determine:

    (*i*) the discharge through the syphon and

    (*ii*) the pressure head of pressure at the summit.

    Neglect minor losses and take coefficient of friction f = 0.005

    (GGSIP University, Delhi, Dec. 2004)

    Ans. (*i*) 87.96 litre/s (*ii*) – 7.399 mm of water or 2.901 m of water absolute

13. Water is discharged from one tank to another with 40 m difference of water levels through a pipe 1200 m long. The diameter for the first 600 m length of pipe is 400 mm and 250 mm for the remaining 600 m length. Find the discharge in litre/s through pipe. Assume coefficient of friction as 0.009 for both the pipes. Neglect minor losses. **Ans.** 140.78 litre/s

14. A pipe of diameter 500 mm and length 4 km is used for the transmission of power by water. The total head at the inlet of the pipe is 520 m. Find the maximum power available at the outlet of the pipe. Take coefficient of friction. f = 0.006.

    Ans. 2808.40 kW

15. The maximum power is to be transmitted through a pipe line. Work-out the condition for maximum transmission of power. It is desired to develop 1200 kW of power at 85% efficiency by supplying water to a hydraulic turbine through a horizontal pipe 600 m long. Determine the necessary flow rate and minimum diameter of pipe to carry that discharge, water is available at head of

200 m. Take $f = 0.0063$ in the formula, $h_f = \dfrac{4\,fl\,V^2}{2\,gd}$

Ans. 0.71955 m³/s, 0.4643 m

16. Power is to be transmitted hydraulically along a distance of 8500 m through a number of 100 mm diameter pipes, laid is parallel. The pressure at the discharge end is maintained constant at 7200 kPa. Find the maximum number of pipes required to ensure an efficiency of at least 91% when the power delivered is 182 kW. Take $f = 0.0061$ for all pipes and neglect losses other than friction.

   Ans. 4

17. A rigid pipe conveying water is 3000 m long. The velocity of flow is 1.3 m/s. Calculate the rise in pressure caused within the pipe due to valve closure in (*i*) 20 seconds, (*ii*) 3 seconds.

   Take bulk modulus of water as 20×10^8 N/m².

   Ans. (*i*) 195 kPa, (*ii*) 1838.47 kPa

18. A water supply consists of three reservoirs A, B and C connected to a common junction J as shown in Fig. 4.60

Pipe	Diameter mm	Length m	Coefficient of friction of
1.	300	200	0.005
2.	300	125	0.004
3.	300	250	0.004

Calculate the discharge in each pipe and the piezometric head at the junction.

[**Ans.** $Q_1 = 0.139$ m³/s, $Q_2 = 0.095$ m³/s

$Q_3 = 0.234$ m³/s, $H_J = 97.447$m]

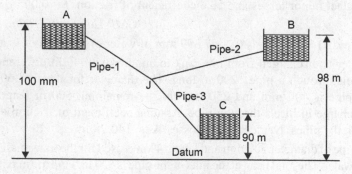

Fig. 4.60: Q.18

5

Pipe Flow Measurement

5.1 INTRODUCTION

After designing a system, it is fabricated and tested to determine whether it is reliable, cost effective and serves the proposed purpose. In engineering, designing, fabrication and testing involve measurements. Measurements signify whether constituent system elements are working as per the design expectations and then evaluate the system's performance as a whole under diverse working conditions. Measurements are vital to measuring different variables. For example, in fluid mechanics, the discharge, pressure, viscosity, density gradients, turbulence, velocity etc. are the essential variables that affects fluids' flow properties. Various instruments/devices are used to measure these variables. This chapter measures two variables, discharge and velocity of the fluid flowing through the pipe.

5.2 VENTURIMETER

Function: It is used for measuring the discharge (*i.e.*, rate of fluid flow) through a pipe.

Principle: A venturimeter* is based on the principle of Bernoulli's equation. When the velocity head increases in an accelerated flow, there is a corresponding reduction in the pressure head.

Main parts of Venturimeter: The following are three main parts of a venturimeter:

1. A short converging part (acts as a nozzle),
2. Throat, and
3. Diverging part (acts as diffuser).

---

* Venturi was an Italian engineer who discovered in 1791 that a pressure difference related to the rate of flow could be created in a pipe by reducing cross-sectional area in a pipe.

© The Author(s) 2023
S. Kumar, *Fluid Mechanics (Vol. 2)*,
https://doi.org/10.1007/978-3-030-99754-0_5

Design Aspects of a Venturimeter:

Let D = diameter of a pipe in which liquid flowing.

and d = diameter of throat

Length of converging part = $2.5\,D$

Length of throat = Diameter of throat

Length of diverging part = $7.5\,D$

$$\frac{d}{D} = 0.25 \text{ to } 0.75$$

Angle of converging cone : $\alpha = 19^\circ$ to 23°

Angle of diverging cone : $\beta = 5^\circ$ to 7° for maximum pressure recovery.

$\qquad\qquad\qquad\qquad\quad = 14^\circ$ for maximum pressure recovery
is not much importance.

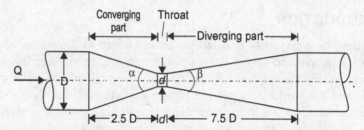

Fig. 5.1: *Design aspects of a Venturimeter.*

Types of Venturimeter:

A Venturimeter consists in three types according to the position (or placement)

1. Horizontal Venturimeter,
2. Vertical Venturimeter, and
3. Inclined Venturimeter.

1. Horizontal Venturimeter:

Consider a venturimeter fitted in a horizontal pipe through which a fluid (say water) is flowing as shown in Fig. 5.2.

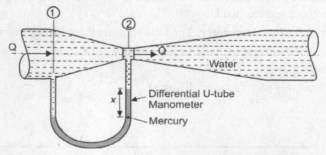

Fig. 5.2: *Horizontal Venturimeter with differential manometer.*

Let D = diameter of a pipe or at section (1)

$$p_1 = \text{pressure of fluid at section (1)}$$
$$V_1 = \text{velocity of fluid at section (1)}$$
$$A = \text{cross-sectional area at section (1)}$$

$$= \frac{\pi}{4}D^2$$

and d, p_2, V_2, $a\left(=\dfrac{\pi}{4}d^2\right)$ are corresponding values at section (2).

Applying the Bernoulli's equation at section (1) and (2), neglecting head losses, we get

$$\frac{p_1}{\rho g}+\frac{V_1^2}{2g}+z_1 = \frac{p_2}{\rho g}+\frac{V_2^2}{2g}+z_2$$

$$\frac{p_1}{\rho g}+\frac{V_1^2}{2g} = \frac{p_2}{\rho g}+\frac{V_2^2}{2g} \qquad [z_1 = z_2, \because \text{Venturimeter}$$

(or pipe) is horizontal.]

or

$$\frac{V_2^2}{2g}-\frac{V_1^2}{2g} - \frac{p_1}{\rho g}-\frac{p_2}{\rho g}$$

$$\frac{V_2^2-V_1^2}{2g} = h$$

where

$$h = \frac{p_1}{\rho g}-\frac{p_2}{\rho g}$$

$$= \text{pressure head difference at sections (1) \& (2)}$$

or $\qquad V_2^2-V_1^2 = 2gh$ $\qquad\qquad$ By continuity equation,

$$\frac{Q^2}{a^2}-\frac{Q^2}{A^2} = 2gh$$

$Q = aV_2$ or $V_2 = \dfrac{Q}{a}$ $\qquad\qquad\qquad$ and $Q = AV_1$, $V_1=\dfrac{Q}{a}$

$$Q^2\left[\frac{1}{a^2}-\frac{1}{A^2}\right] = 2gh$$

$$Q^2\left[\frac{A^2-a^2}{a^2 A^2}\right] = 2gh$$

or $\qquad\qquad\qquad Q^2 = \left[\dfrac{a^2 A^2 \times 2gh}{A^2-a^2}\right]$

$$\text{or} Q = \sqrt{\frac{a^2 A^2 \times 2gh}{A^2 - a^2}}$$

Theoretical discharge : $Q_{th} = \dfrac{aA\sqrt{2gh}}{\sqrt{A^2 - a^2}}$ ...(5.1)

Equation (5.1) is derived under ideal condition. So, it is called theoretical discharge. The actual discharge is always less than theoretical discharge because energy losses involved. The equation of actual discharge is written as

$$Q_{act} = C_d Q_{th}$$

$$Q_{act} = \frac{C_d a A \sqrt{2gh}}{\sqrt{A^2 - a^2}}$$ ...(5.2)

where C_d = coefficient of discharge
which accounts for losses of energy which have been ignored earlier.

Coefficient of discharge: C_d : It is defined as the ratio of actual discharge (Q_{act}) to theoretical discharge (Q_{th}).

Mathematically,

Coefficient of discharge : $C_d = \dfrac{Q_{act}}{Q_{th}}$

The value of C_d is always less than one. For venturimeter, C_d **lies between 0.97 to 0.99.** Average value of $C_d = 0.97$ is taken.

The value of 'h' from equation (5.9.2) is obtained by the differential U-tube manometer

Case I: If the differential U-tube manometer contains a liquid which is heavier than the liquid flowing though the pipe then

$$h = x\left[\frac{\rho_{mano}}{\rho_{pipe}} - 1\right] \quad \text{for} \quad \rho_{mano} > \rho_{pipe}$$ ...(5.3)

also $h = \dfrac{p_1}{\rho g} - \dfrac{p_2}{\rho g}$

where ρ_{mano} = density of the heavier liquid used in U-tube

manometer.

ρ_{pipe} = density of the liquid flowing though pipe.

x = difference of the heavier liquid column in

U-tube manometer.

Let water flow through pipe and mercury used in manometer. Then equation (5.9.3) becomes

$$= x \left[\frac{\rho_{Hg}}{\rho_{H_2O}} - 1 \right]$$

$\because \qquad \rho_{Hg} = 13600 \text{ kg. } \& \ \rho_{H_2O} = 1000 \text{ kg/m}^3$

$$= x \left[\frac{13600}{1000} - 1 \right]$$

$$h = 12.6x \qquad\qquad\qquad\qquad ...(5.4)$$

Case II: If the differential manometer contains a liquid which is lighter than the liquid flowing through the pipe then

$$h = x \left[1 - \frac{\rho_{mano}}{\rho_{pipe}} \right] \text{ for } \rho_{pipe} > \rho_{mano}$$

also $\qquad\qquad h = \frac{p_1}{\rho g} - \frac{p_2}{\rho g}$

Inclined and vertical Venturimeter with differential U-tube manometer:

Case I: If the differential U-tube manometer contains a liquid which is heavier than the liquid flowing though the pipe then

$$h = x \left[\frac{\rho_{mano}}{\rho_{pipe}} - 1 \right] \text{ for } \rho_{mano} > \rho_{pipe}$$

Also $\qquad\qquad h = \left(\frac{p_1}{\rho g} + z_1 \right) - \left(\frac{p_2}{\rho g} + z_2 \right)$

Case II: If the differential U-tube manometer contains a liquid which is lighter than the liquid flowing through the pipe then

$$h = x \left[1 - \frac{\rho_{mano}}{\rho_{pipe}} \right] \text{ for } \rho_{pipe} > \rho_{mano}$$

also $\qquad\qquad h = \left(\frac{p_1}{\rho g} + z_1 \right) - \left(\frac{p_2}{\rho g} + z_2 \right)$

On simplification of equation (5.2)

$$A = \frac{\pi}{4} D^2$$

$$a = \frac{\pi}{4} d^2, \qquad \text{Let} \quad Q_{act} = Q$$

$$\therefore \quad Q = \dfrac{C_d \dfrac{\pi}{4} d^2 \cdot \dfrac{\pi}{4} D^2 \sqrt{2gh}}{\sqrt{\left(\dfrac{\pi}{4} D^2\right)^2 - \left(\dfrac{\pi}{4} d^2\right)^2}}$$

$$= \dfrac{C_d \left(\dfrac{\pi}{4}\right)^2 d^2 D^2 \sqrt{2gh}}{\sqrt{\left(\dfrac{\pi}{4}\right)^2 \left(D^4 - d^4\right)}}$$

$$= \dfrac{C_d \left(\dfrac{\pi}{4}\right)^2 d^2 D^2 \sqrt{2gh}}{\left(\dfrac{\pi}{4}\right)\sqrt{D^4 - d^4}}$$

$$= \dfrac{C_d \left(\dfrac{\pi}{4}\right) d^2 D^2 \sqrt{2gh}}{\sqrt{D^4 - d^4}}$$

$$= \dfrac{\dfrac{\pi}{4} \times \sqrt{2g}\, C_d\, d^2 D^2 \sqrt{h}}{\sqrt{D^4 - d^4}}$$

$$= \dfrac{22}{7 \times 4} \times \sqrt{2 \times 9.81}\, \dfrac{C_d\, d^2 D^2 \sqrt{h}}{\sqrt{D^4 - d^4}} \qquad \left(\text{Take } \pi = \dfrac{22}{7}\right)$$

$$= 3.48\, \dfrac{C_d\, d^2 D^2 \sqrt{h}}{\sqrt{D^4 - d^4}} \qquad\qquad \ldots(5.5)$$

Equation (5.5) applicable only when d & D in m and h in m of liquid flow through pipe then discharge (Q) will be measured in m³/s.

If water flows through pipe and mercury contained in differential U-tube manometer. Then substituting the value of $h = 12.6x$ from equation (5.9.4) in equation (5.5), we get

$$Q = 3.48\, \dfrac{C_d\, d^2 D^2 \sqrt{12.6x}}{\sqrt{D^4 - d^4}}$$

$$Q = 12.35 \dfrac{C_d\, d^2 D^2 \sqrt{x}}{\sqrt{D^4 - d^4}} \qquad\qquad \ldots(5.6)$$

Equation (5.6) is used to determine the discharge for specific case when water flows through pipe and mercury contained in differential U-tube manometer where x in meter of mercury. From equation (5.6), we get discharge (Q) in m³/s when x in m of Hg, d and D in m only.

No other units of x, d & D are applicable in equation (5.6).

Note: Author recommends equation (5.6) for specific case mentioned above, in order to consume less time (*i.e.*, less no. of calculation) and less chance of error involved in the result.

2. Vertical Venturimeter:

Applying the Bernoulli's equation at sections (1) and (2), neglecting head losses, we get

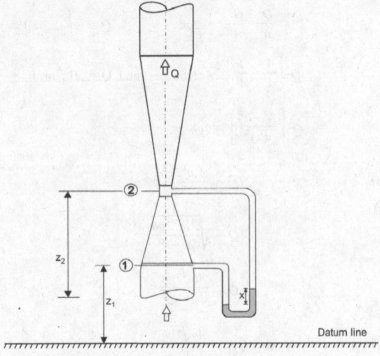

Fig. 5.3: *Vertical Venturimeter with differential manometer*

$$\frac{p_1}{\rho g}+\frac{V_1^2}{2g}+z_1 = \frac{p_2}{\rho g}+\frac{V_2^2}{2g}+z_2$$

or

$$\frac{V_2^2}{2g}-\frac{V_1^2}{2g} = \left(\frac{p_1}{\rho g}+z_1\right)-\left(\frac{p_2}{\rho g}+z_2\right)$$

$$\frac{V_2^2-V_1^2}{2g} = h$$

where
$$h = \left(\frac{p_1}{\rho g} + z_1\right) - \left(\frac{p_2}{\rho g} + z_2\right)$$

= piezometric head difference.

$$= \left(\frac{p_1}{\rho g} - \frac{p_2}{\rho g}\right) + (z_1 - z_2)$$

= pressure head difference

+ datum head difference

$$V_2^2 - V_1^2 = 2gh$$

By continuity equation

$$\frac{Q^2}{a^2} - \frac{Q^2}{A^2} = 2gh \qquad\qquad Q = aV_2 \quad \text{or} \quad V_2 = \frac{Q}{a}$$

$$Q^2\left[\frac{1}{a^2} - \frac{1}{A^2}\right] = 2gh \qquad \text{and } Q = AV_1 \text{ or } V_1 = \frac{Q}{A}$$

$$Q^2\left[\frac{A^2 - a^2}{a^2 A^2}\right] = 2gh$$

or
$$Q^2 = \frac{A^2 a^2 \, 2gh}{A^2 - a^2}$$

$$Q = \sqrt{\frac{A^2 a^2 \, 2gh}{A^2 - a^2}}$$

$$Q = \frac{Aa\sqrt{2gh}}{\sqrt{A^2 - a^2}}$$

Theoretical discharge: $Q_{th} = \dfrac{Aa\sqrt{2gh}}{\sqrt{A^2 - a^2}}$

Actual discharge: $Q_{act} = C_d \, Q_{th}$

$$Q_{act} = \frac{C_d \, Aa\sqrt{2gh}}{\sqrt{A^2 - a^2}}$$

The expression of discharge for vertical venturimeter is similar to that of horizontal Venturimeter. The difference is only that the value of 'h'

For horizontal venturimeter

$$h = \frac{p_1}{\rho g} - \frac{p_2}{\rho g} = \text{pressure heads difference for horizontal}$$

$$= x\left[\frac{\rho_{mano}}{\rho_{pipe}} - 1\right] \qquad\qquad \text{for } \rho_{mano} > \rho_{pipe}$$

$$= x\left[1 - \frac{\rho_{mano}}{\rho_{pipe}}\right] \qquad\qquad \text{for } \rho_{pipe} > \rho_{mano}$$

For vertical venturimeter

$$h = \left(\frac{p_1}{\rho g} + z_1\right) - \left(\frac{p_2}{\rho g} + z_2\right)$$

= piezometric head difference for vertical venturimeter

$$= x\left[\frac{\rho_{mano}}{\rho_{pipe}} - 1\right] \qquad\qquad \text{for } \rho_{mano} > \rho_{pipe}$$

$$= x\left[1 - \frac{\rho_{mano}}{\rho_{pipe}}\right] \qquad\qquad \text{for } \rho_{pipe} > \rho_{mano}$$

3. Inclined Venturimeter

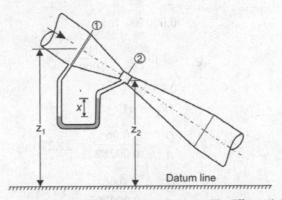

Fig. 5.4: *Inclined Venturimeter with differential manometer*

The expression of inclined venturimeter is same as vertical venturimeter

$$Q_{act} = \frac{C_d \, A a \sqrt{2gh}}{\sqrt{A^2 - a^2}}$$

where

$$h = \left(\frac{p_1}{\rho g} + z_1\right) - \left(\frac{p_2}{\rho g} + z_2\right)$$

piezometric head difference

Problem 5.1: What should be pressure difference between the upstream section and throat of a 60 mm by 30 mm horizontal venturimeter carrying 60 litre/s of water at room temperature.

Solution: Given data:

Diameter at upstream *i.e.*, at section (1)-(1) : D = 60 mm = 0.06 m

$\therefore$

$$\text{Cross-sectional area : } A = \frac{\pi}{4}D^2 = \frac{3.14}{4} \times (0.06)^2$$

$$= 0.00282 \text{ m}^2$$

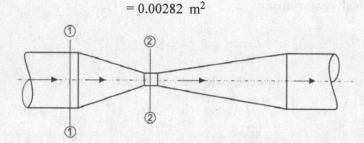

Fig. 5.5: Schematic for Problem 5.1

Diameter at throat *i.e.*, at section (2)–(2) : d = 30 mm = 0.03 m

$\therefore$ Cross-sectional area :

$$a = \frac{\pi}{4}d^2 = \frac{3.14}{4} \times (0.03)^2$$

$$= 0.000706 \text{ m}^2$$

Discharge : $Q = 60 \text{ litre/s} = \dfrac{60}{1000} \text{ m}^3/\text{s}$

$$= 0.06 \text{ m}^3/\text{s}$$

also $Q = AV_1 = aV_2$

$\therefore$ $V_1 = \dfrac{Q}{A} = \dfrac{0.06}{0.00282} = 21.27 \text{ m/s}$

and $V_2 = \dfrac{Q}{a} = \dfrac{0.06}{0.000706} = 84.98 \text{ m/s}$

Applying Bernoulli's equation at section (1)–(1) and (2)–(2), we get

$$\frac{p_1}{\rho g} + \frac{V_1^2}{2g} + z_1 = \frac{p_2}{\rho g} + \frac{V_2^2}{2g} + z_2$$

$$\frac{p_1}{\rho g} + \frac{V_1^2}{2g} = \frac{p_2}{\rho g} + \frac{V_2^2}{2g} \qquad\qquad z_1 = z_2 \text{ because of}$$

venturimeter is horizontal

$$\frac{p_1 - p_2}{\rho g} = \frac{V_1^2 - V_2^2}{2g}$$

or

$$p_1 - p_2 = \frac{\rho\left(V_1^2 - V_2^2\right)}{2g}$$

$$= 1000 \frac{(84.98)^2 - (21.27)^2}{2}$$

$$\because \rho = 1000 \text{ kg/m}^3$$

$$= 3384593.75 \text{ N/m}^2 \quad \text{or} \quad \text{Pa}$$

$$= \mathbf{3.384 \text{ MPa}}$$

Problem 5.2: A 200 mm × 120 mm venturimeter is installed in a horizontal pipe carrying water. If the mercury differential manometer shows a reading of 200 mm, find the discharge through the pipe. Take coefficient of discharge : $C_d = 0.98$

Solution : Given data:

Diameter at inlet : $D = 200$ mm $= 0.2$ m

∴ Cross-sectional at inlet :

$$A = \frac{\pi}{4} D^2 = \frac{3.14}{4} \times (0.2)^2 = 0.0314 \text{ m}^2$$

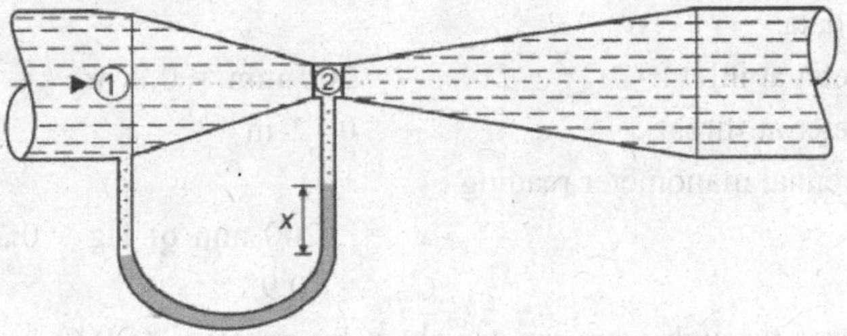

Fig. 5.6: *Horizontal Venturimeter with U-tube differential manometer.*

Diameter at throat : $d = 120$ mm $= 0.12$ m

∴ Cross-sectional area at throat :

$$a = \frac{\pi}{4} d^2 = \frac{3.14}{4} \times (0.12)^2 = 0.0113 \text{ m}^2$$

Differential manometer reading :

$$x = 200 \text{ mm of mercury (Hg)}$$
$$= 0.2 \text{ of Hg}$$

We know manometer reading in term of water : h

$$h = x\left[\frac{\rho_{mano}}{\rho_{pipe}} - 1\right]$$

where ρ_{mano} = density of liquid used in manometer
                            = ρ_{Hg} = 13600 kg/m³
and ρ_{pipe} = density of liquid flow through pipe
                            = ρ_{water} = 1000 kg/m³

$\therefore$ $h = 0.2\left[\dfrac{13600}{1000} - 1\right]$ = 2.52 m of water

We know discharge through venturimeter :

$$Q = \frac{C_d \, Aa\sqrt{2gh}}{\sqrt{A^2 - a^2}}$$

$$= \frac{0.95 \times 0.0314 \times 0.0113\sqrt{2 \times 9.81 \times 2.52}}{\sqrt{(0.0314)^2 - (0.0113)^2}}$$

$$= \frac{2.445 \times 10^{-3}}{0.02929} = 0.083475 \text{ m}^3/\text{s}$$

$$= 0.083475 \times 1000 \text{ litre/s}$$
$$(\because \ 1 \text{ m}^3 = 1000 \text{ litre})$$
$$= \textbf{83.47 litre/s}$$

Note:Given problem 5.2 is also solved by simplified equation (5.6)
Given data:

Diameter at inlet : D =   200 mm = 0.2 m
Diameter at throat : d =   0.12 m
Differential manometer reading :

x =   200 mm of Hg = 0.2 m
C_d =   0.98

Discharge through venturimeter given by equation (5.9.6), we get

$$Q = \frac{12.35 \times 0.98 \times (0.12)^2 \times (0.2)^2\sqrt{0.2}}{\sqrt{(0.2)^4 - (0.12)^4}}$$

$$= 0.08354 \text{ m}^3/\text{s} = \textbf{83.54 litre/s}$$

Problem 5.3: A horizontal venturimeter with inlet diameter 20 cm and throat diameter 10 cm is used to measure the flow of oil of specific gravity 0.8. The discharge of oil through Venturimeter is 60 litre/s. Find the reading of the oil-mercury differential manometer. Take C_d = 0.98
(GGSIP University, Delhi, Dec. 2005)

Solution: Given data:
Inlet diameter of venturimeter :
$$D = 20 \text{ cm} = 0.20 \text{ m}$$

Throat diameter : $d = 10$ cm $= 0.10$ m
Specific gravity of oil $= 0.8$
$\therefore$ density of oil : $\rho = 0.8 \times 1000$
                         $= 800$ kg/m³

Dscharge, Q $= 60$ litre /s $= \dfrac{60}{1000} = 0.06$ m³/s

Coefficient of discharge, $C_d = 0.98$

Cross-sectional area of venturimeter at inlet, $A = \dfrac{\pi}{4}D^2 = \dfrac{3.14}{4} \times (0.20)^2$

$= 0.0314$ m²

Cross-sectional area at throat,

$$a = \frac{\pi}{4}d^2 = \frac{3.14}{4} \times (0.10)^2$$
$$= 0.00785 \text{ m}^2$$

We know,

Discharge, $Q = \dfrac{C_d\, aA\sqrt{2gh}}{\sqrt{A^2 - a^2}}$

$0.060 = \dfrac{0.98 \times 0.00785 \times 0.0314 \times \sqrt{2 \times 9.81 \times h}}{\sqrt{(0.0314)^2 - \sqrt{(0.00725)^2}}}$

or $\sqrt{2 \times 9.81 \times h} = 7.5516$
          $2 \times 9.8 \times h = 57.026$
$\therefore$ Diffrential head, $h = 2.9065$ m of oil

Also we know $h = x\left[\dfrac{\rho_{Hg}}{\rho_{Oil}} - 1\right]$

where $x = $ reading of the oil mercury differential manometer as shown in Fig. 5.7

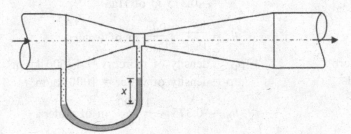

Fig. 5.7: Venturimeter with oil-mercury differential manometer.

$$2.9065 = x\left[\frac{13600}{800} - 1\right]$$

$$= x[17 - 1]$$

($\because$ Density of the mercury,

$\rho_{Hg} = 13600$ kg/m³

$$or \qquad x = \frac{2.9065}{16} = 0.18165 \text{ m of Hg}$$

$$= \mathbf{18.165 \text{ cm of Hg.}}$$

Problem 5.4: A venturimeter is installed in a 300 mm diameter horizontal pipeline. The throat diameter is 1/3 of pipe diameter. Water flows through the installation. The pressure in the pipe line is 13.8 N/cm² (gauge) and vacuum in the throat is 37.5 cm of mercury. Determine the rate of flow of water in the pipe line. Take $C_d = 0.98$.

Solution: Given data :

Diameter of pipe, $\qquad D = 300 \text{ mm} = 0.3 \text{ m}$

∴ Cross-sectional area of pipe,

$$A = \frac{\pi}{4} D^2 = \frac{3.14}{4} \times (0.3)^2 = 0.07065 \text{ m}^2$$

Diameter of throat, $d = \dfrac{1}{3}$ of pipe diameter $= \dfrac{1}{3} \times D$

$$= \frac{1}{3} \times 0.3 = 0.1 \text{ m}$$

∴ Cross-sectional area of throat, $a = \dfrac{\pi}{4} d^2 = \dfrac{3.14}{4} \times (0.1)^2 = 0.00785 \text{ m}^2$

Pressure in the pipe line, $p_1 = 13.8 \text{ N/cm}^2 = 13.8 \times 10^4 \text{ N/m}^2$

∴ Pressure head in the pipeline, $h_1 = \dfrac{p_1}{\rho g} = \dfrac{13.8 \times 10_4}{1000 \times 9.81} = 14.067 \text{ m}$

Vacuum pressure head in throat,

$$h_2 = \frac{p_2}{\rho_{Hg} g} = -37.5 \text{ cm of Hg}$$

$$= -0.375 \text{ m of Hg}$$

$$h_2 = \frac{p_2}{\rho_{Hg} g} \times \frac{\rho_{Hg}}{\rho} \text{ m of water}$$

where $\qquad \rho_{Hg}$ = density of mercury = 13600 kg/m³

and $\qquad \rho$ = density of water = 1000 kg/m³

$$h_2 = -0.375 \times \frac{13600}{1000} \text{ m of water}$$

$$= -5.1 \text{ m of water}$$

(−ve sign shows that pressure in throat is vacuum.

∴   Difference of pressure head,

$$h = \frac{p_1}{\rho g} - \frac{p_2}{\rho g} = h_1 - h_2 = 14.067 - (-5.1)$$

$$= 14.067 + 5.1 = 19.167 \text{ m of water.}$$

$\therefore$ Rate of flow : $Q = \dfrac{C_d \, Aa\sqrt{2gh}}{\sqrt{A^2 - a^2}}$

$$= \dfrac{0.98 \times 0.00785 \times 0.07065 \times \sqrt{2 \times 9.81 \times 19.167}}{\sqrt{(0.07065)^2 - (0.00785)^2}}$$

$$= 0.15011 \ m^3/s = \textbf{150.11 litre/s}$$

Problem 5.5: The inlet and throat diameters of a horizontal venturimeter are 200 mm and 100 mm respectively. The liquid flowing through the venturimeter is water. The pressure at inlet is 120 kPa while the vacuum pressure head at the throat is 150 mm of mercury. Find coefficient of discharge. Assume that 3.33% of the differential head is lost between the inlet and throat. Find also the rate of flow.

Solution: Given data :

Diameter at inlet : $D = 200 \ mm = 0.2 \ m$

$\therefore$ Cross-sectional area at inlet :

$$A = \frac{\pi}{4} D^2 = \frac{3.14}{4} \times (0.2)^2 = 0.0314 \ m^2$$

Diameter at throat : $d = 100 \ mm = 0.1 \ m$

$\therefore$ Cross-sectional area at throat :

$$a = \frac{\pi}{4} d^2 = \frac{3.14}{4} \times (0.1)^2 = 0.00785 \ m^2$$

Pressure at inlet : p_1 $= 120 \ kPa = 120 \times 10^3 \ Pa$ or N/m^2

$\therefore$

Pressure head at inlet : $\dfrac{p_1}{\rho g}$

$$= \frac{120 \times 10^3}{1000 \times 9.81} = 12.23 \ m \ of \ water$$

Vacuum pressure head at throat : $\dfrac{p_2}{\rho g} = -150 \ mm \ of \ Hg$

$$= - \, 0.15 \ m \ of \ Hg$$
$$= 0.015 \times 13.6 \ m \ of \ water$$
$$= - \, 2.04 \ m \ of \ water$$

$\therefore$ Differential head : $h = \dfrac{p_1}{\rho g} - \dfrac{p_2}{\rho g} = 12.23 - (-2.04)$

$$= 14.27 \ m \ of \ water$$

Head lost : $h_f = 3.33\%$ of differential head
$$= 3.33 \ \% \times h = 0.0333 \times 14.27$$
$$= 0.4752 \ m \ of \ water$$

$\therefore$ Coefficient of discharge : $C_d = \sqrt{\dfrac{h - h_f}{h}} = \sqrt{\dfrac{14.27 - 0.4752}{14.27}}$

$$= 0.983$$

Rate of flow : $Q = \dfrac{C_d\, A a \sqrt{2gh}}{\sqrt{A^2 - a^2}}$

$$= \dfrac{0.983 \times 0.0314 \times 0.00785\sqrt{2 \times 9.81 \times 14.27}}{\sqrt{(0.0314)^2 - (0.00785)^2}}$$

$$= 0.13335 \ \text{m}^3/\text{s} = \textbf{133.35 litre/s}$$

Note :Discharge can also calculated by simplified equation (5.9.5)

$$Q = \dfrac{3.48\, C_d\, d^2 D^2 \sqrt{h}}{\sqrt{D^4 - d^4}}$$

where d and D are in metre

$\qquad h =$ differential of pressure head in metre of liquid flow through venturimeter

$\therefore \qquad Q = \dfrac{3.48 \times 0.983 \times (0.1)^2 \times (0.2)^2 \sqrt{14.27}}{\sqrt{(0.2)^4 - (0.1)^4}}$

$$= 0.13346 \ \text{m}^3/\text{s}$$

$$= \textbf{133.46 litre/s}$$

Problem 5.6: If an oil of specific gravity 0.8 is flowing through horizontal venturimeter instead of water in above problem 5.16, remaining data is same.

Solution: Given data:

Specific gravity of oil : $S = 0.8$

$\therefore \qquad$ Density of oil : $\rho = 0.8 \times 1000 = 800 \ \text{kg/m}^3$

Pressure head at inlet : $\dfrac{p_1}{\rho g} = \dfrac{120 \times 10^3}{800 \times 9.81} = 15.29$ m of oil

Vacuum pressure head at throat : $\dfrac{p_2}{\rho g} = -150$ mm of Hg

$$= -0.15 \text{ m of Hg}$$

$$= \dfrac{-0.15 \times 13.6}{0.8} \text{ m of oil}$$

$$= -2.55 \text{ m of oil}$$

$\therefore$ Difference of pressure head : $h = \dfrac{P_1}{\rho g} - \dfrac{P_2}{\rho g} = 15.29 - (-2.55)$

$\qquad = 15.29 + 2.55 = 17.84$ m of oil

Head lost : h_f $\qquad = 3.33\%$ of differential head

$\qquad = 0.0333 \times h$

$\qquad = 0.0333 \times 17.84 = 17.84 = 0.594$ m of oil

$\therefore$ Coefficient of discharge : $C_d = \sqrt{\dfrac{h - h_f}{h}}$

$\qquad = \sqrt{\dfrac{17.84 - 0.594}{17.84}} = \sqrt{0.9667} = 0.983$

Rate of flow : Q $\qquad = \dfrac{C_d\, Aa\sqrt{2gh}}{\sqrt{A^2 - a^2}}$

$\qquad = \dfrac{0.983 \times 0.0314 \times 0.00785\sqrt{2 \times 9.81 \times 17.84}}{\sqrt{(0.0314)^2 - (0.00785)^2}}$

$\qquad = 0.14910$ m^3/s

$\qquad = \mathbf{149.10}$ **litre/s**

Problem 5.7: A venturimeter is used for measurement of discharge of water in a horizontal pipeline. If the ratio inlet diameter to that of throat diameter is 2 : 1, inlet diameter is 300 mm, the difference of pressure head between inlet and throat is 3 m of water and loss of head between inlet and throat is one-eighth of the throat velocity head. Calculate discharge in pipe.

Solution: Given data:

$$\frac{D}{d} = 2$$

Diameter at inlet: $D = 300$ mm $= 0.3$ m

$\therefore$ Cross-sectional area at inlet: $A = \dfrac{\pi}{4}D^2 = \dfrac{3.14}{4} \times (0.3)^2 = 0.07065$ m^2

Diameter at throat: $d = \dfrac{D}{2} = \dfrac{300}{2} = 150$ mm

$\qquad = 0.15$ m

$\therefore$ Cross-sectional area at throat: a

$\qquad = \dfrac{\pi}{4}d^2 = \dfrac{3.14}{4} \times (0.15)^2 = 0.01766$ m^2

Difference of pressure head between inlet and throat:

$$\frac{p_1}{\rho g} - \frac{p_2}{\rho g} = 3 \text{ m of water}$$

Loss of head : $h_L = \frac{1}{8} \times$ throat velocity head

$$= \frac{1}{8} \times \frac{V_2^2}{2g} = \frac{V_2^2}{16g}$$

Now applying modified Bernoulli's equation at inlet & throat, we get

$$\frac{p_1}{\rho g} + \frac{V_1^2}{2g} + z_1 = \frac{p_2}{\rho g} + \frac{V_2^2}{2g} + z_2 + h_L$$

$$(\because \ z_1 = z_2 \text{ for horizontal pipeline})$$

$$\frac{p_1}{\rho g} - \frac{p_2}{\rho g} + \frac{V_1^2}{2g} = \frac{V_2^2}{2g} + \frac{V_2^2}{16g}$$

$$3 + \left(\frac{a}{A} V_2\right)^2 \times \frac{1}{2g} = \frac{V_2^2}{2g} + \frac{V_2^2}{16g}$$

(By continuity Eq. $AV_1 = aV_2$, $V_1 = \frac{a}{A} V_2$)

$$3 + \left(\frac{0.01766}{0.07065}\right)^2 \frac{V_2^2}{2g} = \frac{V_2^2}{2g} + \frac{V_2^2}{16g}$$

$$3 + 0.06248 \frac{V_2^2}{2g} = \frac{V_2^2}{2g} + \frac{V_2^2}{16g}$$

On multiplying 16g both sides, we get

$16g \times 3 + 8 \times 0.06248 \ V_2^2 = 4 \ V_2^2 + V_2^2$

$16 \times 9.81 \times 3 + 0.4998 \ V_2^2 = 5 \ V_2^2$

$470.88 + 0.4998 \ V_2^2 \qquad = 5 \ V_2^2$

$$470.88 = 4.50 \ V_2^2$$

or $$V_2^2 = 104.65$$

$$V_2 = 10.23 \text{ m/s}$$

$\therefore$ \qquad Discharge : $Q = aV_2$

$$= 0.01766 \times 10.23$$

$$= 0.18066 \text{ m}^3/\text{s}$$

$$= \mathbf{180.66 \ litre/s}$$

Problem 5.8: The maximum flow through a 300 mm diameter horizontal pipe line is 300 litre/s. A venturimeter is introduced at a point of the pipe where the pressure head is 5 m of water. Find the diameter of throat so that the pressure at the throat is never negative. Take $C_d = 1$

Solution: Given data:

Diameter at inlet : $D = 300$ mm $= 0.3$ m

∴   Cross-sectional area at inlet :

$$A = \frac{\pi}{4}D^2 = \frac{3.14}{4} \times (0.3)^2 = 0.07065 \text{ m}^2$$

Discharge through pipe : $Q = 300$ litre/s $= 0.3$ m³/s

Pressure heat at inlet : $\dfrac{p_1}{\rho g} = 5$ m of water.

Pressure heat at throat : $\dfrac{p_2}{\rho g} = 0$

∴ Difference of pressure head : h

$$= \frac{p_1}{\rho g} - \frac{p_2}{\rho g} = 5 \text{ m of water}$$

We know,

$$\text{Discharge } Q = \frac{C_d\, Aa\sqrt{2gh}}{\sqrt{A^2 - a^2}}$$

$$0.3 = \frac{1 \times 0.07065 \times a\sqrt{2 \times 9.81 \times 5}}{\sqrt{(0.07065)^2 - (a)^2}}$$

or $$0.4287 = \frac{a}{\sqrt{(0.07065)^2 - (a)^2}}$$

or $0.4287 \times \sqrt{(0.07065)^2 - (a)^2} = a$

Squaring both sides, we get

$(0.4287)^2 \times [(0.07065)^2 - a^2] = a^2$

$0.1837\,[(0.07065)^2 - a^2] = a^2$

$9.169 \times 10^{-4} - 0.1837\,a^2 = a^2$

$9.169 \times 10^{-4} = 0.8163\,a^2$

or $$a^2 = 11.2323 \times 10^{-4} \text{ m}^2$$

$$a = 0.03351 \text{ m}^2$$

Also $$a = \frac{\pi}{4} \times d^2$$

∴ $$0.03351 = \frac{3.14}{4} \times d^2$$

or $$d^2 = 0.04268$$

$$d = 0.20661 \text{ m} = \mathbf{206.61 \text{ mm}}$$

Problem 5.9: A venturimeter is installed in a 300 mm diameter vertical pipe carrying water, flowing in the upward direction. The throat diameter is 150 mm. A differential mercury manometer connected to the inlet and throat gives a reading of 200 mm. Find the rate of flow of water in litre/s. Take $C_d = 0.98$

Solution: Given data:

Diameter of vertical pipe : $D = 300$ mm $= 0.3$ m

$\therefore$ Cross-sectional area of the pipe, $A = \dfrac{\pi}{4} D^2 = \dfrac{3.14}{4} \times (0.3)^2$

$$= 0.07065 \text{ m}^2$$

Diameter of throat, $d = 150$ mm $= 0.150$ m

$\therefore$ Cross-sectional area of throat, $a = \dfrac{\pi}{4} d^2 = \dfrac{3.14}{4} \times (0.150)^2$

$$= 0.01766 \text{ m}^2$$

Reading of differential mercury manometer,

$$x = 200 \text{ mm of Hg}$$
$$= 0.20 \text{ m of Hg}$$

Let $\qquad h = $ manometer reading in terms of water

$\therefore \qquad\qquad h = x \left[\dfrac{\rho_{Hg}}{\rho_{wager}} - 1 \right]$

$$= 0.20 \left[\dfrac{13600}{1000} - 1 \right]$$

$$= 2.52 \text{ m of water}$$

The rate of flow of water,

$$Q = \dfrac{C_d \, Aa\sqrt{2gh}}{\sqrt{A^2 - a^2}}$$

$$= \dfrac{0.98 \times 0.01766 \times 0.07065 \sqrt{2 \times 9.81 \times 2.52}}{\sqrt{(0.07065)^2 - (0.01766)^2}}$$

$$= 0.125683 \text{ m}^3/\text{s}$$
$$= \mathbf{125.683 \text{ litre/s}}$$

Problem 5.10:   A 20×10 cm venturimeter is provided in a vertical pipe line carrying oil of specific gravity 0.8, the flow being upwards. The difference in elevation of the throat and entrance section of the venturimeter is 50 cm. The differential U-tube mercury manometer shows a deflection of 40 cm of mercury. Calculate (i) the discharge of oil and (ii) the pressure difference (in N/cm²) between entrance and throat sections. Assume $C_d = 0.98$ and specific gravity of mercury as 13.0

Solution: Given data:

Diameter at throat, at section (2)–(2), $d = 10$ cm $= 0.1$ m

$\therefore$ Cross-sectional area at throat, $a = \dfrac{\pi}{4}d^2 = \dfrac{3.14}{4} \times (0.1)^2$

$$= 7.85 \times 10^{-3} \text{ m}^2$$

Diameter of pipe at section (1)–(1), $D = 20$ cm

$$= 0.2 \text{ m}$$

$\therefore$ Cross-sectional area at inlet of the venturimeter,

$$A = \frac{\pi}{4}D^2 = \frac{3.14}{4} \times (0.2)^2$$

$$- 0.0314 \text{ m}^2$$

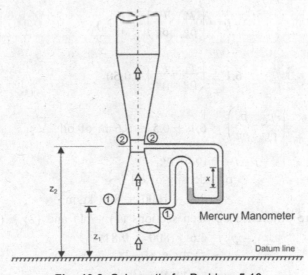

Fig. 12.8: Schematic for Problem 5.10

Specific gravity of oil flow though pipe, $S_{oil} = 0.8$

The difference in elevation of the throat and entrance section of the venturimeter,

$$z_2 - z_1 = 50 \text{ cm} = 0.50 \text{ m}$$

or $z_2 - z_1 = -0.50$ m

Manometer reading, $x = 40$ cm $= 0.4$ m of Hg

Specific gravity of mercury, $S_{Hg} = 13$

The manometer reading in term of oil,

$$h = x \left[\frac{S_{Hg}}{S_{Oil}} - 1 \right]$$

$$= 0.40 \left[\frac{13}{0.8} - 1 \right]$$

$$= 6.1 \text{ m of oil}$$

Coefficient of discharge, $C_d = 0.98$

(i) We know,   discharge: $Q = \dfrac{C_d \, A a \sqrt{2gh}}{\sqrt{A^2 - a^2}}$

$$= \frac{0.98 \times 0.0314 \times 7.85 \times 10^{-3} \sqrt{2 \times 9.81 \times 6.1}}{\sqrt{(0.0314)^2 - (7.85 \times 10^{-3})^2}}$$

$$= 0.086920 \text{ m}^3/\text{s}$$

$$= \textbf{86.92 litre/s}$$

(ii) The manometer reading in term of oil,

$$h = \left(\frac{p_1}{\rho g} - \frac{p_2}{\rho g} \right) + (z_1 - z_2)$$

$$6.1 = \left(\frac{p_1}{\rho g} - \frac{p_2}{\rho g} \right) - 0.50$$

or $\left(\dfrac{p_1}{\rho g} - \dfrac{p_2}{\rho g} \right) = 6.1 + 0.5 = 6.6 \text{ m of oil}$

or $p_1 - p_2 = (6.6) \, \rho g$

where ρ = density of oil

$$= 0.8 \times 1000 = 800 \text{ kg/m}^3$$

∴   The pressure difference between sections (1) – (1) and (2) – (2)

$$p_1 - p_2 = 6.6 \times 800 \times 9.81$$

$$= 51796.8 \text{ N/m}^2$$

$$= \textbf{5.179 N/cm}^2$$

Probme 5.11: The water is flowing through an inclined venturimeter in the upward direction. The inlet and throat diameters of the venturimeter are 200 mm and 100 mm respectively. The pressure of inlet is 19.62 N/cm² (gauge) and at throat is 3.924 N/cm² (vacuum). The length between inlet and throat is 500 mm and venturimeter is inclined at an angle of 60° with horizontal. Find the discharge through venturimeter. Take $C_d = 0.98$

Solution: Given data :

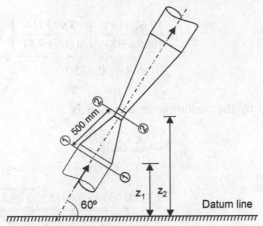

Fig. 5.9: Schematic for Problem 5.11

Diameter at inlet at section (1) – (1) : d_1 = 200 mm = 0.2 m

Diameter at throat at section (2)–(2):d_2 = 100 mm = 0.1 m

∴ Cross-sectional area at section (1)–(1)

$$A_1 = \frac{\pi}{4}d_1^2 = \frac{3.14}{4} \times (0.2)^2$$

$$= 0.0314 \text{ m}^2$$

And cross-sectional area at section (2)–(2) :

$$A_2 = \frac{\pi}{4}d_2^2 = \frac{3.14}{4} \times (0.1)^2$$

$$= 0.00785 \text{ m}^2$$

Pressure at section (1)–(1):p_1 = 19.62 N/cm²

$$= 19.62 \times 10^4 \text{ N/m}^2$$

Pressure at throat *i.e.*, at section (2)–(2) : p_2 = –3.924 N/cm²

$$= 3.924 \times 10^4 \text{ N/m}^2$$

Length between inlet and throat : L = 500 mm = 0.5 m

Difference in elevation between the throat and the inlet

$$= z_2 - z_1 = L \sin \theta$$

$$z_2 - z_1 = 0.5 \times \sin 60°$$

$$= 0.4330 \text{ m}$$

$$h = \left(\frac{p_1}{\rho g} + z_1\right) - \left(\frac{p_2}{\rho g} + z_2\right)$$

$$= \left(\frac{p_1}{\rho g} - \frac{p_2}{\rho g}\right) - (z_2 - z_1)$$

$$= \frac{19.62 \times 10^4}{1000 \times 9.81} - \left(\frac{-3.924 \times 10^4}{1000 \times 9.81} \right) - 0.4330$$

$$= 20 + 4 - 0.4330$$

$$= 23.567 \text{ m}$$

The discharge Q by the venturimeter equation is

$$Q = \frac{C_d\, A_1 A_2 \sqrt{2gh}}{\sqrt{A_1^2 - A_2^2}}$$

$$= \frac{9.81 \times 0.0314 \times 0.00785 \sqrt{2 \times 9.81 \times 23.567}}{\sqrt{(0.0314)^2 - (0.00785)^2}}$$

$$= \frac{2.4156 \times 10^{-4}}{0.030429} \times 21.503$$

$$= 0.17084 \text{ m}^3/\text{s}$$

$$= \mathbf{170.84 \text{ litre/s}}$$

Problem.5.12:   Find the discharge of water flowing a pipe 200 mm diameter placed in an inclined position where a venturimeter is inserted, having throat diameter of 100 mm. The difference of pressure between the main and throat is measured by a liquid of specific gravity 0.75 in an inverted U-tube which gives a reading of 300 mm. The loss of head between the main and throat is 0.3 times the kinetic head of the pipe.

   Solution: Given data:

      Diameter at inlet : $D = 200$ mm $= 0.2$ m

   $\therefore$ Cross-sectional area at inlet :

$$A = \frac{\pi}{4} D^2 = \frac{3.14}{4} \times (0.2)^2 = 0.0314 \text{ m}^2$$

      Diameter at throat : $d = 100$ mm $= 0.1$ m

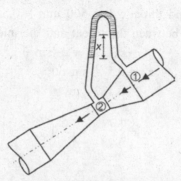

Fig. 5..10 *Inclined Venturimeter with inverted U-tube differential manometer*

∴ Cross-sectional area at throat : $a = \frac{\pi}{4}d^2 = \frac{3.14}{4} \times (0.1)^2 = 0.00785$ m^2

Specific gravity of liquid used in manometer : $S = 0.75$

∴ Density of manometer liquid : $\rho = S \times 1000 = 0.75 \times 1000$
$$= 750 \text{ kg/m}^3$$

Manometer reading : x $= 300$ mm of manometer liquid
$$= 0.3 \text{ m of manometer liquid}$$

Difference of pressure head : $h = x\left[1 - \frac{\rho_{mano}}{\rho_{pipe}}\right]$ for $\rho_{pipe} > \rho_{mano}$

where ρ_{mano} = density of liquid used in manometer

        ρ_{pipe} = density of liquid flow through pipe.

∴ $h = 0.3\left[1 - \frac{750}{1000}\right]$

$$= 0.3 \times 0.25 = 0.075 \text{ m of water}$$

also $h = \left(\frac{p_1}{\rho g} + z_1\right) - \left(\frac{p_2}{\rho g} + z_2\right)$

Loss of head : $h_L = 0.3 \times$ kinetic head of the pipe

$$= 0.3 \times \frac{V_1^2}{2g}$$

Now applying the modified Bernoulli's equation for real fluid at sections (1) and (2), we get

$$\frac{p_1}{\rho g} + \frac{V_1^2}{2g} + z_1 = \frac{p_2}{\rho g} + \frac{V_2^2}{2g} + z_2 + h_L$$

or $\left(\frac{p_1}{\rho g} + z_1\right) - \left(\frac{p_2}{\rho g} + z_2\right) + \frac{V_1^2}{2g} = \frac{V_2^2}{2g} + 0.3\frac{V_1^2}{2g}$

$$0.075 + \frac{V_1^2}{2g} - 0.3\frac{V_1^2}{2g} - \frac{V_2^2}{2g} = 0$$

$$0.075 + 0.7\frac{V_1^2}{2g} - \frac{V_2^2}{2g} = 0 \qquad \qquad ...(i)$$

Applying the continuity equation at sections (1) and (2), we get
$$A_1V_1 = aV_2$$

$$0.0314 \ V_1 = 0.00785 \ V_2$$

or
$$4V_1 = V_2$$
or
$$V_2 = 4V_1$$

Putting the value of V_2 in equation (i), we get

$$0.075 + 0.7\frac{V_1^2}{2g} - \frac{(4V_1)^2}{2g} = 0$$

$$0.075 + 0.7\frac{V_1^2}{2g} - \frac{16V_1^2}{2g} = 0$$

$$0.075 - 15.3\frac{V_1^2}{2g} = 0$$

or
$$0.075 = 15.3\frac{V_1^2}{2g}$$

or
$$\frac{0.075 \times 2 \times 9.81}{15.3} = V_1^{\ 2}$$

or
$$V_1^{\ 2} = 0.096176$$
$$V_1 = 0.310 \ \text{m/s}$$

∴ Discharge : $Q = AV_1$
$$= 0.314 \times 0.310 = 0.09737 \ \text{m}^3/\text{s}$$
$$= \textbf{97.37 litre/s}$$

5.3 ORIFICE METER OR ORIFICE PLATE

Function: It is used for measuring the discharge (*i.e.*, rate of fluid flow) through a pipe.

Principle: Same as that of venturimeter. The orifice is a circular thin metal plate with a small hole on its centre. The plate in held in the pipeline between two flanges called orifice flanges. The plane of orifice plate is perpendicular to the axis of the pipe and hole is concentric with the pipe. The area of flow decreases and then increases in direction of flow. The minimum called cross section of flow between the two is **vena–contracta** as shown in Fig. 12.11.

Design aspects of a orifice meter:
Let
$$D = \text{diameter of pipe.}$$
$$d = \text{diameter of orifice}$$

$$\frac{d}{D} = 0.4 \text{ to } 0.8$$

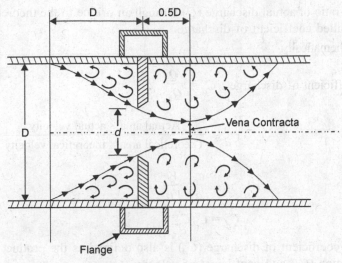

Fig. 5.11: Orifice meter.

HYDRAULIC COEFFICIENTS OF ORIFICE METER

(1)Coefficient of Velocity (C_v):

It is defined as the ratio of actual velocity (V) of the jet at vena contracta to the theoretical (or ideal) velocity (V_{th}) at vena contracta.

Mathematically,

$$\text{Coefficient of velocity} : C_v = \frac{\text{Actual velocity of jet at vena contracta} : V}{\text{Theoretical velocity of jet at vena contracta} : V_{th}}$$

$$C_v = \frac{V}{V_{th}} = \frac{V}{\sqrt{2gh}}$$

The value of C_v lies between 0.95 to 0.99

2.Coefficient of Contraction (C_c)

It is defined as the ratio of the cross-sectional area of the jet at vena-contracta to the cross-sectional area of the orifice.

Mathematically,

$$\text{Coefficient of contraction} : C_c = \frac{\text{Cross-sectional area of jet at vena contract} : a_c}{\text{Cross-sectional area of the orfice} : a}$$

$$C_c = \frac{a_c}{a}$$

The value of C_c lies between 0.61 to 0.69. Average value of C_c taken as 0.64

3.Coefficient of discharge (C_d)

The ratio of actual discharge (Q) through an orifice to the theoretical discharge (Q_{th}) is called coefficient of discharge.

Mathematically,

Coefficient of discharge : $C_d = \dfrac{Q}{Q_{th}}$

$$C_d = \dfrac{\text{Actual area} \times \text{actual velocity}}{\text{Theoretical area} \times \text{theoretical velocity}}$$

$$= \dfrac{a_c}{a} \times \dfrac{V}{Vt_h}$$

$$C_d = C_c C_v$$

The coefficient of discharge (C_d) is also defined as the product of coefficient of contraction (C_c) and coefficient of velocity (C_v).

The value of C_d lies between 0.62 to 0.65.

Derivation of discharge for orifice meter:

Consider an orifice meter fitted in a horizontal pipe through which a fluid is flowing, as shown in Fig. 5.12

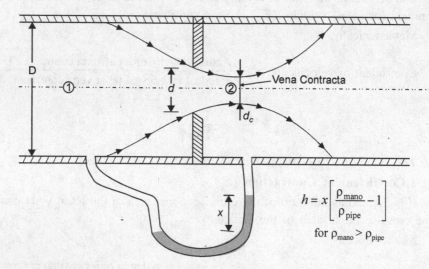

$$h = x\left[\dfrac{\rho_{mano}}{\rho_{pipe}} - 1\right]$$

for $\rho_{mano} > \rho_{pipe}$

Fig. 5.12: *Orifice meter with differential U-tube manometer.*

Let D = diameter of pipe (or at section (1),

d = diameter of orifice

d_c = diameter of vena-contracta.

$A = \dfrac{\pi}{4}D^2$, cross-sectional area of the pipe

(or at sectional (1),

$$a = \frac{\pi}{4}d^2 \text{ , cross-sectional area of the orifice.}$$

$$a_c = \frac{\pi}{4}d_c^{\;2} \text{ , cross-sectional area of vena-contracta.}$$

p_1, v_1, z_1 are pressure, velocity and datum head at section (1) respectively and
p_2, v_2, z_2 are corresponding values at section (2)

Applying the Bernoulli's equation at sections (1) and (2), neglecting head losses, we get

$$\frac{p_1}{\rho g} + \frac{V_1^2}{2g} + z_1 = \frac{p_2}{\rho g} + \frac{V_2^2}{2g} + z_2$$

$$\frac{p_1}{\rho g} + \frac{V_1^2}{2g} = \frac{p_2}{\rho g} + \frac{V_2^2}{2g}$$

$$(z_1 = z_2 \; \because \text{ orificemeter is horizontal})$$

$$\frac{p_1}{\rho g} - \frac{p_2}{\rho g} = \frac{V_1^2}{2g} - \frac{V_2^2}{2g}$$

$$h = \frac{V_1^2}{2g} - \frac{V_2^2}{2g} \qquad\qquad \text{...(5.7)}$$

According to continuity equation, at sections (1) and (2), we get

$$AV_1 = a_c V_2$$

or

$$V_1 = \frac{a_c V_2}{A} \qquad\qquad \text{...(5.8)}$$

and

$$C_c = \frac{a_c}{a} = 1 \text{ for ideal flow}$$

or

$$a_c = a$$

Putting $a_c = a$ in equation (12.8), we get

$$V_1 = \frac{a}{A} V_2$$

Substituting $\qquad\qquad V_1 = \frac{a}{A} V_2$ in equation (5.7), we get

$$h = \frac{V_2^2}{2g} - \frac{a^2 V_2^2}{A^2 2g}$$

or
$$h = \frac{V_2^2}{2g}\left[1 - \frac{a^2}{A^2}\right]$$

or
$$2gh = V_2^2\left[\frac{A^2 - a^2}{A^2}\right]$$

$$V_2^2 = \left[\frac{A^2}{A^2 - a^2}\right] \times 2gh$$

or
$$V_2 = \frac{A\sqrt{2gh}}{\sqrt{A^2 - a^2}}$$

Theoretical (or ideal) velocity at section (2).

i.e.,
$$(V_2)_{\text{th}} = \frac{A\sqrt{2gh}}{\sqrt{A^2 - a^2}}$$

and coefficient of contraction: $C_c = \frac{a_c}{a}$

for real fluid

or
$$a_c = C_c a$$

Putting the value of $a_c = C_c a$ in equation (5.8), we get

$$V_1 = \frac{C_c a V_2}{A}$$

Substituting $V_1 = \frac{C_c a V_2}{A}$ in equation (5.7), we get

$$h = \frac{V_2^2}{2g} - \frac{C_c^2 a^2 V_2^2}{A^2}$$

$$= \frac{V_2^2}{2g}\left[1 - \frac{C_c^2 a^2}{A^2}\right]$$

$$2gh = V_2^2\left[1 - \frac{C_c^2 a^2}{A^2}\right]$$

$$= V_2^2\left[\frac{A^2 - C_c^2 a^2}{A^2}\right]$$

or
$$V_2^2 = \left[\frac{A^2 \times 2gh}{A^2 - C_c^2 a^2}\right]$$

$$V_2 = \frac{A\sqrt{2gh}}{\sqrt{A^2 - C_c^2 a^2}}$$

Actual velocity at section (2)

i.e., $(V_2)_{act} = \dfrac{A\sqrt{2gh}}{\sqrt{A^2 - C_c^2 a^2}}$ ...(5.9)

Coefficient of velocity: $C_v = \dfrac{(V_2)_{act}}{(V_2)_{th}}$

$$= \frac{A\sqrt{2gh}}{\sqrt{A^2 - C_c^2 a^2}} \times \frac{\sqrt{A^2 - a^2}}{A\sqrt{2gh}}$$

$$= \frac{\sqrt{A^2 - a^2}}{\sqrt{A^2 - C_c^2 a^2}}$$

As we know coefficient of discharge : $C_d = C_c C_v$

$$C_d = C_c \frac{\sqrt{A^2 - a^2}}{\sqrt{A^2 - C_c^2 a^2}}$$

or $C_c = \dfrac{C_d \sqrt{A^2 - C_c^2 a^2}}{\sqrt{A^2 - a^2}}$ ...(5.10)

We know discharge : $Q = a_c V_2$

For real fluid $a_c = C_c a$

                        $Q = C_c a V_2$ ...(5.11)

Substituting the value of C_c from equation (5.10) and V_2 from equation (5.11) in equation (5.12), we get

$$Q = \frac{C_d \sqrt{A^2 - C_c^2 a^2}}{\sqrt{A^2 - a^2}} \times a \times \frac{A\sqrt{2gh}}{\sqrt{A^2 - C_c^2 a^2}}$$

$$Q = \frac{C_d a A \sqrt{2gh}}{\sqrt{A^2 - a^2}}$$

where C_d = coefficient of discharge

                        = 0.62 to 0.65

        a = cross-sectional area of the orifice

        A = cross-sectional area of the pipe.

the formula of discharge for orifice meter and venturimeter is same.

i.e.,
$$Q = \frac{C_d a A \sqrt{2gh}}{\sqrt{A^2 - a^2}}$$
... (5.12)

On simplifying equation (5.10.6) like equation (5.11)

$$Q = \frac{3.48 C_d d^2 D^2 \sqrt{h}}{\sqrt{D^4 - d^4}}$$ (general equation)

where
h = pressure difference in metre liquid
flow through pipe
D & d = diameters in metre

Another simplified equation, like equation

$$Q = \frac{12.35 C_d d^2 D^2 \sqrt{x}}{\sqrt{D^4 - d^4}}$$ (specific equation)

where
x = manometer reading in metrr of mercury
D, d = diameters of pipe & orifice
respectively

The above specific equation is used only when water flows through pipe & mercury contains in manometer.

Problem 5.13: An orifice metre consisting of 100 mm diameter orifice in a 250 mm diameter pipe has C_d = 0.65. The pipe delivers oil of specific gravity of 0.8. The pressure difference on the two sides of the orifice plate is measured by a mercury oil differential manometer. If the differential gauge reading is 800 mm of mercury, find the rate of flow in litre/s.

Solution. Given data:

Diameter of pipe : $D = 250$ mm $= 0.25$ m

∴ Cross-sectional area of pipe : $A = \frac{\pi}{4} D^2 = \frac{3.14}{4} \times (0.25)^2$

$= 0.04906$ m^2

Diameter of orifice : d = 100 mm = 0.1 m

∴ Cross-sectional area of orifice :

$$a = \frac{\pi}{4} d^2 = \frac{3.14}{4} \times (0.1)^2$$

$= 7.85 \times 10^3$ m^2

Coefficient of discharge : C_d = 0.65

Specific gravity of oil :

$$S = 0.8$$

∴ Density of oil : $\rho = S \times 1000$

$$= 8 \times 1000 = 800 \text{ kg/m}^3$$

Differential manometer reading : $x = 800$ mm of Hg

$$= 0.8 \text{ m of Hg}$$

Pressure difference : $h = x\left[\dfrac{\rho_{mano}}{\rho_{pipe}} - 1\right]$

$$= 0.8\left[\frac{13600}{800} - 1\right] = 0.8 \times [17.1]$$

$$= 0.8 \times 16 = 12.8 \text{ m of oil}$$

Rate of flow : $Q = \dfrac{C_d\, Aa\sqrt{2gh}}{\sqrt{A^2 - a^2}}$

$$= \frac{0.65 \times 0.0490 \times 7.85 \times 10^{-3}\sqrt{2 \times 9.81 \times 12.8}}{\sqrt{(0.04906)^2 - (7.85 \times 10^{-3})^2}}$$

$$= 0.08191 \text{ m}^3/\text{s}$$
$$= 0.08191 \times 1000 \text{ litre/s}$$
$$= \mathbf{81.916 \text{ litre/s}}$$

5.4 PITOT TUBE

Function: It is used for measuring the velocity of liquid flow at any point in a pipe or a channel.

Principle: If the velocity of flow at a point becomes zero, the pressure is increased there due to the conversion of the kinetic energy into pressure energy. The point at which the velocity of flow becomes zero is called stagnant point or state. The pressure at stagnant state is called total pressure (or head) or stagnation pressure (or head).

Shape: Pitot tube consists of L-shaped glass tube, a tube bent at 90° and with ends unsealed, the horizontal part is called 'body', placed in parallel and opposite to the direction of flow and the other part is called 'stem', vertical and open to atmosphere as shown in Fig. 5.13.

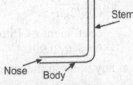

Fig. 5.13: *Pitot Tube*

Consider two points (1) and (2) at the same level in such a way that point (2) is just at the inlet of the Pitot tube and point (1) is for away from the tube at same level of point (2)

Let, p_1, v_1 = pressure, velocity at point (1)

p_2, v_2 = pressure, velocity at point (2)

H = depth of tube in the liquid or depth of points (1) and (2) from the free surface of liquid.

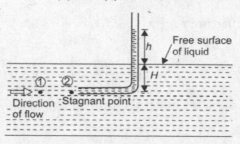

Fig. 5.14: *Pitot tube used in liquid flow through channel.*

h = rise of liquid in the tube above the free surface of liquid.

Applying Bernoulli's equation at point (1) and (2) we get,

$$\frac{p_1}{\rho g} + \frac{V_1^2}{2g} + z_1 = \frac{p_2}{\rho g} + \frac{V_2^2}{2g} + z_2 \qquad \qquad ...(5.13)$$

Datum head, $z_1 = z_2$

Velocity at point (2), $V_2 = 0$

$$\frac{p_1}{\rho g} = H, \qquad \frac{p_2}{\rho g} = H + h$$

Equation (5.13) becomes

$$H + \frac{V_1^2}{2g} = H + h$$

$$\frac{V_1^2}{2g} = h$$

$$V_1 = \sqrt{2gh}$$

This is theoretical velocity at point (1).

Actual velocity at point (1),

$$V_{1act} = C_v \times \text{theoretical velocity at point (1)}$$

$$= C_v \sqrt{2gh}$$

where C_v = coefficient of Pitot tube

= 0.96 to 0.99

∴ The velocity of fluid flow at any point in the channel is measured by Pitot tube by equation $V = C_v \sqrt{2gh}$

Pitot tube is also used for measuring the state head, dynamic head and stagnation head (or total head) of the liquid flow at any point in a pipe or a channel.

Stagnation or total head:

It is sum of static head and dynamic head.

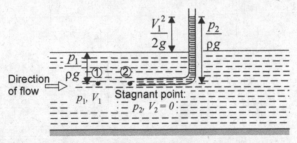

Fig. 5.15: *Pitot tube used in liquid flow through channel.*

Total head :
$$\frac{p_2}{\rho g} = \frac{p_1}{\rho g} + \frac{V_1^2}{2g}$$

or total pressure at point 2, $\quad p_2 = p_1 + \dfrac{\rho V_1^2}{2}$

OR

Total pressure at point 1 = static pressure + dynamic pressure

$$= p_1 + \frac{\rho V_1^2}{2}$$

Static Pressure: It is defined as the pressure of liquid measured when the liquid is static.

When the liquid flows through pipe or in channel, the total pressure at any point in the flow is the sum of static pressure and dynamic pressure. It means the flow of liquid possess both the static pressure and dynamic pressure (pressure due velocity of the liquid).

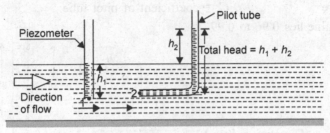

Fig. 5.16: *Liquid flow is channel*

Static pressue head = h_1
Dynamic pressure head = h_2
Total pressure head = $h_1 + h_2$

A simple piezometer tube is used at point 1 in such a way that there is no component of velocity along the axis of the piezometer tube at which liquid enters to it.

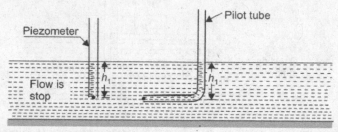

Fig. 5.17: *Liquid at rest.*

$$Static\ pressue\ head = h_1$$
$$Dynamic\ pressure\ head = 0$$
$$Total\ pressure\ head = static\ presure\ head = h_1$$

So the level of the liquid raise in tube is called static pressure head. Say h_1. the liquid raise in piezometer tube upto free surface of the liquid. On the other hand, a Pitot tube is used at another point 1 at same level from the free surface of liquid, the direction of the velocity is parallel to the axis of horizontal point of the Pitot tube. The liquid raise in Pitot tube partly due to static pressure and partly due to dynamic pressure (*i.e.*, pressure due to velocity of the liquid)

i.e., Total pressure head = static pressure + dynamic pressure = $h_1 + h_2$

If the flow of liquid stop in the channel, the level of the liquid in Pitot tube falls and became equal to the level of the liquid in piezometer (or level of free surface of liquid) as shown in Fig. 5.17. It shows that the dynamic pressure (*i.e.*, pressure due to velocity) becomes zero. The level of the liquid in the Pitot tube shows the static pressure head. So the pitot acts as piezometer tube if liquid is at rest.

Note: Total pressure is also known as stagnation pressure.

Pitot tube with Piezometer raised in flow through pipe:

Velocity act point 1, $V_1 = C_v \sqrt{2gh}$

where C_v = coefficient of pitot tube.

Its value lies 0.96 to 0.99

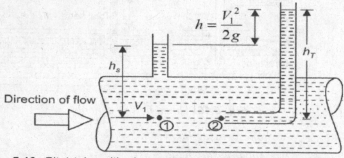

Fig. 5.18: *Pitot tube with piezometer used in flow through pipe*

Pitot tube with differential manometer :

Velocity at point 1, $V_1 = C_v \sqrt{2gh}$

where $h = x \left[\dfrac{\rho_{Hg}}{\rho_{pipe}} - 1 \right]$

x = differential manometer reading in mm of mercury

ρ_{Hg} = density of the mercury.

OR

density of the liquid used in manometer

ρ_{pipe} = density of liquid flow through pipe

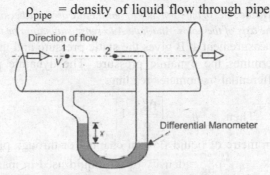

Fig. 5.19: *Pilot tube with differential manometer*

Pitot Static Tube : It consists of two circular concentric tubes one inside the other with some angular spare in between as shown in Fig. 5.20. The holes are added to the sides of the tube and simple Pitot tube and connected to differential manometer

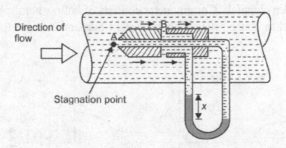

Fig. 5.20: *Pitot-Static tube used in flow through pipe.*

Holes added to the sides of the pilot tube to sence static pressure because there is no component of velocity along the axis of the holes. And hole 'A' to sense stagnaton or total pressure.

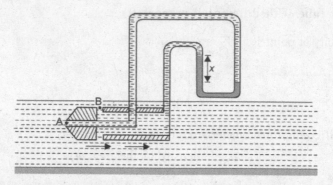

Fig. 5.21: Pitot Static tube used in flow in Channel.

Holes added to the sides of the Pitot tube to sense static pressure because there is no component of velocity along the axis of the holes. And hole A to sense stagnation or total pressure

Pressure measurement at B gives the static pressure and the difference between p_A and p_B determines the dynamic pressure. The dynamic pressure head (x) is measured by differential manometer reading.

$$\text{Then,} \qquad h = x\left[\frac{\rho_{Hg}}{\rho} - 1\right]$$

where h in metre of liquid flow in channel or through pipe.

ρ_{Hg} = density of the liquid used in manometer (normally mercury is used)

ρ = density of the liquid flow through pipe or in channel.

5.5 CURRENT METER

The current meter is a mechanical device used to measure the velocity of water in rivers and in open channels. Current meter consists of hollow hemisphere or cups mounted on spokes so as to cause rotation about a shaft perpendicular to the direction of flow.

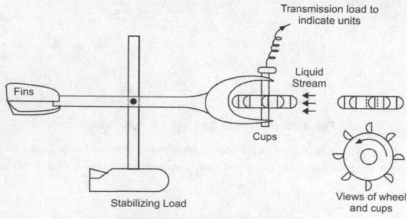

Fig. 5.22: Current Meter.

Table 5.1 : Comparision of Venturimeter, Flow Nozzle meter and Orifice meter

S. No.	Types of Flowmeter	Use	Diagram	Value of C_d	Head Loss	Length	Cost	Accuracy	Typical Installation
1.	Venturimeter	To determine discharge through pipe		0.97 to 0.99	Low	Long	High	Very good	In a pipe of small diameter
2.	Flow Nozzle meter	To determine discharge through pipe		0.93 to 0.98	Intermediate between a Venturimeter and an orifice meter				Size of pipe lies between Venturimeter & orifice pipes
3.	Orifice meter	To determine discharge through pipe		0.62 to 0.65	High	Short	Low	Good	In a pipe of large diameter

The entire setup shown in Fig. 5.22, is lowered into water from a bridge or a boat. The drag on a hollow hemisphere or cups is greater when its open side faces the liquid stream and so there is a net torque on assembly when flow comes from any direction in the plane of rotation. This rotation is converted into an electrical signal by means of a circuit into an electrical signal by means of a circuit into which a set of earphones are plugged in. A fixed number of revolutious of wheel on which cups are mounted, produce a beat that can be heard clearly through the ear phones. The number of beats in a given period of time is a function of fluid velocity.

5.6 ROTAMETER

It is a device used for measuring discharge in pipelines. A rotameter consists of a tapered metering glass tube inside which a rotor or active element (float) is located. The tube is provided with suitable inlet and outlet connections. The float has a specific gravity higher than that of the fluid flowing through the tube. A very small slot cut in the centre of rotor causes it to move up and down about the guide wide and also keep it centered.

With the increase in the flow rate, the float rises in the tube and there occurs an increase in angular area between the float and the tube the float adjust its position in relation to the discharge through the passage i.e., float rides higher or lower depending on the flow rate.

A rotameter has the advantage of simplicity, relatively low cost and ability to handle variety of corrosive fluids.

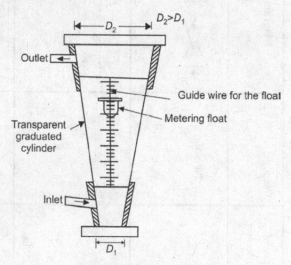

Fig. 5.23: Rotameter

The meter however is to be mounted vertically is limited to small pipe sizes and capacities and is less accurate compared to venture and orifice meter.

5.7 BEND METER

A bend meter is also called elbow meter, makes use of fact that when a liquid flow through a curved pipe (elbow), centrifugal forces cause a pressure difference between the outer and inner sides of the curved pipe. This defference in pressure is utilized to measure discharge. The pressure diference generated by an elbow flowmeter is smaller than that by other pressure differential flow meters.

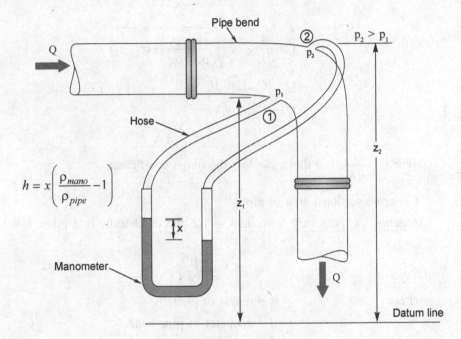

Fig. 5.24: *Bend meter*

In bend meter, the pipe is given two pressure tappings, one at the inner and other at the outer wall of the pipe at bend. These tappings are comected with a U-tube differential menometer as shown in Fig. 5.24. The velocity of flow in pipe is given by following relation.

$$\frac{KV^2}{2g}=h \tag{5.14}$$

where h is piezometric head difference and is given by

$$h=\left(\frac{p_2}{\rho g}+z_2\right)-\left(\frac{p_1}{\rho g}+z_1\right)$$

$$=\left(\frac{p_2}{\rho g}-\frac{p_1}{\rho g}\right)+(z_2-z_1)$$

= pressure head difference + datum head difference.

Here p_2 is pressure at outer wall of the pipe bend

p_1 is pressure at inner wall of the pipe bend

In Eq. (5.14) K is a constant and it depends upon the size and shape of the pipe. Its value ranges between 1.3 to 3.2.

The relationship expressing discharge through bend may be written as:

$$Q = \frac{A}{\sqrt{K}}\sqrt{2g}\sqrt{\left(\frac{p_2}{\rho g} - \frac{p_1}{\rho g}\right) + (z_2 - z_1)}$$

$$Q = AC_d\sqrt{2g}\sqrt{h}$$

$$Q = AC_d\sqrt{2gh}$$

Here $C_d = \frac{1}{\sqrt{K}}$, is a dimensionless discharge coefficient.

A is cross-sectional area of pipe.

In general, C_d can vary from 0.55 to 1.2 and empirically it is given by :

$$C_d = \sqrt{\frac{R}{2d}}$$

where   R = radius of pipe

d = diameter of pipe bend.

In bend meter, as bend itself is a part of piping system, therefore no additional fitting is required for this meter. Also if calibrated properly in place, accurate results can be obtained from bend meter.

Problem 5.14: A Pitot tube was inserted in a river to measure the velocity of water in it. If the water rises in the tube above free surface of water is 300 mm. Find the velocity of water. Take C_v = 0.98

Solution: Given data :

Rise of water above free surface : h = 300 mm = 0.3 m

$$C_v = 0.98$$

We know,

$$V = C_v\sqrt{2gh}$$

$$= 0.98 \times \sqrt{2 \times 9.81 \times 0.3}$$

$$= \textbf{2.37 m/s}$$

Problem 5.15: A Pitot static tube is used to measures the velocity of water in the centre of pipe. The static and total pressures are 6 m and 7.5 m respectively. Find the velocity of flow. Take coefficient of Pitot tube as 0.98.

Solution: Given data:

Static pressure head : $h_{st} = 6$ m

Total pressure head : $h_T = 7.5$ m

Also total pressure head : $h_T = h_{st}$ + dynamic pressure head

$$h_T = h_{st} + h$$
$$7.5 = 6 - h$$

or $\qquad\qquad\qquad h = 1.5$ m of water

$\therefore \qquad$ Velocity of flow : $V = C_v\sqrt{2gh}$

$$= 0.98 \times \sqrt{2 \times 9.81 \times 1.5} = \textbf{5.31 m/s}$$

Problem 5.16: Find the velocity of the flow of an oil through a pipe, when the difference of mercury level in a differential U-tube manometer connected to the two tapping-tube is 200 mm. Take $C_v = 0.98$ and specific gravity of oil is 0.85.

Solution: Given data:

Difference of mercury level : $\quad x = 200$ mm of Hg = 0.2 m of Hg

$$C_v = 0.98$$

Specific gravity of oil : $\qquad S = 0.85$

$\therefore$ Density of oil : $\qquad\qquad \rho = S \times 1000$

$$= 0.85 \times 1000 \text{ kg/m}^3$$
$$= 850 \text{ kg/m}^3$$

We know,

Difference of pressure head : $h = x\left[\dfrac{\rho_{Hg}}{\rho} - 1\right]$

$$= 0.2\left[\dfrac{13600}{850} - 1\right]$$

$$= 3 \text{ m of oil}$$

$\therefore \qquad$ Velocity of oil : $V = C_v\sqrt{2gh}$

$$= 0.98 \times \sqrt{2 \times 9.81 \times 3} = \textbf{7.51 m/s}$$

Problem 5.17: A Pitot static tube placed in the centre of a 250 mm pipe line has one orifice pointing upstream and other perpendicular to it. The mean velocity in the pipe 75% of the central velocity. Find the discharge through the pipe if the pressure difference between the two orifices is 80 mm of water. Take $C_v = 0.99$.

Solution: Given data:

Density of pipe : $D = 250$ mm $= 0.25$ m

Mean velocity : $\overline{V}$ $= 75\%$ central velocity

$= 0.75 \times V$

Difference of pressure head :

$h = 800$ mm of water

$= 0.08$ m of water

Central Velocity : $V = C_v \sqrt{2gh}$

$= 0.99 \times \sqrt{2 \times 9.81 \times 0.08}$

$= 1.24$ m/s

∴   Average velocity : $\overline{V}$ $= 0.75 \times 1.25$

$= 0.9375$ m/s

According to continuity equation :

Discharge : Q = cross-sectional area of pipe $\times$ $\overline{V}$

$= A\overline{V}$ $= \dfrac{\pi}{4} D^2 \times 0.9375$

$= \dfrac{3.14}{4} \times (0.25)^2 \times 0.9375$

$= 0.04599$ m³/s = **45.99 litre/s**

Problem 5.18: A Pitot tube is used in river to determine the following terms at point 1;

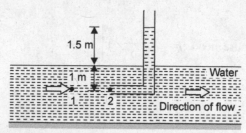

Fig. 5.25: Schematic for Problem 5.18

(*i*) Velocity,

(*ii*) Static pressure,

(*iii*) Dynamic pressure, and

(*iv*) Total (or Stagnation) pressure.

Suitable data are mentioned in Fig. 5.25. Take coefficient of Pitot tube is 0.98.

Solution: Given data:

Static head : $h_{st} = 1$ m

Dynamic head : $h = 1.5$ m

Total head : $h_T = h_{st} + h$

$= 1 + 1.5 = 2.5$ m of water

(i)   Velocity at point 1 : $V = C_v \sqrt{2gh}$

$= 0.98 \times \sqrt{2 \times 9.81 \times 1.5}$

$= \mathbf{5.31}$ **m/s**

(ii) Static pressure : p_{st}

Static pressure head : $h_{st} = \dfrac{p_{st}}{\rho g}$

or $P_{st} = \rho g h_{st} = 1000 \times 9.81 \times 1$

$= 9810$ N/m²

$= \mathbf{9.81}$ **kN/m²**

(iii) Dynamic pressure : p

Dynamic pressure : $h = \dfrac{p}{\rho g}$

or $p = \rho g h = 1000 \times 9.81 \times 1.5$

$= 14715$ N/m²

$= \mathbf{14.715}$ **kN/m²**

(iv)   Total pressure : p_T

Total pressure head : $h_T = \dfrac{p_T}{\rho g}$

or $p_T = \rho g h_T = 1000 \times 9.81 \times 2.5$

$= 24525$ N/m²

$= \mathbf{24.52}$ **kN/m²**

OR

$p_T = p_{st} + p$

$= 9.81 + 14.71$

$= \mathbf{24.52}$ **kN/m²**

SUMMARY

1. Venturimeter: It is used for measuring the discharge through a pipe.

$$\text{Discharge: } Q = \frac{C_d a A \sqrt{2gh}}{\sqrt{A^2 - a^2}}$$

where C_d = coefficient of discharge

A = cross-sectional area at inlet of venturimeter.

a = cross-section area at throat of venturimeter.

h = difference of pressure head in terms of liquid flowing, through the pipe.

$$h = x \left[\frac{\rho_{mano}}{\rho_{pipe}} - 1 \right] \qquad \text{for } \rho_{mano} > \rho_{pipe}$$

where x = differential manometer reading

ρ_{mano} = density of liquid used in manometer

ρ_{pipe} = density of liquid flowing through pipe.

$$h = x \left[1 - \frac{\rho_{mano}}{\rho_{pipe}} \right] \qquad \text{for } \rho_{pipe} > \rho_{mano}$$

Also

$$h = \frac{p_1}{\rho g} - \frac{p_2}{\rho g} \qquad \text{for horizontal venturimeter}$$

$$h = \left(\frac{p_1}{\rho g} + z_1 \right) - \left(\frac{p_2}{\rho g} + z_2 \right)$$

for vertical and inclined venturimeter

2. Discharge through venturimeter

$$Q = \frac{3.48 C_d d^2 D^2 \sqrt{h}}{\sqrt{D^4 - d^4}}$$

where

D = diameter at inlet.

d = diameter at throat.

h = difference of pressure head at inlet and throat in terms of liquid flowing through pipe.

All D, d and h are in m.

3. The specific equation for venturimeter is used when water flow through pipe and mercury is used in manometer :

$$Q = \frac{12.35 C_d d^2 D^2 \sqrt{x}}{\sqrt{D^4 - d^4}}$$

where

x = differential manometer reading in m of Hg

All D, d and h are in m.

Contd...

4. Orifice Meter: It is used for measuring the discharge through a pipe

$$\text{Discharge: } Q = \frac{C_d\, Aa\sqrt{2gh}}{\sqrt{A^2 - a^2}}$$

where
- a = cross-sectional area of the orifice.
- A = cross-sectional area of the pipe.

5. Pitot tube: It is used for measuring the velocity of liquid at any point in a pipe or a channel.

$$\text{Velocity: } V = C_v\sqrt{2gh}$$

where
- C_v = coefficient of pitot-tube.
- h = rise of liquid in the tube above the free surface of liquid\

6. Current Meter: It is a mechanical device used to measure the velocity of water in rivers and in open channels.

7. Rotameter: It is a device used for measuring discharge in pipelines.

8. Bend Meter: It is also called elbow meter, makes use of fact that when a liquid flow through a curved pipe (elbow), centrifugal cause a pressure difference between the outer and inner sides of the curved pipes. This pressure difference is utilized to measure discharge .

ASSIGNMENT - 1

1. Define a venturimeter. Prove that the discharge through a venturimeter is given by the relation.

$$Q = \frac{C_d\, Aa\sqrt{2gh}}{\sqrt{A^2 - a^2}}$$

where
- A = cross-sectional area of pipe.
- a = cross-sectional area of throat.

2. Prove that the discharge through a venturimeter is given by the relation :

$$Q = \frac{3.48 C_d d^2 D^2 \sqrt{x}}{\sqrt{D^4 - d^4}}$$

where
- D = diameter at inlet,
- d = diameter at throat
- h = pressure head difference between, inlet and throat

3. Prove that the discharge through a venturimeter is given by relation, when water flows through pipe and mercury contained in manometer :

$$Q = \frac{12.35 C_d d^2 D^2 \sqrt{x}}{\sqrt{D^4 - d^4}}$$

where x = manometer reading in m of Hg

D, d = diameters at inlet & throat respectively in m.

4. What is a venturimeter ? Derive an expression for the rate of discharge through the venturimeter.

5. Define an orifice meter. Prove that the discharge through an orifice meter is given by the relation

$$Q = \frac{C_d \, Aa \sqrt{2gh}}{\sqrt{A^2 - a^2}}$$

where A = cross-sectional area of pipe in which orifice-meter is filled.

a = area of orifice.

6 What is an orifice plate ? How will the discharge through a pipeline be measured with the help of an orifice plate ? Compare and contrast the use of orificemeter and Venturimeter.   (GGSIP University, Delhi Dec.2008)

7. Describe with a neat sketch, the principle and working of a pitot-tube.

(GGSIP University, Delhi Dec. 2002, Dec.2007)

8. What is pitot ? how is it used to measue the velocity of flow at any point in a pipe or channel.   (GGSIP University, Delhi 2005)

9. What is the difference between pitot tube and pitot static tube ?

10. What is a current meter and how does it work?

11. What is a rotameter?

12. What is a bend meter?

ASSIGNMENT - 2

1. A horizontal venturimeter with inlet and throat diameter 300 mm and 150 mm respectively is used to measure the flow of water. The reading of differential manometer connected to the inlet and the throat is 20 cm of mercury. Find the rate of flow. Take $C_d = 0.98$ [**Ans.** 125.75 litre/s]

2. An oil of specific gravity 0.8 is flowing through a venturimeter having inlet diameter 200 mm and throat diameter 100 mm. The oil mercury differential manometer shows a reading of 250 mm. Find the discharge of oil through the horizontal venturimeter. Take $C_d = 0.98$ [**Ans.** 70.44 litre/s]

3. A horizontal venturimeter with inlet diameter 300 mm and throat diameter 150 mm is used to measure the flow of oil of specific gravity 0.8. The discharge of oil through venturimeter is 100 litre/s. Find the reading of the oil mercury differential manometer. Take $C_d = 0.98$ [**Ans.** 10.48 cm of Hg]

4. A horizontal venturimeter with inlet diameter 300 mm and throat diameter 150

mm is used to measure the flow of water. The pressure at the throat is 250 mm of mercury. Find the discharge of water through venturimeter. Take $C_d = 0.987$
[**Ans.** 363.39 litre/s]

5. The inlet and throat diameters of a horizontal Venturimeter are 300 mm and 100 mm respectively. The liquid flowing through the Venturimeter is water. The vacuum pressure head at the throat is 370 mm of mercury. Find coefficient of discharge. Assume that 4% of the differential head is lost between the inlet and throat. Find also the rate of flow.   [**Ans.** $C_d = 0.989$, Q = 149.39 litre/s]

6. A 300 mm × 150 mm venturimeter is inserted in a vertical pipe carrying water, flowing in upward direction. A differential mercury manometer connected to the inlet and throat gives a reading of 300 mm. Find the discharge. Take coefficient of discharge as 0.98.   /   [**Ans.** 154.04 litre/s]

Hint: Simplified equation is used in this problem:

$$Q = \frac{12.35 C_d d^2 D^2 \sqrt{x}}{\sqrt{D^4 - d^4}}$$

where $x = 0.3$ m of Hg

$D = 0.3$ m

$d = 0.15$ m

7. 200 mm × 100 mm venturimeter is mounted in a vertical pipe carrying water. the flow being upwards. The throat section is 200 mm above the entrance section of the venturimeter. For a certain flow through the meter, the differential gauge between the throat and entrance indicates a gauge of deflection of 250 mm of mercury. Assuming the venturi coefficient as 0.98, find the discharge.
[**Ans.** 62.49 litre/s]

8. Find the throat diameter of a venturimeter, when fitted to a horizontal main 100 mm diameter having a discharge of 20 litre/s. The differential U-tube mercury manometer shows a deflection giving a reading of 0.6 m. Take coefficient of discharge as 0.95.

In case, this venturimeter is introduced in a vertical pipe with the water flowing upwards, find the difference in the readings of mercury gauge. The dimensions of pipe and venturimeter remain unaltered as well as the discharge through the pipe.

[**Ans.** d = 46 mm, the manometer reading will be same, when
Venturimeter is introduced in vertical pipe.]

9. A 300 mm × 150 mm venturimeter is provided in a vertical pipeline carrying oil of specific gravity 0.9, the flow being upwards. The difference in elevations of the throat section and entrance section of the venturimeter is 300 mm. The differential U-tube mercury manometer shows a gauge deflection of 250 mm. Calculate:

    (*i*) *discharge of the oil, and*

    (*ii*) pressure difference between the entrance and throat section.

Take the coefficient of discharge as 0.98 and the specific gravity of the mercury as 13.6.   [**Ans.** 149 litre/s, 3.695 m of oil]

10. An orifice meter consisting of 150 mm of diameter orifice in a 300 mm diameter of pipe has coefficient of discharge to 0.64. The pressure difference on the two sides of the orifice plate is measured by a mercury water differential manometer. If the differential gauge reading is 100 mm of mercury, find the rate of flow in litre.[**Ans.** 58.08 litre/s]

11. A pitot tube was inserted in a pipe to measure the velocity of water in it. If the water rises in the tube is 400 mm, find the velocity of water. Take $C_v = 0.98$

[**Ans.** 2.74 m/s]

12. Find the velocity of the flow of an oil through a pipe, when the difference of mercury level in differential U-tube manometer connected to the two tapping-tube is 150 mm. Take $C_v = 0.98$ and specific gravity of oil is 0.8.

[**Ans.** 6.72 m/s]

13. A pitot static tube is used to measure the velocity of water in the centre of pipe. The static and total pressures are 6.5 m and 8.5 respectively. Find the velocity of flow. Take $C_v = 0.98$.   [**Ans.** 6.13 m/s]

14. A pitot-static tube placed in the centre of a 400 mm pipe line has one orifice pointing upstream and other perpendicular to it. The mean velocity in the pipe is 0.80 of the central velocity. Find the discharge through the pipe if the pressure difference between the two orifices is 60 mm of water. Take $C_v = 0.97$

[**Ans.** 105.74 litre/s]

Orifices and Mouthpieces

6.1 INTRODUCTION

An orifice may be defined as a small opening with a closed perimeter located in the side or bottom of the tank or vessel. The level of the liquid in the vessel must be above the top edge of the orifice. A mouthpiece is a short length of a pipe which is two to three times of diameter of the pipe; fitted in a tank or vessel which contains the liquid. Both the orifices and mouthpieces are used for measuring the discharge of the liquid in the tank or vessel.

The stream of liquid issuing from an orifice is called the jet of liquid or simply 'jet'. The minimum cross-sectional area of the jet is called vena-contracta as shown by section C-C in Fig. 6.1.

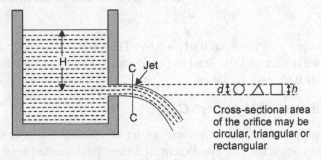

Fig. 6.1: Cross-sectional area of the orifice may be circular, triangular or rectangular.

6.2 TYPES OF ORIFICES

The orifices may be classified as:

    1. According to size:

        (*i*) Small orifice

        (*ii*) Large orifice

   2. According to shape:
      (*i*) Circular orifice
      (*ii*) Rectangular orifice
      (*iii*) Triangular orifice.
   3. According to shape of the edge:
      (*i*) Sharp-edged orifice
      (*ii*) Bell-mouthed orifice
   4. According to nature of discharge:
      (*i*) Fully submerged orifice
      (*ii*) Partially submerged orifice.

6.3 HYDRAULIC COEFFICIENTS

The following three types of coefficients used in an orifice, are called hydraulic coefficients:

   (*i*) Coefficient of contraction: C_c
   (*ii*) Coefficient of velocity: C_v
   (*iii*) Coefficient of discharge: C_d

6.3.1 Coefficient of Contraction: C_c

It is defined as the ratio of the cross-sectional area of the jet at vena-contracta to the cross-sectional area of the orifice. The coefficient of contraction is denoted by C_c.
   Mathematically,

$$\text{Coefficient of contraction: } C_c = \frac{\text{Cross - sectional area at vena - contracta}: a_c}{\text{Cross - sectional area of the orifice}: a}$$

$$C_c = \frac{a_c}{a}$$

   The value of C_c varies from 0.61 to 0.69. The variation in value depends upon size, shape of the orifice and the head of liquid under which the flow takes place. An average value of 0.65 may be taken.

6.3.2 Coefficient of Velocity: C_v

It is defined as the ratio of actual velocity of jet at vena-contracta to the theoretical velocity at vena-contracta. The coefficient of velocity is denoted by C_v.
   Mathematically,

$$\text{Coefficient of Velocity: } C_v = \frac{\text{Actual velocity of jet at vena - contracta}: V}{\text{Theoretical velocity of jet at vena - contracta}: V_{th}}$$

$$\therefore \qquad C_v = \frac{V}{V_{th}}$$

Theoretical velocity of jet at vena-contracta:

$$V_{th} = \sqrt{2gH}, \text{ is also called velocity of spout.}$$

$$\therefore \qquad C_v = \frac{V}{\sqrt{2gH}}$$

or Actual velocity: $\qquad V = C_v\sqrt{2gH}$

The value of C_v varies from 0.95 to 0.99. The variation in value depends upon size, shape of the orifice and the head of liquid under which the flow takes place. An average value of 0.97 may be taken.

6.3.3 Coefficient of Discharge: C_d

It is defined as the ratio of actual discharge through an orifice to the theoretical discharge through an orifice. It is denoted by C_d.

Mathematically,

$$\text{Coefficient of discharge: } C_d = \frac{\text{Actual discharge through an orifice:} Q}{\text{Theoretical discharge through an orifice : } Q_{th}}$$

$$C_d = \frac{Q}{Q_{th}}$$

where $\qquad\qquad Q$ = actual velocity × actual area = $V\, a_c$

and $\qquad\qquad Q_{th}$ = theoretical velocity × theoretical area = $V_{th}\, a$

$\therefore\qquad\qquad$ $$C_d = \frac{V\, a_c}{V_{th}\, a} = \frac{V}{V_{th}} \times \frac{a_c}{a}$$

$$C_d = C_v \times C_c$$

It is also defined as the product of coefficient of velocity and coefficient of contraction. The value of C_d varies from 0.60 to 0.64. An average value of 0.62 may be taken.

6.4 EXPERIMENTAL DETERMINATION OF HYDRAULIC COEFFICIENTS

Consider a tank containing water at a constant level. The constant level of water is maintained by a continuous supply of water as shown in Fig. 6.2.

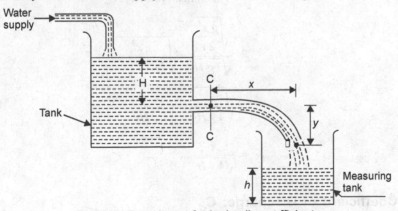

Fig. 6.2: Experiment for hydraulic coefficient.

Let the water be allowed to flow through an orifice fitted in one side of the tank.

Let the section C-C represent the vena-contracta of a jet of water coming out from orifice under constant head H. Consider a water particle which is at vena-contracta at any time and takes the position at point P along the jet in time t.

Let H = constant water head

            x = horizontal distance travelled by the water particle from section C-C to point P in time t.

            y = vertical distance between the centre of vena-contracta and point P.

            V = actual velocity of jet at vena-contracta.

We know that

$$\text{Vertical distance: } y = \frac{1}{2}gt^2 \qquad \qquad ...(6.4.1)$$

and   horizontal distance: $x = V \times t$

or $t = \dfrac{x}{V}$

Substituting the value of t in Eq. (6.4.1), we get

$$y = \frac{1}{2}g\left(\frac{x}{V}\right)^2 = \frac{1}{2}\frac{gx^2}{V^2}$$

or $V^2 = \dfrac{gx^2}{2y}$

or $V = \sqrt{\dfrac{gx^2}{2y}}$

and we know that the theoretical velocity of jet at vena-contracta:

$$V_{th} = \sqrt{2gH}$$

∴ Coefficient of velocity: $C_v = \dfrac{\text{Actual velocity}: V}{\text{Theoretical velocity}: V_{th}}$

$$= \frac{\sqrt{\dfrac{gx^2}{2y}}}{\sqrt{2gH}} = \sqrt{\frac{gx^2}{2y \times 2gH}}$$

$$= \sqrt{\frac{x^2}{4yH}} = \frac{x}{\sqrt{4yH}}$$

$$C_v = \frac{x}{\sqrt{4yH}}$$

6.4.1 Coefficient of Discharge: C_d

The actual discharge through orifice is measured by the volume of water collected in measuring tank per unit time.

Mathematically,

Actual discharge: $Q = \dfrac{\text{Volume of water in collecting tank}}{\text{Time}}$

$$= \frac{\text{Cross - sectional area of collecting tank} \times \text{height of water in measuring tank}}{\text{Time}}$$

$$Q = \frac{A \times h}{t}$$

and theoretical discharge: Q_{th}= area of orifice × theoretical velocity of the jet

$$= a \times V_{th} = a\sqrt{2gH}$$

∴ Coefficient of discharge: $\qquad C_d = \dfrac{Q}{Q_{th}}$

6.4.2 Coefficient of Contraction: C_c

We know that the coefficient of discharge:

$$C_d = C_v \times C_c$$

or $\qquad C_c = \dfrac{C_d}{C_v}$

Problem 6.1: A jet of water issues from an orifice of diameter 20 mm under a head of 1.2 m. Find the coefficient of discharge for the orifice, if actual discharge is 0.94 litre/s.

Solution: Given data:

Diameter of orifice: $\qquad d = 20$ mm $= 0.02$ m

∴ Cross-sectional area of orifice:

$$a = \frac{\pi}{4}d^2 = \frac{3.14}{4} \times (0.02)^2 = 3.14 \times 10^{-4}\ \text{m}^2$$

Head: $\qquad\qquad\qquad\qquad H = 1.2$ m

Actual discharge: $\qquad\qquad Q = 0.94$ litre/s $= 0.94 \times 10^{-3}$ m³/s.

We know that the theoretical discharge: Q_{th}

$$Q_{th} = \text{cross-sectional area} \times \text{theoretical velocity}$$

$$= a \times \sqrt{2gH} = 3.14 \times 10^{-4}\sqrt{2 \times 9.81 \times 1.2}$$

$$= 1.52 \times 10^{-3} \text{m}^3/\text{s}$$

$$C_d = \frac{Q}{Q_{th}} = \frac{0.94 \times 10^{-3}}{1.52 \times 10^{-3}} = \mathbf{0.618}$$

Problem 6.2: A 50 mm diameter orifice is discharging liquid under a head of 8 m. Find the actual discharge and actual velocity of the jet of liquid at vena-contracta. Take $C_d = 0.62$ and $C_v = 0.98$.

Solution: Given data:

Diameter of orifice: $\qquad d = 50$ mm $= 0.05$ m

∴ Cross-sectional area of orifice:

$$a = \frac{\pi}{4}d^2 = \frac{3.14}{4} \times (0.05)^2 = 1.96 \times 10^{-3}\ \text{m}^2$$

Head : $H = 8$ m
$$C_d = 0.62$$
$$C_v = 0.98$$

Theoretical discharge through the orifice: Q_{th}

$$Q_{th} = a \times \sqrt{2gH}$$

$$= 1.96 \times 10^{-3} \sqrt{2 \times 9.81 \times 8} = 0.02455 \, m^3/s$$

$\therefore$ Actual discharge through orifice: Q

$$Q = C_d \times Q_{th} = 0.62 \times 0.02455 = 0.01522 \, m^3/s$$
$$= \mathbf{15.22 \ litre/s}$$

Theoretical velocity at vena-contracta:

$$V_{th} = \sqrt{2gH} = \sqrt{2 \times 9.81 \times 8} = 12.52 \ m/s$$

Actual velocity of jet at vena-contracta: V

$$V = C_v \times V_{th} = 0.98 \times 12.52 = \mathbf{12.269 \ m/s.}$$

Problem 6.3: The head of liquid available at the centre of orifice of diameter 100 mm is 10 m. The liquid coming out from orifice is collected in a measuring tank of diameter 1.5 m. The rise of liquid level in the measuring tank is 0.6 m in 15 seconds. Also the co-ordinates of a point on the jet, measured from vena-contracta are 3.8 m horizontal and 0.4 m vertical. Find the hydraulic coefficients C_d, C_v and C_c.

Solution: Given data:

Diameter of orifice: $d = 100$ mm $= 0.1$ m

$\therefore$ Cross-sectional area of orifice:

$$a = \frac{\pi}{4}d^2 = \frac{3.14}{4} \times (0.1)^2 = 7.85 \times 10^{-3} \, m^2$$

Head: $H = 10$ m

Diameter of measuring tank:

$$D = 1.5 \text{ m}$$

Cross-sectional area of measuring tank:

$$A = \frac{\pi}{4}D^2 = \frac{3.14}{4} \times (1.5)^2 = 1.766 \, m^2$$

Rise of liquid in measuring tank:

$$h = 0.6 \text{ m}$$

Time: $t = 15$ seconds

Horizontal distance: $x = 3.8$ m

Vertical distance: $y = 0.4$ m

Now theoretical velocity: $V_{th} = \sqrt{2gH} = \sqrt{2 \times 9.81 \times 10} = 14$ m/s

$\therefore$ Theoretical discharge: $Q_{th} = a \times V_{th} = 7.85 \times 10^{-3} \times 14 = 0.1099 \, m^3/s.$

Actual discharge: $Q = \dfrac{\text{Volume of liquid in measuring tank}}{\text{Time}}$

$$= \frac{Ah}{t} = \frac{1.766 \times 0.6}{15} = 0.07064 \text{ m}^3/\text{s}$$

$\therefore$ Coefficient of discharge: $C_d = \dfrac{Q}{Q_{th}} = \dfrac{0.07064}{0.1099} = \mathbf{0.642}$

Coefficient of velocity: $\quad C_v = \dfrac{x}{\sqrt{4yH}} = \dfrac{3.8}{\sqrt{4 \times 0.4 \times 10}} = \mathbf{0.95}$

We know that $\quad\quad\quad\quad C_d = C_c \times C_v$

or $\quad\quad\quad\quad\quad\quad C_c = \dfrac{C_d}{C_v} = \dfrac{0.642}{0.94} = \mathbf{0.67}$

Problem 6.4: Water discharge at the rate of 67 litre/s through a 100 mm diameter vertical sharp edge orifice under head of 9.5 m. A point, on the jet, measured from the vena-contracta has coordinates of 4.2 m horizontal and 0.50 m vertical. Find the values of hydraulic coefficients C_v, C_c and C_d.

Solution: Given data:

Actual discharge: $\quad\quad Q = 67$ litre/s $= 0.067$ m^3/s

Diameter of orifice: $\quad\quad d = 100$ mm $= 0.1$ m

$\therefore$ Cross-sectional area of orifice:

$$a = \frac{\pi}{4}d^2 = \frac{3.14}{4} \times (0.1)^2 = 7.85 \times 10^{-3} \text{ m}^2$$

Head: $\quad\quad\quad\quad\quad H = 9.5$ m

$\quad\quad\quad\quad\quad\quad\quad x = 4.2$ m

$\quad\quad\quad\quad\quad\quad\quad y = 0.50$ m

We know that the coefficient of velocity: C_v

$$C_v = \frac{x}{\sqrt{4yH}} = \frac{4.2}{\sqrt{4 \times 0.50 \times 9.5}} = 0.96$$

Theoretical discharge: $\quad Q_{th} = a\sqrt{2gH}$

$$= 7.85 \times 10^{-3} \times \sqrt{2 \times 9.81 \times 9.5}$$
$$= 0.10717 \text{ m}^3/\text{s}$$

We know that the coefficient of discharge: C_d

$$C_d = \frac{Q}{Q_{th}} = \frac{0.067}{0.10717} = 0.62$$

also $\quad\quad\quad\quad\quad C_d = C_v \times C_c$

or $\quad\quad\quad\quad\quad\quad C_c = \dfrac{C_d}{C_v} = \dfrac{0.62}{0.96} = 0.64$

6.5 SMALL AND LARGE ORIFICES

The size of the orifices may be classified as the relation between the head of water in the tank from the centre of the orifice and the height of the orifice.

Fig. 6.3: Orifice

If the head of water in the tank from the centre of the orifice is more than 5 times the height of the orifice, such type of the orifice is referred as a **small orifice**.

Mathematically,

$$H > 5\,H_0$$

where H = head of water in the tank from the centre of the orifice

H_0 = height of the orifice

If the head of water in the tank from the centre of the orifice is less than 5 times of height of the orifice, such type of the orifice is referred as **large orifice**.

Mathematically,

$$H < 5\,H_0$$

6.6 DISCHARGE THROUGH A SMALL RECTANGULAR ORIFICE

In case of a small rectangular orifice, the velocity of water at every point in the cross-section of the jet is considered to be constant. The discharge through an orifice is given by the following relation:

$$Q = C_d\,A\sqrt{2\,g\,H} = C_d\,LH_0\,\sqrt{2\,g\,H}$$

where C_d = coefficient of discharge for the orifice

A = cross-sectional area of the orific

$= L\,H_0$

L = length of the orifice

H_0 = height of the orifice

H = height of the water above the centre of the orifice

6.7 DISCHARGE THROUGH A LARGE RECTANGULAR ORIFICE

In case of a large rectangular orifice, the velocity of water at every point in the cross-section of the jet is not constant, because of a considerable variation of head along the height of the orifice. So, the effect of head variation leads to the variation of velocity.

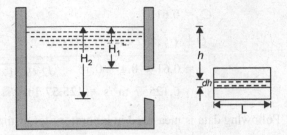

Fig. 6.4: Large rectangular orifice.

Consider a large rectangular orifice is fitted in one side of the tank as shown in Fig. 6.4.

Let H_1 = height of water above the top of the orifice

H_2 = height of water above the bottom of the orifice

L = length of the orifice.

Consider a horizontal strip of thickness dh at depth h from the level of the water as shown in Fig. 6.4.

∴ Area of strip: $dH = L\ dh$

Theoretical velocity of water through the strip = $\sqrt{2gH}$

∴ Discharge flow through strip:

$$dQ = C_d \times dA \times \sqrt{2gH} = C_d\ L\ dh\ \sqrt{2gH}$$

$$= C_d\ \sqrt{2g}\ L\ h^{1/2}\ dh$$

Total discharge flow though the entire orifice:

$$Q = \int_{H_1}^{H_2} dQ$$

$$= \int_{H_1}^{H_2} C_d \sqrt{2g}\ L h^{1/2}\ dh = C_d \sqrt{2g} L \int_{H_1}^{H_2} h^{1/2} dh$$

$$= C_d \sqrt{2g} L \left[\frac{h^{3/2}}{3/2}\right]_{H_1}^{H_2} = \frac{2}{3} C_d \sqrt{2g} L \left[H_2^{3/2} - H_1^{3/2}\right]$$

$$Q = \frac{2}{3} C_d \sqrt{2g} L \left[H_2^{1.5} - H_1^{1.5}\right]$$

Problem 6.5: A small rectangular orifice 150 mm height and 400 mm wide is discharging water under a constant head of 600 mm. Find the discharge through the orifice. Take $C_d = 0.61$.

Solution: Given data:

Height of orifice: $H_0 = 150$ mm $= 0.15$ m

Width of orifice: $L = 400$ mm $= 0.4$ m

Head: $H = 600$ mm $= 0.6$ m

$$C_d = 0.61$$

Discharge: $$Q = C_d A\sqrt{2gH} = C_d LH_0 \sqrt{2gH}$$

$$= 0.61 \times 0.4 \times 0.15 \times \sqrt{2 \times 9.81 \times 0.6}$$

$$= 0.12557 \text{ m}^3/\text{s} = \textbf{125.57 litre/s.}$$

Problem 6.6: Following data is measured in laboratory for a small orifice:

Diameter of orifice = 150 mm

Discharge = 33 litre/s

Head = 450 mm

Find the coefficient of discharge.

Solution: Given data:

Diameter of orifice: $d = 150$ mm $= 0.15$ m

$\therefore$ Cross-sectional area of the orifice:

$$a = \frac{\pi}{4}d^2 = \frac{3.14}{4} \times (0.15)^2 = 0.01766 \text{ m}^2$$

Discharge: $Q = 33$ litre/s $= 0.033$ m^3/s
Head: $H = 450$ mm $= 0.45$ m

We know that the discharge through a small orifice:

$$Q = C_d a\sqrt{2gH}$$

$$0.033 = C_d \times 0.01766 \times \sqrt{2 \times 9.81 \times 0.45}$$

or $$C_d = \textbf{0.628.}$$

Problem 6.7: A rectangular orifice of 1.5 m wide and 0.5 m height is discharging water from a tank. If the water level in the tank is 3 m above the top edge of the orifice, find the discharge through the orifice. Take $C_d = 0.62$.

Solution: Given data:

Width of orifice: $L = 1.5$ m

Height of orifice: $H_0 = 0.5$ m

Level of water above the top edge of the orifice:

$$H_1 = 3 \text{ m}$$

Level of water above the bottom edge of the orifice:

$$H_2 = H_1 + H_0 = 3 + 0.5 = 3.5 \text{ m}$$

Discharge through orifice:

$$Q = \frac{2}{3}C_d\sqrt{2g}L\left[H_2^{1.5} - H_1^{1.5}\right]$$

$$= \frac{2}{3} \times 0.62 \times \sqrt{2 \times 9.81} \times 1.5 \times \left[(3.5)^{1.5} - (3)^{1.5}\right]$$

$$= \textbf{3.71 m}^3/\textbf{s.}$$

6.8 DISCHARGE THROUGH FULLY SUBMERGED ORIFICE

If the level of water in downstream is higher than the upper edge of the orifice then, such type orifice is called fully submerged orifice or wholly drowned orifice. This type of orifice is shown in Fig. 6.5.

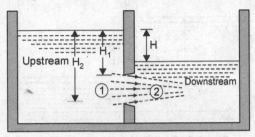

Fig. 6.5: Fully submerged orifice.

Let $\quad\quad\quad$ H_1 = height of water above the top of the orifice in upstream side.

H_2 = height of water above the bottom of the orifice in upstream side.

L = length of the orifice.

H = difference in water level between upstream and downstream.

H_0 = height of the orifice.

$= H_2 - H_1$

Area of the orifice: $\quad a = LH_0$

Applying Bernoulli's equation at points 1 and 2, we get

$$\frac{p_1}{\rho g} + \frac{V_1^2}{\rho g} + z_1 = \frac{p_2}{\rho g} + \frac{V_2^2}{\rho g} + z_2$$

Here $\quad\quad\quad z_1 = z_2$, V_1 is negligible

$\therefore$

$$\frac{p_1}{\rho g} = \frac{p_2}{\rho g} + \frac{V_2^2}{\rho g}$$

or

$$\frac{V_2^2}{\rho g} = \frac{p_1}{\rho g} - \frac{p_2}{\rho g}$$

$$\frac{V_2^2}{\rho g} = H$$

or $\quad\quad\quad V_2^2 = 2 g H$

or $\quad\quad\quad V_2 = \sqrt{2 g H}$

$\therefore$ Discharge through orifice:

$$Q = C_d \times \text{Area} \times \text{Theoretical velocity}$$

$$Q = C_d L H_0 \sqrt{2 g H}$$

where $\quad\quad\quad H_0 = H_2 - H_1$

6.9 DISCHARGE THROUGH PARTIALLY SUBMERGED ORIFICE

If the orifice is partially submerged under downstream water, such type of orifice is called partially submerged orifice.

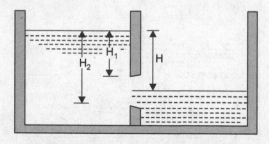

Fig. 6.6: Partially submerged orifice.

The discharge through partially submerged orifice is equal to the discharges through free and the submerged portions.

Mathematically,

Total discharge: $Q = Q_1 + Q_2$

where $Q_1 = \dfrac{2}{3} C_d \sqrt{2g}\, L (H^{1.5} - H_1^{1.5})$, discharge through free

portion

$Q_2 = C_d L (H_2 - H) \sqrt{2g\,H}$, discharge through sub-

merged portion.

where L = length of the orifice.

Problem 6.8:   Find the discharge through fully submerged orifice of width 3 m, if the difference of water levels on both sides of the orifice is 0.4 m. The length of water from top and bottom of the orifice are 2 m and 2.5 m respectively. Take C_d as 0.61.

Solution: Given data:

Width of orifice: $L = 3$ m

Difference of water level: $H = 0.4$ m

Height of water from top of orifice:

$H_1 = 2\text{m}$

Height of water from bottom of orifice: $H_2 = 2.5$ m

$C_d = 0.61$

Height of orifice: $H_0 = H_2 - H_1 = 2.5 - 2 = 0.5$ m

We know that the discharge through fully submerged orifice:

$$Q = C_d L H_0 \sqrt{2g\,H}$$

$$= 0.61 \times 3 \times 0.5 \times \sqrt{2 \times 9.81 \times 0.4} = \mathbf{2.56\ m^3/s.}$$

Problem 6.9: Find the discharge through a fully submerged orifice 2.5 m wide and 1.5 m deep, if the difference of water levels on both the sides of the orifice be 3 m. Take $C_d = 0.60$.

Solution: Given data:

Width of orifice: $L = 2.5$ m

Depth of orifice: $H_0 = 1.5$ m

Difference of water level on both the sides:

$$H = 3 \text{ m}$$
$$C_d = 0.60$$

Discharge: $Q = C_d \, L H_0 \sqrt{2g H}$

$$= 0.60 \times 2.5 \times 1.5 \times \sqrt{2 \times 9.81 \times 3} = \textbf{17.26 m}^3\textbf{/s.}$$

Problem 6.10: A rectangular orifice of 2 m wide and 1.5 m deep is filled in one side of a large tank. The water level on one side of the orifice is 4 m above the top edge of the orifice, while on the other side of the orifice, the water is 0.5 m below its top edge. Find the discharge through the orifice. Take $C_d = 0.62$.

Solution: Given data:

Width of orifice: $L = 2$ m

Depth of orifice: $H_0 = 1.5$ m

Height of water from top edge of orifice:

$$H_1 = 4 \text{ m}$$

Difference of water level on both sides:

$$H = H_1 + 0.5 = 4 + 0.5 = 4.5 \text{ m}$$

Height of water from the bottom edge of orifice:

$$H_2 = H_1 + H_0 = 4 + 1.5 = 5.5 \text{ m}$$

Discharge through free portion: Q_1

$$Q_1 = \frac{2}{3} C_d \sqrt{2g} \, L (H^{1.5} - H_1^{1.5})$$

$$= \frac{2}{3} \times 0.62 \sqrt{2 \times 9.81} \times 2 \times (4.5^{1.5} - 4^{1.5}) = \textbf{5.66 m}^3\textbf{/s.}$$

Discharge through submerged portion: Q_2

$$Q_2 = C_d \, L (H_2 - H) \sqrt{2g H}$$

$$= 0.62 \times 2 \times (5.5 - 4.5) \times \sqrt{2 \times 9.81 \times 4.5} = 11.65 \, \text{m}^3/\text{s}$$

∴ Total discharge through the orifice:

$$Q = Q_1 + Q_2 = 5.66 + 11.65 = \textbf{17.31 m}^3\textbf{/s.}$$

6.10 CLASSIFICATION OF MOUTHPIECES

The mouthpieces are classified as:

    (a) External mouthpieces
    (b) Internal mouthpieces

6.10.1 External Mouthpieces

The external mouthpieces are further classified as:

    (*i*) Cylindrical mouthpieces

    (*ii*) Convergent mouthpieces

    (*iii*) Convergent-divergent mouthpieces

6.10.2 Internal Mouthpiece

The internal mouthpieces are further classified as:

    (*i*) Mouthpieces running full

    (*ii*) Mouthpieces running free.

6.11 DISCHARGE THROUGH EXTERNAL CYLINDRICAL MOUTHPIECE

A piece of pipe connected to an opening at the side of a tank or vessel such that it projects outside of the tank, is known as external mouthpiece as shown in Fig 6.7.

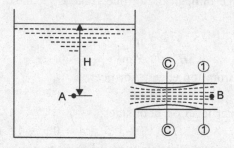

Fig. 6.7: External cylinderical mouthpiece

Let

    H = height of liquid above the centre of mouthpiece

    a = cross-sectional area of the mouthpiece

    a_c = cross-section area of flow at vena-contracta (C)-(C).

    V = velocity of liquid at outlet of mouthpiece or at section (1)-(1).

    V_c = velocity of liquid at vena-contracta

    C_c = coefficient of contraction.

Applying the continuity equation at sections (C)-(C) and (1)-(1), we get

$$V_c\, a_c = V\, a$$

or

$$V_c = V \frac{a}{a_c}.$$

$$V_c = \frac{V}{a_c / a} = \frac{V}{C_c} \qquad \because\ C_c = \frac{a}{a_c} = 0.62\ \text{assume}$$

$$V_c = \frac{V}{0.62}$$

As clear from the Fig. 6.7, the liquid flow from the section (C)-(C) is suddenly enlarged at section (1)-(1). Therefore, due to the sudden enlargement, the loss of head,

$$h_e = \frac{(V_c - V)^2}{2g} = \frac{\left(\dfrac{V}{0.62} - V\right)}{2g} = \frac{0.375V^2}{2g}$$

Applying Bernoulli's equation at points A and B, we get

$$H = \frac{V^2}{2g} + \text{losses} = \frac{V^2}{2g} + \frac{0.375V^2}{2g}$$

$$H = \frac{1.375V^2}{2g}$$

or
$$V^2 = \frac{2gH}{1.375}$$

or
$$V = 0.855\sqrt{2gH}$$

We know that the coefficient of velocity:

$$C_v = \frac{\text{Actual velocity}}{\text{Theoretical velocity}}$$

$$C_v = \frac{V}{\sqrt{2gH}} = \frac{0.855\sqrt{2gh}}{\sqrt{2gh}} = \mathbf{0.855}$$

and
$C_c = 1$ for mouthpiece, because of the cross-sectional area of flow is equal to the cross-sectional area of mouthpiece at outlet.

∴ Coefficient of discharge for mouthpiece: C_d

$$C_d = C_c \times C_v$$
$$C_d = 1 \times 0.855 = 0.855$$

Discharge through mouthpiece:

$$Q = C_d\, a\, \sqrt{2gh}$$

$$Q = \mathbf{0.855\, a\, \sqrt{2gH}}$$

The discharge through mouthpiece is more than the discharge through orifice, because of the value of C_d for mouthpiece is more than the value of C_d for orifice.

6.12 DISCHARGE THROUGH A CONVERGENT MOUTHPIECE

Let H = height of free surface of liquid above the centre of the mouthpiece
        a = cross-sectional area of the orifice at exit *i.e.,* point B.
        V = velocity of jet at point B.

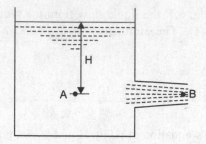

Fig. 6.8: Convergent mouthpiece.

Applying Bernoulli's equation at points A and B, we get

$$\frac{p_A}{\rho g}+\frac{V_A^2}{2g}+z_A = \frac{p_B}{\rho g}+\frac{V_B^2}{2g}+z_B \qquad \text{neglecting head losses}$$

where

$$\frac{p_A}{\rho g} = H,\ V_A = 0,\quad z_A = z_B$$

$$\frac{p_B}{\rho g} = 0,\quad V_B = V_A$$

$$H = \frac{V^2}{2g}$$

or

$$V^2 = 2\,gH$$

$$V = \sqrt{2gH}$$

$\therefore$ Actual discharge: $\quad Q = aV = a\sqrt{2gH}$

and theoretical discharge: $Q_{th} = a\sqrt{2gH}$

$\therefore$ Coefficient of discharge: $C_d = \dfrac{Q}{Q_{th}} = \dfrac{a\sqrt{2gH}}{a\sqrt{2gH}} = 1$

$\therefore$ Actual discharge: $\quad Q = C_d \times Q_{th} = C_d \times a\,\sqrt{2gH} = a\,\sqrt{2gH}$

6.13 DISCHARGE THROUGH A CONVERGENT-DIVERGENT MOUTHPIECE

In this type of mouthpiece, the first section of the mouthpiece is made of convergent and second section is divergent. The minimum cross-sectional area at which two sections meet is called throat or vena-contracta.

Let
     H = height of liquid above the centre of mouthpiece

     a = cross-sectional area of mouthpiece at section (1)-(1)

     V = velocity of liquid at section (1)-(1)

    $a_c,\ V_c$ = cross-sectional area and velocity at vena-contracta respectively.

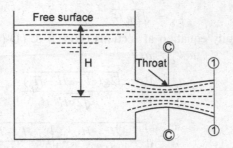

Fig. 6.9: Convergent-divergent mouthpiece.

Applying Bernoulli's equation at sections (C)-(C) and (1)-(1), we get

$$\frac{p_c}{\rho g}+\frac{V_c^2}{2g}+z_c = \frac{p_1}{\rho g}+\frac{V_1^2}{2g}+z_1$$

But $\qquad z_c = z_1,\ \dfrac{p_1}{\rho g} = H_a,$ $\qquad$ atmospheric head

$$\frac{p_c}{\rho g} = H_c$$

$\therefore \qquad\qquad H_c + \dfrac{V_c^2}{2g} = H_a + \dfrac{V_1^2}{2g}$

$$\frac{V_c^2}{2g} = H_a - H_c + \frac{V_1^2}{2g} \qquad\qquad ...(i)$$

Now applying Bernoulli's equation to the free surface of liquid in tank and section (C)-(C), we get

$$\frac{p}{\rho g}+\frac{V^2}{2g}+z = \frac{p_c}{\rho g}+\frac{V_c^2}{2g}+z_c$$

Let the datum line passing through the centre of orifice, we get

$$\frac{p}{\rho g} = H_a,\ V = 0,\quad z = H,\ \frac{p_c}{\rho g} = H_c,\quad z_c = 0$$

$\therefore \qquad\qquad H_a + 0 + H = H_c + \dfrac{V_c^2}{2g} + 0$

$$\frac{V_c^2}{2g} = H_a + H - H_c \qquad\qquad ...(ii)$$

or $\qquad\qquad V_c = \sqrt{2g\,(H_a + H - H_c)} \qquad\qquad ...(iii)$

Equating Eqs. (i) and (ii), we get

$$H_0 + H_c + \frac{V_1^2}{2g} = H_a + H - H_c$$

or $\qquad\qquad \dfrac{V_1^2}{2g} = H$

or $V_1 = \sqrt{2gH}$

Applying continuity equation at sections (C)-(C) & (1)-(1), we get

$$a_c V_c = a_1 V_1$$

or $\dfrac{a_1}{a_c} = \dfrac{V_c}{V_1} = \dfrac{\sqrt{2g(H_a + H - H_c)}}{\sqrt{2gH}}$

$$= \sqrt{\dfrac{H + H_a - H_c}{H}} = \sqrt{1 + \dfrac{H_a - H_c}{H}}$$

Discharge: $Q = C_d\, a_c\, \sqrt{2gH}$

$$Q = a_c\, \sqrt{2gH}$$

$$C_d = 1 \qquad \text{for convergent–divergent mouthpiece.}$$

Problem 6.11: Find the discharge from a 150 mm diameter external mouthpiece, fitted in one side of a large tank, if the head over the mouthpiece is 3 m.

Solution: Given data:

Diameter of the mouthpiece:

$$d = 150 \text{ mm} = 0.15 \text{ m}$$

∴ Cross-sectional area of the mouthpiece:

$$a = \frac{\pi}{4}d^2 = \frac{3.14}{4} \times (0.15)^2 = 0.01766 \text{ m}^2$$

Head: $H = 3$ m

We know that the external mouthpiece:

$$Q = 0.855\, a\, \sqrt{2gH}$$

$$= 0.855 \times 0.01766 \times \sqrt{2 \times 9.81 \times 3} = 0.11584 \text{ m}^3/\text{s}$$

$$= \textbf{115.84 litre/s.}$$

Problem 6.12: A convergent mouthpiece is discharging water under a constant head of 4 m. Find the discharge, if diameter of the mouthpiece at exit is 80 mm.

Solution: Given data:

Head: $H = 4$ m

Diameter at exit: $d = 80$ mm $= 0.08$ m

∴ Cross-sectional area of exit:

$$a = \frac{\pi}{4}d^2 = \frac{3.14}{4} \times (0.08)^2 = 5.024 \times 10^{-3} \text{ m}^2$$

We know that the discharge through a convergent mouthpiece:

$$Q = a\, \sqrt{2gH} = 5.024 \times 10^{-3}$$

$$\times \sqrt{2 \times 9.81 \times 4} = 0.04450 \text{ m}^3/\text{s} = \textbf{44.50 litre/s.}$$

Problem 6.13: A convergent-divergent mouthpiece has 60 mm throat diameter and is discharging water under a constant head of 2 m. Find the discharge through the mouthpiece.

Solution: Given data:

Diameter at throat or vena-contracta:

$$d_c = 60 \text{ mm} = 0.06$$

∴ Cross-sectional area at throat or vena-contracta:

$$a_c = \frac{\pi}{4}d^2 = \frac{3.14}{4} \times (0.06)^2 = 2.826 \times 10^{-3} \text{ m}^2$$

Head: $H = 2$ m

We know that the discharge through the convergent-divergent mouthpiece:

$$Q = a_c \sqrt{2gH}$$

$$= 2.862 \times 10^{-3} \times \sqrt{2 \times 9.81 \times 2} = 0.01770 \text{ m}^3/\text{s}$$

$$= \textbf{17.70 litre/s.}$$

6.14   DISCHARGE THROUGH AN INTERNAL MOUTHPIECE (RE-ENTRANT OR BORDA'S MOUTHPIECE)

A piece of pipe connected to an opening at the side of a tank or vessel such that it projects inside the tank, is known as internal mouthpiece as shown in Figs 6.10, 6.11. Following are the two types of internal mouthpieces, depending upon their nature of discharge:

(*i*) Borda's mouthpiece running free, and

(*ii*) Borda's mouthpiece running full.

6.14.1 Borda's Mouthpiece Running Free

If the length of the pipe is equal to its diameter, the jet of liquid, after contraction, does not touch the sides of the mouthpiece. Such type of mouthpiece is called running free as shown in Fig. 6.10.

Let H = height of liquid above the centre of the mouthpiece.

a = cross-sectional area of the mouthpiece.

a_c = cross-sectional area of flow at vena-contracta.

V = velocity of liquid through mouthpiece.

We know that the pressure of the liquid on the mouthpiece:

$$p = \rho g H$$

∴ Force acting on the mouthpiece = pressure × area = $p \times a$

$$= \rho g H a \qquad\qquad\qquad\qquad ...(i)$$

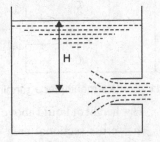

Fig 6.10: Borda's mouthpiece running free.

and mass flow rate through mouthpiece:
$$m = \rho a_c V$$
∴ Momentum of the flowing liquid
$$= mV = \rho a_c V^2$$
Since the water is initially at rest, therefore initial momentum = 0

∴ Change of momentum $\quad = \rho a_c V^2 - 0 = \rho a_c V^2 \qquad\qquad ...(ii)$

According to Newton's 2nd law of motion, the force is equal to the rate of change of momentum. Therefore, equating Eqs. (i) and (ii), we get:
$$\rho g H a = \rho a_c V^2$$
$$g H a = a_c V^2$$

$$a = a_c \frac{V^2}{gH} \quad \because V = \sqrt{2gH} \text{ or } V^2 = 2gH \text{ or } \frac{V^2}{gH} = 2$$

$$a = a_c \times 2$$

or
$$\frac{a_c}{a} = \frac{1}{2} = 0.5$$

∴ Coefficient of contraction: $C_c = \dfrac{a_c}{a} = 0.5$

Since, there is no loss of head, coefficient of velocity: $C_v = 1$

∴ Coefficient of discharge: $C_d = C_c \times C_v = 0.5 \times 1 = 0.5$

∴ Discharge through the mouthpiece running free:

$$Q = C_d\, a\, \sqrt{2gH} = 0.5\, a\, \sqrt{2gH}$$

6.14.2 Borda's Mouthpiece Running Full

If the length of the pipe is more than 2.5 times its diameter, the jet of liquid, after contraction, expands and fills up the whole mouthpiece. Such type of mouthpiece is called running full as shown in Fig. 6.11.

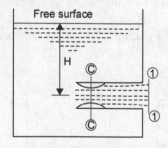

Fig. 6.11: Borda's mouthpiece running fall.

Let $\qquad\qquad\qquad H$ = height of liquid above the centre of the mouthpiece

$\qquad\qquad a$ = cross-sectional area of the mouthpiece

$\qquad\qquad a_c$ = cross-sectional area of flow at vena-contracta

V = velocity of the liquid at the outlet *i.e.*, at section (1)-(1).

V_c = velocity of the liquid at the vena-contracta.

Applying continuity equation at sections (C)-(C) and (1)-(1), we get

$$a_c V_c = aV$$

or $$\frac{a_c}{a} V_c = V$$

$$C_c \, V_c = V \qquad\qquad\qquad\qquad \because C_c = 0.5$$

$$0.5 \, V_c = V$$

or $$V_c = 2V$$

As clear from the Fig. 6.11, the jet of liquid passing through section (C)-(C) suddenly enlarges at section (1)-(1). Therefore, there will be a loss of head due to sudden enlargement. We know that the loss of head due to sudden enlargement (h_e) is given by:

$$h_e = \frac{(V_c - V)^2}{2g} = \frac{(2V - V)^2}{2g} = \frac{V^2}{2g} \qquad\qquad \because V_c = 2V$$

Applying Bernoulli's equation to the free surface of liquid in the tank and section (1)-(1), we get:

$$\frac{p_t}{\rho g} + \frac{V_t}{2g} + z_t = \frac{p_t}{\rho g} + \frac{V_1^2}{2g} + z_1 + h_e$$

Let datum line passing through the centre of the mouthpiece at the free surface of liquid in tank,

$$\frac{p_t}{\rho g} = 0, \quad \frac{V_t}{2g} = 0, \quad z_t = H$$

and at section (1)-(1),

$$\frac{p_1}{\rho g} = 0, \; z_1 = 0, \; V_1 = V$$

$\therefore$

$$0 + 0 + H = 0 + \frac{V^2}{2g} + 0 + h_e$$

or $$H = \frac{V^2}{2g} + \frac{V^2}{2g} = \frac{V^2}{2g} \qquad\qquad \because h_e = \frac{V^2}{2g}$$

or $$V = \sqrt{gH}$$

$\therefore$ Actual discharge: $\qquad Q = a\sqrt{gH}$

We know that theoretical discharge:

$$Q_{th} = a\sqrt{2gH}$$

∴ Coefficient of discharge:

$$C_d = \frac{\text{Actual discharge}}{\text{Theoretical discharge}}$$

$$C_d = \frac{2\sqrt{gH}}{a\sqrt{2gH}} = \frac{1}{\sqrt{2}} = 0.707$$

∴ Discharge: $Q = C_d\, a\sqrt{2gH} = 0.707\, a\sqrt{2gH} = a\sqrt{gH}$.

Problem 6.14: A Borda's mouthpiece of 50 mm diameter is fitted on one side of a tank containing liquid upto a height of 3 m above the centre line of the mouthpiece. Find the discharge through the mouthpiece, if the mouthpiece is running free.

Solution: Given data:

Diameter of mouthpiece:

$$d = 50 \text{ mm} = 0.05 \text{ m}$$

∴ Cross-sectional area of the mouthpiece:

$$a = \frac{\pi}{4}d^2 = \frac{3.14}{4}\times(.05)^2 = 1.962\times10^{-3}\,m^2$$

Head: $H = 3$ m

We know that the discharge through the mouthpiece running free:

$$Q = 0.5\ a\sqrt{2gH} = 0.5 \times 1.962 \times 10^{-3}\ \sqrt{2\times9.81\times3}$$
$$= 0.00752 \text{ m}^3/s = \textbf{7.52 litre/s.}$$

Problem 6.15: A Borda's mouthpiece of 50 mm diameter is fitted on one side of a tank containing liquid up to a height of 4 m above the centre line of the mouthpiece. Find the discharge through the mouthpiece, if the mouthpiece is running full.

Solution: Given data:

Diameter of mouthpiece: $d = 50$ mm $= 0.05$ mm

∴ Cross-sectional area of the mouthpiece:

$$a = \frac{\pi}{4}d^2 = \frac{3.14}{4}\times(0.05)^2 = 1.962\times10^{-3}\,m^2$$

Head: $H = 4$ m

We know that the discharge through the mouthpiece running full:

$$Q = a\sqrt{gH} = 1.962 \times 10^{-3} \times \sqrt{9.81\times4}$$
$$= 0.01229 \text{ m}^3/s = \textbf{12.29 litre/s.}$$

Problem: 6.16 A rectangular tank with vertical sides is provided with an office of diameter 50 mm at the bottom of the tank. During the inflow into the tank at a uniform rate from an external source and outflow from the orifice it was observed that the level of water rose from 2 m to 2.5 m in 100 seconds and from 3.5 to 3.75 m in 120 seconds. Assuming C_d for the office is 0.62, determine the rate of inflow and the cross-sectional area of the tank

Solution: Given data Diameter of the orifice: $d = 50$ mm $= 0.050$ m

$$\therefore \text{ Cross-sectional area of the office: } a = \frac{\pi}{4}d^2 = \frac{3.14}{4} \times (0.050)^2$$

$$= 1.962 \times 10^{-3} \text{ m}^2$$

At an average head:
$$h = \frac{2.5+2}{2} = 2.25 \text{ m}$$

$$dh = 2.5 - 2 = 0.5 \text{ m}$$

and
$$dt = 100 \text{ s}$$

Similarly, at an average head,
$$h = \frac{3.5+3.75}{2} = 3.625 \text{ m}$$

$$dh = 3.75 - 3.5 = 0.25 \text{ m}$$

and
$$dt = 120 \text{ s}$$

$$dt = \frac{Adh}{Q-\left(C_d a\sqrt{2gh}\right)}$$

or
$$A\frac{dh}{dt} = Q - C_d\, a\sqrt{2gh}$$

when $h = 2.25$ m

$$A \times \frac{0.5}{100} = Q - 0.62 \times 1.962 \times 10^{-3} \times \sqrt{2 \times 9.81 \times 2.25}$$

$$\frac{A}{200} = Q - 8.08 \times 10^{-3}$$

or
$$A = 200\, Q - 200 \times 8.08 \times 10^{-3}$$

$$A = 200\, Q - 1.616 \qquad \qquad ...(i)$$

when $h = 3.625$ m

$$A \times \frac{0.25}{120} = Q - 0.62 \times 1.962 \times 10^{-3} \times \sqrt{2 \times 9.81 \times 3.625}$$

$$\frac{A}{480} = Q - 1.025 \times 10^{-2}$$

or
$$A = 480\, Q - 480 \times 1.025 \times 10^{-2}$$

$$A = 480\, Q - 4.92 \qquad \qquad ...(ii)$$

Equating Eqs. (i) and (ii), we get

$$480\, Q - 4.92 = 200\, Q - 1.616$$

or
$$280\, Q = 3.304$$

or
$$Q = \textbf{0.0118 m}^3\textbf{/s}$$

Substituting the value of Q in Eq. (ii), we get

$$A = 480 \times 0.0118 - 4.92 = \textbf{0.744 m}^2$$

SUMMARY

1. An orifice may be defined as a small opening with a closed perimeter located in the side or bottom of the tank or vessel. The level of the liquid in the vessel must be above the top edge of the orifice. A mouthpiece is a short length of a pipe which is two to three times of diameter of the pipe, fitted in a tank or vessel which contains the liquid. Both the orifices and mouthpieces are used for measuring the discharge of the liquid in the tank or vessel.

2. The stream of liquid issuing from an orifice is called the jet of liquid or simply 'jet'.

3. Hydraulic coefficients:

   (i) Coefficient of contraction: C_c

   $$C_c = \frac{\text{Cross-sectional area at vena-contracta}: a_c}{\text{Cross-sectional area of the orifice}: a}$$

   $$C_c = \frac{a_c}{a}$$

   The value of C_c varies from 0.61 to 0.69.

   (ii) Coefficient of velocity:

   $$C_v = \frac{\text{Actual velocity of jet at vena-contracta}: V}{\text{Theoretical velocity of jet at vena-contracta}: V_{th}}$$

   $$C_v = \frac{V}{V_{th}}$$

   The value of C_v varies from 0.95 to 0.99.

   (iii) Coefficient of discharge:

   $$C_d = \frac{\text{Actual discharge through an orifice}: Q}{\text{Theoretical discharge through an orifice}: Q_{th}}$$

   $$C_d = \frac{Q}{Q_{th}}$$

   also $C_d = C_v \times C_c$.

   The value of C_d varies from 0.60 to 0.64.

4. Theoretical velocity of jet at vena-contract: V_{th}

   $$V_{th} = \sqrt{2gH}$$

5. The minimum cross-sectional area of the jet is called vena-contracta.

6. Experimental determination of hydraulic coefficient.

   Coefficient of velocity:

   $$C_v = \frac{x}{\sqrt{4yH}}$$

Contd...

where x = horizontal distance travelled by the water particle from section $C–C$ to point P in time t.

y = vertical distance between the centre of vena-contracta and point P.

7. If the head of water in the tank from the centre of the orifice is more than 5 times the height of the orifice, such type of the orifice is referred as a small orifice.

Mathematically,

$$H > 5 \, H_0$$

where H = head of water in the tank from the centre of the orifice.

H_0 = height of the orifice.

If the head of water in the tank from the centre of the orifice is less than 5 times the height of the orifice, such type of the orifice is referred as large orifice.

Mathematically,

$$H < 5 \, H_0$$

8. Discharge through a small rectangular orifice:

$$Q = C_d L H_0 \sqrt{2gH}$$

where C_d = coefficient of discharge for the orifice

L = length of the orifice

H_0 = height of the orifice

H = height of the water above the centre of the orifice.

9. Discharge through a large rectangular orifice:

$$Q = \frac{2}{3} C_d \sqrt{2g} L \left[H_2^{1.5} - H_1^{1.5} \right]$$

where H_1 = height of the water above the top of the orifice.

H_2 = height of water above the bottom of the orifice.

L = length of the orifice.

10. Discharge through fully submerged orifice:

$$Q = C_d L H_0 \sqrt{2gH}$$

11. Discharge through partially submerged orifice.

$$Q = \frac{2}{3} C_d L \sqrt{2g} \, (H^{1.5} - H_1^{1.5}) + C_d L (H_2 - H) \sqrt{2gH}$$

12. Discharge through external cylindrical mouthpiece:

$$Q = 0.855 \, a \sqrt{2gH}$$

where a = cross-sectional area of the mouthpiece.

Contd...

H = height of liquid above the centre of mouthpiece.

13. Discharge through a convergent mouthpiece:

$$Q = a\sqrt{2gH}$$

where a = cross-sectional area at exist of the mouthpiece

14. Discharge through a convergent-divergent mouthpiece

$$Q = a_c\sqrt{2gH}$$

where a_c = cross-sectional area at the throat or at vena-contracta.

15. Discharge through Borda's mouthpiece running free:

$$Q = 0.5\,a\sqrt{2gH}$$

where a = cross-sectional area of the mouthpiece.

16. Discharge through Borda's mouthpiece running full: $Q = a\sqrt{gH}$

ASSIGNMENT - 1

1. Define an orifice and a mouthpiece.
2. What is the difference between an orifice and a mouthpiece?
3. Define the following hydraulic coefficients:
   (a) Coefficient of contraction.
   (b) Coefficient of velocity.
   (c) Coefficient of discharge.
4. What do you mean by the term vena-contracta and how does it occur?
5. Prove that: Coefficient of velocity:

$$C_v = \frac{x}{2yH}$$

   where x = horizontal distance travelled by water particle from section C-C to any point P in time t.

   y = vertical distance between the centre of vena-contracta and point P.

   H = constant water head.

5. Prove that: $C_d = C_c \times C_v$
6. What is the difference between a large and a small orifice?
7. Obtain an expression for discharge through a large rectangular orifice.
8. What do you understand by the terms fully submerged orifice and partially submerged orifice?
9. Prove the expression for discharge through fully submerged orifice is given by:

$$Q = C_d L H_0 \sqrt{2gH}$$

where C_d = coefficient of discharge.
         L = length of the orifice.
         H_0 = height of the orifice.
         H = difference in water level between upstream and downstream.

10. Prove that the expression for discharge through an external mouthpiece is given by:

$$Q = 0.855\, a\sqrt{2gH}$$

where a = cross-sectional area of the mouthpiece.
         H = height of liquid above the centre of mouthpiece.

11. What is a convergent-divergent mouthpiece? Prove that the expression for discharge through a convergent-divergent mouthpiece is given by:

$$Q = a_c\,\sqrt{2gH}$$

where a_c = cross-sectional area at vena-contracta.
         H = height of liquid above the centre of mouthpiece.

12. Differentiate between:
   (*i*) External mouthpiece and internal mouthpiece.
   (*ii*) Mouthpiece running full and mouthpiece running free.

13. Obtain an expression for discharge through a Borda's mouthpiece running free.

14. Prove that the expression for discharge through a Borda's mouthpiece running full is given by:

$$Q = a\sqrt{gH}$$

where a = cross- sectional area of the mouthpiece.
         H = height of liquid above the centre of the mouthpiece.

ASSIGNMENT - 2

1. A jet of water issues from an orifice of diameter 20 mm under a head of 1 m. Find the coefficient of discharge for the orifice, if actual discharge is 0.9 litre/s.   **Ans.** 0.64

2. The head of liquid over an orifice of diameter 40 mm is 10 m. Find the actual discharge and actual velocity of the jet of liquid at vena-contracta, if $C_d = 0.6$ and $C_v = 0.98$.   **Ans.** 10.55 litre/s, 13.72 m/s

3. The head of liquid available at the centre of orifice of diameter 100 mm is 10 m. The liquid coming out from orifice is collected in a measuring tank of diameter 1.5 m. The rise of liquid level in the measuring tank is 1 m in 25 seconds. Also the coordinates of a point on the jet, measure from vena-contracta are 4.3 m horizontal and 0.5 m vertical. Find the hydraulic coefficients C_d, C_v and C_c.   **Ans.** 0.64, 0.96, 0.666

4. The head of water over an orifice of diameter 40 mm is 10 m. What is the actual discharge and actual velocity of the jet?
   Take $C_d = 0.6$ and $C_v = 0.98$.   **Ans.** 10.6 litre/s; 13.73 m/s

5. In experiment, water issues horizontally from an orifice under a head of 160 mm. Determine the coefficient of velocity of the jet, if the horizontal distance travelled by a point on the jet is 320 mm and vertical distance is 170 mm. **Ans.** 0.97

6. A jet of water issues from an orifice 1250 mm^2 in area under a constant head of 1.125 m. It falls vertically 1 m before striking the ground at a distance of 2 m measured horizontally from the vena-contracta. If the flow rate of water through the orifice is 3.653 litre/s, calculate the coefficients of discharge, velocity and contraction. **Ans.** 0.622; 0.943; 0.66

7. A small rectangular orifice 100 mm height and 500 mm wide is discharging water under a constant head of 400 mm. Find the discharge through the orifice. Take $C_d = 0.62$. **Ans.** 86.84 litre/s

8. Following data is measured in laboratory for a small orifice:

$$\text{Diameter of orifice} = 100 \text{ mm}$$
$$\text{Discharge} = 13.6 \text{ litre/s}$$
$$\text{Head} = 400 \text{ mm}$$

Find the coefficient of discharge. **Ans.** 0.618

9. A rectangular orifice 2 m wide and 0.5 m height is discharging water from a tank. If the water level in the tank is 2 m above the top edge of the edge of the orifice. Find the discharge through the orifice.

Take $C_d = 0.61$. **Ans.** 4.05 m^3/s

10. Find the discharge through fully submerged orifice of width 2 m, if the difference of water levels on both sides of the orifice is 0.5 m. The height of water from top and bottom of the orifice are 2 m and 2.5 m respectively.

Take $C_d = 0.62$. **Ans.** 1.94 litre/s

11. Find the discharge through a fully submerged orifice 2 m wide and 1 m deep, if the difference of water levels on both sides of the orifice be 4 m.

Take $C_d = 0.61$. **Ans.** 10.80 m^3/s

12. Find the discharge from a 100 mm diameter external mouthpiece, fitted in one side of a large tank, if the head over the mouthpiece is 2.5 m. **Ans.** 47 litre/s

13. A convergent-divergent mouthpice has 50 mm throat diameter is discharging water under a constant head of 2 m. Find discharge through the mouthpiece. **Ans.** 12.29 litre/s

14. A Borda's mouthpiece of 60 mm diameter is fitted on one side of a tank containing liquid upto a height of 4 m above the centre line of the mouthpiece. Find the discharge through the mouthpiece, if the mouthpiece is running free. **Ans.** 12.51 litre/s

15. A Borda's mouthpiece of 60 mm diameter is fitted on one side of a tank containing liquid upto height of 5 m above the centre line of the mouthpiece. Find the discharge through the mouthpiece, if the mouthpiece is running full. **Ans.** 19.79 litre/s

□□□

Flow Past Submerged Bodies

7.1 INTRODUCTION

When a solid body is placed in the flowing fluid, force is exerted by the fluid on the body. According to Newton's third law of motion, an equal and opposite force is exerted by the solid body on the fluid. Examples are flow of air over buildings, chimneys and towers.

In another case, when the fluid is at rest and the solid body moving through it, force is exerted by the solid body on the fluid. For examples: (*i*) Automobiles moving through air (*ii*) Airplanes moving through air (*iii*) Ships and submarines moving through water.

This chapter deals with the study of force exerted by the fluid on the stationary submerged body.

7.2 DRAG AND LIFT

When a solid body is placed in the flow of real fluid.

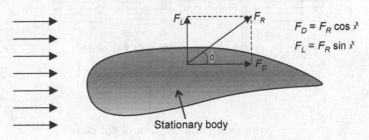

$$F_D = F_R \cos \lambda$$
$$F_L = F_R \sin \lambda$$

Fig. 7.1: Drag and Lift on a stationary body.

The fluid will exert the resultant force F_R on the stationary body making an angle ϕ in the direction of motion. The resultant force can be resolved in two components, one in the direction of flow and other normal to the flow.

Drag

The force in the direction of flow exerted by the fluid on the solid body is called drag force or drag. (*i.e.*, the component of the resultant force in the direction of flow) It is denoted by F_D.

Lift

The force exerted by the fluid in the direction perpendicular to the flow is called lift (*i.e.*, the component of the resultant force in the direction perpendicular to the flow) It is denoted by F_L.

7.3 TYPES OF DRAG FORCE

The pressure difference over the surface of the object causes a pressure force on the object. In addition to the force due to pressure difference, there is another force which is everywhere trangential to the surface of the object and due to action of viscous or wall shear stress (τ_0).

The component of these forces parallel to the flow direction is called drag force or simply "drag" and the component normal to drag is called 'lift' from above text it is clear that drag force has two types:

(*i*) Friction drag or wall drag

(*ii*) Pressure drag or form drag

(*i*) **Friction Drag:** When the surface of the object is parallel with the direction of flow, as in case of the thin flat plate as shown in Fig. 7.2, the fluid exerts a force on the surface due to viscous action. The friction force acts in the direction of flow is called friction drag. It is also called wall or skin drag.

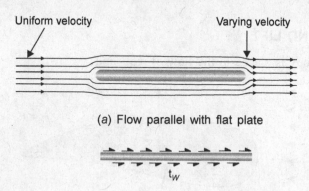

(a) Flow parallel with flat plate

(b) Free body diagram of flat plate

Fig. 7.2: Friction drag.

(*ii*) **Pressure Drag:** When the surface of the object is not parallel with the direction of flow, an additional drag force due to pressure difference on the surface of object acts. This is called pressure drag or form drag.

As shown in Fig. 7.3, the thin flat plate is perpendicular to the direction of flow, then only pressure drag is exerted on the plate.

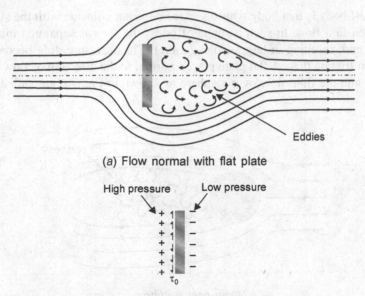

(a) Flow normal with flat plate

(b) Free body diagram of flat plate

Fig. 7.3: Pressure drag.

When the surface of the object is neither parallel to the flow nor perpendicular to the flow *i.e.*, the surface of the object is inclined with the direction of flow, then fluid exerts both friction and pressure drag on the body. The total drag is equal to the sum of friction drag and pressure drag.

Mathematically:

Total drag,

$$F_D = \text{friction drag } (F_{Df}) + \text{pressure drag } (F_{Dp})$$
$$F_D = F_{Df} + F_{Dp}$$

7.3.1 Streamlined and Bluff Bodies

A streamlined is that body whose surface coincides with the stream-line, when body is placed in a flow. In such a shape of body the boundary layer separation occurs towards the rear most part (*i.e.*, at trailing edge) of the body and small eddies formation (wake) takes place at rear end. Small eddies formation gives rise to a small pressure drag. Thus, the function drag makes a major contribution to the total drag in a streamlined body. The fluid flow over an airfoil is called the streamlined body.

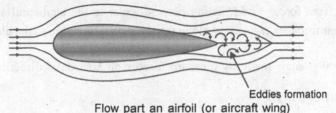

Flow part an airfoil (or aircraft wing)

Fig. 7.4: Streamlined body.

A bluff body is that body whose surface does not coincide with the stream-line when placed in a flow. In such a shape of body, the flow is separated much ahead of its rear end, resulting in large eddies (wake). Thus pressure drag becomes much greater than friction drag. A flat plate placed normal to the direction of flow as shown in Fig. 7.3 and the fluid flow over a sphere as shown in Fig. 7.5 are called the bluff bodies.

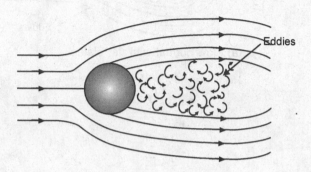

Flow past a sphere

Fig. 7.5: Bluff body.

7.4 EXPRESSION FOR DRAG AND LIFT

Consider a solid body immerged in a real fluid. The surface of the body is neither parallel nor perpendicular to the direction of flow as shown in Fig. 7.6.

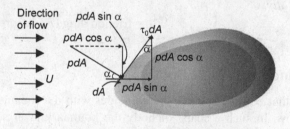

Fig. 7.6: Flow past immerged solid.

Let dA = small elemental area on the surface of the body.

The forces acting on the small element:

  (*i*) pressure force = *pdA*, acting perpendicular to the elemental area *dA*.

  (*ii*) shear force = $\tau_0 dA$, acting along the tangential direction to the elemental area *dA*.

Let θ = angle made by pressure force with horizontal direction.

7.4.1 Drag Force: F_D

Small drag force on elemental area;

$$dF_D = \text{force due to pressure in the direction of flow} + \text{force due to shear stress in the direction of flow}$$

$$dF_D = pdA \cos \alpha + \tau_0 \, dA \sin \alpha$$

$$\therefore \quad \text{Total drag: } F_D = \int pdA \cos \alpha + \int \tau_0 dA \sin \alpha \qquad \qquad ...(7.4.1)$$

$$= F_{Dp} + F_{Df}$$

where $\qquad \qquad F_{Dp} = \int pdA \cos \alpha,$ \qquad \qquad pressure drag

$$F_{Df} = \int \tau_0 dA \sin \alpha,$$ \qquad \qquad friction drag

7.4.2 Lift Force: F_L

Small lift force on elemental area,

$dF_L = $ force due to pressure in the direction perpendicular to the direction of flow + force due to shear stress in the direction perpendicular to the direction of flow.

$$dF_L = -pdA \sin \alpha + \tau_0 \, dA \cos \alpha$$

where negative sign show that the pressure force acting in downward direction

$$= \tau_0 \, dA \cos \alpha - pdA \sin \alpha$$

$$\therefore \quad \text{Total lift: } F_L = \int \tau_0 dA \cos \alpha - \int pdA \sin \alpha \qquad \qquad ...(7.4.2)$$

Equations (7.4.1) and (7.4.2) are used to predict the total drag and lift forces on the bodies.

The process to obtain both the total drag and total lift in actual fluids is very complicated. Equations (7.4.1) and (7.4.2) are not practical since the detailed distributions of pressure and shear forces are difficult to obtain by measurements especially for irregular shapes. To overcome this problem, the resultant drag force and lift acting on the irregular shape of body are most easily determined by direct experiment. In case of sphere and other regular shape at low fluid velocity, the lift and drag forces may be estimated by using Eqs. (7.4.1) and (7.4.2).

7.4.3 Co-efficient of Drag: C_D

It is defined as the ratio of the drag force per unit projected area $\left(\dfrac{F_D}{A}\right)$ to the dynamic pressure $\left(\dfrac{1}{2}\rho U^2\right)$ of the uniform flow stream.

Mathematically

$$\text{Co-efficient of drag: } C_D = \frac{F_D / A}{\rho U^2 / 2} = \frac{2F_D}{A\rho U^2}$$

or Drag force: $\qquad \qquad F_D = \dfrac{C_D A\rho U^2}{2} \qquad \qquad ...(7.4.3)$

where A is the projected area. The projected area of the solid body is defined as the area obtained by projecting the object on a plane perpendicular to the direction of flow.

For sphere, $A = \dfrac{\pi}{4}D^2$ where L_D is diameter of the sphere

For cylinder, $A = L_D$,   axis of cylinder is perpendicular to the direction of flow.

$A = \dfrac{\pi}{4}D^2$, axis of cylinder is parallel to the direction of flow, the same as for sphere of the same diameter.

7.4.4 Co-efficient of Lift: C_L

It is defined as the ratio of the lift force per unit projected area $\left(\dfrac{F_L}{A}\right)$ to the dynamic

pressure $\left(\dfrac{1}{2}\rho U^2\right)$ of the uniform flow stream.

Mathematically:

$$\text{Co-efficient of lift: } C_L = \frac{F_L / A}{\rho U^2 / 2} = \frac{2F_L}{A\rho U^2}$$

or   Lift force: $F_L = \dfrac{C_L A\rho U^2}{2}$ ...(7.4.4)

where A is the projected area. Define the projected area of the solid body as the area obtained by projecting the body from above in a direction normal to the body (the projected area is top view of solid body).

For sphere, $A = \dfrac{\pi}{4}D^2$

For cylinder, $A = \dfrac{\pi}{4}D^2$, axis of cylinder is horizontal and perpendicular to the direction of flow

$= LD$,   axis of cylinder is parallel to the direction of flow.

The resultant force on the body,

$$F_R = \sqrt{F_D^2 + F_L^2}$$

where F_D, F_L are determined by using Eqs. (7.4.3) and (7.4.4) respectively.

Problem 7.1:   A circular disc of 2 m in diameter is held normal to a 30 m/s wind of density 1.2 kg/m³. What force is required to hold it at rest ? Assume co-efficient of drag of disc is 1.2.

Solution: Given data:

Diameter of disc: $D = 2$ m

$$\therefore \text{Area:} \qquad A = \frac{\pi}{4}D^2 = \frac{3.14 \times (2)^2}{4} = 3.14 \text{ m}^2.$$

Velocity of wind: $\qquad\qquad U = 30$ m/s

Density of wind: $\qquad\qquad \rho = 1.2$ kg/m^3

Co-efficient of drag: $\qquad\qquad C_D = 1.2$

The force required to hold the disc at rest is equal to the drag force exerted by wind on the disc,

$$\text{We know, Drag force:} \qquad F_D = C_D \frac{A\rho U^2}{2} = \frac{1.2 \times 3.14 \times 1.2 \times (30)^2}{2}$$

$$= \textbf{2034.72 N}$$

Problem 7.2: A man weighing 735.75 N descends to the ground from an aeroplane with the help of a parachute against the resistance of air. The shape of the parachute is hemispherical of 3 m diameter. Find the velocity of the parachute with which it comes down. Assume $C_D = 0.5$ and density for air is 1.2 kg/m^3.

Solution: Given data:

Weight of the main: $\qquad\qquad W = 735.75$ N

$\therefore$ Drag force: $\qquad\qquad F_D = W = 735.75$ N

Diameter of the parachute: $\qquad D = 3$ m

$$\therefore \text{Projected area:} \qquad A = \frac{\pi}{4}D^2 = \frac{3.14}{4} \times (3)^2 = 7.065 \text{ m}^3$$

$$C_D = 0.5$$

Density of air: $\qquad\qquad \rho = 1.2$ kg/m^3

Let $\qquad\qquad\qquad U = $ velocity of the parachute

We know that the drag force:

$$F_D = \frac{C_D A \rho U^2}{2}$$

$$735.75 = \frac{0.5 \times 7.065 \times 1.2 \times U^2}{2}$$

or $\qquad\qquad\qquad U^2 = 347.133$

$$U = \textbf{18.63 m/s}$$

Problem 7.3: A flat plate 2 m × 2 m moves at 40 km/hr in a stationary air of density 1.2 kg/m^3. If the co-efficients of drag and lift are 0.1 and 0.5 respectively, find

(i) the lift force

(ii) the drag force

(iii) the resultant force, and

(iv) the power required to keep the plate in motion.

Solution: Given data:

Area of the plate: $\quad\quad\quad\quad\quad\quad A = 2 \times 2 = 4 \ \text{m}^2$

Velocity of the plate: $\quad\quad\quad U = 40 \ \text{km/hr} = \dfrac{40 \times 1000}{3600} \ \text{m/s} = 11.11 \ \text{m/s}$

Density of air: $\quad\quad\quad\quad\quad \rho = 1.2 \ \text{kg/m}^3$

Co-efficient of drag : $\quad\quad\quad C_D = 0.1$

Co-efficient of lift: $\quad\quad\quad\quad C_L = 0.5$

(i) Lift force: $\quad\quad\quad F_L = \dfrac{C_L A \rho U^2}{2} = \dfrac{0.5 \times 4 \times 1.2 \times (11.11)^2}{2}$

$$= \mathbf{148.11 \ N}$$

(ii) Drag force: $\quad\quad\quad F_D = \dfrac{C_D A \rho U^2}{2} = \dfrac{0.1 \times 4 \times 1.2 \times (11.11)^2}{2}$

$$= \mathbf{29.62 \ N}$$

(iii) Resultant force: $\quad F_R = \sqrt{F_D^2 + F_L^2} = \sqrt{(29.62)^2 + (148.11)^2}$

$$= \sqrt{877.34 + 21936.57} = \mathbf{151.04 \ N}$$

(iv) Power required to keep the plate in motion:

$$P = \text{force in the direction of motion} \times \text{velocity}$$
$$= F_D \times U = 29.62 \times 11.11 = \mathbf{329.07 \ W}$$

Problem 7.4: Experiments were conducted in a wind tunnel with a wind speed 60 km/hr on a flat plate of size 3 m long and 1.5 m wide. The density of air is 1.22 kg/m². The co-efficients of lift and drag are 0.80 and 0.20 respectively.

Find: (i) the lift force
      (ii) the drag force
      (iii) the resultant force
      (iv) direction of resultant force, and
      (v) power exerted by air on the plate.

Solution: Given data:

Speed of wind: $\quad\quad\quad U = 60 \ \text{km/hr} = \dfrac{60 \times 1000}{3600} \ \text{m/s} = 16.66 \ \text{m/s}$

Length of flat plate: $\quad\quad L = 3 \ \text{m}$

Width of flat plate: $\quad\quad b = 1.5 \ \text{m}$

$\therefore$ Area of plate: $\quad\quad\quad A = L \times b = 3 \times 1.5 = 4.5 \ \text{m}^2$

Density of air: $\quad\quad\quad\quad \rho = 1.22 \ \text{kg/m}^3$

$\quad\quad\quad\quad\quad\quad\quad\quad C_L = 0.80$

$\quad\quad\quad\quad\quad\quad\quad\quad C_D = 0.20$

(i) Lift force: F_L

We know, $\quad\quad\quad\quad F_L = \dfrac{C_L A \rho U^2}{2} = \dfrac{0.80 \times 4.5 \times 1.22 \times (16.66)^2}{2}$

$$= \mathbf{609.51 \ N}$$

(*ii*) Drag force: F_D

we know, $$F_D = \frac{C_D A \rho U^2}{2} = \frac{0.2 \times 4.5 \times 1.22 \times (16.66)^2}{2}$$

$$= \mathbf{152.37 \ N}$$

(*iii*) Resultant force: F_R

We know, $$F_R = \sqrt{F_D^2 + F_L^2} = \sqrt{(152.37)^2 + (609.51)^2}$$

$$= \mathbf{628.26 \ N}$$

(*iv*) Direction of the resultant force: θ

we know, $$\tan \theta = \frac{F_L}{F_D} = \frac{609.51}{152.37}$$

$$\tan \theta = 4$$

or $$\theta = \tan^{-1} 4 = \mathbf{75.96^\circ}$$

(*v*) Power exerted by air on the plate: P

we know, $$P = F_D \times U = 152.37 \times 16.66 \ W = \mathbf{2538.48 \ W}$$

Problem 7.5: A truck having a projected area of 10 m² travelling at 60 km/hr has a total resistance of 2500 N. Of this 20% is due to rolling friction and 15% is due to surface friction. The rest is due to form drag. Find the co-efficient of form drag. Take $\rho = 1.22$ kg/m³ for air.

Solution: Given data

Projected area: $A = 10$ m²

Speed of truck: $U = 60$ km/hr $= \dfrac{60 \times 1000}{3600}$ m/s $= 16.66$ m/s

Total resistance: $F_T = 2500$ N

Resistance due to rolling friction: $F_r = 20\%$ of $F_T = 0.20 \times 2500 = 500$ N

Resistance due to surface friction: $F_s = 15\%$ of $F_T = 0.15 \times 2500 = 375$ N

∴ Form drag: $F_D = F_T - F_r - F_s = 2500 - 500 - 375$

$$= 1625 \ N$$

Density of air: $\rho = 1.22$ kg/m³

also we know $$F_D = \frac{C_D A \rho U^2}{2}$$

$$1625 = \frac{C_D \times 10 \times 1.22 \times (16.66)^2}{2}$$

or $$C_D = \mathbf{0.959}$$

Problem 7.6: A kite of dimension 0.9 m × 0.9 m and weight 7.9 N is maintained in air at an angle 10° to the horizontal. The string attached to the kite makes an angle of 45° to the horizontal. The pull on the string is 32 N when the wind is flowing at a speed 30 km/hr. Find the co-efficients of drag and lift. Take density of air as 1.2 kg/m³.

Solution: Given data:

Projected area of kite: $A = 0.9 \times 0.9 = 0.81$ m²

Weight of kite: $W = 7.9$ N

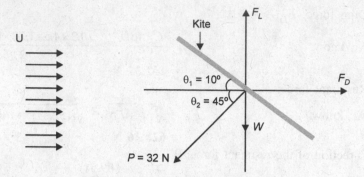

Fig. 7.7: Schematic for Problem 7.6

Angle made by kite with horizontal: $\theta_1 = 10°$

Angle made by string with horizontal: $\theta_2 = 45°$

Pull on the string: $P = 32$ N

Speed of wind: $U = 32$ km/hr $= \dfrac{32 \times 1000}{3600}$ m/s $= 8.888$ m/s

Density of air: $\rho = 1.2$ kg/m^3

Drag force: F_D = force exerted by wind on the kite in the direction of wind

= component of pull P along x-axis

= $P \cos 45° = 32 \times \cos 45° = 22.62$ N

and Lift force: F_L = force exerted by wind on the kite perpendicular to the direction of wind

= component of P in vertical downward direction + weight of kite.

= P sin 45° + $W = 32 \times \sin 45° + 7.9$

= 30.527 N

(*i*) Drag co-efficient: C_D

We know, $$F_D = \frac{C_D A \rho U^2}{2}$$

$$22.62 = \frac{C_D \times 0.81 \times 1.2 \times (8.888)^2}{2}$$

or $C_D = \mathbf{0.589}$

(*ii*) Lift co-efficient: C_L

we know, $$F_L = \frac{C_L A \rho U^2}{2}$$

$$30.527 = \frac{C_L \times 0.81 \times 1.2 \times (8.888)^2}{2}$$

or $C_L = \mathbf{0.795}$

Problem 7.7: A body of length 2.5 m has a projected area 2 m² normal to the direction of its motion. The body is moving through water, which is having dynamic viscosity = 0.01 poise. Find the drag on the body if it has a drag co-efficient 0.5 for Reynold's number of 8×10^6.

Solution: Given data:

Length of body: $L = 2.5$ m

Projected area: $A = 2$ m²

Dynamic viscosity of water: $\mu = 0.01$ poise $= \dfrac{0.01}{10}$ Ns/m² $= 0.001$ Ns/m²

Drag co-efficients: $C_D = 0.05$

Reynold's number: $Re = \dfrac{\rho L U}{\mu}$

$$8 \times 10^6 = \dfrac{1000 \times 2.5 \times U}{0.001} \quad (\because \rho = 1000 \text{ kg/m}^3 \text{ for water})$$

or $U = 3.2$ m/s

We know,   Drag force: $F_D = C_D \dfrac{A \rho U^2}{2} = \dfrac{0.5 \times 2 \times 1000 \times (3.2)^2}{2}$

$$= 5120 \text{ N}$$

Problem 7.8: A jet plane which weighs 30 kN and having a wing area of 20 m² flies at a velocity of 850 km/hr, when the engine delivers 8 MW power, 70% of the power is used to overcome the drag resistance of the wing. Find the co-efficients of lift and drag for the wing. Take density of the atmospheric air as 1.2 kg/m³.

Solution: Given data:

Weight of plane: $W = 20$ kN $= 30 \times 10^3$ N

Wing area: $A = 20$ m²

Speed of plane: $U = 850$ km/hr $= \dfrac{850 \times 1000}{3600}$ m/s $= 236.11$ m/s

Engine power: $P = 8$ MW $= 8 \times 10^6$ W

Power used to overcome drag resistance:

$$= 70\% \text{ of } P = 0.70 \times 8 \times 10^6 = 5.6 \times 10^6 \text{ W}$$

Density of air: $\rho = 1.2$ kg/m³

The lift force should be equal to weight of the plane

i.e., $F_L = W = 30 \times 10^3$ N

also $F_L = \dfrac{C_L A \rho U^2}{2}$

$$30 \times 10^3 = \dfrac{C_L \times 20 \times 1.2 \times (236.11)^2}{2}$$

or $C_L = \mathbf{0.0448}$

Power used to overcome drag resistance $= F_D \times U$

$$5.6 \times 10^6 = F_D \times 236.11$$

or $F_D = 23717.75 \text{ W}$

also $F_D = \dfrac{C_D A \rho U^2}{2}$

$$23717.75 = \dfrac{C_D \times 20 \times 1.2 \times (236.11)^2}{2}$$

or $C_D = \mathbf{0.03545}$

7.5 DRAG ON A SPHERE

Consider a smooth sphere immersed in a flowing fluid.

Let $U =$ velocity of free steam

$D =$ diameter of sphere

$\rho =$ density of fluid, and

$\mu =$ dynamic viscosity of fluid

We know, Reynold's number: $Re = \dfrac{\text{Inertia force: } F_i}{\text{Viscous force: } F_V} = \dfrac{\rho D U}{\mu}$

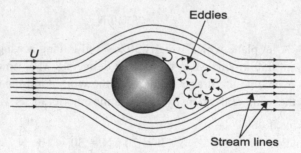

Fig. 7.8: Flow past immersed sphere.

If Reynold's number, $Re < 0.2$: In this case, viscous force is more predominant than the inertia force. So, the drag on a sphere, is entirely due to friction only and is prescribed by Stokes law:

$$F_D = 3\pi D \mu U \qquad \qquad ...(7.5.1)$$

Also we know $F_D = \dfrac{1}{2} C_D \rho A U^2 \qquad \qquad ...(7.5.2)$

Equating Eqs. (7.5.1) and (7.5.2), we get

$$3\pi D \mu U = \dfrac{1}{2} C_D \rho A U^2$$

or $\dfrac{6\pi D \mu}{\rho A U} = C_D$

where $A = \dfrac{\pi}{4} D^2$

$$\therefore \qquad \frac{6\pi D\mu}{\rho \dfrac{\pi}{4} D^2 U} = C_D$$

$$\frac{24\mu}{\rho DU} = C_D$$

or $$C_D = \frac{24}{Re} \qquad\qquad ...(7.5.3)$$

Equation (7.5.3) is called Stokes law. It is applicable for sphere when Reynold's number is less than 0.2.

Oseen has modified the stokes law as

$$C_D = \frac{24}{Re}\left(1 + \frac{3}{16Re}\right) \qquad\qquad ...(7.5.4)$$

Equation (7.5.4) is applicable for sphere when Reynold's number lies between 0.2 and 5.

$$\begin{aligned} C_D &= 0.4 && \text{for } 5 \le Re < 10^3 \\ &= 0.5 && \text{for } 10^3 \le Re < 10^5 \\ &= 0.2 && \text{for } Re \ge 10^5 \end{aligned}$$

7.6 DRAG ON A CYLINDER

Consider cylinder of diameter D and length L when the cylinder is placed in the fluid such that its axis is horizontal and perpendicular to the direction of flow as shown in Fig. 7.9, we know that

Reynolds number: $Re = \dfrac{\rho DU}{\mu} = \dfrac{DU}{\nu}$

where, U = velocity of free stream, ν = kinematic viscosity.

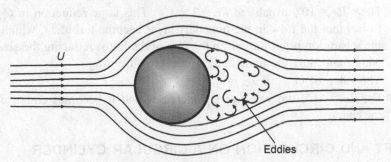

Fig. 7.9: Laminar boundary layer separation with eddies formation; flow over circular cylinder at Re = 2000.

At very low upstream velocity ($Re \le 1$), the fluid completely wraps around the cylinder and the fluid on the top and the bottom meet on the rear side of the cylinder in an orderly manner. Thus, the fluid follows the curvature of the cylinder.

At higher velocities, the fluid still hugs cylinder on the frontal side, but it is too fast to remain attached to the surface as it approaches the top or bottom of the cylinder. As a result, the boundary layer detaches from the surface, forming a separation region behind the cylinder and results in eddies formation. Thus, the pressure at the rear side is much lower than the pressure in front side.

The nature of the flow across a cylinder strongly affects the drag co-efficient C_D. The value of C_D changes with the variation of Reynolds number has been observed by following experiments:

(i) For $Re \leq 1$, the drag co-efficient decreases with increasing Reynolds number.

(ii) At about $Re = 10$, separation starts occurring on the rear of the body and starting vortex at about $Re = 90$. The region of separation increases with increasing Reynolds number upto about $Re = 10^3$.

i.e., C_D = 4.5   at $Re = 10$

$= 2$ at $Re = 90$

$= 0.95$ at $Re = 10^3$

The drag co-efficient continues to decrease with increasing Reynold's number in this range of $10 \leq R \leq 10^3$. A decrease in the drag co-efficient does not necessarily indicate a decrease in drag. The drag force is proportional to the square of the velocity, and the increase in velocity at higher Reynold's number usually more than offsets the decrease in the drag co-efficient.

(iii) In the moderate range of $10^3 \leq Re \leq 10^5$, the drag co-efficient remains relatively constant. Thus behaviour is characteristic of blunt bodies. The flow in the boundary layer is laminar in this range, but the flow in the separated region past the cylinder is highly turbulent with wide turbulent eddies.

The value of co-efficient of drag, C_D = 0.95   at $Re = 10^3$

$= 1.3$ at $Re = 10^4$

$= 1.1$ at $Re = 10^5$

(iv) There is a sudden drop in the drag co-efficient some where in the range of $10^5 < Re < 10^6$, usually at $Re = 3 \times 10^5$. This large reduction in C_d is due to fact that the flow in the boundary layer become turbulent, which moves the separation point further on the rear of the body, reducing the size of the eddies and thus the magnitude of the pressure drag.

The value of $C_D = 0.3$ at $Re = 3 \times 10^5$

(v) For $Re > 10^6$, the value of C_D increases and it becomes equal to 0.7 at the end.

7.7 LIFT AND CIRCULATION ON A CIRCULAR CYLINDER

Lift was defined earlier as the component of the net force (due to viscous and pressure forces) perpendicular to direction of flow. It is zero, when a body is placed in fluid in such a way that axis of body is parallel to the direction of flow and body is symmetrical. In this case only drag force acts on the body.

Lift will act on the body when the axis of symmetrical body is inclined to the direction of flow or body is unsymmetrical. Consider the case of a circular cylinder whose axis is parallel to the direction of flow and cylinder is stationary. So in this the lift will be zero. But if the cylinder is rotated, the axis of the cylinder is not maintained parallel to the direction of flow and hence, lift will act on the rotating cylinder. Consider the following cases:

(*i*) Ideal fluid flow over stationary cylinder.

Consider an ideal fluid flow over stationary cylinder as shown in Fig. 7.5.

Let U = velocity of free stream fluid

    R = radius of the cylinder

    θ = angle made by any point C on the circumference of the cylinder with direction of flow.

The velocity at any point C on the cylinder is given by:

$$u_\theta = 2U \sin \theta \qquad\qquad ...(7.7.1)$$

The velocity distribution over the upper half and lower half of the cylinder from the axis AB of the cylinder are identical and hence, the pressure distributions will also be same. Hence, the lift acting on the cylinder will be zero.

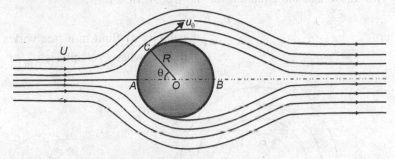

Fig. 7.10: Ideal fluid flow over stationary cylinder.

(*ii*) Circulation

Consider a closed curve ABC of fluid flow around the cylinder. The circulation around such a curve is defined as the summation of product of velocity component along the element (such as ds) of the curve and the elemental length ds. It is denoted by Γ (Greek capital letter 'gamma')

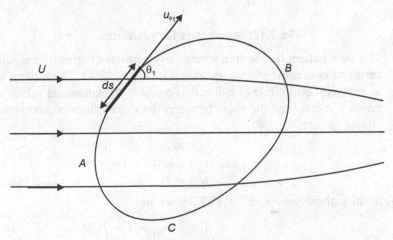

Fig. 7.11: Circulation.

Let $\quad u_\theta$ = component of free stream velocity (U) along the tangent on element ds

$\quad = U \cos \theta_1$

$\quad \theta_1$ = angle made between the length of element ds and free stream velocity U

$\therefore$ Circulation along the closed curve:

$$\Gamma = \oint \text{velocity component along the element}$$
$$\times \text{length of element}$$
$$= \oint u\theta_1 . \, ds = \oint U \cos \theta_1 . \, ds \quad ...(7.7.2)$$

Circulation for the flow field in a free-vortex flow:

We know that the equation for the free vortex flow,

$$u_{\theta 1} R = C$$

where $\quad u_{\theta 1}$ = velocity of fluid in a free vortex flow

$\quad R$ = radius, where velocity is $u_{\theta 1}$

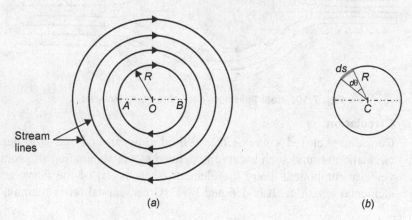

(a)   (b)

Fig. 7.12: Stream lines for free vortex.

The flow pattern for the free vortex flow consists of stream lines which are series of concentric circles as shown in Fig. 7.12(a). The stream velocity at any point on a circle of radius R is equal to the tangential velocity at that point. It means that the angle between the stream lines and tangent on the stream is zero.

$\therefore$

$$U = u_{\theta 1}$$
$$\cos \theta_1 = \cos 0° = 1$$
$$ds = Rd\theta \quad\quad\quad \text{From Fig. 7.12(b)}$$

Substituting these values in Eq. (7.7.2), we get

$$\Gamma = \oint u_{\theta_1} \times 1 \times Rd\theta$$
$$= \oint u_{\theta_1} \, Rd\theta \quad\quad (\because u_{\theta_1} R = C)$$

$$= C \oint d\theta = \oint C d\theta$$

$$= C \times 2\pi \qquad\qquad (\because \oint d\theta = 2\pi)$$

$$\Gamma = u_{\theta 1} \ R \times 2\pi$$

or $\qquad\qquad u_{\theta_1} = \dfrac{\Gamma}{2\pi R}$ ...(7.7.3)

Flow over a rotating cylinder due to constant circulation:

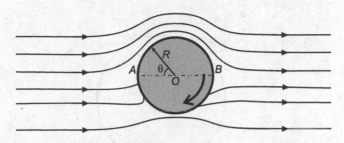

Fig. 7.13: Flow over a rotating cylinder.

The flow pattern over a rotating cylinder to which constant circulation is imparted is obtained by combining the flow patterns as shown in Fig. 7.10 and 7.12(a). The resultant flow pattern is shown in Fig. 7.13. The velocity at any point on the surface of the cylinder is obtained by adding Eqs. (7.7.1) and (7.7.3) as

$$u = u_\theta + u_{\theta 1}$$

$$u = 2U \sin\theta + \frac{\Gamma}{2\pi R} \qquad\qquad ...(7.7.4)$$

For upper half portion of the cylinder (*i.e.*, above *AB*), θ varies from 0° to 180° and hence the velocity component $2\,U \sin\theta$ is positive.

For lower half portion of the cylinder (*i.e.*, below *AB*), θ lies between 180° and 360° and hence the velocity component $2U \sin\theta$ will be negative. It means, the velocity on the upper half portion of the cylinder will be more than the velocity on the lower half portion. According to Bernoulli's equation, the surface where velocity is less, pressure will be more and vice-versa. Hence, the lower half portion of the cylinder, where velocity is less, pressure will be more than the pressure on the upper half portion of the cylinder. Due to this pressure difference on the two portions of the cylinder, a force will be acting on the cylinder in a direction perpendicular to the direction of flow, is called lift. The lift can be exerted even if the cylinder is rotating at constant speed in a uniform flow.

(*iii*) **Expression for lift force and lift co-efficient acting on rotating cylinder:**
Consider a small element of the surface of the cylinder shown in Fig. 7.14.
Let $\qquad\qquad\qquad\qquad$ ds = length of a small element

p_s = pressure on the surface of the element

R = radius of cylinder

$d\theta$ = angle made by the length ds at the centre of the cylinder

p = pressure of the free stream of fluid

U = free stream velocity of fluid

u_s = velocity of fluid on the surface of the cylinder

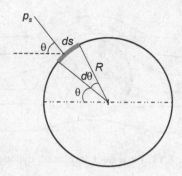

Fig. 7.14: Lift on a rotating cylinder.

Applying Bernoulli's equation to a point away from cylinder in free stream to a point laying on the surface of cylinder such that both the points are on the same horizontal line, we have

$$\frac{p}{\rho g} + \frac{U^2}{2g} = \frac{p_s}{\rho g} + \frac{u_s^2}{2g}$$

$$\therefore \qquad \frac{p_s}{\rho g} = \frac{p}{\rho g} + \frac{U^2}{2g} - \frac{u_s^2}{2g}$$

$$= \frac{p}{\rho g} + \frac{U^2}{2g}\left[1 - \frac{u_s^2}{U^2}\right] \qquad \qquad ...(7.7.5)$$

The value of u_s from Eq. (7.7.4), we get

$$u_s = u = 2U\sin\theta + \frac{\Gamma}{2\pi R}$$

Substituting the value of u_s in eq.(7.7.5), we get

$$\frac{p_s}{\rho g} = \frac{p}{\rho g} + \frac{U^2}{2g}\left[1 - \frac{\left(2U\sin\theta + \dfrac{\Gamma}{2\pi R}\right)^2}{U^2}\right]$$

$$= \frac{p}{\rho g} + \frac{U^2}{2g}\left[1 - \frac{4U^2\sin^2\theta + \dfrac{\Gamma^2}{4\pi^2 R^2} + 4U\sin\theta\dfrac{\Gamma}{2\pi R}}{U^2}\right]$$

$$p_s = p + \frac{\rho U^2}{2}\left[1 - 4\sin^2\theta - \frac{\Gamma^2}{4\pi^2 R^2 U^2} - \frac{4\Gamma\sin\theta}{2\pi UR}\right] \qquad \ldots(7.7.6)$$

The lift force acting on the small length ds on the element, due to pressure p_s is

$$dF_L = \text{component of ps in the direction perpendicular to flow} \times \text{area of the element}$$

$$= -p_s \sin\theta \times ds \, L$$

−ve sign shown that the pressure force component acts downward direction.

where,　　L = length of the cylinder

$$ds = Rd\theta$$

∴　　$dF_L = -p_s \sin\theta \times d\theta \,.\, L$

The total lift force is obtained by integrating above equation, so

∴
$$F_L = \int_0^{2\pi} -p_s \sin\theta \times Rd\theta.L$$

$$= \int_0^{2\pi} -p_s RL \sin\theta.d\theta$$

Substituting the value of p_s from Eq.(7.7.6), we get

$$= \int_0^{2\pi} -\left[p + \frac{\rho U^2}{2}\left(1 - 4\sin^2\theta - \frac{\Gamma^2}{4\pi^2 R^2 U^2} - \frac{4\Gamma\sin\theta}{2\pi UR}\right)\right] RL\sin\theta.d\theta$$

$$= -RL\int_0^{2\pi}\left[p\sin\theta + \frac{\rho U^2}{2}\left(\sin\theta - 4\sin^3\theta - \frac{\Gamma^2\sin\theta}{4\pi^2 R^2 U^2} - \frac{4\Gamma\sin^2\theta}{2\pi UR}\right)\right]d\theta$$

But
$$\int_0^{2\pi}\sin\theta.d\theta = \int_0^{2\pi}\sin^3\theta.d\theta = 0$$

∴
$$F_L = -RL\int_0^{2\pi}\frac{\rho U^2}{2}\left(-\frac{4\Gamma\sin\theta}{2\pi UR}\right)d\theta$$

$$= RL\frac{\rho U^2}{2}\times\frac{4\Gamma}{2\pi UR}\int_0^{2\pi}\sin^2\theta d\theta$$

$$= \frac{L\rho U\Gamma}{\pi}\int_0^{2\pi}\sin^2\theta d\theta$$

But
$$\int_0^{2\pi}\sin^2\theta d\theta = \left|\frac{\theta}{2} - \frac{\sin^2 2\theta}{4}\right|_0^{2\pi} = \left(\frac{2\pi}{2} - \frac{\sin 4\pi}{4}\right) = \pi$$

$$\therefore \qquad F_L = \frac{L\rho U\Gamma}{\pi} \times \pi$$

$$F_L = \rho L U\Gamma \qquad\qquad\qquad ...(7.7.7)$$

Equation (7.7.7) is known as **Kutta-Joukowaski equation.**

also
$$F_L = C_L \frac{A\rho U^2}{2}$$

$$\therefore \qquad \rho L U\Gamma = C_L \frac{A\rho U^2}{2}$$

or
$$C_L = \frac{2L\Gamma}{AU}$$

where $\qquad A = 2RL$, projected area.

From Eq. (7.7.3),

$$u_{\theta 1} = \frac{\Gamma}{2\pi R}$$

$$\frac{T}{R} = 2\pi u_{\theta 1}$$

$$\therefore \qquad C_L = \frac{2L\Gamma}{2RLU} = \frac{\Gamma}{RU}$$

$$C_L = \frac{2\pi u_{\theta 1}}{U} \qquad\qquad \left(\because \frac{\Gamma}{R} = 2\pi u_{\theta 1}\right)$$

where $u_{\theta 1}$ = velocity of rotation of the cylinder in the tangential direction.

(*iv*) **Drag force acting on a rotating cylinder:**
The drag force acting on the rotating cylinder in a uniform flow is zero because the velocity and pressure distribution on symmetrical about the vertical axis of the cylinder.

(*v*) **Location of stagnation points for a rotating cylinder in a uniform flow:**
Stagnation point is defined as the point at which the velocity of flow is zero. The cylinder is rotating in the uniform velocity as shown in Fig. 7.13. The velocity at any point on the surface of the cylinder is given by Eq. (7.7.4) as

$$= U \cos \theta_1$$

$$u = 2U \sin\theta + \frac{\Gamma}{2\pi R}$$

For stagnation point, $\qquad u = 0$

$$\therefore \qquad 0 = 2U \sin\theta + \frac{\Gamma}{2\pi R}$$

or
$$2U \sin\theta = \frac{\Gamma}{2\pi R}$$

or $\qquad\qquad\qquad\qquad sin\ \theta\ =\ -\dfrac{\Gamma}{2\pi UR}$ ...(7.7.8)

Equation (7.7.8) gives the location of stagnation points on the surface of the cylinder. There are two values of θ, which satisfy the Eq. (7.7.8). As sin θ is −ve in Eq. (7.7.8), it means θ is more than $180°$ but less than $360°$. The two values of θ are such that one value is between $180°$ to $270°$ and other value is between $270°$ and $360°$.

For a single stagnation point, $\theta = 270°$ and then Eq. (7.7.8) becomes as

$$sin\ 270°\ =\ -\dfrac{\Gamma}{2\pi UR}$$

or $\qquad\qquad\qquad\qquad -1\ =\ -\dfrac{\Gamma}{2\pi UR}$

or $\qquad\qquad\qquad\qquad \Gamma\ =\ 4\pi UR$ ...(7.7.9)

Equation (7.7.9) shows that the condition of single stagnation point in terms of circulation Γ.

From Eqs. (7.7.3) we get

$$u_{\theta 1}\ =\ \dfrac{\Gamma}{2\pi R}$$

or $\qquad\qquad\qquad\qquad \Gamma\ =\ 2\pi R u_{\theta 1}$ ...(7.7.10)

Equating Eqs. (7.7.9) and (7.7.10) we get

$$4\pi UR\ =\ 2\pi R u_{\theta 1}$$
$$2U\ =\ u_{\theta 1}$$

or $\qquad\qquad\qquad\qquad u_{\theta 1}\ =\ 2U$

7.8 MAGNUS EFFECT: LIFT GENERATED BY SPINNING

When a cylinder or sphere in a flow is rotated at a sufficiently high rate, a lift is produced on the cylinder. The phenomenon of producing lift by the rotation of a solid body is called the Magnus effect after the German scientist Heinrich Magnus (1802-1870), who was the first to study the lift of rotating bodies, which is shown in Fig. 7.15 for the simplified case of ideal flow. When the ball is not spinning, the lift is zero because of top-bottom symmetry. But when the cylinder is rotated about its axis, the cylinder drags some fluid around because of the no-slip condition and the flow field reflects the superposition of the spinning and non-spinning flows. The stagnation points shift down, and the flow is no longer symmetric about the horizontal plane that passes through the centre of the cylinder. The average pressure on the upper half is less than the average pressure at the lower half because of the Bernoulli's effect, and thus there is a net upward force (lift) acting on the cylinder.

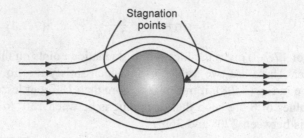

(a) Irrotational flow over a stationary cylinder

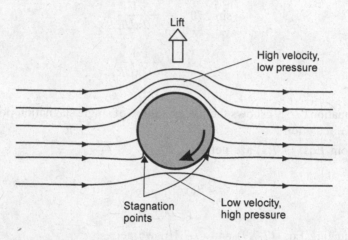

(b) Irrotational flow over a rotating cylinder

Fig. 7.15: Lift on a rotating circular for the case of ideal flow.

Problem 7.9:   A metallic sphere of specific gravity of 6 falls in an oil of specific gravity 0.85. The diameter of the sphere is 10 mm and it attains a terminal velocity of 0.05 m/s. Find the viscosity of the oil.

Solution: Given data:

Specific gravity of metallic sphere: $S_s = 6$

$\therefore$ Density of metallic sphere: $\rho_s = S_s \times \rho_{water} = 6 \times 1000 \text{ kg/m}^3 = 6000 \text{ kg/m}^3$

Specific gravity of oil: $S_o = 0.85$

$\therefore$ Density of oil: $\rho_o = S_o \times \rho_{water} = 0.85 \times 1000 \text{ kg/m}^3$
$= 850 \text{ kg/m}^3$

Diameter of sphere: $D = 10 \text{ mm} = 10 \times 10^{-3} \text{ m}$

Terminal velocity of sphere: $U = 0.05 \text{ m/s}$

Weight of sphere: $W = Mg$

$$= \rho_s \times \text{volume of sphere} \times g = \rho_s \times \frac{\pi}{6} D^3 \times g$$

$$= 6000 \times \frac{3.14}{6} \times (10 \times 10^{-3}) \times 9.81 = 0.0308 \text{ N}$$

Buoyant force on sphere: $F_B = \rho_o g V$

where

ρ_o = density of oil

V = volume of sphere

$\therefore \qquad F_B = 850 \times 9.81 \times \dfrac{\pi}{6} D^3$

$\qquad\qquad = 850 \times 9.81 \times \dfrac{3.14}{6} (10 \times 10^{-3})^3 = 0.00436$ N

Drag force on the sphere: F_D

$F_D = 3\pi\mu DU \qquad\qquad$ (by Stokes law)

$\qquad = 3 \times 3.14 \times \mu \times 10 \times 10^{-3} \times 0.05$

$\qquad = 0.00471\ \mu$

For equilibrium condition,

Equating downward force = upward forces

$W = F_D + F_B$

$0.0308 = 0.00471\ \mu + 0.00436$

or $\quad 0.00471\ \mu = 0.02644$

or $\qquad \mu = 5.613$ Ns/m^2 = 5.613 × 10 poise

$\qquad\quad = 56.13$ poise

The expression for drag force: $F_D = 3\pi\mu DU$,

Stokes law is valid only upto Reynolds number less than 0.2. Hence, it is necessary to calculate Reynolds number for the flow.

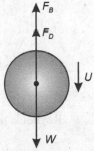

$\therefore\ Re = \dfrac{\rho U D}{\mu} = \dfrac{850 \times 0.05 \times 10 \times 10^{-3}}{5.613} = 0.0757$

Hence, $Re < 0.2$ and so the expression for drag force,

$F_D = 3\pi\mu DU$ is valid.

Fig. 7.16: Schematic for Problem 7.9

Problem 7.10: Find the velocity of fall of rain drop of 0.3 mm diameter in atmospheric air having density 1.2 kg/m^3 and kinematic viscosity 0.15 stokes.

Solution: Given data:

Diameter of rain drop: $\qquad\qquad\ D = 0.3$ mm = 0.3 × 10^{-3} m

Density of air: $\qquad\qquad\qquad \rho = 1.2$ kg/m^3

Kinematic viscosity of air: $\qquad \nu = 0.15$ stokes = 0.15 × 10^{-4} m^3/s

Weight of rain drop: $\qquad\qquad W = Mg = \rho_w V g$

where $\qquad\qquad\qquad\qquad V = \dfrac{\pi}{6} D^3$, volume of rain drop

$\therefore \qquad\qquad\qquad\qquad W = \rho_w \times \dfrac{\pi}{6} D^3 \times g$

$\qquad\qquad\qquad\qquad\quad = 1000 \times \dfrac{3.14}{6} \times (0.3 \times 10^{-3}) \times 9.81$

$$= 1.386 \times 10^{-7} \text{ N}$$

Drag force on rain drop is given by stokes law:

$$F_D = 3\pi\,\mu\,DU$$

where U = uniform velocity of rain drop

$\mu = \rho\nu$, viscosity of the air

$\therefore$ $F_D = 3\pi\,\rho\nu DU$

$$= 3 \times 3.14 \times 1.2 \times 0.15 \times 10^{-4} \times 0.3 \times 10^{-3}\, U$$

$$= 5.086 \times 10^{-8}\, U$$

Buoyant force: $F_D = \rho gV = \rho g \times \dfrac{\pi}{6} D^3$

$$= 1.2 \times 9.81 \times \frac{3.14}{6} \times (0.3 \times 10^{-3})^3$$

$$= 1.66 \times 10^{-10} \text{ N}$$

For equilibrium condition,

Equating downward force = upward force

$$W = F_D + F_B$$

$$1.386 \times 10^{-7} = 5.086 \times 10^{-8}\, U$$

$$+ 1.66 \times 10^{-10}$$

or $5.086 \times 10^{-8}\, U = 1.384 \times 10^{-7}$

$$U = 2.72 \text{ m/s}$$

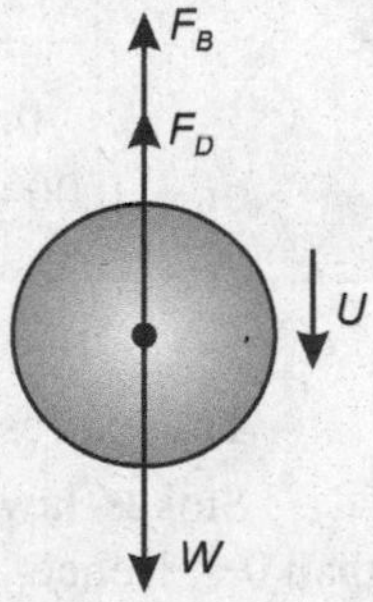

Reynolds number: $Re = \dfrac{\rho UD}{\mu} = \dfrac{UD}{\nu}$

Fig. 7.17: Schematic for Problem 7.10

$$= \frac{2.72 \times 0.3 \times 10^{-3}}{0.15 \times 10^{-4}} = 54.4$$

Thus, $Re > 0.2$ and therefore the Stokes law cannot be applicable.

2nd approach:

Drag force: $F_D = \dfrac{C_D \rho U^2 A}{2}$

where C_D = co-efficient of drag

$$= 0.4 \quad \text{for } 5 < Re < 1000$$

$$F_D = 0.4 \times 1.2 \times U^2 \times \frac{\pi}{4} D^2$$

$$= 0.4 \times 1.2 \times U^2 \times \frac{3.14}{4}(0.3 \times 10^{-3})^2$$

$$= 3.39 \times 10^{-8}\, U^2$$

Now $W = F_D + F_B$

$$1.386 \times 10^{-7} = 3.39 \times 10^{-8}\, U^2 + 1.66 \times 10^{-10}$$

or $3.39 \times 10^{-8}\, U^2 = 1.38 \times 10^{-7}$

or
$$U^2 = 4.07$$
$$U = 2.01 \text{ m/s}$$

Reynolds number:
$$Re = \frac{\rho UD}{\mu} = \frac{UD}{v} = \frac{2.01 \times 0.3 \times 10^{-3}}{0.15 \times 10^{-4}} = 40.2$$

The value of Reynolds number lies between 5 and 1000. So, assumed values of $C_D = 0.4$ is right.

Hence, the velocity of fall of rain drop = **2.01 m/s**.

Problem 7.11: During the flood a very fine silt of diameter 0.002 mm and specific gravity 2.8, enters the reservoir 40 m deep. Find the time required for all the silt to settle down. Take viscosity for water as 0.01 poise.

Solution: Given data:

Diameter of the silt:
$$D = 0.002 \text{ mm} = 0.002 \times 10^{-3} \text{ m}$$
$$= 2 \times 10^{-6} \text{ m}$$

Specific gravity of the silt: $S = 2.8$

∴ Density of the silt: $\rho = S \times \rho_w = 2.8 \times 1000 \text{ kg/m}^3 = 2800 \text{ kg/m}^3$

Depth of the reservoir: $H = 40 \text{ m}$

Viscosity of water: $\mu = 0.01 \text{ poise} = 0.001 \text{ Ns/m}^2$

According to Stokes law, the drag force on the silt:
$$F_D = 3\pi\mu DU = 3 \times 3.14 \times 0.001 \times 2 \times 10^{-6} \times U$$
$$= 1.884 \times 10^{-8} \ U$$

Buoyant force:
$$F_B = \rho_w gV = \rho_w g \frac{\pi}{6} D^3$$
$$= 1000 \times 9.81 \times \frac{3.14}{6} \times (2 \times 10^{-6})^3$$
$$= 4.107 \times 10^{-14} \text{ N}$$

Weight of the silt:
$$W = Mg = \rho Vg = \rho \times \frac{\pi}{6} D^3 \times g$$
$$= 2800 \times \frac{3.14}{6} \times (2 \times 10^{-6})^3 \times 9.81$$
$$= 1.149 \times 10^{-13} \text{ N}$$

Now equating the forces *i.e.*,

Downward force = sum of upward forces
$$W = F_D + F_B$$
$$1.149 \times 10^{-13} = 1.884 \times 10^{-8} \ U$$
$$+ 4.107 \times 10^{-14}$$

or $$1.884 \times 10^{-8} \ U = 7.383 \times 10^{-14}$$

or $$U = 3.91 \times 10^{-6} \text{ m/s}$$

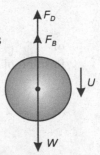

Fig. 7.18: Schematic
for Problem 7.11

also
$$U = \frac{\text{distance travel: } H}{\text{time take: } t}$$

or
$$3.91 \times 10^{-6} = \frac{40}{t}$$

or
$$t = \frac{40}{3.91 \times 10^{-6}} \text{ seconds} = 10.23 \times 10^6 \text{ seconds}$$

$$= \frac{10.23 \times 10^6}{60 \times 60 \times 24} \text{ days} = \textbf{118.40 days}$$

Problem 7.12: The air having a velocity of 50 m/s is flowing over a cylinder of diameter 1 m and length 8 m, when the axis of the cylinder is horizontal and perpendicular to the direction of flow. The cylinder is rotated about its axis and a lift of 6500 N per metre length of the cylinder is developed. Find the speed of rotation and location of the stagnation points. Take density of air as 1.22 kg/m³.

Solution: Given data:

Velocity of air: $U = 50$ m/s

Diameter of the cylinder: $D = 1$ m

Length of the cylinder: $L = 8$ m

Lift per metre length: $\dfrac{F_L}{L} = 6500$ N/m

Density of air: $\rho = 1.22$ kg/m³

We know, $F_L = \rho L U \Gamma$

or $\dfrac{F_L}{L} = \rho U \Gamma$

or $6500 = 1.22 \times 50 \times \Gamma$

∴ Circulation: $\Gamma = 106.55$ m³/s

We know from Eq. (7.7.3) the speed of rotation corresponding to circulation,

$$u_\theta = \frac{\Gamma}{2\pi R} = \frac{\Gamma}{\pi D} = \frac{106.55}{3.14 \times 1} = 33.93 \text{ m/s}$$

also
$$u_\theta = \frac{\pi D N}{60}$$

∴
$$33.93 = \frac{3.14 \times 1 \times N}{60}$$

or $N = \textbf{648.34 rpm}$

Position of stagnation points are given by Eq. (7.7.8), as

$$\sin \theta = -\frac{\Gamma}{4\pi U R} = -\frac{\Gamma}{2\pi U D} = -\frac{106.55}{2 \times 3.14 \times 50 \times 1}$$

$$\sin \theta = -0.3393$$

$$\theta = -19.83°$$

(−ve sin θ lies in 3rd and 4th quadrants)

$$\theta = 180° + 19.83, 360° - 19.83°$$
$$= \textbf{199.83°, 340.17°}$$

Problem 7.13: A cylinder rotates at 120 rpm with its axis perpendicular in an air stream which is having uniform velocity 20 m/s. The cylinder is 0.8 m in diameter and 5 m long.

Find: (*i*) the circulation, (*ii*) lift force and (*iii*) position of stagnation points.

Take density of air as 1.22 kg/m³.

Solution: Given data:

Rotation of the cylinder: $N = 120$ rpm

Velocity of air: $U = 20$ m/s

Diameter of the cylinder: $D = 0.8$ m

Length of the cylinder: $L = 5$ m

Density of air: $\rho = 1.22$ kg/m³

Tangential velocity of the cylinder:

$$u_\theta = \frac{\pi DN}{60} = \frac{3.14 \times 0.8 \times 120}{60} = 5.02 \text{ m/s}$$

also, $u_\theta = \dfrac{\Gamma}{2\pi R}$ From Eq. (7.7.3)

$$u_\theta = \frac{\Gamma}{2\pi D}$$

$$5.02 = \frac{\Gamma}{3.14 \times 0.8}$$

or $\Gamma = \textbf{12.61 m}^2\textbf{/s}$

(*ii*) Lift force: $F_L = \rho LU\Gamma = 1.22 \times 5 \times 20 \times 12.61$
$$= \textbf{1538.42 N}$$

(*iii*) Position of stagnation points are given by Eq. (7.7.8), as

$$\sin\theta = -\frac{\Gamma}{4\pi UR}$$

$$\sin\theta = -\frac{\Gamma}{2\pi UD}$$

$$\sin\theta = -\frac{12.61}{2 \times 3.14 \times 20 \times 0.8}$$

$$\sin\theta = -0.12549$$

$$\theta = -7.20°$$

(−*ve* sin θ lies in 3rd and 4th quadrants)

∴ $\theta = 180° + 7.20°, 360° - 7.20°$
$$= \textbf{187.2°, 352.8°}$$

7.9 LIFT ON AN AIRFOIL

An airfoil is a stream line body. It has a rounded leading edge and profile is gradually tapered towards the trailing edge. Such a geometrical configuration produces high lift and low drag.

The profile of an airfoil is used in aircraft wings (also known as airfoils), blades of propellers, turbines and pumps etc.

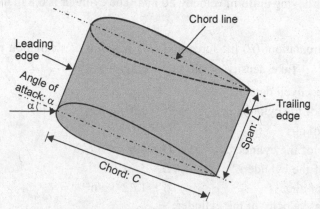

Fig. 7.19: Configuration of an airfoil

Definitions of various terms associated with an airfoil:

(*i*) *Leading and trailing edges*: The leading of an airfoil is usually a circular arc at which the fluid first strikes. The trailing edge is of zero radius is also known as rear of airfoil.

(*ii*) *Chord line or chord*: It is a line joining the leading and trailing edges of the airfoil.

(*iii*) *Profile or centre line*: It is a line obtained by joining the mid points of the profile.

(*iv*) *Angle of attack*: It is angle between the chord line and the direction of free stream. It is denoted by α.

(*v*) *Wingspan or span*: It is the length of leading or trailing edges.

(*vi*) *Stall*: Beyond a certain value of the angle of attack, the separation point moves toward the leading edge and the lift produced by the airfoil starts diminishing. The airfoil is then said to be operating under stalled condition. The value of the angle of attack at which lift reaches its maximum value is known as the stalling angle.

(*vii*) *Aspect ratio*: It is a ratio of the span L to the chord length C.
Mathematically,

Aspect ratio: $\qquad AR = \dfrac{\text{Span}}{\text{Chord length}} = \dfrac{L}{C}$

The condition of flow at the trailing edge depends upon magnitude of circulation. The circulation required to produce a tangential stream line at the trailing edge is given by

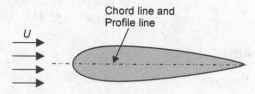

(a) Symmetrical airfoil (Chord line coincides with profile line)

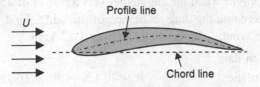

(b) Non-symmetrical airfoil (Chord line does not coincide with profile line)

Fig. 7.20

$$\Gamma = \pi \, CU \sin \alpha$$

where,
$$C = \text{chord}$$
$$U = \text{velocity of airfoil}$$
$$\alpha = \text{angle of attack}$$

According to Kutta-Joukowaki's Eq. (14.7.7), we get

$$F_L = \rho \, U L \, \Gamma$$
$$= \rho \, U L \times \pi \, CU \sin \alpha \, (\because \Gamma = \pi C U \sin \alpha)$$
$$= \pi \, \rho \, C U^2 \, L \sin \alpha \qquad \qquad ...(7.9.1)$$

Also we know that

$$F_L = C_L \frac{A \rho U^2}{2}$$

where
$$C_L = \text{co-efficient of lift}$$
$$A = C \times L, \text{projected area}$$

$$\therefore \qquad F_L = C_L \frac{C L \rho U^2}{2} \qquad \qquad ...(7.9.2)$$

Equating Eqs. (14.9.1) and (14.9.2), we get

$$\pi \, \rho \, C U^2 \, L \sin \alpha = C_L \frac{C L \rho U^2}{2}$$

$$\pi \sin \alpha = \frac{C_L}{2}$$

or
$$\boldsymbol{C_L = 2\pi \sin \alpha} \qquad \qquad ...(7.9.3)$$

Thus, it is clear from above Eq. (7.9.3), the co-efficient of lift (C_L) depends upon the angle of attack (α).

7.9.1 Steady State of a Flying Object

When a flying object, for example, airplane is in a steady state, the weight (W) of the airplane is equal to the lift force (F_L) and thrust developed by the engine is equal to the drag force.

Mathematically,

Weight:
$$W = F_L = C_L \frac{\rho A U^2}{2}$$

and thrust by the engine:
$$F_t = F_D = \frac{C_D \rho A U^2}{2}$$

Problem 7.14: A jet plane which weights 30 kN and has a wing area of 20 m² flies at a velocity of 950 km/hr when the engine delivers a power of 7500 kW, 65% of the power is used to overcome the drag resistance of the wing. Find the co-efficients of lift and drag for the wing. Take density of air as 1.22 kg/m³.

Solution: Given data:

Weight of the plane: $\qquad W = 30\ \text{kN} = 30 \times 10^3\ \text{N} = F_L$, lift force

Area of the wing: $\qquad A = 20\ \text{m}^2$

Speed of the plane: $\qquad U = 950\ \text{km/hr} = \dfrac{950 \times 1000}{3600}\ \text{m/s}$

$$= 263.88\ \text{m/s}$$

Power delivered by the engine: $\quad P = 7500\ \text{kW} = 7500 \times 10^3\ \text{W}$

Drag power: $\qquad P_D = 0.65\ P = 0.65 \times 7500 \times 10^3\ \text{W}$

$$= 4875 \times 10^3\ \text{W}$$

also $\qquad\qquad P_D = F_D \cdot U$

∴ $\qquad\qquad 4875 \times 10^3 = F_D \times 263.88$

or $\qquad\qquad F_D = 18474.30\ \text{N}$

We know, Lift force: $\qquad F_L = C_L \dfrac{A \rho U^2}{2}$

$$30 \times 10^3 = \dfrac{C_L \times 20 \times 1.2 \times (263.88)^2}{2}$$

or $\qquad\qquad C_L = \mathbf{0.0359}$

Drag force: $\qquad\qquad F_D = \dfrac{C_D A \rho U^2}{2}$

$$18474.30 = \dfrac{C_D \times 20 \times 1.2 \times (263.88)^2}{2}$$

or $\qquad\qquad C_D = \mathbf{0.0221}$

Problem 7.15: An airfoil of chord length 2 m and of span 15 m has an angle of attack as 6°. The airfoil is moving with a velocity of 360 km/hr in air whose density is 1.2 kg/m³. Find the weight of the airfoil and the power required to drive it. The values of co-efficients of drag and lift corresponding to angle of attack are given as 0.035 and 0.52 respectively.

Solution: Given data:

Chord length: $\qquad\qquad C = 2\ \text{m}$

Span: $\qquad\qquad L = 15\ \text{m}$

Angle of attack: $\qquad\qquad \alpha = 6°$

Velocity of airfoil: $\qquad\qquad U = 360\ \text{km/hr} = \dfrac{360 \times 1000}{3600}\ \text{m/s} = 100\ \text{m/s}$

Density of air:	$\rho = 1.2$ kg/m^3
Co-efficient of drag:	$C_D = 0.035$
Co-efficient of lift:	$C_L = 0.52$
Area of wind:	$A = C \times L = 2 \times 15 = 30$ m^2

Weight of airfoil:

$$W = F_L = \frac{C_L \rho A U^2}{2} = \frac{0.52 \times 1.2 \times 30 \times (100)^2}{2}$$

$$= \mathbf{93600\ N}$$

Drag force:

$$F_D = \frac{C_D \rho A U^2}{2} = \frac{0.035 \times 1.2 \times 30 \times (100)^2}{2}$$

$$= 6300\ N$$

Power required:

$$P_D = F_D \cdot U = 6300 \times 100\ W = 630 \times 10^3\ W$$

$$= \mathbf{630\ kW}$$

Problem 7.16: An airplane weighing 40 kN is flying in a horizontal direction at 350 km/hr and has a wing surface area of 30 m^2. Find the lift co-efficient and the power required to drive the plane. Assume drag co-efficient is 0.03 and density of air is 1.2 kg/m^3.

Solution: Given data:

Weight of the plane: $W = 40$ kN $= 40 \times 10^3$ N

Speed of the plane: $U = 350$ km/hr $= \dfrac{350 \times 1000}{3600}$ m/s $= 97.22$ m/s

Area of wing: $A = 30$ m^2
Co-efficient of drag: $C_D = 0.03$
Density of air: $\rho = 1.2$ kg/m^3

For equilibrium in vertical direction, lift is equal to the weight of airplane.

i.e.,

$$F_L = W = \frac{C_L \rho A U^2}{2}$$

$$40 \times 10^3 = \frac{C_L \times 1.2 \times 30 \times (97.22)^2}{2}$$

or

$$C_L = \mathbf{0.235}$$

Drag force:

$$F_D = \frac{C_D \rho A U^2}{2} = \frac{0.03 \times 30 \times 1.2 \times (97.22)^2}{2}$$

$$= 5103.93\ N$$

Power required to drive the plane:

$$P_D = F_D \times U = 5103.93 \times 97.22\ W$$

$$= 496204.07\ W = \mathbf{496.204\ kW}$$

SUMMARY

1. The force in the direction of flow exerted by the fluid on the solid body is called drag force or drag.

2. The force exerted by the fluid in the direction perpendicular to the flow is called lift.

3. Types of drag force:

   (i) Friction drag or wall drag

   (ii) Pressure drag or form drag

4. Expression for the drag and lift forces are:

$$\text{Drag force: } F_D = \frac{C_D \rho A U^2}{2}$$

$$\text{Lift force: } F_L = \frac{C_L \rho A U^2}{2}$$

where,

C_D = co-efficient of drag

C_L = co-efficient of lift

A = projected area of the body

ρ = density of fluid

U = free stream velocity of fluid

Resultant force exerted by fluid on solid body,

$$F_R = \sqrt{F_D^2 + F_L^2}$$

5. A body whose surface coincides with the stream-lines, when the body is placed in a flow is called stream-lined body. In such types of body, the friction drag is much greater than the pressure drag.

6. A body whose surface does not coincides with the stream-lines, when the body is placed in a flow is called bluff body. In such types of body, the pressure drag is much greater than the friction drag.

7. The drag force on the sphere:

$$F_D = 3\pi \, D\mu U$$

for $Re < 0.2$

Co-efficient of drag for sphere: C_D

$$C_D = \frac{24}{Re} \qquad\qquad \text{for } Re < 0.2$$

$$= \frac{24}{Re}\left(1 + \frac{3}{16Re}\right) \qquad \text{for } 0.2 < Re < 5$$

$$= 0.4 \qquad\qquad \text{for } 5 \le Re < 10^3$$

$$= 0.5 \qquad\qquad \text{for } 10^3 \le Re < 10^5$$

$$= 0.2 \qquad\qquad \text{for } Re \ge 10^5$$

Contd...

8. The velocity at any point C on the cylinder is prescribed by
$$u_\theta = 2U\sin\theta$$
where, u_θ = tangential velocity on the surface of the cylinder.

U = velocity of free stream fluid

θ = angle made by any point C on the circumference of the cylinder with direction of flow.

9. The circulation around the fluid flow curve is defined as the summation of product of velocity component along the element of the curve and the element length.

Circulation for free-vortex flow at any radius is given by
$$\Gamma = 2\pi R\, u_{\theta 1}$$

10. The resultant velocity on a circular cylinder which rotates at constant speed in uniform flow is given by
$$u = 2U\sin\theta + \frac{\Gamma}{2\pi R}$$

11. Lift acting on a rotating cylinder in uniform flow is given by
$$F_L = \rho\, L\, U\, \Gamma$$
where
ρ = density of fluid

L = length of the cylinder

U = free stream velocity

Γ = circulation

Above equation is known as Kutta-Joukowaski equation.

12. The co-efficient of lift for a rotating cylinder in a uniform flow is given by
$$C_L = \frac{\Gamma}{RU}$$
(in terms of circulation)
$$= \frac{2\pi u_{\theta 1}}{U}$$
(in terms of tangential velocity)

13. Stagnation point is defined as the point at which the velocity of flow is zero.

14. The location of stagnation points for a rotating cylinder in a uniform flow is given by
$$\sin\theta = -\frac{\Gamma}{4\pi UR}$$
where R = radius of cylinder

Contd...

U = free stream velocity

15. For a single stagnation point, the condition is
$$\Gamma = 4\pi U R$$
(in terms of circulation)

or $\qquad u_{\theta 1} = 2U$
(in terms of tangential velocity)

16. When a cylinder or sphere in a flow is rotated at a sufficiently high rate, a lift is produced on the cylinder. The phenomenon of producing lift by the rotation of a solid body is called the Magnus effect.

17. Circulation developed on the airfoil is given by
$$\Gamma = \pi C U \sin\alpha$$
where $\qquad C$ = chord

U = velocity of airfoil

α = angle of attack

18. The co-efficient of lift for an airfoil is given by
$$C_L = 2\pi \sin\alpha$$

19. When an airplane is in a steady state, then

Weight of airplane: $\qquad W$ = lift force
$$W = F_L = \frac{C_L \rho A U^2}{2}$$
and thrust by the engine: $\qquad F_t$ = Drag force
$$F_t = F_{D_1} = \frac{C_D \rho A U^2}{2}$$

ASSIGNMENT - 1

1. Define the terms,
    (i) Drag, and $\qquad$ (ii) Lift
2. Define the following terms:
    (i) Wall drag $\qquad$ (ii) Form drag
    (iii) Total drag
3. What is the expression for the drag on a sphere, when Reynold's number of the flow is up to 0.2 ? Hence, prove that the co-efficient of drag for sphere for this range of the Reynold's number is given by
$$C_D = \frac{24}{Re}, \quad \text{where, } Re = \text{Reynold's number}$$
4. Explain the variation of co-efficient of drag with Reynold's number, when fluid flows over a stationary cylinder.
5. What is circulation ? Find an expression for circulation for a free vortex of radius R.

6. Prove that the lift force acting on rotating cylinder in a uniform flow is given by

$$F_L = \rho L U \Gamma$$

where
L = length of the cylinder
U = free stream velocity
Γ = circulation

7. Prove that the co-efficient of lift for a rotating cylinder placed in a uniform flow is given by

$$C_L = \frac{\Gamma}{RU}$$

where
Γ = circulation
R = radius of the cylinder
U = free steam velocity

8. Define stagnation points. How the position of the stagnation points for a rotating cylinder in a uniform flow is determined ? What is the condition for single stagnation point ?

9. What is Magnus effect ? Why is it known as Magnus effect ?

10. Define the following terms:
   (i) Airfoil   (ii) Leading and trailing edges
   (iii) Chord   (iv) Angle of attack
   (v) Aspect ratio

11. Prove that the co-efficient of lift for airfoil is given by

$$C_L = 2\pi \sin\alpha$$

When the circulation developed on an airfoil is equal to $\pi C U \sin\alpha$.
where, α = angle of attack.

ASSIGNMENT - 2

1. A circular disc of 2.5 m in diameter is held normal to a 25 m/s wind velocity 1.2 kg/m³. What force is required to hold it at rest ? Assume co-efficient of drag of disc is 1.15.   **Ans.** 2115.82 kW

2. A man weighing 784.8 N descends to the ground from an aeroplane with the help of a parachute against the resistance of air. The shape of the parachute is hemispherical of 2 m diameter. Find the velocity of the parachute with which it comes down. Assume $C_D = 0.6$ and density for air is 1.25 kg/m³.
                                                                Ans. 25.81 m/s

3. A flat plate 1.5 m × 1.5 m moves at 50 km/hr in stationary air of density 1.15 kg/m³. If the co-efficient of drag and lift are 0.15 and 0.75 respectively, find:
   (i) the lift force   (ii) the drag force
   (iii) the resultant force, and   (iv) the power required to keep the plate in motion

   Ans. (i) 187.20 N, (ii) 37.44 N, (iii) 190.90 N, (iv) 520.04 kW

4. Experiments were conducted in a wind tunnel with speed of 50 km/hr on flat plate of size 2 m long and 1 m wide. The density of air is 1.15 kg/m^3. The co-efficients of lift and drag are 0.75 and 0.15 respectively. Find:

(*i*) the lift force   (*ii*) the drag force

(*iii*) the resultant force   (*iv*) direction of resultant force and

(*v*) power exerted by air on the plate.

Ans. (*i*) 166.40 N, (*ii*) 33.28 N, (*iii*) 169.69 N, (*iv*) 78.69°, (*v*) 462.259 W

5. A truck having a projected area of 7 m^2 travelling at 60 km/hr has a total resistance of 2000 N; of this 20% is due to rolling friction and 10% is due to surface friction. The rest is due to form drag. Find the co-efficient of form drag. Take ρ = 1.25 kg/m^3.   **Ans.** 1.15

6. A kite of dimension 0.8 m × 0.8 m and weighing 4 N assumes an angle of 15° to the horizontal. The string attached to the kite makes an angle of 45° to the horizontal. The pull on the string is 24 N when the wind is flowing at a speed of 30 km/hr. Find the co-efficient of drag and lift. Take density of air as 1.22 kg/m^3.   **Ans.** C_D = 0.626, C_L = 0.774

7. A submarine which may be supposed to approximate a cylinder 4 m in diameter and 20 m long travels submerged at 1.4 m/s in sea-water. Find the drag exerted on it, if the drag co-efficient for Reynolds number greater than 10^5 may be taken as 0.75. The density of sea-water is given as 1025 kg/m^3 and kinematic viscosity as 0.02 stokes.   **Ans.** 60.27 kN

8. A body of length 2.5 m has a projected area 1.8 m^2 normal to the direction of its motion. The body is moving through water with a velocity such that the Reynold's number is 6 × 10^6 and the drag co-efficient is 0.5. Find the drag on the body. Take viscosity of water as 0.01 poise.   **Ans.** 2592 N

9. A cylinder rotates at 200 rpm with its axis perpendicular in an air stream which is having uniform velocity of 20 m/s. The cylinder is 2 m in diameter and 8 m long. Assuming ideal fluid theory, find (*i*) the circulation, (*ii*) lift force, and (*iii*) position of stagnation points. Take density of air as 1.25 kg/m^3.

Ans. (*i*) 131.57 m^2/s, (*ii*) 26.309 kN, (*iii*) 211.56° and 328.44°

10. A metallic sphere of specific gravity 8 falls in an oil of specific gravity 0.8. The diameter of the sphere is 10 mm and it attains a terminal velocity of 0.05 m/s. Find the velocity of the oil in poise.   **Ans.** 78.48 poise

11. A jet plane which weighs 19620 N has a wing area of 25 m^2. It is flying at a speed of 200 km/hr. When the engine develops 588.6 kW, 70% of this power to overcome the drag resistance of the wing. Find the co-efficient of lift and co-efficient of drag for the wing. Take density of air as 1.25 kg/m^3.

Ans. C_L = 0.4068, C_D = 0.1538

12. An airplane weighing 50 kN in flying in a horizontal direction at 400 km/hr and has a wing surface area of 40 m^2. Find the lift co-efficient and the power required to drive the plane. Assume drag co-efficient is 0.03 and density of air is 1.2 kg/m^3.   **Ans.** C_L = 0.1687, P = 987.624 kW

□□□

Flow Through Open Channels

8.1 INTRODUCTION

The flow of liquid with a free surface is called **open channel flow.** The upper surface of the liquid is subject to the atmospheric pressure is known as free surface. Flows in rivers, natural stream, artificial canals, navigation channels, drainage channels and sewers are some examples of open channel flow. Pipelines or tunnels which are not completely full of liquid also treated as open channel flow. All the open channels have a bottom slope and the component of the weight of the liquid along the slope acts as the driving force. The boundary resistance at the perimeter acts as the resisting force.

8.2 GEOMETRICAL TERMINOLOGIES: FLOW THROUGH OPEN CHANNELS

(*a*) **Depth of Flow: H.** The depth of flows H at any section is the vertical distance from the free surface of liquid to the bed of the channel at that section.

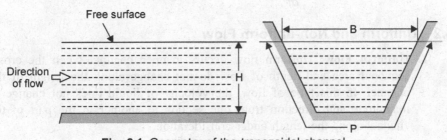

Fig. 8.1 Geometry of the trapezoidal channel.

(*b*) **Top Breadth: B.** It is the breadth of the channel section at the free surface.
(*c*) **Water Area: A.** The water area is flow cross-sectional area perpendicular in the direction of flow.

© The Author(s) 2023
S. Kumar, *Fluid Mechanics (Vol. 2)*,
https://doi.org/10.1007/978-3-030-99754-0_8

(*d*) **Wetted Perimeter:** *P.* The wetted perimeter is the perimeter of the solid boundary in contact with the liquid.

(*e*) **Hydraulic Radius:** *m.* It is defined as the ratio of the water area to the wetted perimeter. It is also called hydraulic mean depth.

Mathematically,

Hydraulic radius: $\qquad m = \dfrac{\text{Water area or cross-sectional area of flow}}{\text{Wetted perimeter}}$

$$m = \frac{A}{P}$$

8.3 TYPES OF FLOW IN OPEN CHANNELS

The flow in open channel is classified as:

    (*a*) Steady and unsteady flow

    (*b*) Uniform and non-uniform flow

    (*c*) Laminar and turbulent flow, and

    (*d*) Sub-critical, critical and super critical flow.

8.3.1 Steady and Unsteady Flow

Flow is termed steady or unsteady depending upon the velocity at a point in the channel is invariant with time or not. If the velocity of the flow at any point does not change with time, such type of flow is called **steady flow.** If the velocity of the flow changes with time, such type of flow is called **unsteady flow.**

    Mathematically,

$$\frac{\partial V}{\partial t} = 0 \quad \text{for steady flow}$$

$$\frac{\partial V}{\partial t} \neq 0 \quad \text{for unsteady flow}$$

8.3.2 Uniform and Non-uniform Flow

1. **Uniform Flow:** Uniform flow occurs in open channel when the cross-sectional area and depth of flow do not change along the length of the channel. In this type of flow, the velocity of liquid does not change in magnitude and direction from one section to another in the part of the channel along the length under consideration.

2. **Non-uniform Flow:** Non-uniform flow occurs in open channel when the cross-sectional area and depth of flow change along the length of the channel. This flow is also called varied flow because the depth of flow continuously varies from one section to another. Thus flow occurs in a channel which is shaped irregularly when depth and the velocity of flow vary.

Non-uniform flow is further classified into two types given below:

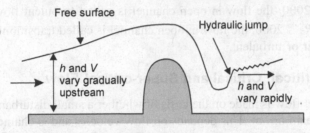

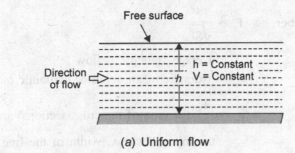

(b) Non-uniform flow–gradually and rapidly Varied flow

(a) Uniform flow

Fig. 8.2 Uniform and non-uniform flow in open channel.

(a) Gradually varied flow, and

(b) Rapidly varied flow

(a) **Gradually Varied Flow:** As the depth of flow changes gradually over a length of the channel, flow is called gradually varied flow. In this non-uniform flow, the degree of non-uniformity is small and gradual as shown in Fig. 8.2(b).

(b) **Rapidly Varied Flow:** As the depth of flow changes rapidly over a small length of the channel, flow is called rapidly varied flow. In this non-uniform flow, the degree of non-uniformity is large and rapidly as shown in Fig. 8.2(b).

8.3.3 Laminar and Turbulent Flow

The flow in open channel is said to be laminar or turbulent depending upon the value of Reynold's number.

Reynold's number: $\quad$ Re $= \dfrac{\rho V y_m}{\mu}$ for flow in open channel

where
$\qquad \rho$ = density of the liquid

$\qquad \mu$ = viscosity of the liquid

$\qquad V$ = average velocity of flow

$\qquad y_m$ = hydraulic mean depth or hydraulic radius

$$y_m = \frac{A}{P}$$

$\qquad A$ = cross-section of flow normal to the direction of flow

$\qquad P$ = wetted perimeter

If   Re < 500, the flow in open channel is called laminar flow.

If   Re > 2000, the flow in open channel is called turbulent flow.

If $500 \leq Re \leq 2000$, the flow in open channel is called transition flow and may be either laminar or turbulent.

8.3.4   Sub-critical, Critical and Super-critical Flow

An open channel flow is made on the basis of whether a small disturbance in the flow can travel upstream or not. This depends on flow velocity and is characterised by the value of Froude's number.

Froude's number: $F_r = \dfrac{V}{\sqrt{gy_m}}$

where V = average velocity of flow

y_m = hydraulic mean depth or hydraulic radius

$y_m = \dfrac{A}{B}$, the ratio of the cross-sectional area of flow to

the top width (*i.e.*, width of the free surface.)

The magnitude of Froude's number determines the type of flow carried by the channel.

(*i*) If $F_r < 1$, the flow is said to be sub-critical or tranquil. Any small disturbance can travel against the flow and affects the upstream condition.

(*ii*) If $F_r = 1$, the flow is called critical flow. The velocity of flow is just equal to the velocity of an elementary wave.

(*iii*) If $F_r > 1$, the flow is called super critical flow. It is also referred to as rapid or shooting flow. Any disturbance occurring downstream cannot travel upstream and consequently no change is affected in the upstream conditions.

8.4 CHEZY'S FORMULA

We know that the Chezy's formula;

$$V = C\sqrt{im}$$

where V = average velocity of flow

C = Chezy's constant. It has the dimensions $L^{1/2} T^{-1}$

$= \sqrt{\dfrac{\rho g}{f_1}}$ where f_1 is coefficient depending on the roughness of the channel.

i = slope of the bed of the channel

$= \tan \theta = \dfrac{\sin \theta}{\cos \theta}$

where θ is small angle, $\cos \theta = 1$ and $\sin \theta \approx \theta$

$\therefore$ $i = \theta$

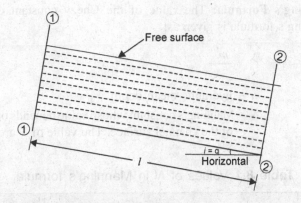

Fig. 8.3 Uniform flow in open channel.

m = hydraulic mean depth or hydraulic radius

$= \dfrac{A}{P}$, it is ratio of the cross-sectional area of flow to wetted perimeter.

The discharge of water through the open channel:

Q = cross-sectional area of flow × velocity of flow

$Q = AV$

$$Q = AC\sqrt{im}$$

8.5 EMPIRICAL RELATIONS FOR DETERMINATION OF CHEZY CONSTANT

We known that the Chezy's formula;

$$V = C\sqrt{im}$$

or

$$C = \frac{V}{\sqrt{im}}$$

where

$V = LT^{-1}$

$m = L$

and

i is dimensionless.

∴

$$C = \frac{LT^{-1}}{\sqrt{L}}$$

$$C = L^{1/2}\,T^{-1}$$

The Chezy constant C is not a dimensionless quantity, it has the dimension $L^{1/2}T^{-1}$ and so its numerical value depends upon the units of the system. The followings are the empirical relations are used to determine the value of the Chezy constant C.

(*a*) **Manning's Formula:** The value of the Chezy constant C according to Manning's formula is given as:

$$C = \frac{m^{1/6}}{N}$$

where m = hydraulic mean depth

N = Manning's constant whose value depends on the roughness of the channel surface. The value of N are given is Table 8.1.

Table 8.1 Values of *N* in Manning's formula

S. No.	Nature of the Surface of Channel	N
1.	Smooth surface of concrete	0.012
2.	Rubble masonry	0.018
3.	Earthen channel of ordinary surface	0.027
4.	Earth channel of rough surface	0.030

(*b*) **Bazin Formula:** The value of the Chezy constant C according to Bazin formula is given as

$$C = \frac{157.6}{1.81 + \dfrac{K}{\sqrt{m}}}$$

where K = Bazin's constant and depends upon the roughness of the surface of the channel. The value of K are given in Table 8.2

Table 8.2 Values of *K* in Bazin's formula

S. No.	Nature of the Surface of Channel	K
1.	Smooth cement plaster	0.11
2.	Smooth brick, stone or wood	0.29
3.	Rubble masonry or poor brick work	0.83
4.	Earthen channel of very good surface	1.54
5.	Earthen channel of ordinary surface	2.36
6.	Earthen channel of rough surface	3.17

Problem 8.1:   A rectangular channel is 3 m wide. The slope of its bed is 1 in 800. Determine the discharge when depth of water 1.2 m. Take C = 60.

Solution: Given data:

Width of channel: $L = 3$ m

Slope of bed: $i = \dfrac{1}{800}$

Depth of water: $H = 1.5$ m

Chezy's constant: $C = 60$
Cross-sectional area of flow:
$$A = LH = 3 \times 1.5 = 4.5 \text{ m}^2$$
Wetted perimeter: $P = L + 2H = 3 + 2 \times 1.5 = 3 + 3 = 6 \text{ m}$

$\therefore$ Hydraulic mean depth: $m = \dfrac{A}{P} = \dfrac{4.5}{6} = 0.75 \text{ m}$

Discharge through channel: $Q = AC\sqrt{im} = 4.5 \times 60\sqrt{\dfrac{1}{800}} \times 0.75 = \textbf{8.267 m}^3\textbf{/s}$

Problem 8.2: Determine the slope of the bed of a rectangular channel of width 4 m when depth of water is 1.6 m and rate of flow is given as 10 m³/s. Take Chezy's constant as 50.

Solution: Given data:
    Width of channel: $L = 4 \text{ m}$
    Depth of water: $H = 1.6 \text{ m}$
    Rate of flow: $Q = 10 \text{ m}^3/\text{s}$
    Chezy's constant: $C = 50$
    Area of flow: $A = LH = 4 \times 1.6 = 6.4 \text{ m}^2$
    and wetted perimeter: $P = L + 2H = 4 + 2 \times 1.6 = 7.2 \text{ m}$

$\therefore$ Hydraulic mean depth: $m = \dfrac{A}{P} = \dfrac{6.4}{7.2} = 0.888 \text{ m}$

Rate of flow: $Q = AC\sqrt{im}$

$\therefore$ $10 = 6.4 \times 50 \times \sqrt{i \times 0.888}$

or $0.03125 = \sqrt{i \times 0.888}$

Squaring both sides, we get
$$(0.03125)^2 = i \times 0.888$$

or $i = \dfrac{(0.03125)^2}{0888} = \dfrac{1}{909.31}$

$\therefore$ **Slope of the bed is 1 in 909.31**

Problem 8.3: Determine the discharge through a rectangular channel 3 m wide, depth of water 2 m and bed slope as 1 in 5000 by using (i) Chezy's formula for which C = 50 (ii) Manning's formula for which N = 0.012.

Solution: Given data:
    Length of the channel: $L = 3 \text{ m}$
    Depth of water: $H = 2 \text{ m}$

    Bed slope: $i = \dfrac{1}{5000}$

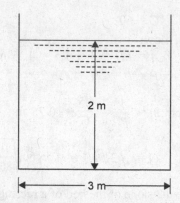

Fig. 8.4: Schematic for Problem 8.3

Area of flow: $A = HL = 2 \times 3 = 6$ m²

Wetted perimeter: $P = L + 2H = 3 + 2 \times 2 = 7$ m

$\therefore$ Hydraulic mean depth: $m = \dfrac{A}{P} = \dfrac{6}{7}$ m

(i) By Chezy's formula;

    Chezy's constant: $C = 50$

    Discharge: $Q = AC\sqrt{im}$

$$= 6 \times 50 \times \sqrt{\frac{1}{5000} \times \frac{6}{7}} = \textbf{3.927 m}^3\textbf{/s}$$

(ii) By Manning's formula;

    Manning's constant: $N = 0.019$

$\therefore$ Chezy's constant: $C = \dfrac{m^{1/6}}{N}$

$$= \frac{\left(\dfrac{6}{7}\right)^{1/6}}{0.019} = 51.30$$

$\therefore$ Discharge: $Q = AC\sqrt{im} = 6 \times 51.30 \times \sqrt{\dfrac{1}{5000} \times \dfrac{6}{7}} = \textbf{4.03 m}^3\textbf{/s}$

Problem 8.4:   Determine the discharge through a trapezoidal channel of width 6 m and side slope of 1 horizontal to 3 vertical. The depth of flow of water is 3 m and Chezy's constant, $C = 60$. The slope of the bed of the channel is given 1 in 5000.

Solution: Give data:

    Width of the bed: $L = 6$ m

    Side slope $= 1$ horizontal and 3 vertical

    Depth of flow: $H = 3$ m

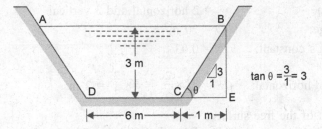

Fig. 8.5: Schematic for Problem 8.4

Chezy's constant: $C = 60$

Slope of the bed: $i = 1$ in 5000

Length of horizontal: $CE = H \times \dfrac{1}{3} = 3 \times \dfrac{1}{3} = 1$ m

$\therefore$ Width of the free surface:
$$AB = DC + 2\,CE = 6 + 2 \times 1 = 6 + 2 = 8 \text{ m}$$

Area of flow: $A = \left(\dfrac{DC + AB}{2}\right) \times H = \left(\dfrac{6+8}{2}\right) \times 3 = 21 \text{ m}^2$

and wetted perimeter: $P = AD + DC + CB = DC + 2CB \qquad \because \quad AD = CB$

$$= 6 + 2 \times \sqrt{(3)^2 + (1)^2} = 6 + 2 \times \sqrt{9+1}$$

$$= 6 + 2 \times \sqrt{10} = 12.32 \text{ m}$$

$\therefore$ Hydraulic mean depth: $m = \dfrac{A}{P} = \dfrac{21}{12.32} = 1.70$ m

Discharge: $Q = AC\sqrt{im} = 21 \times 60 \sqrt{\dfrac{1}{5000}} \times 1.7 = \mathbf{23.23 \ m^3/s}$

Problem 8.5: Determine the bed of slope of trapezoidal channel of bed width 4 m, depth of water 3 m and side slope of 2 horizontal to 3 vertical, when the discharge through the channel is 20 m^3/s. Take manning's constant, N = 0.03.

Solution: Give data:

Width of the bed: $L = 4$ m

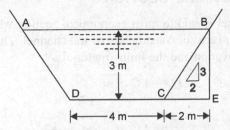

Fig. 8.6: Schematic for Problem 8.5

Depth of flow: $H = 3$ m

Side slope   = 2 horizontal and 3 vertical

Discharge: $Q = 20 \ \text{m}^3/s$

Manning's constant: $N = 0.03$

Length of horizontal: $CE = H \times \dfrac{2}{3} = 3 \times \dfrac{2}{3} = 2$ m

∴ Width of the free surface:
$$AB = DC + 2 \ CE = 4 + 2 \times 2 = 8 \text{ m}$$

Area of flow: $A = \left(\dfrac{DC + AB}{2}\right) \times H = \left(\dfrac{4+8}{2}\right) \times 3 = 18 \text{ m}^2$

and Wetted perimeter: $P = DC + 2CB$ ∵ $AD = CB$

$$= 4 + 2 \times \sqrt{(3)^2 + (2)^2} = 11.21 \text{ m}$$

∴ Hydraulic mean depth: $m = \dfrac{A}{P} = \dfrac{18}{11.21} = 1.60$ m

Chezy's constant: $C = \dfrac{m^{1/6}}{N} = \dfrac{(1.60)^{1/6}}{0.03} = 36.04$

Discharge: $Q = A C \sqrt{im}$

$$20 = 18 \times 36.04 \times \sqrt{i \times 1.60}$$

$$0.0308 = \sqrt{i \times 1.60}$$

Squaring both sides, we get
$$0.00095 = i \times 1.60$$

or $i = 5.940 \times 10^{-4} = \dfrac{1}{1683}$

Slope of the bed 1 in 1683.

8.6 MOST ECONOMICAL SECTION

A section of the channel is said to most economical section when we get maximum discharge with minimum cost of construction of the channel. The cost of construction depends upon the excavation and the lining material.

Discharge: $Q = A C \sqrt{i m}$

$$= A C \sqrt{i \dfrac{A}{P}} = \dfrac{A^{3/2} C \sqrt{i}}{\sqrt{P}}$$

For a given A, C and i, the above equation becomes

$$Q = C_1 \frac{1}{\sqrt{P}} \quad \text{where} \quad C_1 = A^{3/2} C \sqrt{i}$$

or

$$Q \propto \frac{1}{\sqrt{P}}$$

Hence the discharge Q will be maximum, when the wetted perimeter P is minimum.

This condition will be used for the construction of the most economical section of the channel for a given cross-sectional area. Although a semi-circular channel has the maximum hydraulic mean radius and it is built from prefabricated sections, the semi-circular shape is impractical for other forms of construction.

The condition for the most economical channel is used for the following channel:

(*i*) Rectangular channel, and (*ii*) Trapezoidal channel.

8.6.1 Most Economical Rectangular Channel

For a given channel slope, surface roughness and area of flow, the most economical is one which has the minimum wetted perimeter. Consider a rectangular channel as shown in Fig. 8.7

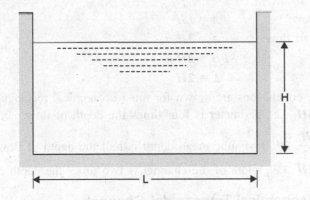

Fig. 8.7 Rectangular channel.

Let L = width of the channel

H = depth of the flow in the channel

∴ Area of flow: $A = LH$

and wetted perimeter: $P = L + 2H$

$$P = \frac{A}{H} + 2H$$

$$\because A = LH$$
$$\text{or } L = \frac{A}{H}$$

For most economical section, P should be minimum and $\dfrac{dP}{dH} = 0$, keeping A as constant.

$$\frac{dP}{dH} = 0$$

$$\frac{d}{dH}\left(\frac{A}{H}+2H\right) = 0$$

$$\frac{d}{dH}(AH^{-1}+2H) = 0$$

$$-AH^{-2}+2 = 0$$

or $$\frac{-A}{H^2}+2 = 0$$

or $$A = 2H^2$$

and Wetted perimeter: $$P = \frac{A}{H}+2H$$

$$P = \frac{2H^2}{H}+2H \qquad\qquad \because A = 2H^2$$

$$P = 2H + 2H$$

$$\boldsymbol{P = 4H}$$

$\therefore$ Hydraulic mean depth: $$m = \frac{A}{P} = \frac{2H^2}{4H}$$

$$\boldsymbol{m = \frac{H}{2}}$$

Area of flow: $A = LH$

Also area of flow: $A = 2H^2$

$\therefore$ $LH = 2H^2$

or $\boldsymbol{L = 2H}$

Hence the conclusions are drawn for most economical rectangular channel as

(a) $\boldsymbol{P = 4H}$ *i.e.*, perimeter is four times the depth of flow.

(b) $\boldsymbol{P = \dfrac{H}{2}}$ *i.e.*, hydraulic mean depth is half the depth of flow.

(c) $\boldsymbol{P = 2H}$ *i.e.*, width of the channel is two times the depth of flow.

8.6.2 Most Economical Trapezoidal Channel

The trapezoidal section of a channel will be most economical, when its wetted perimeter is minimum. Consider a trapezoidal channel as shown in Fig. 8.8

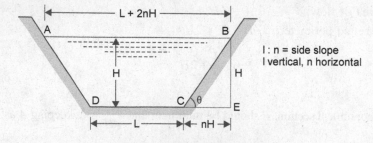

Fig. 8.8 Trapezoidal section.

Let
$$L = \text{width of the channel at bottom}$$
$$H = \text{depth of the flow in the channel}$$
$$\theta = \text{angle made by the sides with horizontal}$$

$$\frac{1}{n} = \text{slope of the slanting side}$$

$$= \tan\theta$$

$$= \frac{BE}{CE}$$

Width at free surface: $AB = L + 2nH$

$\therefore$ Area of flow:
$$A = \frac{(L+L+2nH)H}{2}$$

$$= \frac{(L+L+2nH)H}{2} = \left(\frac{2L+2nH}{2}\right)H$$

$$A = (L + nH)\,H$$

or
$$\frac{A}{H} = L + nH$$

or
$$L = \frac{A}{H} - nH \qquad\qquad\qquad …(i)$$

and wetted perimeter:
$$P = AD + DC + CB$$
$$= DC + 2CB \qquad\qquad\qquad \because AD = CB$$
$$= L + 2H\sqrt{n^2+1} \qquad\qquad\qquad …(ii)$$

Substituting the value of L from Eq. (i) in above Eq. (ii), we get

$$P = \frac{A}{H} - nH + 2H\sqrt{n^2+1} \qquad\qquad …(iii)$$

1. Minimum perimeter:

For most economical section, differentiating P w.r.t. H and equating to zero, we get

$$\frac{d}{dH}\left[\frac{A}{H} - nH + 2H\sqrt{n^2+1}\right] = 0$$

$$-\frac{A}{H^2} - n + 2\sqrt{n^2+1} = 0$$

or
$$\frac{-A}{H^2} - n = -2\sqrt{n^2+1}$$

or
$$\frac{A}{H^2} + n = 2\sqrt{n^2+1}$$

$$\frac{(L+nH)}{H^2} + n = 2\sqrt{n^2+1} \qquad\qquad \because A = (L+nH)H$$

$$\frac{L + nH}{H} + n = 2\sqrt{n^2 + 1}$$

$$\frac{L + nH + nH}{H} = 2\sqrt{n^2 + 1}$$

$$\frac{L + 2nH}{H} = 2\sqrt{n^2 + 1}$$

$$\frac{L + 2nH}{2} = H\sqrt{n^2 + 1} \qquad \qquad ...(iv)$$

$$OB = BC$$

Hence, the discharge through a trapezoidal channel with a given area will be a maximum when half of its width at free surface is equal to the sloping side.

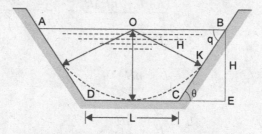

Fig. 8.9 Most economical trapezoidal section.

$$\frac{L + 2nH}{2} = H\sqrt{n^2 + 1}$$

2. Hydraulic mean depth:

Hydraulic mean depth: $$m = \frac{A}{P}$$

where $$A = (L + nH)H$$

and $$P = L + 2H\sqrt{n^2 + 1}$$

From Eq. (iii) $$2H\sqrt{n^2 + 1} = L + 2nH$$

∴ $$P = L + L + 2nH$$
$$P = 2L + 2nH = 2\,(L + nH)$$

∴ Hydraulic mean depth: $$m = \frac{(L + nH)H}{2(L + nH)}$$

$$m = \frac{H}{2}$$

Hence for most economical trapezoidal channel, hydraulic mean depth is equal to half the depth of flow.

3. **The discharge with a given cross-section will be a maximum when the sides of the channel are tangential to a semi-circle, drawn with radius H from the centre O.**

From O i.e., centre of the free surface AB, draw a perpendicular on the sloping sides AD and BC

From $\triangle$ BOK, $\sin \theta = \dfrac{OK}{OB}$

and from $\triangle$ BCE, $\sin \theta = \dfrac{BE}{BC} = \dfrac{H}{H\sqrt{n^2 + 1}}$

$$\sin \theta = \dfrac{1}{\sqrt{n^2 + 1}}$$

$\therefore$ $\dfrac{OK}{OB} = \dfrac{1}{\sqrt{n^2 + 1}}$

where OB = half of the free surface

$$= \dfrac{L + 2nH}{2}$$

also $OB = H\sqrt{n^2 + 1}$

$\therefore$ $OK = \dfrac{H\sqrt{n^2 + 1}}{\sqrt{n^2 + 1}}$

or   **OK = H**

Thus, the discharge with a given cross-section will be a maximum when the sides of the channel are tangential radius H from the centre O.

4. **Best side slope:**

For the most economical tropezoidal section, the depth of flow and flow area are constant. Then only n is the variable. Best side slop will be when the section is most economical, differentiating P w.r.t n and equating to zero, we get

$$\dfrac{dP}{dn} = 0$$

$$\dfrac{d}{dn}\left(\dfrac{A}{H} - nH + 2H\sqrt{n^2 - 1}\right) = 0$$

$$0 - H + \dfrac{1}{2} \times 2H(n^2 - 1)^{-\frac{1}{2}} \times 2n = 0$$

$$-H + \dfrac{2nH}{\sqrt{n^2 + 1}} = 0$$

or $\dfrac{2nH}{\sqrt{n^2 + 1}} = H$

or $\dfrac{2n}{\sqrt{n^2 + 1}} = 1$

or $2n = \sqrt{n^2 + 1}$

Squaring both sides, we get

$$4n^2 = n^2 + 1$$
$$3n^2 = 1$$

$$n^2 = \frac{1}{3}$$

or $n = \dfrac{1}{\sqrt{3}}$

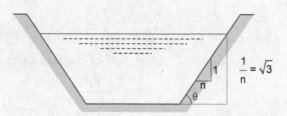

Fig. 8.10: Most economical rectangular channel

If the sloping sides makes an angle θ with the horizontal, then

$$\tan \theta = \frac{1}{n} = \frac{1}{1/\sqrt{3}}   \qquad \because n = \frac{1}{\sqrt{3}}$$

$$\tan \theta = \sqrt{3}$$
$$\theta = 60°$$

Hence, the best side slope with horizontal is 60°.

The conclusions are drawn for most economical rectangular channel as:

(a) $\dfrac{L + 2nH}{2} = H\sqrt{n^2 + 1}$ *i.e.*, half width of the free surface is equal to the sloping side.

(b) $\mathbf{m} = \dfrac{H}{2}$ *i.e.*, hydraulic mean depth is equal to half the depth of flow.

(c) A semi-circle drawn from the centre O of free surface with radius equal to the depth of flow will touch the sides of the channel.

(d) Best side slope with horizontal = 60° *i.e.*, $\theta = 60°$.

Problem 8.6:   A rectangular channel 6 m wide has a bed slope of 1 in 2000. Determine the channel Take C = 50.

Solution: Give data:

Width of channel: $L = 6$ m

Bed slope: $i = \dfrac{1}{2000}$

Chezy's constant: $C = 50$

We know the conditions for maximum discharge flow through a rectangular channel:

(i) Depth of water: $\quad H = \dfrac{L}{2} = \dfrac{6}{2} = 3$ m

(ii) Hydraulic mean depth: $m = \dfrac{H}{2} = \dfrac{3}{2} = 1.5$ m

$\therefore$ Area of most economical rectangular channel:
$$A = LH = 6 \times 3 = 18 \text{ m}^2$$

Maximum discharge: $\quad Q = AC\sqrt{im} = 18 \times 50 \times \sqrt{\dfrac{1}{2000}} \times 1.5 = \textbf{24.64 m}^3\textbf{/s}$

Problem 8.7: Determine the most economical dimensions of the rectangular channel when discharge flow through the channel is 500 litre/s and bed slope of 1 in 2000. Take Chezy's constant as 50.

Solution: Give data:

Discharge: $\qquad Q = 500 \text{ litre/s} = \dfrac{500}{1000} \text{ m}^3/s \qquad \because \text{ litre} - \dfrac{1}{1000}\text{m}^3$

$\qquad\qquad = 0.5 \text{ m}^3/s$

Bed slope: $\qquad i = \dfrac{1}{2000}$

We know the conditions for the most economical rectangular channel:

(i) Depth of water: $\qquad\qquad\qquad\qquad H = \dfrac{L}{2}$

(ii) Hydraulic mean depth: $\qquad\qquad\qquad m = \dfrac{H}{2}$

Area of flow: $\qquad A = LH = 2H \times H \qquad\qquad\qquad \because L = 2H$

$\qquad\qquad\qquad = 2H^2$

Discharge: $\qquad Q = AC\sqrt{im}$

$$0.5 = 2H^2 \times 50 \times \sqrt{\dfrac{1}{2000}} \times \dfrac{H}{2}$$

$$0.5 = 2H^2 \times 50 \times \sqrt{\dfrac{1}{4000}} \ H^{1/2}$$

$$0.5 = 1.58 \ H^{5/2}$$

or $\qquad\qquad H^{5/2} = 0.3164$

or $\qquad\qquad \textbf{\textit{H} = 0.631m}$

also $\qquad\qquad\qquad\qquad \dfrac{L}{2} = 0.631$

or $\qquad\qquad\qquad\qquad L = 0.631 \times 2 = \textbf{1.26 m}$

Problem 8.8: A trapezoidal channel has side slopes 3 horizontal to 4 vertical and slope of its bed is 1 in 1500. Determine the most economical dimensions of the channel, if it is to carry water at 10 m ^{3}/s. Take C = 60.

Solution: Give data:

Side slopes: $\qquad\qquad\qquad n = \dfrac{3}{4}$

Slope of bed: $\qquad\qquad\qquad i = \dfrac{1}{1500}$

Discharge: $\qquad\qquad\qquad Q = 10$ m ^{3}/s

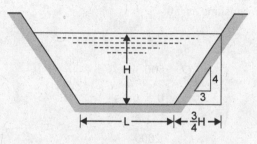

Fig. 15.11

Chezy's constant: $C = 60$

We know that the conditions for the most economical channel:

(i) $\dfrac{L + 2nH}{2} = H\sqrt{n^2 + 1}$

(ii) $m = \dfrac{H}{2}$

From condition (i), we get

$$\dfrac{L + 2 \times \dfrac{3}{4} H}{2} = H \sqrt{\left(\dfrac{3}{4}\right)^2 + 1}$$

$$\dfrac{L + 1.5\,H}{2} = H \times \sqrt{\dfrac{25}{16}}$$

$$\dfrac{L + 1.5\,H}{2} = H \times \dfrac{5}{4}$$

or
$$L + 1.5H = H \times \frac{5}{4} \times 2$$

$$L + 1.5H = 2.5H$$

or
$$L = H$$

Area of flow:
$$A = \left[\frac{L + (L + 2nH)}{2} \right] \times H = \left[\frac{L + L + 2nH}{2} \right] \times H$$

$$= (L + nH)\, H = \left(H + \frac{3}{4} \times H \right) H = \frac{7}{4} H^2 - 1.75 H^2$$

Discharge:
$$Q = AC\sqrt{i\,m}$$

$$10 = 1.75 H^2 \times 60 \times \sqrt{\frac{1}{1500}} \times \frac{H}{2}$$

$$10 = 105 H^2 \times \sqrt{\frac{1}{300}} H^{1/2}$$

$$10 = 1.917\ H^{5/2}$$

or
$$H^{5/2} = 5.216$$
or
$$H = (5.216)^{2/5} = 1.936 \text{ m}$$

From Eqs. (*i*), we get
$$L = H = 1.936 \text{ m}$$

∴ The most economical dimensions of the channel are $L = H = \mathbf{1.936}$ **m.**

SUMMARY

1. The flow of liquid with a free surface is called open channel flow. The upper surface of the liquid is subjected to the atmospheric pressure is known as free surface.

2. **Depth of flow: H.** The depth of flow H at any section is the vertical distance from the free surface of liquid to the bed of the channel at that section.

3. **Top breadth: B.** It is the breadth of the channel section at the free surface.

4. **Water area:** A. The water area is flow cross-sectional area perpendicular in the direction of flow.

5. **Wetted perimeter: P.** The wetted perimeter is the perimeter of the solid boundary in contact with the liquid.

6. **Hydraulic radius:** *m.* It is defined as the ratio of the water area to the wetted perimeter. It is also called hydraulic mean depth.

Contd...

Mathematically,

Hydraulic radius: $m = \dfrac{A}{P}$

7. **Steady and unsteady flow:** If the velocity of the flow at any point does not change with time, such type of flow is called steady flow. If the velocity of flow changes with time, such type of flow is called unsteady flow.

8. **Uniform and non-uniform flow:** If the velocity of flow does not change in magnitude and direction from one section to another in the part of the channel along the length, is called uniform flow. If the velocity of flow changes from one section to another in the part of the channel along the length, is called non-uniform flow.

9. **Gradually varied flow:** It is the type of non-uniform flow in which the depth of flow changes gradually over a length of the channel.

10. **Rapidly varied flow:** It is the type of non-uniform flow in which the depth of flow changes rapidly over a small length of the channel.

11. **Laminar and turbulent flow:** The flow in open channel is said to be laminar or turbulent depending upon the value of Reynolds number.

$$\text{Re} = \frac{\rho V y_m}{\mu}$$

where ρ = density of the liquid
  μ = viscosity of the liquid
  V = average velocity of flow
  y_m = hydraulic mean depth

   (*i*) If $\text{Re} < 500$, the flow in open channel is called laminar flow.

  (*ii*) If $\text{Re} > 2000$, the flow in open channel is called turbulent flow

 (*iii*) If $500 \le \text{Re} \le 2000$, the flow in open channel is called transition flow.

12. **Sub-critical, critical and super critical flow:**

Froude's number: $F_r = \dfrac{V}{\sqrt{gy_m}}$

   (*i*) If $F_r < 1$, the flow is said to be sub-critical or tranquil.

  (*ii*) If $F_r = 1$, the flow is called critical flow.

 (*iii*) If $F_r > 1$, the flow is called super-critical.

13. **Chezy's formula:**

Average velocity: $V = C\sqrt{im}$
Discharge: $Q = AV$
where A = area of flow
                                        C = Chezy's constant.

Contd...

It the has the dimension $L^{1/2}T^{-1}$

i = slope of the bed of the channel

$$m = \frac{A}{P}, \text{ hydraulic mean depth}$$

14. Empirical Relations for Determination of Chezy constant:

(i) $$C = \frac{m^{1/6}}{N}$$

Manning's formula

where m = hydraulic mean depth

N = Manning's constant

(ii) $$C = \frac{157.6}{1.81 + \dfrac{K}{\sqrt{m}}}$$

Bazin formula

where m = hydraulic mean depth

K = Bazin's constant

15. Most Economical Section: A section of the channel is said to most economical section when we get maximum discharge with minimum cost of construction of the channel. The cost of construction depends upon the excavation and the lining material.

16. Condition for Most Economical Section:

(a) Rectangular section;

    (i) $P = 4H$ *i.e.*, perimeter is four times the depth of flow

    (ii) $m = \dfrac{H}{2}$ *i.e.*, hydraulic mean depth is half the depth of flow.

    (iii) $L = 2H$ *i.e.*, width of the channel is two times the depth of flow.

(b) Trapezoidal channel:

    (i) $\dfrac{L + 2nH}{2} = H\sqrt{n^2 + 1}$, *i.e* half width of the free surface is equal to the sloping side.

    (ii) $m = \dfrac{H}{2}$, *i.e* hydraulic mean depth is equal to half the depth of flow.

    (iii) A semi-circle drawn from the centre O of free surface with radius equal to the depth of flow will touch the sides of the channel.

(iv) Best side slope with horizontal = 60° *i.e.*, $\theta = 60°$

Contd...

ASSIGNMENT - 1

1. Define an open channel. How does it differ from a pipe running full ?
2. Why is it necessary to provide longitudinal slope to the bed of channel ?
3. Differentiate between:
    (*i*) Steady and unsteady flow
    (*ii*) Uniform and non-uniform flow.
    (*iii*) Laminar and turbulent flow in open channel.
4. What do you understand by critical, sub-critical and super-critical flow in open channel ?
5. Explain the terms:
    (*i*) Water area
    (*ii*) Wetted perimeter
    (*iii*) Hydraulic mean depth.
6. Explain the terms:
    (*i*) Rapidly varied flow, and
    (*ii*) Gradually varied flow.
7. Describe the Chezy's formula for flow through open channel.
8. How the value of Chezy's constant is determined by different empirical relations ?
9. What is the relation between Manning's constant and Chezy's constant ?
10. What is the dimensions of Chezy's constant ?
11. What is meant by an economical section of a channel ?
12. Derive the conditions for the economical conditions for:
    (*i*) Rectangular Channel
    (*ii*) Trapezoidal Channel
13. Derive the condition for the best side slope of the most economical trapezoidal channel.

ASSIGNMENT - 2

1. A rectangular channel is 4 m wide. The slope of its bed is 1 in 1000. Determine the discharge when depth of water 2 m. Take $C = 60$.   **[Ans. 15.17m^3/s**
2. Determine the slope of the bed of a rectangular channel of width 3 m when depth of water is 2 m and rate of flow is given as 12 m^3/s. Take Chezy's constant as 51.   **[Ans. 1 in 557.33]**
3. Determine the discharge through a trapezoidal channel of width 5 m and side slope of 2 horizontal to 3 vertical. The depth of flow of water is 3 m and Chezy's constant, $C = 60$. The slope of the bed of the channel is given 1 to 4000.   **[Ans. 26.12 m^3/s]**

4. Determine the bed slope of trapezoidal channel of bed width 6 m, depth of water is 3m and side slope of 3 horizontal to 4 vertical, when the discharge through the channel is 30 m³/s.Take C = 70.   **[Ans. 1 in 6133]**

5. Determine the discharge through a trapezoidal channel of width 8 m and side slope of 1 horizontal to 3 vertical. The depth of flow of water is 2.4 m and bed slope as 1 in 4000. Take the Chezy's constant as 50.**[Ans. 21.23 m³/s]**

6. A rectangular channel is 4 m wide has a bed slope of 1 in 1500. Determine the discharge the channel. Take C = 50.   **[Ans. 10.32 m³/s]**

7. Determine the most economical dimensions of the rectangular channel when discharge flow through the channel is 600 litre/s and bed slope of 1 in 1500. Take Chezy's constant as 54.   **[Ans. H = 0.6213, L = 1.24 m]**

8. A trapezoidal channel has side slopes of 1 horizontal to 2 vertical and the slope of the bed is 1 in 1500. The area of the section is 40 m². Determine the most economical dimensions of the section. Determine also the discharge of the most economical section. Take C = 50.

   [Ans. H = 4.8 m, L = 5.93 m, Q = 80 m³/s]

9. A trapezoidal channel has side slopes of 3 horizontal to 4 vertical and the slope of the bed is 1 in 2000. Determine the most economical dimension of the channel, if it is to carry water at 0.5 m³/s. Take Chezy's constant as 80.

   [Ans. H = 0.55 m, L = 0.55 m]

Notches and Weirs

9.1 INTRODUCTION

Notches and Weirs are used for measuring the rate of flow of liquid through open channel. Both are having same function *i.e.*, to measure the rate of flow. Difference between the two is their size and the quantity of discharge. **Notches** are small in size, hence used to measure small discharge through tank in laboratory. **Weirs** are big in size, thus measure large discharge in dams, rivers, *etc.*

A notch may be defined as an opening provided in the side of tank (or channel) in such a way that the liquid surface in the tank or channel is below the top edge of the opening. Normally, there is no need of upper edge of a notch. Liquid flows over a notch or weir while it passes through an orifice. The stream of liquid issuing from the orifice is called a jet while the stream of liquid issuing from a notch or weir is called a **nappe** or **vein**. The upper surface of the notch or weir over which the liquid flows is called the **crest** or **sill** of the weir.

9.2 DIFFERENCE BETWEEN NOTCH AND ORIFICE

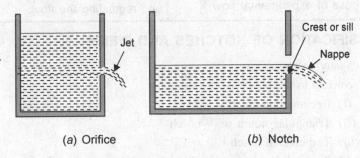

(a) Orifice   (b) Notch

Fig. 9.1 Orifice and notch.

© The Author(s) 2023
S. Kumar, *Fluid Mechanics (Vol. 2)*,
https://doi.org/10.1007/978-3-030-99754-0_9

S.No.	Orifice	Notch
1.	An orifice is an opening provided in the side or bottom of the tank or vessel in a such a way that the level of liquid in the tank is above the top edge of the opening as shown in Fig. 9.1 (a).	A notch is an opening provided in the side of tank in a such a way that the level of liquid in the tank is below the top edge of the opening as shown in Fig. 9.1 (b).
2.	It can be provided on the side or at the bottom of a tank.	It can only be provided on the side of the tank.
3.	The liquid flowing out of orifice is called a jet.	The liquid flowing out over a notch is called nappe or vein.
4.	An orifice is used to measure discharge through the tank or vessel which contains liquid.	A notch is used to measure discharge through small open channels.
5.	In an orifice, the free surface of liquid is always above the top edge of orifice.	In a notch, the free surface of flowing liquid is always below the top edge of the notch.

9.3 DIFFERENCE BETWEEN A NOTCH AND A WEIR

The following are main difference between a notch and a weir.

S.No.	Notch	Weir
1.	A notch is an opening made of metallic plate.	A weir is made of masonary or concrete.
2.	A notch is small in size, hence used to large discharge.	A weir is big in size, hence used to measure large discharge.
3.	The edges of a notch are thin and sharp	The edges of a weir are much wider as compared to notch.
4.	It is used in tanks or small open channels in laboratory for the purpose of experimental flow.	It is used in an ancient or spillway, canal or river for the purpose of storing and regulating the flow.

9.4 CLASSIFICATION OF NOTCHES AND WEIRS

Notches may be classified:
1. According to the shape of the opening:
   (i) Rectangular notch
   (ii) Triangular notch or V-notch
   (iii) Trapezoidal notch
   (iv) Stepped notch.
2. According to the effect of the sides on the nappe:
   (i) Notch with end contractions
   (ii) Notch without end contractions or suppressed notch.

Weirs may be classified:

1. According to the shape of the opening:
   - (*i*) Rectangular weir
   - (*ii*) Triangular weir
   - (*iii*) Trapezoidal weir.
2. According to the effect of the sides of the nappe.
   - (*i*) Weir with end contractions
   - (*ii*) Weir without end contractions or suppressed weir.
3. According to the shape of the crest:
   - (*i*) Sharp-crested weir
   - (*ii*) Broad-crested weir
   - (*iii*) Narrow-crested weir
   - (*iv*) Ogee-shaped weir

9.5 DISCHARGE OVER A RECTANGULAR NOTCH OR WEIR

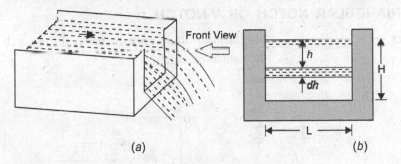

Fig. 9.2: Rectangular notch or weir.

Consider a rectangular weir from which the liquid is flowing as shown in Fig. 9.2.

Let L = length of the weir,

H = head of liquid over the crest of the weir.

Let us consider a horizontal strip of liquid of thickness dh at the depth h from the free surface of liquid as shown in Fig. 16.2 (*b*).

$\therefore$ Area of strip: $dA = Ldh$

and the theoretical velocity of liquid flowing through the elemental strip $= \sqrt{2gh}$

$\therefore$ Theoretical discharge through elemental strip:

$$dQ_{\text{th}} = \text{area of strip} \times \text{theoretical velocity}$$

$$= Ldh \times \sqrt{2gh} \qquad \qquad ...(9.5.1)$$

The total theoretical discharge, over the weir, may be determined by integrating the above Eq. (16.5.1) within the limits 0 and H, we get

$$Q_{\text{th}} = \int_{0}^{H} Ldh \times \sqrt{2gh}$$

$$= L\sqrt{2g}\int_0^H \sqrt{h}\,.dh = L\sqrt{2g}\left[\frac{h^{3/2}}{3/2}\right]_0^H$$

$$= \frac{2}{3}L\sqrt{2g}\left[H^{3/2} - 0\right]$$

$$Q_{th} = \frac{2}{3}\sqrt{2g}LH^{3/2} = \frac{2}{3}\sqrt{2g}LH^{1.5}$$

Actual discharge: $\qquad Q = C_d\,Q_{th}$

where $\quad C_d$ = coefficient of discharge.

$$Q = \frac{2}{3}C_d\sqrt{2g}LH^{1.5}$$

Note: The expression for discharge over a rectangular notch or weir is the same.

$$Q = \frac{2}{3}C_d\sqrt{2g}LH^{1.5}$$

9.6 TRIANGULAR NOTCH OR V-NOTCH

Let $\qquad\qquad\qquad$ H = head of liquid over the apex of the notch

$\qquad\qquad\qquad\quad$ θ = angle of notch

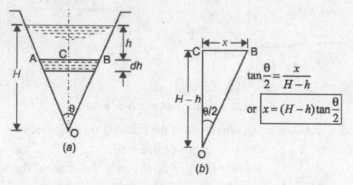

Fig. 9.3: Triangular notch.

Let us consider a horizontal elemental strip of liquid of thickness dh at the depth h from the free surface of liquid as shown in Fig. 9.3 (a).

Width of the elemental strip:

$$AB = AC + CB = 2\,CB \qquad\qquad \because AC = CB$$

$$= 2x = 2(H - h)\tan\frac{\theta}{2}$$

$\therefore$ Area of elemental strip:

$$dA = AB \times dh = 2(H - h)\tan\frac{\theta}{2} \times dh$$

$$= 2(H - h)\,dh\,\tan\frac{\theta}{2}$$

and the theoretical velocity of liquid flowing through the elemental strip $= \sqrt{2gh}$

$\therefore$ Theoretical discharge through elemental strip:

$$dQ_{th} = \text{area of strip} \times \text{theoretical velocity}$$

$$= 2(H - h)\, dh \, \tan\frac{\theta}{2} \times \sqrt{2gh}$$

$$= 2\sqrt{2g} \, \tan\frac{\theta}{2}(H - h)\sqrt{h}\, dh$$

$$= 2\sqrt{2g} \, \tan\frac{\theta}{2}(Hh^{1/2} - h^{3/2})\, dh \qquad \ldots(9.6.1)$$

The total theoretical discharge, over the notch, may be determined by integrating the above Eq. (9.6.1) with the limits 0 and H, we get

$$Q_{th} = \int_0^H 2\sqrt{2g} \, \tan\frac{\theta}{2}(Hh^{1/2} - h^{3/2})\, dh$$

$$= 2\sqrt{2g} \, \tan\frac{\theta}{2}\left[\frac{Hh^{3/2}}{3/2} - \frac{h^{5/2}}{5/2}\right]_0^H$$

$$= 2\sqrt{2g} \, \tan\frac{\theta}{2}\left[\frac{2}{3}H^{5/2} - \frac{2}{5}H^{5/2}\right]$$

$$= \frac{8}{15}\sqrt{2g} \, \tan\frac{\theta}{2}H^{5/2} = \frac{8}{15}\sqrt{2g} \, \tan\frac{\theta}{2}H^{2.5}$$

Actual discharge: $\qquad Q = C_d \, Q_{th}$

where $\qquad C_d = $ coefficient of discharge.

$$Q = \frac{8}{15}C_d\sqrt{2g}\tan\frac{\theta}{2}H^{2.5} \qquad \ldots(9.6.2)$$

The expression for discharge in above Eq. (9.6.2) over a triangular notch or weir is the same.

$$Q_\nabla = \frac{8}{15}C_d\sqrt{2g}\tan\frac{\theta}{2}H^{2.5}$$

Problem 9.1: Find the discharge over a rectangular notch of 4 m long when the head of liquid over the crest in 0.5 m. Take $C_d = 0.60$.

Solution: Given data for a rectangular notch:

Length of notch: $\qquad L = 4$ m

Head of liquid over the crest:

$$H = 0.5 \text{ m}$$
$$C_d = 0.60$$

We know that the discharge over a rectangular notch:

$$Q_\square = \frac{2}{3}C_d\sqrt{2g}LH^{1.5} = \frac{2}{3} \times 0.60 \times \sqrt{2 \times 9.81} \times 4 \times (0.5)^{1.5}$$

$$= \mathbf{2.50 \text{ m}^3/\text{s}}$$

Problem 9.2: The discharge over a rectangular notch is 120 litre/s when the water level is 300 mm above the sill. Find the length of a notch if the coefficient of discharge is 0.62.

Solution: Given data for a rectangular notch:

Discharge:
$$Q = 120 \text{ litre/s} = 0.12 \text{ m}^3/\text{s}$$
$$H = 300 \text{ mm} = 0.3 \text{ m}$$
$$L = ?$$
$$C_d = 0.62$$

We know that
$$Q = \frac{2}{3} C_d \sqrt{2g} L H^{1.5}$$

$$0.12 = \frac{2}{3} \times 0.60 \times \sqrt{2 \times 9.81} \times L \times (0.3)^{1.5}$$

or
$$L = 0.39888 \text{ m} = \textbf{398.88 mm.}$$

Problem 9.3: A rectangular weir of 5 m long is used to measure the rate of flow of water. The head of water over the weir is 800 mm. If the available height of waterfall is 22 m, find the power of the waterfall. Take $C_d = 0.60$.

Solution: Given data for rectangular weir:
$$L = 5 \text{ m}$$
$$H = 800 \text{ mm} = 0.8 \text{ m}$$

Available height of water fall:
$$H_1 = 22 \text{ m}$$
$$C_d = 0.60$$

Discharge:
$$Q = \frac{2}{3} C_d \sqrt{2g} L H^{1.5} = \frac{2}{3} \times 0.60 \times \sqrt{2 \times 9.81} \times 5 \times (0.8)^{1.5}$$
$$= 6.339 \text{ m}^3/\text{s}$$

Power of the waterfall:
$$P = \rho Q g H_1$$
$$= 1000 \times 6.339 \times 9.81 \times 22 \text{ W} = 1368082.9 \text{ W}$$
$$= \textbf{1.368 MW.}$$

Problem 9.4: The maximum flow through a rectangular channel 1.5 m deep and 2 m wide is 1.5 m³/s. It is proposed to install a full width, sharp-edged rectangular weir across the channel to measure the flow. Find the maximum height at which the crest of the weir must be placed in order that water may not overflow the sides of the channel. Take $C_d = 0.62$.

Solution: Given data for rectangular channel and rectangular weir:

Depth of flow in the channel:
$$Z = 1.5 \text{ m}$$

Width of the channel = length of weir

i.e., $L = 2$ m

Discharge through channel:

$$Q = 1.5 \text{ m}^3/\text{s}$$
$$C_d = 0.62$$

Discharge: $Q = \dfrac{2}{3} C_d \sqrt{2g} LH^{1.5}$

$$1.5 = \frac{2}{3} \times 0.62 \times \sqrt{2 \times 9.81} \times 2 \times H^{1.5}$$

or $H^{1.5} = 0.40964$

or $H = 0.5516$ m

Let h = height of the crest of the weir above the bottom of the channel.

∴ $Z = H + h$

$$1.5 = 0.5516 + h$$

or $h = \mathbf{0.9484}$ **m.**

Problem 9.5: The daily rainfall over a catchment area was found to 2.5×10^8 litre. It was observed that 25% of the rain water is lost due to evaporization and the remaining reaches the reservoir which passes over a rectangular weir: Find the length of the weir, if water over the weir will never rise more than 500 mm. Take coefficient of discharge as 0.62.

Solution: Given data:

2.5×10^8 litre rainfall in 24 hrs.

∴ Discharge: $Q_1 = \dfrac{2.5 \times 10^8}{24}$ litre/hr

$$= \frac{2.5}{24} \times \frac{10^8 \times 1}{1000} \text{ m}^3/\text{hr} = \frac{2.5}{24} \times 10^5 \text{ m}^3/\text{hr}$$

$$= \frac{2.5}{24 \times 3600} \times 10^5 \text{ m}^3/\text{s} = 2.89 \text{ m}^3/\text{s}$$

25% water is lost due to evaporization and the remaining discharge reaches the reservoir is 75%.

i.e., $Q = 0.75 \ Q_1 = 0.75 \times 2.89 \text{ m}^3/\text{s} = 2.167 \text{ m}^3/\text{s}.$

Head of water over the sill of weir:

$$H = 500 \text{ mm} = 0.5 \text{ m}$$
$$C_d = 0.62$$

Discharge: $Q = \dfrac{2}{3} C_d \sqrt{2g} \ LH^{1.5}$

$$2.89 = \frac{2}{3} \times 0.62 \times \sqrt{2 \times 9.81} \times L(0.5)^{1.5}$$

or $L = \mathbf{4.46}$ **m.**

Problem 9.6: A right-angled V-notch is used to measure the discharge of a pump. If the head of water over the sill is 300 mm, find the discharge over the notch in litre/s. Take $C_d = 0.61$.

Solution: Given data for V-notch:

Angle of notch: $\theta = 90°$ $\qquad\qquad\qquad$ ∵ Right-angled V-notch

$H = 300$ mm $= 0.3$ m

$C_d = 0.61$

We know that the discharge over the V-notch: Q

$$Q = \frac{8}{15} C_d \sqrt{2g} \tan\frac{\theta}{2} H^{2.5}$$

$$= \frac{8}{15} \times 0.61 \times \sqrt{2 \times 9.81} \times \tan\left(\frac{90°}{2}\right) \times (0.3)^{2.5}$$

$$= 0.01863 \text{ m}^3/\text{s} = \textbf{18.63 litre/s.}$$

Problem 9.7: During an experiment in a laboratory, 140 litres of water flowing over a right-angled V-notch was collected in 30 seconds. If the head of water over the sill is 100 mm, find the coefficient of discharge of the notch.

Solution: Given data for V-notch:

140 litres of water flowing over a V-notch in 30 seconds.

i.e., $\qquad$ Discharge: $\qquad Q = \dfrac{140}{30}$ litre/s $= 4.66$ litre/s $= 0.00466$ m^3/s

$\theta = 90°$

$H = 100$ mm $= 0.1$ m

We know that the discharge over the V-notch: Q

$$Q = \frac{8}{15} C_d \sqrt{2g} \tan\frac{\theta}{2} H^{2.5}$$

$$0.00466 = \frac{8}{15} \times 0.61 \times \sqrt{2 \times 9.81} \times \tan\left(\frac{90}{2}\right) \times (0.1)^{2.5}$$

$$0.00466 = 0.00747 \, C_d$$

or $\qquad\qquad C_d = \textbf{0.623}$

Problem 9.8: Water flows over a rectangular notch of 1m length over a head of water 200 mm. Then, the same discharge over a right-angled triangular notch. Find the height of water above the sill of the notch. Take C_d for the rectangular and triangular notches as 0.60 and 0.61 respectively.

Solution: Given data:

For rectangular notch	For right-angled triangular notch
$L = 1$ m	$\theta = 90°$
$H = 200$ mm $= 0.2$ m	$C_d = 0.61$
$C_d = 0.60$	

Discharge: $$Q_{\square} = \frac{2}{3} C_d \sqrt{2g} L H^{1.5}$$

$$= \frac{2}{3} \times 0.60 \times \sqrt{2 \times 9.84} \times 1 \times (0.2)^{1.5} = 0.1584 \text{ m}^3/\text{s}$$

Given condition:

$$Q_{\square} = Q_{\nabla} = 0.1584 \text{ m}^3/\text{s}$$

We know that

$$Q_{\nabla} = \frac{8}{15} C_d \sqrt{2g} \tan\frac{\theta}{2} H^{2.5}$$

$$0.1584 = \frac{8}{15} \times 0.61 \times \sqrt{2 \times 9.81} \times \tan\left(\frac{90}{2}\right) \times H^{2.5}$$

or $$H^{2.5} = 0.1099$$

or $$H = 0.41342 \text{ m} = \mathbf{413.42 \text{ mm}}.$$

Problem 9.9 Water flows through a triangular right-angled weir first and then over a rectangular weir of 1 m width. The discharge coefficients of the triangular and rectangular weirs are 0.6 and 0.7 respectively. If the depth of water over the triangular weir is 360 mm, find the depth of water over the rectangular weir.

Solution: Given data :

Right-angled triangular weir	**Rectangular weir width**
$\theta = 90°$	Width: $L = 1$ m
$C_d = 0.6$	$C_d = 0.7$
$H = 360$ mm	
$= 0.36$ m	

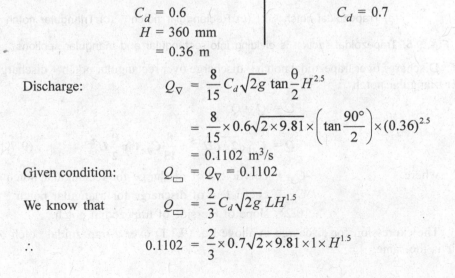

Discharge: $$Q_{\nabla} = \frac{8}{15} C_d \sqrt{2g} \tan\frac{\theta}{2} H^{2.5}$$

$$= \frac{8}{15} \times 0.6\sqrt{2 \times 9.81} \times \left(\tan\frac{90°}{2}\right) \times (0.36)^{2.5}$$

$$= 0.1102 \text{ m}^3/\text{s}$$

Given condition: $$Q_{\square} = Q_{\nabla} = 0.1102$$

We know that $$Q_{\square} = \frac{2}{3} C_d \sqrt{2g} L H^{1.5}$$

$\therefore$ $$0.1102 = \frac{2}{3} \times 0.7\sqrt{2 \times 9.81} \times 1 \times H^{1.5}$$

or $\qquad\qquad H^{1.5} = 0.053312$

or $\qquad\qquad H = 0.14167$ m = **141.67 mm**

9.7 DISCHARGE OVER A TRAPEZOIDAL NOTCH OR WEIR

A trapezoidal notch is a combination of triangular notch and rectangular notch. Therefore, discharge over a trapezoidal notch is sum of discharge over triangular notch and discharge over rectangular notch.

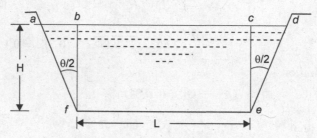

Fig. 9.4: Trapezoidal notch or weir.

Consider a trapezoidal notch $a\ b\ c\ d\ e\ f$ as shown in Fig. 9.4. This trapezoidal section is dividing into rectangular and triangular sections. The rectangular section $b\ c\ e\ f$ is a rectangular notch of length L and the height of liquid over notch H. Two triangular sections $a\ b\ f$ and $c\ d\ e$ are having angle of notch equal to $\theta/2$ and height of liquid is H. The discharge through two triangular sections is equal to the discharge through single triangular notch with angle of notch θ and same height of liquid H.

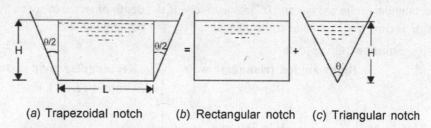

(a) Trapezoidal notch   (b) Rectangular notch   (c) Triangular notch

Fig. 9.5: Trapezoidal section is dividing into rectangular and triangular sections

Discharge over trapezoidal notch = discharge over rectangular notch + discharge over triangular notch

$$Q = Q_\square + Q_\nabla$$

$$Q = C_{d1}\sqrt{2g}\,LH^{1.5} + \frac{8}{15}C_{d2}\tan\frac{\theta}{2}H^{2.5} \qquad ...(9.7.1)$$

where $\qquad C_{d1}$ = coefficient of discharge for rectangular notch

$\qquad\qquad\quad C_{d2}$ = coefficient of discharge for triangular notch

$\qquad\qquad\quad \theta/2$ = slope of the side of trapezodial notch.

The expression for discharge in above Eq. (9.7.1) over a trapezoidal notch or weir is the same.

Problem 9.10: A trapezoidal weir of 3 m wide at the top and 2 m at the bottom is 800 mm high. Find the discharge over the weir, if the head of water is 500 mm. Take $C_d = 0.61$.

Solution: Given data for trapezoidal weir:

Width of weir at the top $= 3$ m

Width of weir at the bottom: $L = 2$ m

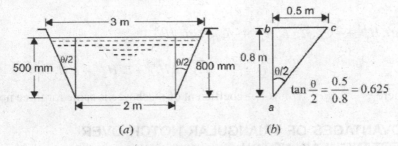

(a)   *(b)*

Fig. 9.6: Schematic for Problem 9.10

Height of weir $= 800$ mm $= 0.8$ m

Head of water: $H = 500$ mm $= 0.5$ m

$C_d = 0.61$

From triangular section on one side of the weir as shown in Fig. 9.6 *(b)*.

$$\tan\frac{\theta}{2} = \frac{0.5}{0.8} = 0.625$$

Discharge

$$Q_{\bigtriangledown} = Q_{\Box} + Q_{\triangledown}$$

$$= \frac{2}{3}C_d\sqrt{2g}LH^{1.5} + \frac{8}{15}C_d\sqrt{2g}\tan\frac{\theta}{2}H^{2.5}$$

$$= \frac{2}{3}\times0.61\times\sqrt{2\times9.81}\times2\times(0.5)^{1.5}$$

$$+\frac{8}{15}\times0.61\times\sqrt{2\times9.81}\times0.625\times(0.5)^{2.5}$$

$$= 1.273 + 0.159 = \mathbf{1.432\ m^3/s}$$

9.8 DISCHARGE OVER A STEPPED NOTCH

A stepped notch is a combination of rectangular notches. The discharge over a stepped notch is equal to the sum of the discharges over separate notches.

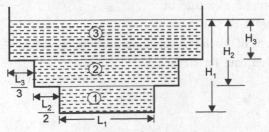

Fig. 9.7: Stepped notch.

Consider three steps notch as shown in Fig. 9.7.

Let H_1 = height of liquid in notch 1

L_2 = length of notch 1

Similarly H_2, L_2 and H_3, L_3 are corresponding values for notches 2 and 3 respectively.

∴ Total discharge $Q = Q_1 + Q_2 + Q_3$

=

$$\frac{2}{3}C_d\sqrt{2g}L_1H_1^{1.5} + \frac{2}{3}C_d\sqrt{2g}\,L_2\,H_2^{1.5} + \frac{2}{3}C_d\sqrt{2g}L_3\,H^{1.5}$$

$$Q = \frac{2}{3}C_d\sqrt{2g}\left[L_1H_1^{1.5} + L_2\,H_2^{1.5} + L_3H_3^{1.5}\right]$$

where C_d = coefficient of discharge is same for three notches.

9.9 ADVANTAGES OF TRIANGULAR NOTCH OVER RECTANGULAR NOTCH

Advantages of triangular notch over rectangular notch are listed below:

1. For low discharge, a triangular notch gives more accurate discharge than a rectangular notch. This is because, a triangular notch provides a greater head than the rectangular notch for same low discharge. Hence, head measurement can be done more accurately over the triangular notch than over the rectangular notch.

2. The coefficient of discharge for a triangular notch is independent of the head (*i.e.*, C_d = constant, for wide range of liquid head over triangular notch). Whereas in a rectangular notch, the coefficient of discharge is not constant (*i.e.*, $C_d = f(H)$).

3. No need of ventilation for the nappe of a triangular notch. But in a rectangular notch is necessary.

4. For a right angle V-notch, the expression for discharge becomes very simple:

    i.e., $\theta = 90°$, $C_d = 0.6$, $g = 9.81$ m/s^2

    then, $Q = 1.417\ H^{2.5}$.

Problem 9.11:   Find the discharge over a stepped weir of the following dimensions:

Top section: 2 m × 0.5 m

Middle section: 1.5 m × 0.25 m

Bottom section: 1 m × 0.15 m

Take coefficient of discharge for three sections as 0.62.

Solution: Given data for stepped weir:

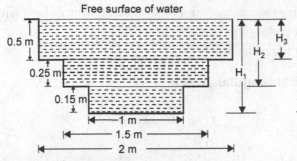

Fig. 9.8: Schematic for Problem 9.11

For bottom section: $L_1 = 1$ m
                       $H_1 = 0.5 + 0.25 + 0.15 = 0.9$ m
For middle section: $L_2 = 1.5 - L_1 = 1.5 - 1 = 0.5$ m
                       $H_2 = 0.5 + 0.25 = 0.75$ m
For top section: $L_3 = 2 - 1.5 = 0.5$ m
                       $H_3 = 0.5$ m

We know that the discharge through stepped weir:

$$Q = \frac{2}{3} C_d \sqrt{2g} \left[L_1 H_1^{1.5} + L_2 H_2^{1.5} + L_3 H_3^{1.5} \right]$$

$$= \frac{2}{3} \times 0.62 \times \sqrt{2 \times 9.81}$$

$$\left[1 \times (0.9)^{1.5} + 0.5 \times (0.75)^{1.5} + 0.5 \times (0.5)^{1.5} \right]$$

$$= 1.83 \times [0.8538 + 0.3247 + 0.1767]$$

$$= 1.83 \times 1.355 = \mathbf{2.479 \ m^3/s.}$$

9.10   EFFECT ON THE DISCHARGE OVER A NOTCH DUE TO AN ERROR IN THE MEASUREMENT OF HEAD

We know that the discharges over a rectangular and triangular notches are:

$$Q = \frac{2}{3} C_d \sqrt{2g} L H^{1.5} \qquad \text{for rectangular notch}$$

$$= \frac{8}{15} C_d \sqrt{2g} \tan\frac{\theta}{2} H^{2.5} \qquad \text{for triangular notch}$$

$Q \propto H^{1.5}$ for rectangular notch
$Q \propto H^{2.5}$ for triangular notch.

Thus, the accurate measurement of the head (H) of the water above the sill of the notch is very essential to know the accurate discharge over the notch. A small error in the measurement of head (H), will affect in the calculation of the discharge over the notch, and is generally expressed as the percentage of the discharge.

The following two cases of error in the measurement of head will be considered:
1. Over a rectangular notch,
2. Over a triangular notch.

9.10.1 For a Rectangular Notch

$$Q = \frac{2}{3}C_d\sqrt{2g}LH^{1.5}$$
$$Q = K_1\, H^{1.5} \qquad\qquad ...(9.10.1)$$

where $$K_1 = \frac{2}{3}C_d\sqrt{2g}L$$

Differentiating above Eq. (12.10.1) with respect to H, we get
$$dQ = K_1 \times 1.5\, H^{0.5}\, dH \qquad\qquad ...(9.10.2)$$
Dividing Eq. (12.10.2) by Eq. (12.10.1), we get
$$\frac{dQ}{Q} = \frac{K_1 \times 1.5 H^{0.5} dH}{K_1\, H^{1.5}}$$

$$\frac{dQ}{Q} = 1.5\frac{dH}{H} \qquad\qquad ...(9.10.3)$$

Equation (9.10.3) shows that an error of 1% in measuring head (H) will produce 1.5% error in discharge over a rectangular notch (same as in case of weir).

9.10.2 For a Triangular Notch

$$Q = \frac{8}{15}C_d\sqrt{2g}\,\tan\frac{\theta}{2}H^{2.5}$$
$$Q = K_2\, H^{2.5} \qquad\qquad ...(9.10.4)$$

where $$K_2 = \frac{8}{15}C_d\sqrt{2g}\,\tan\frac{\theta}{2}$$

Differentiating above Eq. (9.10.4) with respect to H, we get
$$dQ = K_2 \times 2.5\, H^{1.5}\, dH \qquad\qquad ...(9.10.5)$$
Dividing Eq. (9.10.5) by Eq. (9.10.4), we get
$$\frac{dQ}{Q} = \frac{K_2 \times 2.5\, H^{1.5} dH}{K_2\, H^{2.5}}$$

$$\boxed{\frac{dQ}{Q} = 2.5\frac{dH}{H}} \qquad\qquad ...(9.10.6)$$

Equation (9.10.6) shows that an error of 1% in measuring head (H) will produce 2.5% error in discharge over a triangular notch (same as in case of weir).

Problem 9.12: A rectangular weir of 500 mm long is used for measuring a distance of 140 litre/s. An error of 2.5 was made, while measuring the head over the weir. Find the percentage error in the discharge. Take $C_d = 0.61$.

Solution: Given data for rectangular weir:

Length of weir: $L = 500$ mm $= 0.5$ m

Discharge: $Q = 140$ litre/s $= \dfrac{140}{1000}$ m³/s $= 0.14$ m³/s

Error in head: $dH = 2.5$ mm $= 0.0025$ mm

$C_d = 0.61$

Discharge over a rectangular weir:

$$Q = \frac{2}{3} C_d \sqrt{2g}\, LH^{1.5}$$

$$0.14 = \frac{2}{3} \times 0.61 \times \sqrt{2 \times 9.81} \times 0.5\, H^{1.5}$$

or $H^{1.5} = 0.15544$

or $H = 0.2891$ m

Percentage error in discharge:

$$\frac{dQ}{Q} = 1.5 \frac{dH}{H} \times 100 = 1.5 \times \frac{0.0025}{0.2891} \times 100 = \mathbf{1.29\%.}$$

Problem 9.13: A discharge of 60 litre/s was measured over a right-angled triangular notch. While measuring the head over the notch, an error of 1.5 mm was made. Find the percentage of error in the discharge, if coefficient of discharge is 0.62.

Solution: Given data for right-angled triangular notch:

$$Q = 60 \text{ litre/s} = 0.06 \text{ m}^3/\text{s}$$

$$dH = 1.5 \text{ mm} = 0.0015 \text{ m}$$

$$\theta = 90°$$

$$C_d = 0.62$$

Discharge: $Q = \dfrac{8}{15} C_d \sqrt{2g}\, \tan\dfrac{\theta}{2} H^{2.5}$

$$0.06 = \frac{8}{15} \times 0.62 \times \sqrt{2 \times 9.81} \times \tan\frac{90°}{2} \times H^{2.5}$$

or $H^{2.5} = 0.0496$

or $H = 0.3$ m

Error in discharge: $\dfrac{dQ}{Q} = 2.5 \dfrac{dH}{H} = 2.5 \times \dfrac{0.0015}{0.3} = 0.0125 = \mathbf{1.25\%}$

9.11 CIPOLLETTI WEIR

We know that a discharge over rectangular weir is given by

$$Q = \frac{2}{3} C_d \sqrt{2g}\, LH^{1.5}$$

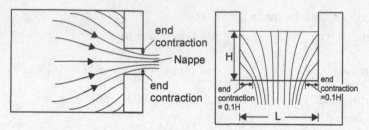

(a) No. of end contractions: $n = 2$ (b) Effective length = L – 0.2H
Fig. 9.9: Discharge over rectangular weir with ends contraction.

,   When the length (L) of the weir is less than the width of the channel in which liquid is flowing. Then, there will be end contraction at each end as shown in Fig. 9.9. According to Francis, each end contraction is equal to 0.1 H.

If the actual length of weir is L, then the effective length of the weir will be (L – 0.2H). Hence, the discharge is decreases due to formation of end contraction.

∴ Discharge through rectangular weir with end contraction:

$$Q = \frac{2}{3}C_d\sqrt{2g}(L-0.2H)H^{1.5}$$

$$= \frac{2}{3}C_d\sqrt{2g}LH^{1.5} - \frac{0.4}{3}C_d\sqrt{2g}\,H^{2.5}$$

$$Q = Q - Q_1$$

where $Q = \frac{2}{3}C_d\sqrt{2g}LH^{1.5}$,

discharge through rectangular weir without end contractions.

$$Q_1 = \frac{0.4}{3}C_d\sqrt{2g}\,H^{2.5},$$

reduction in discharge due to formation of end contraction.

This reduction in discharge Q_1 over rectangular weir is to be compensated by the addition discharge due to additional area at same base length L as shown in Fig. 9.10(b), is called cipolletti weir.

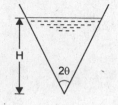 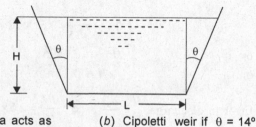

(a) Additional area acts as   (b) Cipoletti weir if θ = 14°
      triangular weir if   2θ = 28°

Fig. 9.10: Cipolletti weir

Deduction in discharge Q_1 over rectangular weir = gain in discharge by
                                                                                                adding area which acts
                                                                                                as triangular notch

$$Q_1 = Q_v$$

$$\frac{0.4}{3} C_d \sqrt{2g} H^{2.5} = \frac{8}{15} C_d \sqrt{2g} \tan\theta \, H^{2.5}$$

$$\frac{0.4}{3} = \frac{8}{15} \tan\theta$$

or $\qquad\qquad \tan\theta = 0.25$

or $\qquad\qquad \theta = \tan^{-1}(0.25) \approx \mathbf{14°}$

Cipolletti weir is a specific type of trapezoidal weir in which sloping side makes an angle of 14° (*i.e.*, $\theta = 14°$) with vertical on each side. In other words sloping sides have an inclination of 1 horizontal to 4 vertical. The cipolletti weir was invented by an Italian engineer Cipoletti.

The discharge over cipolletti weir is equal to discharge over rectangular weir without end contraction at same base length L.

$$Q = Q = \frac{2}{3} C_d \sqrt{2g} L H^{1.5}$$

For experimental analysis, Cipolletti proposed following formula for discharge over cipolletti weir.

$$Q = \mathbf{1.86 \ LH^{1.5}}.$$

9.12   FRANCIS'S FORMULA FOR RECTANGULAR WEIR WITH END CONTRACTIONS

Francis has given empirical formula for calculation of discharge

$$Q = 1.84 \, (L - 0.2 \, H) H^{1.5}$$

for single weir or number of contractions: $n = 2$

$$Q = 1.84 \, (L - 0.1 \, nH) H^{1.5}$$

where $\qquad\qquad n$ = number of contractions

$\qquad\qquad L$ = total length of weir

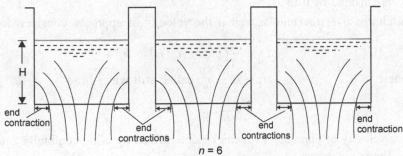

$$n = 6$$

Fig. 9.11: Three rectangular weir with six end contraction.

Francis's formula for rectangular weir without end contractions or for cipolleti weir:

$$Q = \mathbf{1.84 \ LH^{1.5}}$$

9.13 VELOCITY OF APPROACH

In the previous articles, the discharge equations are derived on the assumption that the water on the upstream side of the weir is not in motion. Therefore, the head of water considered in deriving the equations of discharge was taken as the height of the free surface of water above the sill of the weir. But in actual practice, the weir is provided across a river or a stream and the water approaching the weir which gets a certain velocity is called velocity of approach.

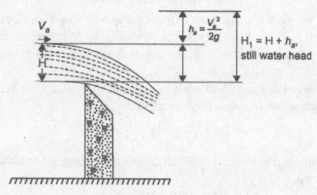

Fig. 9.12: Velocity of approach

Let V_a = velocity of approach

$$\frac{V_a^2}{2g} = h_a, \text{ head due to the velocity of approach.}$$

Because of velocity of approach, the available total head, upstream the weir, is not the height of the free surface above the sill, but is equal to height of the free surface above the sill plus head due to the velocity of approach.

i.e.,   Total head: $H_1 = H + h_a$.

Therefore, the limits of integration for the discharge over a rectangular weir will be h_a to H_1 instead of 0 to H.

Discharge over rectangular weir if the velocity of approach considered:

$$Q_{va} = \frac{2}{3} C_d \sqrt{2g} \, L \, (H_1^{1.5} - h_a^{1.5})$$

where $H_1 = H + h_a$, called still water head

$$h_a = \frac{V_a^2}{2g}$$

The velocity of approach (V_a) can be determined by using continuity equation.

Discharge: $Q = A \, V_a$

or   Velocity of approach:

$$V_a = \frac{Q}{A}$$

where Q = discharge over the weir is determined by neglecting the velocity of approach.

A = cross-sectional area of flow of water in the stream on the upstream side of the weir.

Discharge over rectangular weir with end contraction.

$$Q = \frac{2}{3} C_d \sqrt{2g}\ (L - 0.1 n H_1)\ (H_1^{1.5} - h_a^{1.5})$$

where n = number of end contractions

$H_1 = H + h_a$, total head over the sill of weir.

Francis's formula for rectangular weir with end contraction.

$$Q = 1.84\ (L - 0.1\ n\ H_1)\ (H_1^{1.5} - h_a^{1.5})$$

Problem 9.14: Water is flowing over a Cipolletti weir 5 m long under a head of 1.5 m. Find the discharge, if the coefficient of discharge for the weir is 0.60.

Solution: Given data for a Cipoletti weir:

Length: $L = 5$ m

Head: $H = 1.5$ m

$C_d = 0.60$

Discharge: $Q = \frac{2}{3} C_d \sqrt{2g}\ L H^{1.5} = \frac{2}{3} \times 6.0 \times \sqrt{2 \times 9.81} \times 5 \times (1.5)^{1.5}$

= **16.27 m³/s.**

Problem 9.15: Find the length of a Cipolletti weir required for a flow of 500 litre/s, if the head of water is not to exceed one-tenth of its length. Use Francis's formula for the weir.

Solution: Given data for Cipoletti weir:

Discharge: $Q = 500$ litre/s = 0.5 m³/s

$$H = \frac{1}{10} L$$

or $L = 10H$

Francis's formula for a Cipoletti weir:

$$Q = 1.84\ L\ H^{1.5} = 1.84 \times 10\ H \cdot H^{1.5}$$
$$0.5 = 1.84 \times 10 \times H^{2.5}$$

or $H^{2.5} = 0.02717$

$H = 0.2364$ m

and $L = 10\ H = 10 \times 0.2364 =$ **2.364 m.**

Problem 9.16: Find the discharge over a rectangular weir 20 m in length with a head of 2 m. Take the velocity of approach as 1.2 m/s and $C_d = 0.59$.

Solution: Given data for a rectangular weir:

$L = 20$ m

$H = 2$ m

$V_a = 1.2$ m/s

$$\therefore \qquad h_a = \frac{V_a^2}{2g} = \frac{(1.2)^2}{2 \times 9.81} = 0.073 \, \text{m}$$

$\therefore$ Total head: $H_1 = H + h_a = 2 + 0.073 = 2.073$ m.

$C_d = 0.59$

Discharge over the rectangular weir,

$$Q = \frac{2}{3} C_d \sqrt{2g} \; L \, [H_1^{1.5} - h_a^{1.5}]$$

$$= \frac{2}{3} \times 0.59 \times \sqrt{2 \times 9.81} \times 20 \times [(2.073)^{1.5} - (0.073)^{1.5}]$$

$$= 34.84 \times [2.984 - 0.019] = \textbf{103.30 m}^3\textbf{/s}$$

Problem 9.17:   A weir of 25 m long is divided into 10 equal bays by vertical posts each of 500 mm width. Find the discharge over the weir, if the head over the crest is 1.5 m and the velocity of approach is 2 m/s.

Solution: Given data for a rectangular weir:

Total length of the weir = 25 m

Number of bays = 10

Width of each post = 500 mm = 0.5 m

$H = 1.5$ m

$V_a = 2$ m/s

We know that the number of end contractions,

$n = 10 \times 2 = 20$ ($\because$ each bay contains two end contractions)

For 10 bays, number of vertical posts = 9

Length of the weir: $L = 25 - 9 \times 0.5 = 20.5$ m

Head due to velocity of approach,

$$h_a = \frac{V_a^2}{2g} = \frac{(2)^2}{2 \times 9.81} = 0.203 \, \text{m}$$

$\therefore$ Total head: $H_1 = H + h_a = 1.5 + 0.203 = 1.703$ m

$\therefore$ Discharge over the weir according to Francis's formula,

$$Q = 1.84 \, (L - 0.1 \; nH_1) \; [H_1^{1.5} - h_a^{1.5}]$$

$$= 1.84 \, (20.5 - 0.1 \times 20 \times 1.703)$$

$$[(1.703)^{1.5} - (0.203)^{1.5}]$$

$$= 31.45 \times [2.222 - 0.091] = \textbf{72.74 m}^3\textbf{/s.}$$

Problem 9.18:   A sharp crested rectangular weir of 1.4 m height extends across a rectangular channel of 4 m width. If the head of water over the weir is 0.5 m, find the discharge over the weir. Consider velocity of approach and take coefficient of discharge as 0.61.

Solution: Given data for rectangular weir:

Height of weir from base of channel:
$$h = 1.4 \text{ m}$$

Length of weir: $\qquad L = 4$ m

Head of water over the weir:
$$H = 0.5 \text{ m}$$

Height of water in the channel:
$$= h + H = 1.4 + 0.5 = 1.9 \text{ m}$$
$$C_d = 0.61$$

For calculation of the velocity of approach.

Discharge of rectangular weir without considering velocity of approach,

$$Q = \frac{2}{3} C_d \sqrt{2g} \; L H^{1.5} \; = \frac{2}{3} \times 0.61 \times \sqrt{2 \times 9.81} \times 4 \times (0.5)^{1.5}$$

$$= 2.547 \text{ m}^3/\text{s}.$$

Wetted cross-sectional area of the channel:

$$A = \text{width} \times \text{heights of water in the channel}$$
$$= 4 \times 1.9 = 7.6 \text{ m}^2$$

$\therefore$ Velocity of approach: $V_a = \dfrac{Q}{A} = \dfrac{2.547}{7.6} = 0.335$ m/s

and head due to velocity of approach,

$$h_a = \frac{V_a^2}{2g} = \frac{(0.335)^2}{2 \times 9.81} = 0.0057 \text{ m}$$

$\therefore$ Total head: $\qquad H_1 = H + h_a = 0.5 + 0.0057 = 0.5057$ m

Now the discharge over rectangular weir with velocity of approach,

$$Q_{va} = \frac{2}{3} C_d \sqrt{2g} \; L \; (H_1^{1.5} - h_a^{1.5})$$

$$= \frac{2}{3} \times 0.61 \times \sqrt{2 \times 9.81} \times 4 \times [(0.5057)^{1.5} - (0.0057)^{1.5}]$$

$$= \textbf{2.588 m}^3/\textbf{s}.$$

9.14 VENTILATION OF WEIRS

Some cases, a suppressed weir (*i.e.,* the length of weir is equal to the width of the channel) is provided in a channel to measure the discharge of water. Then, the nappe touches the side walls of the channel.

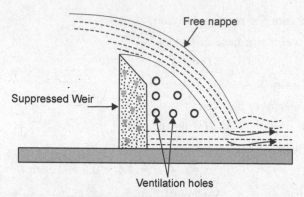

Fig. 9.13: Ventilation of weir

Because of this, a hollow space is maintained among the nappe, the weir, bottom and side walls of the channel in downstream of the weir as shown in Fig. 9.13. This space is occupied by air. The air, thus entrapped is slowly carried away by the flowing water, thus creating a vacuum pressure below nappe. This vacuum pressure below nappe draws the lower nappe towards the downstream surface of the weir. Because of the vacuum pressure (*i.e.,* pressure below the atmospheric pressure) below the nappe, the discharge over weir slightly increases.

In order to keep the atmospheric pressure in the space below the nappe, holes are provided in the side walls of the channel in the space below the lower nappe, so that the space is connected to free atmosphere and air is supplied continuously for the amount of air carried away by flowing water. The number of holes act like ventilators and are called ventilation holes, and the weir where ventilation holes are provided on the side walls of the channel in downstream of the weir, is called ventilated weir.

If the weir is well ventilated and the atmospheric pressure is maintained below the nappe, then it is called free nappe as shown in Fig. 9.13.

If a weir is not properly ventilated, the amount of air which is carried away by water is more than the quantity of air supplied through ventilation holes. Then, partial vacuum is created below the nappe. As a result, the nappe is gradually drawn towards the downstream surface of a weir. Such a nappe is called depressed nappe as shown in Fig. 9.14. It has been observed that the discharge over weir with depressed nappe is increased by 6 to 7% as compared with free nappe.

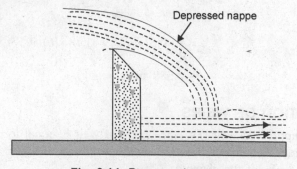

Fig. 9.14: Depressed nappe.

If there is no air left between the lower nappe and the down stream surface of the weir, then the lower nappe adheres the down stream surface of the weir. Such nappe which adheres to the down stream surface of the weir is called adhering or clinging nappe as shown in Fig. 9.15. The discharge over clinging nappe is 20 to 30% more than the free nappe.

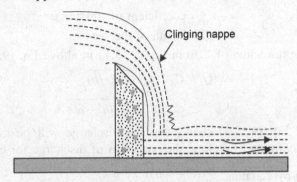

Fig. 9.15: Clinging nappe

9.15 DISCHARGE OVER A BROAD CRESTED WEIR

A weir with a wide crest is known as broad crested weir. The width of the crest B is greater than 0.5 H. This type of weir is shown in Fig. 9.16.

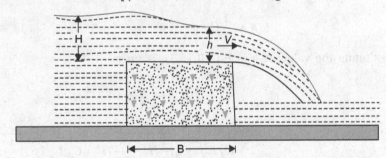

Fig. 9.16: Broad crested weir.

Let $\qquad$ H = head of water on the upstream side of the weir.

h = head of water on the downstream side of the weir.

V = velocity of water on the downstream side of the weir.

Applying the Bernoulli's equation to the crest of the weir on the upstream side and downstream side:

$$H + 0 + 0 = h + \frac{V^2}{2g} + 0$$

or

$$H = h + \frac{V^2}{2g}$$

or

$$\frac{V^2}{2g} = H - h$$

or $\qquad\qquad\qquad V = \sqrt{2g(H-h)}$ $\qquad\qquad\qquad$ ...(9.15.1)

$\therefore$ Discharge over the weir:

$$Q = C_d \times \text{area of flow} \times \text{velocity}$$
$$= C_d \times Lh \times V = C_d \, L \, hV \qquad\qquad ...(9.15.2)$$

where $\qquad\qquad\qquad C_d = \text{coefficient of discharge}$

$\qquad\qquad\qquad\qquad L = \text{length of the weir}$

Substituting the value of V from Eq. (9.15.1) in above Eq. (9.15.2) we get

$$Q = C_d \, L \, h\sqrt{2g(H-h)}$$

$$= C_d \, L \sqrt{2g} \sqrt{(Hh^2 - h^3)} \qquad\qquad ...(9.15.3)$$

The above Eq. (9.15.3) shows that the discharge will be maximum, when $(Hh^2 - h^3)$ is maximum. Therefore, the condition of discharge for constant head H may be obtained by equating $\dfrac{d}{dh}(Hh^2 - h^3)$ to zero.

$\therefore \qquad\qquad \dfrac{d}{dh}(Hh^2 - h^3) = 0$

$\qquad\qquad\qquad 2Hh - 3h^2 = 0$

$\qquad\qquad\qquad 2H - 3h = 0$

or $\qquad\qquad\qquad h = \dfrac{2}{3}H$

Substituting the value of h in Eq. (9.15.3), we get

$$Q_{\max} = C_d \, L \sqrt{2g} \; \sqrt{H\left(\frac{2}{3}H\right)^2 - \left(\frac{2}{3}H\right)^2}$$

$$= C_d \, L \sqrt{2g} \; \sqrt{\frac{4}{9}H^3 - \frac{8}{27}H^3} = C_d \, L \sqrt{2g} \; \sqrt{\frac{4}{27}H^3}$$

$$= C_d \, L \sqrt{2g} \times \frac{2}{3}\sqrt{\frac{H^3}{3}} = \frac{2}{3} C_d \, L \sqrt{2g}\,\frac{H^{3/2}}{\sqrt{3}}$$

$$= \frac{2}{3\sqrt{3}} C_d \, L \sqrt{2g} \, H^{1.5} = \frac{2}{3\sqrt{3}} \times \sqrt{2 \times 9.81} \; C_d \, L H^{1.5}$$

$$Q_{\max} = 1.71 \; C_d \, L \, H^{1.5}$$

9.16 DISCHARGE OVER A SUBMERGED WEIR

When the water level on the down stream side is above the crest of the weir, then the weir is said to be submerged or drowned weir. It is used for large discharge capacity. The discharge over submerged weir may be divided into two portions as discussed below:

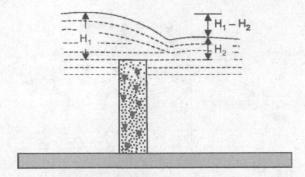

Fig. 9.17: Submerged weir.

Let H_1 = height of water on the upstream side of the weir,
and

H_2 = height of water on the downstream side of the weir.

The portion between upstream and downstream water surface may be treated as a freely discharge weir for the available head equal to $(H_1 - H_2)$.

∴ The portion of discharge over the freely discharge weir:

$$Q_1 - \frac{2}{3} C_d \sqrt{2g}\, L (H_1 - H_2)^{1.5}$$

The portion between downstream water surface and the crest of the weir may be considered as a submerged orifice.

∴ The portion of discharge through a submerged orifice.

$$Q_2 = C_d L H_2 \sqrt{2g(H_1 - H_2)}$$

where C_d = coefficient of discharge
L = length of the weir.

∴ Total discharge over submerged weir:

$$Q = Q_1 + Q_2$$

$$= \frac{2}{3} C_d \sqrt{2g}\, L (H_1 - H_2)^{1.5} + C_d L H_2 \sqrt{2g(H_1 - H_2)}$$

9.17 OGEE WEIR

In case of a sharp crested weir, the nappe as it leaves the crest, rises slightly at the lower surface. The space below the bottom surface of the nappe is filled with masonry or concrete. In this manner a new weir formed beside a sharp crested weir is called an ogee weir. Thus in an ogee weir, the solid boundary of the weir remains in contact with the bottom surface of the nappe of the sharp crested weir under designed head.

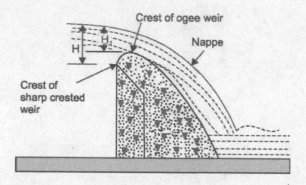

Fig. 9.18: An ogee weir.

Discharge over an ogee weir:

$$Q = \frac{2}{3} C_d \sqrt{2g} \, L H^{1.5} \qquad\qquad ...(9.17.1)$$

According to Francis's formula;

$$Q = 1.84 \, L \, H^{1.5} \qquad\qquad ...(9.17.2)$$

where H = head over the sharp crest weir.

Eqs. (9.17.1) and (9.17.2) same as discharge over rectangular weir.

If H_1 = head above the crest of ogee weir be considered,
                                       then the discharge is given by,

$$Q = 2.20 \, L H_1^{1.5}.$$

Problem 9.19: A submerged sharp crested weir 0.5 m high stands clear across a channel having vertical sides and a width of 2 m. The depth of water in the channel is 1.2 m. The depth of water is 0.8 m in downstream and 5 m from the weir. Find the discharge over the weir. Assume $C_d = 0.6$.

Solution: Given data:

Length of weir: $L = 2$ m

$C_d = 0.6$

$H_1 = 1.2 - 0.5 = 0.5$ m

$H_2 = 0.8 - 0.5 = 0.3$ m

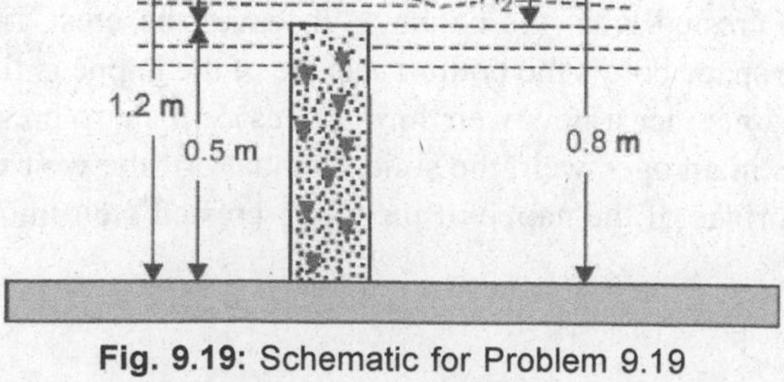

Fig. 9.19: Schematic for Problem 9.19

Discharge over the freely discharge weir:

$$Q_1 = \frac{2}{3} C_d \sqrt{2g} \, L (H_1 - H_2)^{1.5}$$

$$= \frac{2}{3} \times 0.6 \times \sqrt{2 \times 9.81} \times 2(0.5 - 03)^{1.5} = 0.3169 \ \text{m}^3/\text{s}$$

Discharge through a submerged orifice:

$$Q_2 = C_d \, L H_2 \sqrt{2g(H_1 - H_2)}$$

$$= 0.6 \times 2 \times 0.3 \times \sqrt{2 \times 9.81(0.5 - 03)} = 0.7131 \ \text{m}^3/\text{s}$$

∴ Total discharge over submerged weir:

$$Q = Q_1 + Q_2 = 0.3169 + 0.7131 = \textbf{1.03 m}^3/\textbf{s.}$$

Problem 9.20: An ogee weir 5 m long has 400 mm head of water. Find the discharge over the weir. Take $C_d = 0.61$.

Solution: Given data:

$$L = 5 \ \text{m}$$
$$H = 0.4 \ \text{m}$$
$$C_d = 0.61$$

We know that the discharge over an ogee weir:

$$Q = \frac{2}{3} C_d \sqrt{2g} \, L H^{1.5} = \frac{2}{3} \times 0.61 \times \sqrt{2 \times 9.81} \times 5 \times (0.4)^{1.5}$$

$$= \textbf{2.278 m}^3/\textbf{s}$$

SUMMARY

1. Notches and weirs are used for measuring the rate of flow of liquid through open channel. Both are having same function *i.e.*, to measure the rate of flow. Difference between the two is their size and the quantity of discharge. Notches are small in size, hence used to measure small discharge through tank in laboratory. Weirs are big in size, thus measure large discharge in dams, rivers, *etc.*

2. Classification of notches:
   (*a*) According to the shape of the opening:
      (*i*) Rectangular notch
      (*ii*) Triangular notch or V-notch
      (*iii*) Trapezoidal notch
      (*iv*) Steeped notch.

Contd...

(b) According to the effect of the sides on the nappe:

    (i) Notch with end contractions

    (ii) Notch without end contractions or suppressed weir.

3. Classification of weirs:

  (a) According to the shape of the opening:

    (i) Rectangular weir

    (ii) Triangular weir

    (iii) Trapezoidal weir.

  (b) According to the effect of the sides of the nappe.

    (i) Weir with end contractions

    (ii) Weir without end contractions or suppressed weir.

  (c) According to the shape of the crest:

    (i) Sharp-crested weir

    (ii) Broad-crested weir

    (iii) Narrow-crested weir

    (iv) Ogee-shaped weir

4. The discharge over a rectangular notch or weir:

$$Q = \frac{2}{3} C_d \sqrt{2g}\, L H^{1.5}$$

where C_d = coefficient of discharge.

      L = length of the notch or weir.

      H = head of liquid over the crest of the notch or weir.

5. The discharge over a triangular notch or weir.

$$Q = \frac{8}{15} C_d \sqrt{2g}\, \tan\frac{\theta}{2} H^{2.5}$$

where θ = angle of notch or weir

       H = head of liquid over the apex of the notch or weir.

6. The discharge over a trapezoidal notch or weir.

$$Q = \frac{2}{3} C_{d1} \sqrt{2g}\, L H^{1.5} + \frac{8}{15} C_{d2} \tan\frac{\theta}{2} H^{2.5}$$

where C_{d1} = coefficient of discharge for rectangular notch.

      C_{d2} = coefficient of discharge for triangular notch.

      $\dfrac{\theta}{2}$ = slope of the side of trapezoidal notch.

Contd...

7. The effect on the discharge over a notch due to an error in the measurement of head:

$$\frac{dQ}{Q} = 1.5\frac{dH}{H} \qquad \text{for a rectangular notch or weir}$$

$$= 2.5\frac{dH}{H} \qquad \text{for a triangular notch or weir.}$$

For a rectangular notch or weir, an error of 1% in measuring head (H) will produce 1.5% error in discharge.

For a triangular notch or weir, an error of 1% in measuring head (H) will produce 2.5% error in discharge.

8. Cipoletti weir is a specific type of trapezoidal weir in which sloping side makes an angle of 14° (*i.e.,* θ = 14°) with vertical on each side. In the other words sloping sides have an inclination of 1 horizontal to 4 vertical. The cipoletti weir was invented by an Italian engineer Cipoletti.

Discharge over a Cipoletti weir: $Q = \frac{2}{3}C_d\sqrt{2g}\,LH^{1.5}$

9. Francis's formula for rectangular weir with end contractions

$$Q = 1.84\,(L - 0.2H)\,H^{1.5} \quad \text{for } n = 2$$
$$= 1.84\,(L - 0.1\,nH)\,H^{1.5}$$

where n = number of contractions,
         L = total length of weir,

10. Francis's formula for rectangular weir without end contractions or for Cipolletti weir:

$$Q = 1.84\,LH^{1.5}$$

11. The discharge over rectangular weir if the velocity of approach considered:

$$Q_{av} = \frac{2}{3}C_d\sqrt{2g}\,L(H_1^{1.5} - h_a^{1.5})$$

where $H_1 = H + h_a$, called still water head.

$$h_a = \frac{V_a^2}{2g}$$

V_a = velocity of approach

ASSIGNMENT - 1

1. Define the following terms:
   (*i*) Notch
   (*ii*) Crest
   (*iii*) Nappe.
2. What is a weir? Differentiate between a notch and weir.
3. Differentiate between a notch and a orifice.
4. Derive an expression for the discharge over a rectangular weir.
5. Prove that the discharge over a triangular notch is given by:

$$Q = \frac{8}{15} C_d \sqrt{2g} \tan\frac{\theta}{2} H^{2.5}$$

   where H = head of water over the sill of a notch.
                θ = angle of notch.

6. Prove that the discharge over a rectangular weir is given by:

$$Q = \frac{2}{3} C_d \sqrt{2g} \, L H^{1.5}$$

   where C_d = coefficient of discharge
                L = length of weir
                H = head of water over the sill of a weir.

7. What are the advantages of triangular weir over rectangular weir?
8. Prove that the error in discharge due to the error in the measurement of head over a rectangular weir is given by

$$\frac{dQ}{Q} = 1.5\frac{dH}{H}$$

   where Q = discharge over a rectangular weir, and
                H = head over the crest of a weir.

9. Prove that the error in discharge due to the error in the measurement of head over a triangular weir is given by:

$$\frac{dQ}{Q} = 2.5\frac{dH}{H}$$

   where Q = discharge over a rectangular weir, and
                H = head over the crest of a weir.

10. What is a Cipolletti weir? How does it differ from the rectangular weirs?
11. What do you mean by end contractions of a rectangular weir? How can the loss of discharge due to end contractions be compensated?
12. Define the velocity of approach.
13. How does the velocity of approach affect the expression for discharge over a weir?

14. What are the ventilation holes?

15. What is a nappe of weir? Describe the free, depressed and clinging nappes with the help of sketches. State how do they effect the discharge measurement in case of weir.

16. Explain clearly the following weirs:

    (i) Broad crested   (ii) Submerged   (iii) Ogee.

ASSIGNMENT - 2

1. Find the discharge over a rectangular notch 5 m long when the head of liquid over the sill is 0.4 m. Take $C_d = 0.62$.   **Ans. 2.315 m³/s**

2. The discharge over a rectangular notch is 90 litre/s when the water level is 250 mm above the crest. Find the length of a notch if the coefficient of discharge is 0.61.   **Ans. 399.71 mm**

3. A rectangular weir of 4 m long is used to measure the rate of flow of water. The head of water over the weir is 500 mm. If the available height of waterfall is 12 m, find the power of the waterfall. Take $C_d = 0.62$. **Ans. 304.78 kW**

4. The maximum flow through a rectangular channel 1 m deep and 1.5 m wide is 0.9 m³/s. It is proposed to install a full width, sharp-edged rectangular weir across the channel to measure the flow. Find the maximum height at which the crest of the weir must be placed in order that the water may not overflow the sides of the channel. Take $C_d = 0.6$.   **Ans. 514.12 mm**

5. A right-angled V-notch is used to measure the discharge in a small channel. If the depth of water at V-notch is 200 mm, find the discharge over the notch in litre per second. Take $C_d = 0.62$.   **Ans. 26.20 litre/s**

6. Water flows over a rectangular notch of 1.2 m length over a head of water 300 mm. Then, the same discharge over a right-angled triangular notch. Find the height of water above the sill of the notch. Take C_d for the rectangular and triangular notches as 0.60 and 0.62 respectively.   **Ans. 563.65 mm**

7. A trapezoidal weir 4 m wide at the top and 3 m at the bottom is 1 m high. Find the discharge over the weir, if the head of water is 600 mm.

    Take $C_d = 0.60$.   **Ans. 2.667 m³/s**

8. Find the discharge over a stepped notch of the following dimensions:

    Top section: 1000 mm × 150 mm

    Middle section: 800 mm × 100 mm

    Bottom section: 600 mm × 800 mm

    Take coefficient of discharge for three sections as 0.62. **Ans. 275.28 litre/s**

9. A rectangular notch 400 mm long is used for measuring a discharge of 30 litre/s. An error of 1.5 mm was made, while measuring the head over the sill of the notch. Find the percentage error in the discharge. Take coefficient of discharge as 0.60.   **Ans. 1.85 %**

10. A right-angled V-notch is used for measuring a discharge of 30 litre/s. An error of 2 mm was made while measuring the head over the notch. Find the percentage error in the discharge.

    Take coefficient of discharge as 0.62.   **Ans.** 2.36%

11. Water is flowing over a Cipoletti weir 4 m long under a head of 1 m. Find the discharge, if the coefficient of discharge for the weir is 0.62. **Ans.** 7.32 m³/s

12. Find the length of a Cipoletti weir required for a flow of 500 litre/s, if the head of water does not exceed one-tenth of its length. Use Francis's formula for the weir.   **Ans.** 2.36 m

13. Water is flowing in a rectangular channel 1m wide and 0.75 m deep. Find the discharge over a rectangular weir of sill length 600 mm if the head of water over the sill of weir is 200 mm and water from channel flows over the weir. Take C_d = 0.62, neglecting end contraction, but considering the velocity of approach.   **Ans.** 98.81 litre/s

14. A rectangular weir is constructed across a channel of 770 mm width with a head of 390 mm and the sill 600 mm above the bed of the channel. Find the discharge over a rectangular weir neglecting the end construction and considering the velocity of approach. Assume C_d = 0.62. **Ans.** 0.3555 m³/s

15. A submerged sharp crested weir 800 mm high stands clear across a channel having vertical sides and a width of 3 m. The depth of water in the channel is 1.25 m. The width of water is 1 m in the downstream and 10 m from the weir. Find the discharge over the weir. Take C_d = 0.6.   **Ans.** 0.6644 m³/s

16. An ogee weir of 3 m long has 800 mm head of water. Find the discharge over the weir. Take C_d = 0.62.   **Ans.** 3.93 m³/s

10

Compressible Flow

10.1 INTRODUCTION

In the previous chapters, we considered the incompressible flow which means that the density of the fluid is unchanged in the flow. But in this chapter, we will study the behaviour of the compressible flow. In compressible flow, the density of the fluid changes from point to point in the fluid flow. Gases (like air, carbon dioxide *etc.*) are compressible fluid.

So in this chapter, we will discuss the equation of state, thermodynamics process, steady and unsteady flow, uniform and non-uniform flow, compressible and incompressible flow, rate of flow, continuity equation, steady flow energy equation, stagnation state, velocity of sound in compressible fluids, nozzle and diffuser, Rayleigh and Fanno flow.

10.2 EQUATION OF STATE

The equation involving temperature, pressure and volume (or specific volume) is used to describe the condition or state of the system is called an **equation of state**.

For a perfect (or ideal) gas, the equation of state is

$$pV = mRT \qquad \qquad ...\,(10.2.1)$$

where
p = absolute pressure of a perfect gas in kPa.
V = volume of a perfect gas in m^3
m = mass of a perfect gas in kg
R = gas constant in kJ/kgK
$\quad$ = 0.287 kJ/kgK for air
T = absolute temperature *i.e.*, temperature in K (kelvin)

$$\frac{pV}{m} = RT$$

$$pv = RT \qquad \qquad ...\,(10.2.2)$$

© The Author(s) 2023
S. Kumar, *Fluid Mechanics (Vol. 2)*,
https://doi.org/10.1007/978-3-030-99754-0_10

where $v = \dfrac{V}{m}$, specific volume in m³/kg

also $v = \dfrac{1}{\rho}$, specific volume is the reciprocal of the density and its SI unit is m³/kg

∴ $\dfrac{p}{\rho} = RT$

or $p = \rho RT$ ...(10.2.3)

Gas constant : R. It is defined as the ratio of the universal gas constant $(\bar{R})$ to the molecular weight of a perfect gas.

Mathematically,

$$\text{Gas constant: } R = \dfrac{\text{universal gas constant} : \bar{R}}{\text{molecular weight} : M}$$

$$R = \dfrac{\bar{R}}{M}$$

where $\bar{R} = 8.314$ kJ/k mol K

Substituting $R = \dfrac{\bar{R}}{M}$ in Eq. (16.2.1), we get

$$pV = m\dfrac{\bar{R}}{M}T$$

$$pV = n\bar{R}T$$ ...(10.2.4)

S. No.	Ideal Gas	Molecular Weight : M kg/k mol	Universal Gas Constant : $\bar{R}$ kJ/k mol K	Gas Constant : R kJ/kg K
1.	Air	28.92	8.314	$\dfrac{8.314}{28.92} = 0.287$
2.	CO_2	44	8.314	$\dfrac{8.314}{44} = 0.189$
3.	O_2	32	8.314	$\dfrac{8.314}{32} = 0.259$
4.	N_2	28	8.314	$\dfrac{8.314}{28} = 0.297$
5.	$H_2O(v)$ (superheated steam)	18	8.314	$\dfrac{8.314}{18} = 0.462$

where $n = \dfrac{m}{M}$, number of moles. It is defined as the ratio of the mass of a gas to the molecular weight of a gas. Eqs. (10.2.1), (10.2.2), (10.2.3) and (10.2.4) are all form of the equation of state. These equations are applicable at any state of system for an ideal gas. For example: If we know V, m, R and T, then p find out by using Eq. (10.2.1.).

10.3 THERMODYNAMIC PROCESSES

Thermodynamic processes are used to study the thermodynamics behaviour of a series of changes in state of the system. Following are the main thermodynamic processes.

(*i*) Isothermal process
(*ii*) Isobaric process
(*iii*) Isochoric process
(*iv*) Adiabatic process
(*v*) Polytropic process

10.3.1 Isothermal Process [$T = c$]

Isothermal process takes place at constant temperature. The volume (or specific volume) of the gas varies inversely with pressure at constant temperature in this type of process.

From the equation of state

$$pv = RT$$

For any process, R is constant and T is constant for isothermal process.

∴ $$pv = c$$

or $$\dfrac{pV}{m} = c \qquad\qquad | \because v = V/m$$

or $$pV = c$$

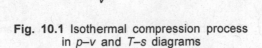

Fig. 10.1 Isothermal compression process
in *p–v* and *T–s* diagrams

Hence, the product of absolute pressure and volume (or specific volume) is constant.

For isothermal process 1 – 2

$$p_1 v_1 = p_2 v_2$$

or
$$\frac{p_2}{p_1} = \frac{v_1}{v_2}$$

Isothermal process follows the Boyle's law which state that the specific volume of a perfect gas is inversely proportional to the absolute pressure when the temperature is kept constant.

Mathematically,

$$v \propto \frac{1}{p} \quad \text{at} \quad T = c$$

or
$$pv = c$$

10.3.2 Isobaric Process [$p = c$]

Isobaric process takes place at constant pressure. The volume (or specific volume) of the gas varies directly with temperature at constant pressure in this type of process.

From the equation of state

$$pv = RT$$

or
$$\frac{v}{T} = \frac{R}{p}$$

$$\frac{v}{T} = c$$

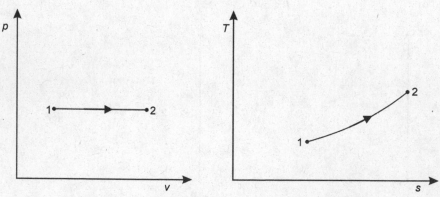

Fig. 10.2 Isobaric process in p–v and T–s diagrams.

For isothermal process 1 – 2

$$\frac{v_1}{T_1} = \frac{v_2}{T_2}$$

or

$$\frac{T_2}{T_1} = \frac{v_2}{v_1}$$

Isobaric process follows the Charles' law which state that the specific volume of a perfect gas is directly proportional to the absolute temperature when the pressure is kept constant.

Mathematically,

$$v \propto T \quad \text{at} \quad p = c$$

or

$$\frac{v}{T} = c$$

10.3.3 Isochoric Process (or Isometric Process) [$V = c$]

Isochoric process takes place at constant volume. The pressure of the gas varies directly with temperature at constant volume in this type of process.

From the equation of state

$$pv = RT$$

$$\frac{p}{T} = \frac{R}{v}$$

$$\frac{p}{T} = c$$

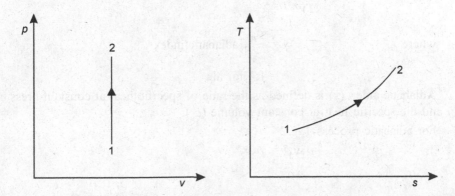

Fig. 10.3 Isochoric process in *p–v* and *T–s* diagrams.

For isochoric process 1 – 2

$$\frac{p_1}{T_1} = \frac{p_2}{T_2}$$

or

$$\frac{T_2}{T_1} = \frac{p_2}{p_1}$$

Isochoric process follows the Gay-Lussac's law which state that the absolute pressure of a perfect gas is directly proportional to the absolute temperature when the volume is kept constant.

Mathematically,

$$p \propto T \quad \text{at} \quad v = c$$

or

$$\frac{p}{T} = C$$

10.3.4 Adiabatic Process [$pv^\gamma = c$]

When there is no heat transfer between the system and surroundings during a process, it is known as an adiabatic process. The properties like a pressure, temperature and volume of the system will vary during this process. Normally, this process is considered as ideal process and also called reversible adiabatic (or adiabatic isentropic). In this process, the entropy of the system will remain constant without transfer of heat to or from the surroundings.

This process can be represented by the equation

$$pv^\gamma = c$$

$$\frac{RT}{v} v^\gamma = c'$$

$$Tv^{\gamma-1} = c$$

$$pv = RT$$

or $$p = \frac{RT}{v}$$

where

$$\gamma = \frac{c_p}{c_v}, \text{ adiabatic index}$$

$$= 1.4 \text{ for air}$$

Adiabatic index (γ) is defined as the ratio of specific heat at constant pressure (c_p) and the specific heat at constant volume (c_v).

For adiabatic process 1 – 2

$$p_1 v_1^\gamma = p_2 v_2^\gamma$$

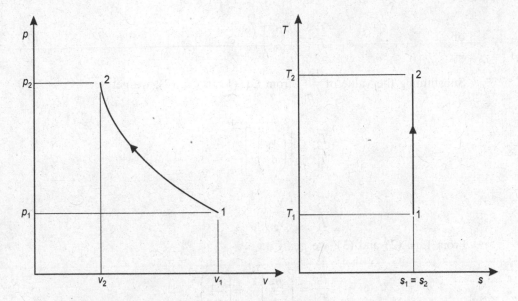

Fig. 10.4 Adiabatic process (*i.e.*, adiabatic compression) in *p–v* and *T–s* diagrams

$$\frac{p_2}{p_1} = \left(\frac{v_1}{v_2}\right)^{\gamma} = \left(\frac{v_2}{v_1}\right)^{-\gamma} \qquad \qquad ...(1)$$

From the equation of state

$$pv = RT$$

$$\frac{pv}{T} = R = c$$

For process 1 – 2

$$\frac{p_1 v_1}{T_1} = \frac{p_2 v_2}{T_2}$$

or

$$\frac{v_1}{v_2} = \frac{p_2}{p_1} \times \frac{T_1}{T_2}$$

Substituting the value of $\dfrac{v_1}{v_2}$ in Eq. (1), we get

$$\frac{p_2}{p_1} = \left(\frac{p_2}{p_1} \times \frac{T_1}{T_2}\right)^{\gamma} = \left(\frac{p_2}{p_1}\right)^{\gamma} \times \left(\frac{T_1}{T_2}\right)^{\gamma}$$

or

$$\left(\frac{T_2}{T_1}\right)^{\gamma} = \left(\frac{p_2}{p_1}\right)^{\gamma-1}$$

or
$$\frac{T_2}{T_1} = \left(\frac{p_2}{p_1}\right)^{\frac{\gamma-1}{\gamma}} \qquad \qquad \dots(2)$$

Substituting the value of $\dfrac{p_2}{p_1}$ from Eq. (1) in Eq. (2), we get

$$\frac{T_2}{T_1} = \left[\left(\frac{v_1}{v_2}\right)^{\gamma}\right]^{\frac{\gamma-1}{\gamma}}$$

$$\frac{T_2}{T_1} = \left(\frac{v_1}{v_2}\right)^{\gamma-1} \qquad \qquad \dots(3)$$

From Eqs. (2) and (3), we get

$$\frac{T_2}{T_1} = \left(\frac{p_2}{p_1}\right)^{\frac{\gamma-1}{\gamma}} = \left(\frac{v_1}{v_2}\right)^{\gamma-1}$$

also
$$\frac{T_2}{T_1} = \left(\frac{p_2}{p_1}\right)^{\frac{\gamma-1}{\gamma}} = \left(\frac{v_1}{v_2}\right)^{\gamma-1} = \left(\frac{\rho_2}{\rho_1}\right)^{\gamma-1}$$

From adiabatic process:

(a) Relation between p and T

$$\frac{T_2}{T_1} = \left(\frac{p_2}{p_1}\right)^{\frac{\gamma-1}{\gamma}}$$

(b) Relation between p and v

$$\left(\frac{p_2}{p_1}\right)^{\frac{\gamma-1}{\gamma}} = \left(\frac{v_1}{v_2}\right)^{\gamma-1}$$

or
$$\frac{p_2}{p_1} = \left(\frac{v_1}{v_2}\right)^{\gamma}$$

(c) Relation between T and v

$$\frac{T_2}{T_1} = \left(\frac{v_1}{v_2}\right)^{\gamma-1} = \left(\frac{v_2}{v_1}\right)^{1-\gamma}$$

(d) Relation between T and ρ

$$\frac{T_2}{T_1} = \left(\frac{\rho_2}{\rho_1}\right)^{\gamma-1}$$

10.3.5 Polytropic Process [$pv^n = c$]

Polytropic process is a real process followed by practically compression and expansion processes. For examples : Actual expansion of gases in the nozzles, turbines, IC engines and actual compression in the compressors and *IC* engines are followed the polytropic process. This process can be represented by the equation.

$$pv^n = c$$

where n = polytropic index

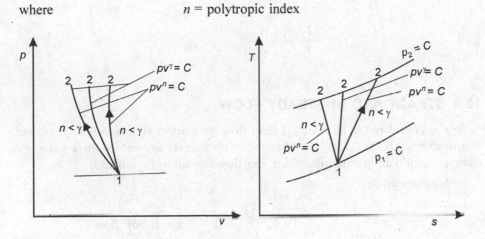

Fig. 10.5 Polytropic process in *p–v* and *T–s* diagrams.

Characteristics of polytropic process:

(*a*) Entropy of the process changes.

(*b*) Both heat and work transfer take place.

The relations between *T, p* and *v* are obtained by replacing *n* instead of γ in relations for adiabatic process:

$$\frac{T_2}{T_1} = \left(\frac{p_2}{p_1}\right)^{\frac{n-1}{n}} = \left(\frac{v_1}{v_2}\right)^{n-1}$$

For polytropic process:

(*a*) Relation between *p* and *T*

$$\frac{T_2}{T_1} = \left(\frac{p_2}{p_1}\right)^{\frac{n-1}{n}}$$

(*b*) Relation between *p* and *v*

$$\frac{p_2}{p_1} = \left(\frac{v_1}{v_2}\right)^{n}$$

(c) Relation between T and v

$$\frac{T_2}{T_1} = \left(\frac{v_1}{v_2}\right)^{n-1} = \left(\frac{v_2}{v_1}\right)^{1-n}$$

(d) Relation between T and ρ

$$\frac{T_2}{T_1} = \left(\frac{\rho_2}{\rho_1}\right)^{n-1}$$

10.4 STEADY AND UNSTEADY FLOW

A flow is considered to be steady if fluid flow parameters such as velocity, pressure, temperature *etc.* at any point do not change with time. If any one of these parameters changes with time at particular point, the flow is said to be unsteady.

Mathematically,

$$\frac{\partial V}{\partial t} = 0, \ \frac{\partial p}{\partial t} = 0 \qquad \text{for steady flow}$$

$$\frac{\partial V}{\partial t} \neq 0, \ \frac{\partial p}{\partial t} \neq 0 \qquad \text{for unsteady flow}$$

10.5 UNIFORM AND NON-UNIFORM FLOW

A type of flow in which velocity (V), pressure (p), density (ρ), temperature (T) etc. at any given time do not change with respect to space (*i.e.,* length of direction of the flow) is called uniform flow.

Mathematically,

$$\frac{\partial V}{\partial s} = 0, \ \frac{\partial p}{\partial s} = 0 \qquad \text{for uniform flow}$$

In case of non-uniform flow, velocity, pressure, density *etc.* at given time change with respect to space.

Mathematically,

$$\frac{\partial V}{\partial s} \neq 0, \ \frac{\partial p}{\partial s} \neq 0 \qquad \text{for non-uniform flow}$$

10.6 COMPRESSIBLE AND INCOMPRESSIBLE FLOW

10.6.1 Compressible Flow [$\rho \neq c$]

If the density of the fluid changes from point to point in the fluid flow, it is referred to as compressible flow.

Mathematically,

Density: $\rho \neq c$

For example: Gases (like air, carbon dioxide *etc.*) are compressible:

10.6.2 Incompressible Flow ($\rho = c$).

If the density of the fluid remains constant at every point in the fluid flow, it is referred to as an incompressible flow.

Mathematically,

Density: $\rho = c$

For example: Liquids are generally incompressible in nature (Solids are also incompressible in nature).

The compressible or incompressible flows are also defined on basis of Mach number (M). It is index of the ratio of inertia and elastic forces.

$$\dot{M}^2 = \frac{\text{inertia force}}{\text{elastic force}}$$

$$M^2 = \frac{\rho L^2 V^2}{K L^2}$$

$$M^2 = \frac{\rho V^2}{K}$$

where $\quad K = $ bulk modulus of the fluid

$$M^2 = \frac{V^2}{K/\rho}$$

$$M^2 = \frac{V^2}{a^2}$$

where $\quad \dfrac{K}{\rho^2} = a^2$

$\therefore \qquad\qquad M = \dfrac{V}{a}$

where $\quad a = $ speed of sound in the flowing medium

This relation given another important definition of the Mach number (M) as the ratio of the fluid velocity to the velocity of sound in the flowing medium.

If M > 0.2, the gases are considered to be compressible.

If M < 0.2, the gases are assumed to be incompressible.

10.7 RATE OF FLOW

The quantity of the fluid flowing per unit time through a section of flow is called **rate of flow**. The rate of flow is expressed as the weight of the fluid flowing per second across the section, is called **weight flow rate.** It is denoted by $\dot{W}$ and its unit is N/s. The rate of flow is expressed as the mass of the fluid flowing across the section per second, is called **mass flow rate.** It is denoted by m and its unit is kg/s. The rate of flow is expressed as the volume of the fluid flowing across the section per second, is called **volume flow rate** or commonly known as **discharge.** It is denoted by Q and its unit is m³/s.

Mathematically,

Discharge: $Q = AV = mv$

where A is the cross-sectional area, m²

V is average velocity of flow across the section, m/s

m is mass flow rate, kg/s

v is specific volume, m³/kg

or $AV = mv$

or $m = \dfrac{AV}{v}$ $\left| \because v = \dfrac{1}{\rho} \right.$

$m = \rho AV$

10.8 CONTINUITY EQUATION

The equation based on the law of conservation of mass is called continuity equation. It means that mass of fluid can neither be created nor destroyed. If there is no accumulation of mass within the control volume, the mass flow rate entering the system must equal to the mass flow rate leaving the system.

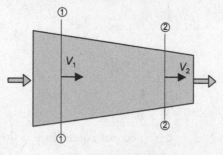

Fig. 10.6

According to law of conservation of mass, mass rate of flow at section (1) – (1) = mass rate of flow at section (2) – (2) (Fig. 10.6)

$$m_1 = m_2 \qquad\qquad \left| \because m = \dfrac{AV}{v} \right.$$

$$\frac{A_1V_1}{v_1} = \frac{A_2V_2}{v_2} \qquad\qquad ...(10.8.1)$$

where $\qquad\qquad m_1 = \dfrac{A_1V_1}{v_1}$ and $m_2 = \dfrac{A_2V_2}{v_2}$

or $\qquad\qquad \rho_1A_1V_1 = \rho_2A_2V_2 \qquad\qquad ...(10.8.2)$

Eqs. (10.8.1) and (10.8.2) are used for compressible fluid where $\rho_1 \neq \rho_2$ or $v_1 \neq v_2$. If fluid is incompressible, then $v_1 = v_2$ or $\rho_1 = \rho_2$, the Eqs. (1) and (2) become

$$A_1V_1 = A_2V_2$$

$$\rho_1A_1V_1 = \rho_2A_2V_2 \qquad \text{for compressible fluid}$$
$$A_1V_1 = A_2V_2 \qquad \text{for incompressible fluid}$$

10.9 STEADY FLOW ENERGY EQUATION [SFEE]

Steady flow means to the flow in which its properties like m, p, T, v, V do not change with respect to time at any point in the system.

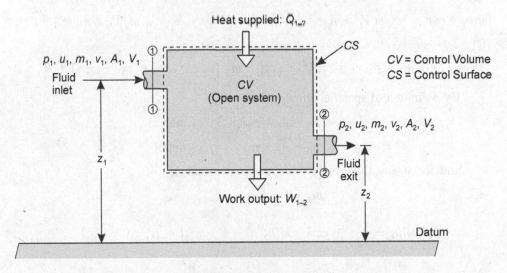

Fig. 10.7 Steady flow system

Consider an open system as shown in Fig. 10.7. Let p_1, u_1, m_1, v_1, A_1 and V_1 are pressure, specific internal energy, mass flow rate, specific volume, cross-sectional area, velocity at section (1) – (1) inlet to the system respectively.

and p_2, u_2, m_2, v_2, A_2 and V_2 are pressure, specific internal energy, mass flow rate, specific volume, cross-sectional area, velocity at section (2) – (2) outlet to the system respectively.

z_1, z_2 are datum heads at inlet and exit respectively

$\underset{1-2}{Q}$ is heat supplied to the system.

$\dfrac{W}{1\text{-}2}$ is work output from the system.

[**Note :** We assumed both heat and work are positive *i.e.*, heat supplied to the system is **+ve** and work output by the system is **+ve**]

Following assumptions are set up to derive the steady flow energy equation:

(*i*) Mass flow rate at the inlet and outlet is same

(*ii*) All the parameters of fluid and flow properties like p, u, v, and V do not change with respect to time at inlet and outlet of the system.

(*iii*) Both heat and work do not change with respect to time.

According to law of conservation of energy, "energy can neither be created nor destroyed"

∴   Net inlet energy to the system =net output energy by the system

Energy at section (1) – (1) + heat supplied = energy at section (2) – (2) + work output

[Internal energy + flow work + $KE - PE]_1 + \dfrac{Q}{1\text{-}2}$

$$= [\text{internal energy} + \text{flow work} + \text{KE} + \text{PE}]_2 + \dfrac{W}{1\text{-}2}$$

$$[m_1 u_1 + m_1 p_1 v_1 + \tfrac{1}{2} m_1 V_1^2 + m_1 g z_1] + \dfrac{Q}{1\text{-}2} = [m_2 u_2 + m_2 p_2 v_2 + \tfrac{1}{2} m_2 V_2^2 + m_2 g z_2] + W_{1\text{-}2}$$

$$m_1 [u_1 + p_1 v_1 + \tfrac{1}{2} V_1^2 + g z_1] = m_2 [u_2 + p_2 v_2 + \dfrac{V_2^2}{2} + g z_2] + W_{1\text{-}2}$$

By definition of specific enthalpy,

$$h_1 = u_1 + p_1 v_1 \qquad \text{at inlet}$$
$$h_2 = u_2 + p_2 v_2 \qquad \text{at outlet}$$

and for steady flow, mass flow rate is constant *i.e.*,

$$m_1 = m_2 = m$$

∴ $$m[h_1 + \dfrac{V_1^2}{2} + g z_1] = m[h_2 + \dfrac{V_2^2}{2} + g z_2] + W_{1\text{-}2}$$

or $$h_1 + \dfrac{V_1^2}{2} + g z_1 + \dfrac{Q_{1\text{-}2}}{m} = h_2 + \dfrac{V_2^2}{2} + g z_2 + \dfrac{W_{1\text{-}2}}{m}$$

or $$h_1 + \dfrac{V_1^2}{2} + g z_1 + q_{1\text{-}2} = h_2 + \dfrac{V_2^2}{2} + g z_2 + w_{1\text{-}2} \qquad \ldots (10.9.1)$$

where $q_{1\text{-}2} = \dfrac{Q_{1\text{-}2}}{m}$, specific heat transfer, which is defined as the rate of heat transfer per unit mass flow rate. Its SI unit is kJ/kg and $w_{1\text{-}2} = \dfrac{W_{1\text{-}2}}{m}$, specific work, which is defined as the rate of work done per unit of mass flow rate. Its SI unit is kJ/kg. Eq. (10.9.1) is called steady flow energy equation. In this equation, the specific enthalpy, kinetic energy, potential energy, specific heat transfer and specific work do not change with respect to time in open system.

10.10 STAGNATION STATE

Stagnation state is the state of flowing fluid at which the flow come to stop. This state is achieved by decelerating the fluid adiabatic isentropically to zero velocity. The concept of a reference state of the gas in a compressible flow is very useful. The stagnation state of a gas is often used as a reference state. It is commonly designated with the subscript zero. The enthalpy, temperature, pressure, density at stagnation state are called stagnation enthalpy h_0, stagnation temperature T_0, stagnation pressure p_0 and stagnation density ρ_0 respectively.

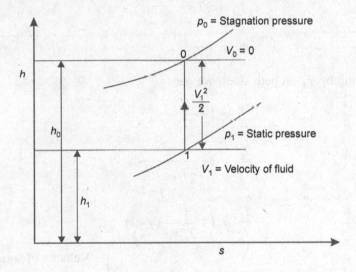

Fig. 10.8 Decelerating of flow

Applying the steady flow energy equation at state 1 and stagnation stage 0, we get

$$h_1 + \frac{V_1^2}{2} + gz_1 + q_{1-0} = h_0 + \frac{V_0^2}{2} + gz_0 + w_{1-0}$$

where $z_1 = z_2$, flow is horizontal

$q_{1-0} = 0$, adiabatic process

$w_{1-0} = 0$, no work done

$V_0 = 0$, at stagnation point

$$\therefore \qquad h_1 + \frac{v_1^2}{2} = h_0$$

or Stagnation enthalpy: $\boxed{h_0 = h_1 + \frac{V_1^2}{2}}$ \qquad\qquad\qquad\qquad (10.10.1)

Stagnation temperature: T_0

For an ideal gas, a stagnation temperature is defined through stagnation enthalpy. From equation (10.10.1)

$$h_0 = h_1 + \frac{V_1^2}{2}$$

$$c_p T_0 = c_p T_1 + \frac{V_1^2}{2}$$

or $\qquad\qquad\qquad T_0 = T_1 + \frac{V_1^2}{2c_p}$

Dividing by T_1 on both sides, we get

$$\frac{T_0}{T_1} = 1 + \frac{V_1^2}{2T_1 c_p}$$

$$\frac{T_0}{T_1} = 1 + \frac{V_1^2}{2T_1 \times \dfrac{\gamma R}{\gamma - 1}} \qquad\qquad \left| \because c_p = \frac{\gamma R}{\gamma - 1} \right.$$

$$\frac{T_0}{T_1} = 1 + \frac{V_1^2}{2T_1 \gamma R}(\gamma - 1)$$

$$\frac{T_0}{T_1} = 1 + \frac{V_1^2(\gamma - 1)}{2a^2} \qquad\qquad
\begin{array}{l} \because \text{Velocity of sound:} \\ \qquad a = \sqrt{\gamma R T_1} \text{ for ideal gas} \\ \text{or } a^2 = \gamma R T_1 \end{array}$$

$$\frac{T_0}{T_1} = 1 + \frac{(\gamma - 1)}{2}\frac{V_1^2}{a^2} = 1 + \left(\frac{\gamma - 1}{2}\right).M^2$$

where $\qquad\qquad\qquad M = \frac{V_1}{a}$, Mach number of the fluid flow at state 1.

$$\boxed{\frac{T_0}{T_1} = 1 + \frac{(\gamma - 1)}{2}.M^2} \qquad\qquad\qquad ...(10.10.2)$$

Stagnation pressure: p_0

The pressure of a gas or fluid which is obtained by decelerating it in a adiabatic isentropically to zero velocity is known as the stagnation pressure.

For adiabatic process 1 – 0,

$$\frac{T_o}{T_1} = \left(\frac{p_o}{p_1}\right)^{\frac{\gamma-1}{\gamma}} \qquad\qquad ...(10.10.3)$$

Equating Eqs. (10.10.3) and (10.10.2), we get

$$\left(\frac{p_o}{p_1}\right)^{\frac{\gamma-1}{\gamma}} = 1 + \frac{(\gamma-1)}{2}M^2$$

or

$$\boxed{\frac{p_o}{p_1} = \left[1 + \frac{(\gamma-1)}{2}M^2\right]^{\frac{\gamma}{\gamma-1}}}$$

Stagnation density: ρ_0

The density of a gas at stagnation state is called the stagnation density.

For adiabatic process:

$$\frac{T_o}{T_1} = \left(\frac{v_1}{v_0}\right)^{\gamma-1} = \left(\frac{\rho_1}{\rho_0}\right)^{\gamma-1} \qquad\qquad ...(10.10.4)$$

Equating Eqs. (10.10.4) and (10.10.2), we get

$$\left(\frac{\rho_1}{\rho_0}\right)^{\gamma-1} = 1 + \frac{(\gamma-1)}{2}M^2$$

or

$$\boxed{\frac{\rho_o}{\rho_1} = \left[1 + \frac{(\gamma-1)}{2}M^2\right]^{\frac{1}{\gamma-1}}}$$

10.11 VELOCITY OF SOUND WAVE IN COMPRESSIBLE FLUIDS

In the first chapter, we discussed, solids are incompressible and liquids are also considered as a incompressible because of the compressibility in liquid is very small or supposed to be neglected. However, gases are purely compressible in nature. In solids and liquids, there is compactness and closeness of molecules. Because of this, if a minor disturbance caused on one end is instantaneously transferred to the other end of liquids or solids. So in liquid, and solids the disturbance travels at infinite velocity. But in case of gases, the molecules are spaced at a considerable distance between molecules. Because of this, if disturbance is applied on gases, it takes time in transferring the distance from one set of molecules to the other. Hence, the velocity of disturbance transfer in compressible fluids like gases is finite and its value depends upon the elastic properties such as bulk modulus of elasticity and density of the gases.

Following assumptions are set up to derive an expression for velocity of sound wave in a fluid.

(i) Pipe should be rigid.

(*ii*) Uniform cross-sectional area.

(*iii*) Fluid is compressible.

(*iv*) Pressure waves travel with velocity of sound in one-direction.

Let the pipe is filled with a compressible fluid, which is at rest initially. The piston is moved towards right and a disturbance is created in the fluid. This disturbance is in the form of pressure wave, which travels in the fluid with a velocity of sound wave.

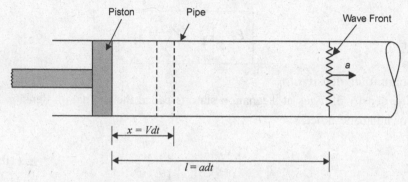

Fig. 10.9 Propagation of pressure wave.

Let V = velocity of piston.

a = velocity of pressure wave (or velocity of sound in the fluid)

p = pressure of fluid in pipe before the movement of the piston.

ρ = density of fluid before the movement of the piston

dt = small interval of time with which piston is moved

Distance travelled by the piston in time dt:

$$x = \text{velocity of piston} \times dt = Vdt$$

Distance travelled by the pressure wave in time dt:

$$l = \text{velocity of sound} \times dt = a\, dt$$

$\therefore$ $a\, dt \gg V\, dt$ | $\because a \gg V$

or $l \gg x$

The length of pipe when fluid compressed in time dt,

$$= l - x$$

Due to compression of the fluid, the pressure and density of fluid will change.

Let $p + dp$ = pressure after compression

$\rho + d\rho$ = density after compression

Mass of fluid for length l before compression,

$$= \rho Al = \rho A a\, dt \qquad\qquad\qquad ...(10.11.1)$$

Mass of fluid for length $(l - x)$ after compression,

$$= (\rho + d\rho)\, A\, (l - x)$$
$$= (\rho + d\rho)\, A\, (a\, dt - V\, dt)$$
$$= (\rho + d\rho)\, (a - V)\, A\, dt \qquad\qquad ...(10.11.2)$$

According to continuity equation, the mass flow rate is remain constant before and after compression. So, equating the Eqs (10.11.1) and (10.11.2), we get

$$\rho A \, a \, dt = (\rho + d\rho) \, (a - V) \, A \, dt$$

or
$$\rho a = (\rho + d\rho) \, (a - V)$$

$$\rho a = \rho a - \rho V - V d\rho + a d\rho$$

or
$$-\rho V - V d\rho + a d\rho = 0$$

$$a d\rho = \rho V + V d\rho$$

where V = velocity of position is very small as compared with a,

∴ $V d\rho \approx 0$, product of small quantity is neglected.

∴ $a d\rho = \rho V$

or $\rho V = a d\rho$ ...(10.11.3)

When the piston is moved with a velocity V in time dt, the fluid which is at rest initially (*i.e.*, velocity of fluid is zero) and then velocity of the fluid is equal to the velocity of the piston. Also the pressure of the fluid increases from p to $p + dp$ due to the movement of the piston:

According to impulse momentum equation, the net force exerted on the fluid is equal rate of change of momentum in the direction of force.

Mathematically,

Net force on the fluid = rate of change of momentum in the direction of force

$$(p + dp) \, A - pA = M \frac{dV}{dt}$$

$$dp \, A = \frac{\rho A l}{dt} [V - 0]$$

Initial velocity : $V_1 = 0$
and final velocity : $V_2 = V$

$$dp \, A = \frac{\rho A l}{dt} V$$

$$dp = \frac{\rho l}{dt} V$$

∵ $l = a \, dt$

or $\dfrac{l}{dt} = a$

∴ $dp = \rho \, aV$

or $\rho V = \dfrac{dp}{a}$ ....(10.11.4)

Equating Eqs. (10.11.3) and (10.11.4), we get,

$$a d\rho = \frac{dp}{a}$$

or
$$a^2 = \frac{dp}{d\rho}$$

or
$$a = \sqrt{\frac{dp}{d\rho}}$$
...(10.11.5)

Equation (10.11.5) gives the velocity of sound which is the square root of the ratio of change of pressure to the change of density of a fluid.

10.11.1 Velocity of Sound (a) in terms of Bulk Modulus of Elasticity (K)

Bulk modulus of elasticity (K) defined as the ratio of the increase in pressure to volumetric strain.

Mathematically,

Bulk modules of elasticity:

$$K = \frac{\text{Increase in pressure} : dp}{\text{Volumetric strain} : \varepsilon_v}$$

or
$$\frac{dp}{d\rho} = \frac{K}{\rho}$$

Substituting $\dfrac{dp}{d\rho} = \dfrac{K}{\rho}$ in Eq. (10.12.5), we get

$$a = \sqrt{\frac{K}{\rho}}$$
...(10.11.6)

Equation (10.11.6) gives the velocity of sound which is the square root of the ratio of the bulk modulus of elasticity to the density of fluid. This equation is applicable for liquids and gases.

$$K = \frac{dp}{\varepsilon_v}$$

where $\varepsilon_v = -\dfrac{dV}{V}$, change in volume to original volume. The −ve sign shows volume decreases with increase in pressure.

We know that the mass of a fluid is constant. Hence

$$\text{Mass} = \text{density} \times \text{volume}$$

or
$$m = \rho V$$

or
$$\rho V = m$$

Taking $\log_e$ both sides, we get

$$\log_e \rho + \log_e V = \log_e m$$

On differentiating the above equation, we get

$$\frac{d\rho}{\rho} + \frac{dV}{V} = 0$$

or $$\frac{d\rho}{\rho} = -\frac{dV}{V}$$

∴ Volumetric strain: $$\varepsilon_v = \frac{d\rho}{\rho}$$

∴ Bulk modulus of elasticity: $$K = \frac{dp}{d\rho/\rho}$$

10.12 VELOCITY OF SOUND IN AN IDEAL GAS

(i) Adiabatic isentropic process

Adiabatic isentropic follows the law:

$$pv^\gamma = C$$

$$p\frac{1}{\rho^\gamma} = C \qquad \qquad |\because \text{Specific volume:} \, v = \frac{1}{\rho}$$

or $$p\rho^{-\gamma} = C$$

Taking $\log_e$ both sides, we get

$$\log_e p - \gamma \log_e \rho = \log_e C$$

On differentiating the above equation, we get

$$\frac{dp}{p} - \gamma\frac{d\rho}{\rho} = 0$$

$$\frac{dp}{p} = \gamma\frac{d\rho}{\rho}$$

$$\frac{dp}{\rho} = \gamma\frac{p}{\rho}$$

∴ Velocity of second :

$$a^2 = \gamma\frac{p}{\rho} \qquad\qquad a = \sqrt{\frac{dp}{d\rho}}$$

$$a^2 = \gamma \, pv \qquad\qquad |\because \, pv = RT$$

or $$a^2 = \gamma \, RT$$

or $$\boxed{a = \sqrt{\gamma RT}}$$

where γ = adiabatic index

= 1.4 for air

R = gas constant

= 0.287 kJ/kgK for air = 287 J/kg K

T = absolute temperature of the air *i.e.*, temperature in

K (kelvin).

Hence, the sound velocity is directly proportional to the square root of absolute temperature

i.e., $a \propto \sqrt{T}$ for an ideal gas.

For air at constant values of $\gamma = 1.4$

and $R = 287$ J/Kg K

Velocity of sound in air:

$$a = \sqrt{\gamma RT} \quad \text{at } 0°C$$

$$= \sqrt{1.4 \times 287 \times 273} = 331.20 \text{ m/s}$$

$$= \sqrt{1.4 \times 287 \times 288} \quad \text{at } 15°C$$

$$= 340.17 \text{ m/s}$$

$$= \sqrt{1.4 \times 287 \times 298} \quad \text{at } 25°C$$

$$= 346 \text{ m/s}$$

The Mach number of the flow defined as the ratio of the velocity of fluid to the velocity of sound in same fluid. It is denoted by letter M. Mathematically,

Mach number : $M = \dfrac{V}{a}$

If $0.9 < M < 1.1$, the flow is transonic

$M < 1$, the flow is subsonic,

$M = 1$, the flow is sonic,

$1 < M < 7$, the flow is supersonic

$7 \leq M < 10$, the flow is hypersonic, and

$M \geq 10$, the flow is high hypersonic.

(ii) For isothermal process

Isothermal process follows the law:

$$pv = c$$

$$\frac{p}{\rho} = c \qquad \qquad \left| \because v = \frac{1}{\rho} \right.$$

$$p\rho^{-1} = c$$

Taking $\log_e$ both sides, we get

$$\log_e p - \log_e \rho = \log_e c$$

On differentiating the above equation, we get

$$\frac{dp}{p} - \frac{d\rho}{\rho} = 0$$

or
$$\frac{dp}{p} = \frac{d\rho}{\rho}$$

or
$$\frac{dp}{d\rho} = \frac{p}{\rho} = pv$$

$$a^2 = pv$$

$$a^2 = RT$$

$$\boxed{a = \sqrt{RT}}$$

$$\because \ a = \sqrt{\frac{dp}{d\rho}} \ \text{and} \ pv = RT$$

10.13 PROPAGATION OF PRESSURE WAVES (OR DISTURBANCES IN A COMPRESSIBLE FLUID)

Whenever any disturbance is produced in a compressible fluid due to movement of any projectile in fluid, the disturbance propagates in all directions with a velocity of sound (*i.e.*, *a*). The nature of propagation of disturbance depends upon the Mach number.

Consider a small projectile moving in a stationary fluid. Let the movement of projectile is from left to right in a straight line. Due to movement of projectile, disturbances will be created in fluid and move with velocity *a* in all directions.

So consider $V =$ velocity of the projectile

$a =$

velocity
o f
disturbances
o r
(pressure
wave)
created
in the
fluid.

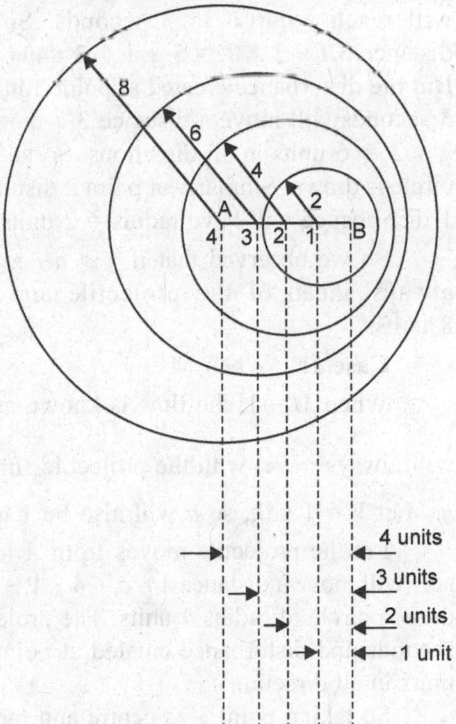

Fig. 17.10 $M < 1$

Let us study the nature of disturbance for different Mach numbers.

Case I : $M < 1$

When Mach number is less than 1.0, the flow is called subsonic flow. Consider M = 0.5 which is less than 1.0.

For $M = 0.5$, $\dfrac{V}{a} = 0.5$ or $\dfrac{V}{a} = \dfrac{1}{2}$ *i.e.*, if $V = 1$ unit then a will be 2 units. Let the projectile is at point A and is moving towards point B. Let in 4 seconds the projectile reaches at point B. As shown in Fig. 10.10, at point A, the point 4 is also marked.

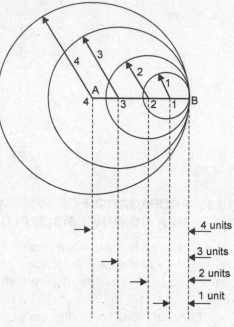

Fig. 17.11 $M = 1$

The position of the projectile after 1 sec, 2 sec, 3 sec and 4 sec along the straight line is shown by points 3, 2, 1 and B respectively. The projectile moves from A to B in 4 seconds. So distance AB $= 4 \times V = 4 \times 1 = 4$ units. The disturbance acted at point A in 4 seconds will move a distance $4 \times a = 4 \times 2 = 8$ units in all directions. So taking A as centre and radius = 8 units a circle as shown in Fig. 10.10 is drawn. This will give the position of disturbance after 4 seconds.

When the projectile is at point 3, it will reach point B in 3 seconds. So distance $3B = 3 \times V = 3 \times 1 = 3$ units. But the disturbance created at point 3 in 3 seconds will move a distance $3 \times a = 3 \times 2 = 6$ units in all directions. So taking point 3 as centre and radius = 6 units, a circle is drawn. Similarly at point 2 disturbance will have radius = 4 units and at point 1 disturbance will have radius = 2 units.

So we observed that if $V < a$ i.e. $M < 1$, the pressure wave or disturbance is always ahead of the projectile and point B is inside the sphere of radius 8 units.

Case II : When M = 1.

When $M = 1$, the flow is known as sonic flow. In this case, the pressure wave will always travel with the projectile shown in Fig.10.10. For $M = 1$, $\dfrac{V}{a} = 1$ or $V = a$. Let $V = 1$ unit, so a will also be 1 unit.

Let the projectile moves from A to B in 4 seconds so the disturbance created at A will move a distance $4 \times a = 4 \times 1 = 4$ units in all directions. So taking A as centre draw a circle of radius 4 units. The projectile from point 3 will move to point B in 3 seconds and disturbance created at point B will move a distance $3 \times a = 3 \times 1 = 3$ units in all directions.

So taking point 3 as centre and radius = 3 units a circle is drawn. Similarly at point 2 and point 1 the disturbance created at these points having radius 2 and 1 unit respectively in all directions. So we observed that the disturbance always travels with the projectile.

Case III : When $M > 1$

When $M > 1$, flow is called supersonic flow. For $M > 1$, consider $M = 2$ which means $\dfrac{V}{a} = 2$ so let $V = 1$ unit and $a = 0.5$ unit so that $M = \dfrac{V}{c} = \dfrac{1}{0.5} = 2$ units.

Again consider that projectile moves from point A to B in 4 seconds. Hence distance travelled by projectile in 4 seconds $= 4 \times V = 4 \times 1 = 4$ units. So take $AB = 4$ units. The disturbance created at point A will move a distance $= 4 \times a = 4 \times 0.5 = 2$ units in all directions. So taking A as centre draw a circle of radius $= 2$ units as shown in Fig. 10.11.

After 1 second, the projectile will be at point 3 and it will reach to point B in 3 seconds. So distance $3B = 3 \times V = 3 \times 1 = 3$ units. Hence, the disturbance created from point 3 will move a distance having radius $= 3 \times a = 3 \times 0.5 = 1.5$ units in add directions.

Similarly the radius of disturbance at point 2 and at point 1 will be $2 \times a = 2 \times 0.5 = 1$ unit and $1 \times a = 1 \times 0.5 = 0.5$ unit respectively.

So we observe that the propagation of disturbance always lags behind the projectile movement. It we draw a tangent to different circles representing the propagation waves on both sides, we will get a cone with vertex at B. This cone is called Mach cone.

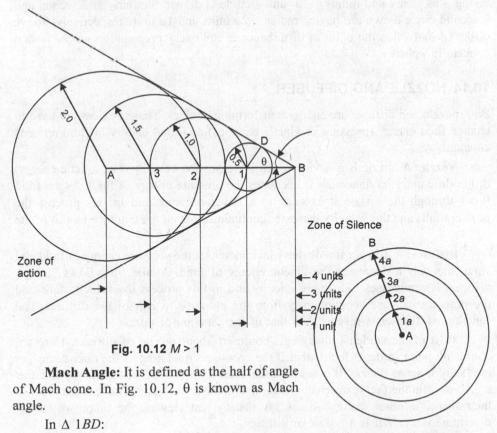

Fig. 10.12 $M > 1$

Mach Angle: It is defined as the half of angle of Mach cone. In Fig. 10.12, θ is known as Mach angle.

In $\Delta\ 1BD$:

$$1B = 1 \times V = V = $$ velocity of projectile

Fig. 10.13 Point of disturbance is stationary

$$1D = 1 \times a = a = \text{velocity of sound wave}$$

Hence we have

$$\sin \theta = \frac{1D}{1V} = \frac{a}{V} = \frac{1}{V/a} = \frac{1}{M}$$

Zone of Action: When $M > 1$, the disturbances are confined to the cone, the area within the Mach cone is called zone of action.

Zone of Silence: When $M > 1$, the disturbance felt only inside the cone, so the area outside the cone is called zone of silence.

Case IV: If the point of disturbance is stationary, then the wave fronts (propagation waves) are concentric spheres as shown in Fig. 10.13.

Consider the point of disturbance is at point A and is stationary. so $V = 0$.

Consider disturbance created from point A at 4th second had moved a distance having radius = $4 \times a = 4a$ units so taking A as centre

and radius = $4a$ units a circle is drawn. The disturbance created at point A in 3rd second have moved a distance having radius = $3 \times a = 3a$ units in all directions. So taking A as centre and radius = $3a$ unit a circle is drawn. Similarly at 2 second and 1 second circle drawn are having radius = $2a$ units and $1a$ units respectively. So we observed that when the point of disturbance is stationary propagation waves from a concentric sphere.

10.14 NOZZLE AND DIFFUSER

Both nozzle and diffuser are energy transforming devices. These devices are used to change fluid energy (pressure or kinetic energies) and total energy in fluid remains constant.

Nozzle: A nozzle is a device which increases the velocity (*i.e.*, kinetic energy) of the fluid and simultaneously it decreases the pressure energy of fluid. As the fluid flows through the nozzle it expands to a lower pressure and in this process the pressure falls and the velocity increases continuously from the entrance to exit of the nozzle.

Diffuser: A diffuser is a device which increases the pressure energy of fluid and simultaneously it decreases the kinetic energy of fluid. As the fluid flows through diffuser, it compresses to a lower velocity and in this process the velocity falls and the pressure increases continuously from the entrance to exit of the diffuser. The function of diffuser is reverse than that of the function of nozzle.

Most of the students always get confused about nozzle of diffuser. Here the confusing point students feel is that if the cross-sectional area of the duct decreases gradually it act as nozzle. If cross-sectional area of duct increases gradually it act as a diffuser. But the fact is not that. Nozzle and diffuser are not defined on the basis of their shape. In other words, we can say that by just viewing the shape we cannot determine whether it is a nozzle or diffuser.

Actually these are defined on the basis of inlet condition of fluids or Mach number (M).

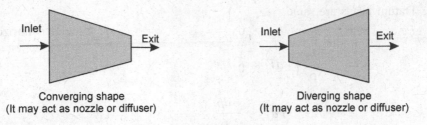

Converging shape
(It may act as nozzle or diffuser)

Diverging shape
(It may act as nozzle or diffuser)

Fig. 10.14: Nozzle and diffuser

Designing of nozzle and diffuser.

Assumption: The fluid flowing through the nozzle and diffuser is steady and isentropic in behaviour.

According to continuity equation:

Discharge flow through nozzle or diffuser,

$$Q = mv = AV$$

where m = mass flow rate, kg/s,

v = specific volume, m^3/kg,

A = cross-sectional area, m^2,

V = velocity of fluid, m/s

$\therefore$ $mv = AV$

$$m = \frac{AV}{v} \qquad \left[\because \text{Specific volume: } v = \frac{1}{\rho} \right]$$

$$m = \rho AV \qquad \qquad ...(10.14.1)$$

The above Eq. (10.14.1) is also called *continuity equation*. It states that mass flow rate at any section in the flow is constant.

For all applications, the mass flow rate is taken constant and hence the Eq. 10.14.1 can be written as

$$\rho AV = \text{constant} = C$$

Taking $\log_e$ both sides, we get

$$\log_e (\rho AV) = \log_e C$$

$$\log_e \rho + \log_e A + \log_e V = \log_e C$$

On differentiating both sides, we get

$$\frac{d\rho}{\rho} + \frac{dA}{A} + \frac{dV}{V} = 0$$

$$\frac{dA}{A} = -\left[\frac{dV}{V} + \frac{d\rho}{\rho} \right] \qquad \qquad ...(10.14.2)$$

According to Euler's equation [Momentum equation]

$$\frac{dp}{\rho} + V \, dV + g \, dz = 0$$

Datum heads are same

i.e., $dz = 0$ [$\therefore$ Nozzle is horizontal]

$\therefore$ $\dfrac{dp}{\rho} + VdV = 0$

$$VdV = -\dfrac{dp}{\rho}$$

$$dV = -\dfrac{dp}{\rho V} \qquad \qquad ...(10.14.3)$$

Substituting the value of dV in Eq. (10.14.2), we get

$$\dfrac{dA}{A} = -\dfrac{dV}{V}\left[1 + \dfrac{d\rho}{\rho}\dfrac{V}{\left(-\dfrac{dp}{\rho V}\right)}\right] = -\dfrac{dV}{V}\left[1 - \dfrac{d\rho}{dp}V^2\right]$$

$$= -\dfrac{dV}{V}\left[1 - \dfrac{V^2}{\dfrac{dp}{d\rho}}\right] \qquad \qquad ...(10.14.4)$$

Sonic-velocity for isentropic process:

$$a = \sqrt{\dfrac{dp}{d\rho}} \qquad \text{for compressible fluid}$$

or $a^2 = \dfrac{dp}{d\rho}$

or $\dfrac{dp}{d\rho} = a^2$

Substituting the value of $\dfrac{dp}{d\rho} = a^2$ in above Eq. (10.14.4), we get

$$\dfrac{dA}{A} = -\dfrac{dV}{V}\left[1 - \dfrac{V^2}{a^2}\right]$$

By definition of Mach number,

Mach number : $M = \dfrac{\text{Velocity of fluid}}{\text{Velocity of sound in same fluid}}$

$$M = \dfrac{V}{a}$$

$$\frac{dA}{A} = -\frac{dV}{V}\left[1-M^2\right]$$

$$\boxed{\frac{dA}{A} = \frac{dV}{V}[M^2-1]} \qquad\qquad ...(10.14.5)$$

The above Eq. (10.14.5) is used to design the nozzle.

From Eq. (10.14.3)

$$dV = -\frac{dp}{\rho V}$$

Substituting the value of dV in above Eq. (10.14.5), we get

$$\boxed{\frac{dA}{A} = -\frac{dp}{\rho V^2}[M^2-1]} \qquad\qquad ...(10.14.6)$$

The above Eq. (10.14.6) is used to design the diffuser.

Nozzles : The following conclusion can be drawn from Eq. (10.14.5)

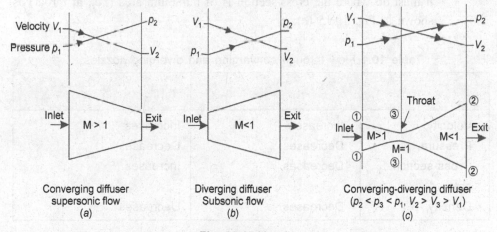

Converging diffuser
supersonic flow
(a)

Diverging diffuser
Subsonic flow
(b)

Converging-diverging diffuser
$(p_2 < p_3 < p_1, V_2 > V_3 > V_1)$
(c)

Fig. 10.15 Nozzles

(*i*) If Mach number, $M < 1$, *i.e.*, $V < a$ (for subsonic flow)

From Eq. (10.14.5), we get

$$\frac{dA}{A} = -\frac{dV}{V}$$

or $\qquad\qquad \dfrac{dV}{V} = -\dfrac{dA}{A}$

As dA and dV must be opposite in sign *i.e.*, decrease in cross-sectional area causes an increase in velocity. This takes place in converging nozzle as shown in Fig. (10.15 (*a*))

i.e., for converging nozzle; $M < 1$ or $V < a$. The flow of fluid in converging nozzle is called *sub-sonic* ($\because$ Velocity of fluid is less than sonic velocity *i.e.*, velocity of sound).

(*ii*) If Mach number $M > 1$, *i.e.*, $V > a$ (for supersonic flow)

From Eq. (10.13.5), we get

$$\frac{dA}{A} = \frac{dV}{V}$$

As dA and dV are of same sign. An increase of cross-sectional area, then causes an increase of velocity. This takes place in diverging nozzle as shown in Fig. 10.15 (*b*) *i.e.*, for diverging nozzle; $M > 1$ or $V > a$. The flow of fluid in diverging nozzle is called *supersonic flow* ($\because$ Velocity of fluid is more than sonic velocity).

(*iii*) If Mach number, $M = 1$ *i.e.*, $V = a$ (For sonic flow).

As dA must be zero and since the second derivative is positive, area 'A' must be minimum. Thus if the velocity of flow equals the sonic velocity anywhere, it must do where the cross-section is of minimum area (*i.e.*, at throat) as shown in Fig. 10.15 (*c*).

Table 10.1 Flow through converging and diverging nozzles.

Flow Parameter	Converging Nozle (Subsonic Flow, M < 1)	Diverging Nozzle (Supersonic Flow, M > 1)
Velocity, V	Increases	Increases
Pressure, p	Decreases	Decreases
Cross-sectional area, A	Decreases	Increases
Density, ρ	Decreases	Decreases

Diffusers: The following conclusion can be drawn from Eq. (17.14.6)

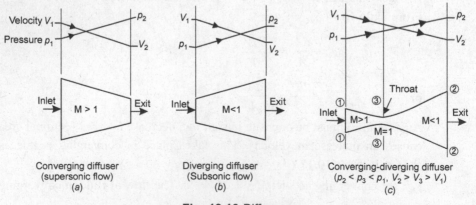

Converging diffuser (supersonic flow) (a) — Diverging diffuser (Subsonic flow) (b) — Converging-diverging diffuser ($p_2 < p_3 < p_1$, $V_2 > V_3 > V_1$) (c)

Fig. 10.16 Diffusers

(i) If Mach number, $M > 1$, $i.e.$, $V > a$ (For supersonic flow)

From Eq. (10.14.6), $dA = -dp$

or $dp = -dA$

As dA and dp must be opposite in sign $i.e.$, decrease in cross-sectional area causes increase in pressure. This takes place in converging diffuser as shown in Fig. 10.16 (a), i.e., for converging diffuser; $M = 1$ or $V > a$.

The flow of fluid in converging diffuser is called *supersonic flow*.

(ii) If Mach number $M > 1$ $i.e.$, $V < a$ (For subsonic flow)

From Eq. (10.14.6) $dA = dp$

As dA and dp are of same sign. An increase of cross-sectional area causes an increase in pressure. This takes place in diverging diffuser as shown in Fig. 10.16 (b), $i.e.$, for diverging diffuser, $M < 1$ or $V < a$. The flow of fluid in diverging diffuser is *sub-sonic flow.*

Table 10.2: Flow through converging and diverging diffusers.

Flow Parameter	Converging Diffusers (Super-sonic Flow, $M > 1$)	Diverging Diffusers (Sub-sonic Flow, $M < 1$)
Velocity, V	Decreases	Decreases
Pressure, p	Increases	Increases
Cross-sectional area, A	Decreases	Increases
Density, ρ	Increases	Increases

(iii) If Mach number, $M = 1$ $i.e.$, $V = a$ (For sonic flow). As dA must be zero and since the second derivative is positive, area A must be minimum. Thus, if the velocity of flow equals the sonic velocity anywhere, it must do where the cross-section is of minimum area ($i.e.$, at throat) as shown in Fig. 10.16 (c).

10.15 FLOW THROUGH NOZZLE

Let us consider the compressible fluid flow through a convergent-divergent nozzle as shown in Fig. 10.17.

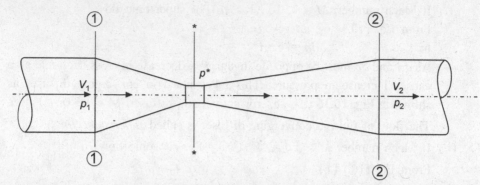

Fig. 10.17 Flow through converging-diverging

Let $\qquad\qquad p_1$, V_1 = pressure and velocity at inlet of nozzle respectively.
$\qquad\qquad\qquad\quad p^* $ = pressure at the throat. $\qquad\cdot$
$\qquad\qquad\qquad p_2$, V_2 = pressure and velocity at exit of nozzle respectively.
Applying the steady flow energy equation at sections 1–1 and 2–2, we get

$$h_1 + \frac{V_1^2}{2} + gz_1 + \underset{1-2}{q} = h_2 + \frac{V_2^2}{2} + gz_2 + \underset{1-2}{w}$$

If the flow is adiabatic: $\underset{1-2}{q} = 0$

No work interaction involved: $\underset{1-2}{w} = 0$, and $z_1 = z_2$, because of nozzle is horizontal.

Therefore, the above is reduced to

$$h_1 + \frac{V_1^2}{2} = h_2 + \frac{V_2^2}{2}$$

If the inlet velocity V_1 is small and neglected, then

$$h_1 = h_2 + \frac{V_2^2}{2}$$

or $\qquad\qquad\qquad \dfrac{V_2^2}{2} = h_1 - h_2 \qquad\qquad\qquad\qquad\qquad ...(10.15.1)$

or $\qquad\qquad\qquad V_2^2 = 2(h_1 - h_2)$

or $\qquad\qquad\qquad V_2 = \sqrt{2(h_1 - h_2)}$ m/s

where $\qquad h_1$ and h_2 in J/kg.

or $\qquad\qquad\qquad V_2 = \sqrt{2 \times 10^3 (h_1 - h_2)}$ m/s

where $\qquad h_1$ and h_2 in kJ/kg.

$$V_2 = 44.72\sqrt{(h_1 - h_2)}$$

We know that the specific enthalpy :

$$h = u + pv$$

On differentiating, we get

$$dh = du + pdv + vdp$$
$$dh = \delta q + vdp \qquad\qquad | \because \delta q = du + pdv$$

or
$$dh = vdp \qquad\qquad | \because \delta q = 0 \text{ for adiabatic flow}$$

On integrating above equation, we get

or
$$\int_1^2 dh = \int_1^2 vdp$$

$$h_2 - h_1 = \int_1^2 C^{1/\gamma} \frac{dp}{p^{1/\gamma}}$$

$$| \text{ For adiabatic process } pv^\gamma = C \text{ or } V = \frac{C^{1/\gamma}}{p^{1/\gamma}}$$

$$= C^{1/\gamma} \int_1^2 p^{-1/\gamma} \, dp$$

$$= C^{1/\gamma} \left[\frac{p^{\frac{-1}{\gamma}+1}}{-\frac{1}{\gamma}+1} \right]_1^2 = C^{1/\gamma} \left[\frac{p_2^{\frac{\gamma-1}{\gamma}} - p_1^{\frac{\gamma-1}{\gamma}}}{\frac{\gamma-1}{\gamma}} \right]_1^2$$

$$= \frac{\gamma}{\gamma-1} \left[C^{1/\gamma} p_2^{\frac{\gamma-1}{\gamma}} \; C^{1/\gamma} p_1^{\frac{\gamma-1}{\gamma}} \right]$$

$$= \frac{\gamma}{\gamma-1} \left[(p_2 v_2^\gamma)^{1/\gamma} p_2^{\frac{\gamma-1}{\gamma}} - (p_1 v_1^\gamma)^{1/\gamma} p_1^{\frac{\gamma-1}{\gamma}} \right]$$

$$= \frac{\gamma}{\gamma-1} \left[p_2^{1/\gamma} v_2 p_2^{\frac{\gamma-1}{\gamma}} - p_1^{1/\gamma} v_1 p_1^{\frac{\gamma-1}{\gamma}} \right]$$

$$h_2 - h_1 = \frac{\gamma}{\gamma-1} [p_2 v_2 - p_1 v_1]$$

or
$$h_1 - h_2 = \frac{\gamma}{\gamma-1} [p_1 v_1 - p_2 v_2] \qquad\qquad ...(10.15.2)$$

Substituting the value of $(h_1 - h_2)$ from Eq. (10.15.2) in Eq. (10.15.1), we get

$$\frac{V_2^2}{2} = \frac{\gamma}{\gamma-1} [p_1 v_1 - p_2 v_2] \qquad\qquad | \; p_1 v_1 - p_2 v_2$$

or
$$V_2{}^2 = \frac{2\gamma}{\gamma-1}\left[p_1 v_1 - p_2 v_2\right]$$

or
$$V_2 = \sqrt{\frac{2\gamma}{\gamma-1}\left[p_1 v_1 - p_2 v_2\right]}$$

$$V_2 = \sqrt{\frac{2\gamma}{\gamma-1}p_1 v_1\left[1 - \frac{p_2 v_2}{p_1 v_1}\right]}$$

$$V_2 = \sqrt{\frac{2\gamma}{\gamma-1}p_1 v_1\left[1 - \frac{p_2}{p_1}\times\left(\frac{p_2}{p_1}\right)^{\frac{-1}{\gamma}}\right]} \qquad \left| \therefore \frac{v_2}{v_1} = \left(\frac{p_2}{p_1}\right)^{\frac{-1}{\gamma}} \right.$$

$$V_2 = \sqrt{\frac{2\gamma}{\gamma-1}p_1 v_1\left[1 - \left(\frac{p_2}{p_1}\right)^{\frac{\gamma-1}{\gamma}}\right]} \qquad \qquad ...(10.15.3)$$

Applying continuity equation at section 2–2, we get
$$m v_2 = A_2 V_2$$

or
$$\frac{m}{A_2} = \frac{V_2}{v_2}$$

Substituting the value of V_2 in above equation from Eq. (10.15.3), we get

$$\frac{m}{A_2} = \frac{1}{v_2}\sqrt{\frac{2\gamma}{\gamma-1}p_1 v_1\left[1 - \left(\frac{p_2}{p_1}\right)^{\frac{\gamma-1}{\gamma}}\right]}$$

$$= \sqrt{\frac{2\gamma}{\gamma-1}\frac{p_1 v_1}{v_2^2}\left[1 - \left(\frac{p_2}{p_1}\right)^{\frac{\gamma-1}{\gamma}}\right]}$$

$$= \sqrt{\frac{2\gamma}{\gamma-1}\frac{p_1 v_1^2}{v_1 v_2^2}\left[1 - \left(\frac{p_2}{p_1}\right)^{\frac{\gamma-1}{\gamma}}\right]}$$

$$= \sqrt{\frac{2\gamma}{\gamma-1}\frac{p_1}{v_1}\left(\frac{p_2}{p_1}\right)^{\frac{2}{\gamma}}\left[1 - \left(\frac{p_2}{p_1}\right)^{\frac{\gamma-1}{\gamma}}\right]}$$

$$\frac{m}{A_2} = \sqrt{\frac{2\gamma}{\gamma-1}\frac{p_1}{v_1}\left[\left(\frac{p_2}{p_1}\right)^{\frac{2}{\gamma}} - \left(\frac{p_2}{p_1}\right)^{\frac{\gamma+1}{\gamma}}\right]}$$

By analogy, changing

$$p_1 \text{ to } p_0$$
$$v_1 \text{ to } v_0$$
$$A_2 \text{ to } A$$
$$p_2 \text{ to } p$$

$$\therefore \qquad \frac{m}{A} = \sqrt{\frac{2\gamma}{\gamma-1} \frac{p_o}{v_o} \left[\left(\frac{p}{p_o}\right)^{\frac{2}{\gamma}} - \left(\frac{p}{p_o}\right)^{\frac{\gamma+1}{\gamma}} \right]} \qquad \qquad ...(10.15.4)$$

Equation (10.15.4) shows that for given values of p_0 and v_0, the mass flow rate

of the fluid depends upon the pressure ratio $\dfrac{p}{p_o}$. There is a certain pressure ratio for

which mass flow rate per unit area $\left(\dfrac{m}{A}\right)$ is maximum. Therefore, by differentiating

the term in square brackets and equating it to zero.

$$\frac{d}{d(p/p_o)} \left[\left(\frac{p}{p_o}\right)^{\frac{2}{\gamma}} - \left(\frac{p}{p_o}\right)^{\frac{\gamma+1}{\gamma}} \right] = 0$$

$$\frac{2}{\gamma} \left(\frac{p}{p_o}\right)^{\frac{2}{\gamma}-1} - \left(\frac{\gamma+1}{\gamma}\right) \left(\frac{p}{p_o}\right)^{\frac{\gamma+1}{\gamma}-1} = 0$$

$$2 \left(\frac{p}{p_o}\right)^{\frac{2-\gamma}{\gamma}} - (\gamma+1) \left(\frac{p}{p_o}\right)^{\frac{1}{\gamma}} = 0$$

or $$\qquad 2 \left(\frac{p}{p_o}\right)^{\frac{2-\gamma}{\gamma}} = (\gamma+1) \left(\frac{p}{p_o}\right)^{\frac{1}{\gamma}}$$

or $$\qquad \frac{2}{\gamma+1} = \frac{\left(\dfrac{p}{p_o}\right)^{\frac{1}{\gamma}}}{\left(\dfrac{p}{p_o}\right)^{\frac{2-\gamma}{\gamma}}} = \left(\frac{p}{p_o}\right)^{\frac{1}{\gamma} - \left(\frac{2-\gamma}{\gamma}\right)}$$

$$\frac{2}{\gamma+1} = \left(\frac{p}{p_o}\right)^{\frac{1-2+\gamma}{\gamma}}$$

$$\frac{2}{\gamma+1} = \left(\frac{p}{p_o}\right)^{\frac{\gamma-1}{\gamma}}$$

$$\frac{p}{p_o} = \left(\frac{2}{\gamma+1}\right)^{\frac{\gamma}{\gamma-1}}$$

or
$$\frac{p}{p_o} = \left(\frac{2}{\gamma+1}\right)^{\frac{\gamma}{\gamma-1}}$$

or
$$\frac{p^*}{p_o} = \left(\frac{2}{\gamma+1}\right)^{\frac{\gamma}{\gamma-1}} \qquad\qquad ...(10.15.5)$$

Equation (10.15.5) is the critical pressure ratio. It is defined as thze ratio of the pressure at the throat to the inlet pressure when the Mach number is unity at the throat. It is standard practice in the analysis of compressible flow to let the superscript asterisk (*) represent the critical values *i.e.*, p^* = critical pressure at which $M = 1$ (at throat).

The critical pressure is also defined as the pressure at which discharge (or mass flow rate per unit area) is maximum. This is occurred at the throat when Mach number is unity.

For air : Adiabatic index : $\gamma = 1.4$

$$\therefore \text{ Critical pressure : } \frac{p^*}{p_o} = \left(\frac{2}{1.4+1}\right)^{\frac{1.4}{1.4-1}} = (0.8333)^{3.5} = 0.528$$

$$\frac{p^*}{p_o} = 0.528$$

where p_0 = pressure at inlet of the nozzle *i.e.*, at section 1–1.

10.16 NOZZLES OPERATING IN THE OFF-DESIGN CONDITION

(*i*) **Converging nozzle:** As we explained in section 16.14, that in converging nozzle flow velocity increases in the direction of flow. The Mach number obtained at the exit of converging nozzle is unity (*i.e.*, we get flow velocity equivalent to sound velocity at the exit of the nozzle).

By nozzle operating in the off-design condition means it is not possible to achieve $M = 1$ at the exit of the nozzle, reason being that the back pressure is not maintained in the right range.

Let us consider a convergent nozzle as shown in Fig. 10.18, which also shows the pressure ratio p/p_0 varying along the length of the nozzle. The inlet condition of an ideal gas is the stagnation state at p_0, T_0, which is assumed to be constant.

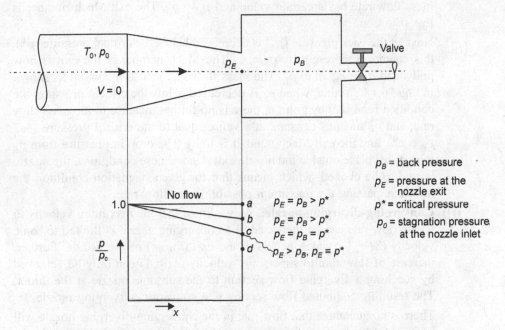

p_B = back pressure
p_E = pressure at the nozzle exit
p^* = critical pressure
p_0 = stagnation pressure at the nozzle inlet

No flow

a	$p_E = p_B > p^*$
b	$p_E = p_B > p^*$
c	$p_E = p_B = p^*$
d	$p_E > p_B, p_E = p^*$

Fig. 10.18: Pressure ratio as a function of back pressure for a convergent nozzle

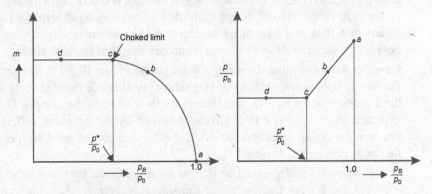

Fig. 10.19: Mass flow rate and exit pressure as function of back pressure for a convergent nozzle.

Let p_E = pressure of the gas at nozzle exit.

p_B = back pressure (*i.e.*, pressure at outside the nozzle exit, that can be varied by the valve)

As the back pressure p_B is varied, the mass flow rate m and the pressure ratio p_E/p_0 also vary, as shown in Fig. 10.19.

When $p_B = p_0$ (*i.e.*, $\dfrac{p_B}{p_0} = 1$), there is no flow and also $p_E = p_B = p_0$, as designated by point *a*. If the back pressure is now decreased to represent by point *b*, so that p_B/p_0 is greater than the critical pressure ratio p^*/p_0. The mass flow rate has a certain value and $p_E = p_B$. The exit Mach number is less than one.

Next let the back pressure (p_B) is decreased below the critical pressure (p^*), this condition represent by point c. The Mach number at the exit is now unity, and p_E is equal to p_B. The mass flow rate is also increased as shown in Fig. 16.18. Further, when p_B is decreased below the critical pressure, the condition represent by point d, there is no further increase in the mass flow rate, and p_E remains constant at a value equal to the critical pressure (*i.e.*, $p_E = p^*$), and the exit Mach number is unity. The drop in pressure from p_E to p_B takes place outside the nozzle exit. Under these conditions, the nozzle is said to be choked, which means that for given stagnation condition the nozzle is passing the maximum possible mass flow rate.

(*ii*) **Converging-diverging nozzle:** As we know that the maximum velocity to which flow rate can be increased in a converging nozzle is limited to sonic velocity ($M = 1$), which occurs at the exit (throat) of the nozzle. Further, increase of flow rate to supersonic velocity ($M > 1$) can only be achieved by attaching a diverging flow section to the subsonic nozzle at the throat. The resulting combined flow section is a converging-diverging nozzle.

There is no guarantee that flow rate in the converging-diverging nozzle will be increased to supersonic velocity. In fact, flow rate may decelerate in the diverging section instead of accelerating, if the back pressure is not maintained in the right range. By off-design condition of converging-diverging nozzle means that flow rate cannot be accelerated to supersonic velocity (or not possible to achieve $M > 1$) due to improper range of back pressure.

Consider the converging-diverging nozzle shown in Fig. 10.20. A fluid enters the nozzle with a low velocity at stagnation pressure p_0. Point a designates the condition when $p_B = p_0$ and thee is no flow through the nozzle. This is expected since the flow in a nozzle is driven by the pressure difference between the nozzle inlet and the exit. Now let us examine what happens as the back pressure is lowered.

(*a*) When p_B is decreased to the pressure indicated by point b, so that p_B/p_0 is less than one but considerably greater than the critical pressure ratio (p^*/p_0), the velocity increases in the converging section, but $M < 1$ at the throat. Therefore, the diverging section acts as a subsonic diffuser (*i.e.*, $M < 1$) in which the pressure increases and velocity decreases.

(*b*) Further, when p_B is decreased to the pressure indicated by point c and $p_B = p_E < p_0$, the throat pressure become the critical pressure (p^*) and the fluid achieves sonic velocity at the throat (*i.e.*, $M = 1$). But the diverging section of the nozzle still acts as a sub-sonic diffuser (*i.e.*, $M < 1$) in which the pressure increases and velocity decreases.

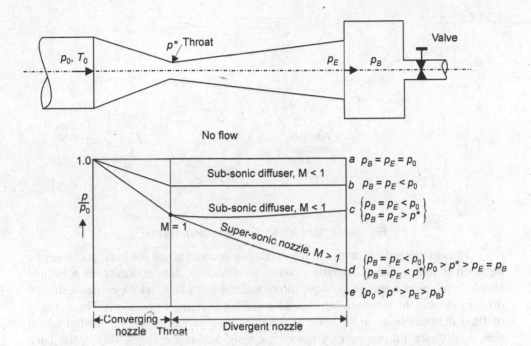

Fig. 10.20 Pressure distribution in a convergent-divergent nozzle

(c) When p_B is decreased and equal to the designing of exit pressure p_E, this condition indicated by point d, this condition permits isentropic flow, and in this case the diverging section acts as a super-sonic nozzle (*i.e.*, $M > 1$), with a decrease in pressure and an increase in velocity. When back pressure p_B is decreased below that designated by point d, the exit pressure p_E remain constant, and drop in pressure from p_E to p_B takes place outside the nozzle. This is designated by point e.

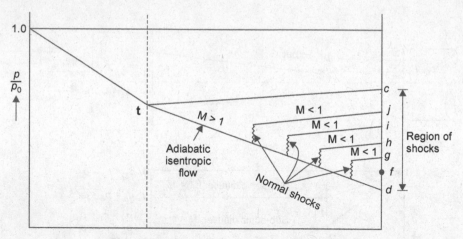

Fig. 10.21 Normal shock in diverging nozzle.

The flow is not isentropic in the diverging section, when the back pressures lie between c and d. The acceleration comes to a sudden stop, however, as a normal shock develops at a section between throat and the exit plane, which causes a sudden drop in velocity to sub-sonic levels and a sudden increase in pressure. The velocity of fluid then continues to decrease in the remaining part of diverging section which acts as diffuser. Properties vary discontinuously across the shock. When the back pressure is as indicated by point f in Fig. 10.21, the flow throughout the nozzle is isentropic, with pressure continuity decreasing and velocity increasing, but a shock appears just at the exit of the nozzle. When the back pressure is increased from f to g, the shock moves upstream as shown in Fig. 10.21. When the back pressure is further increased, the shock moves upstream and disappears at the nozzle throat where the back pressure corresponds to c. Since the flow thought is sub-sonic and no shock is possible.

Normal shock : It is defined as the shock waves that occurs due to increase in back pressure above the designing exit pressure of the converging-diverging nozzle in the diverging section normal to the direction of flow, called normal shock waves. The flow process through the shock wave is highly irreversible.

10.17 NORMAL SHOCKS

Normal shock waves are compression waves that is seen in converging-diverging nozzles, turbomachinery blade passages, flow through duct *etc*. In the first two examples, normal shock usually occurs under off-design operating condition or during start-up. The compression process across the shock wave is highly irreversible and so it is highly undesirable in such cases. In the last example, normal shock is designed to achieve extremely fast compression and heating of a gas with the aim of studying highly transient phenomena. The term normal is used to denote the fact that the shock wave is perpendicular to the flow direction, before and after passage through the shock waves. This latter fact implies that there is no change in flow direction as a result of passing through the shock wave.

10.17.1 Flow of Perfect Gases with Heat-transfer (Rayleigh Flow)

This flow is based upon the conservation of mass and momentum equations. Combining the conservation of mass and momentum equations into a single equation and plotting it on the h–s diagram yield a curved called the Rayleigh line.

Assumptions:

(a) One-dimensional flow with heat transfer.

(b) Constant cross-sectional area and frictionless duct.

(c) Perfect gas with constant specific heat and molecular weight.

Governing equations:

(i) Conservation of mass:

$$\rho_1 A_1 V_1 = \rho_2 A_2 V_2 \qquad |\because A_1 = A_2$$

or

$$\boxed{\rho_1 V_1 = \rho_2 V_2} \qquad ...(10.17.1)$$

where state points 1 and 2 represent the conditions at the entrance and the exit of the duct.

(ii) Conservation of momentum

$$A_1(p_1 - p_2) = m\ (V_2 - V_1)$$

$$p_1 - p_2 = \frac{m}{A}(v_2 - v_1)$$

We know that the continuity equation :

$$mv = AV$$

$$\frac{m}{A} = \frac{V}{v} = \rho V$$

or

$$\frac{m}{A} = \rho V$$

$\therefore$

$$p_1 - p_2 = \rho V\ (V_2 - V_1)$$

$$\boxed{p_1 - p_2 = \rho(V_2^2 - V_1^2)} \qquad ...(10.17.2)$$

(iii) Conservation of energy:

$$h_1 + \frac{V_1^2}{2} + q_{1-2} = h_2 + \frac{V_2^2}{2}$$

where q_{1-2} is the heat interaction per unit mass and is positive when heat is added to the flow and negative when heat is removed. Upon using the calorically perfect gas assumption and the definition of the stagnation temperature:

$$T_0 = T + \frac{V^2}{2c_p},$$

we get, $$\boxed{T_{02} - T_{01} = \frac{q}{c_p}}$$ ...(10.17.3)

Equation (10.17.3) shows that addition of heat to a flow increases the stagnation temperature, while heat removal decreases it.

(*iv*) 2nd law:

$$ds \geq 0$$

where $$ds = c_v \log_e \frac{p_2}{p_1} + c_p \log_e \frac{v_2}{v_1}$$ ...(10.17.4)

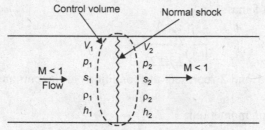

Control volume   Normal shock

V_1 V_2

M < 1 p_1 p_2 M < 1
Flow s_1 s_2
        ρ_1 ρ_2
        h_1 h_2

Control volume for flow across normal shock wave

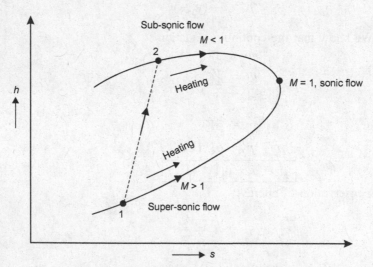

Sub-sonic flow
M < 1

2

Heating   M = 1, sonic flow

h

Heating

M > 1

1   Super-sonic flow

s

Fig. 10.22 Normal shock and Rayleigh line

Let state 1 corresponds to the state before the shock and state 2 corresponds to the state after the shock. Note that the flow is super-sonic before the shock and sub-sonic afterward. Therefore, the flow must change from super-sonic to sub-sonic if a shock is to occur. The larger the Mach number before the shock, the stronger the shock will be. In the limiting case of $M = 1$, the shock wave simply becomes a sound wave.

By simplifying above equations, we get

$$dM = \left(\frac{1+\gamma M^2}{2}\right)M\frac{dV}{V}$$

$$dT_0 = dT + \frac{VdV}{c_p}$$

$$ds = c_v\gamma(1-M^2)\frac{dV}{V}$$

$$dT = (1-\gamma M^2)T\frac{dV}{V}$$

$$d\rho = -\rho\frac{dV}{V}$$

$$dp = m\frac{dV}{A}$$

Table 10.3 Change in Properties for a Given Change in Velocity V

	V ↑	V ↓
M < 1	ρ↓ p↓ T↑ s↑ M↑ T_0↑	ρ↑ p↑ T↓ s↓ M↓T_0↓
M > 1	ρ↓ p↓ T↓ s↓ M↑T_0↓	ρ↑ p↑ T↑ s↑ M↓T_0↑

Using symbols ↑ and ↓ to indicate increasing and decreasing trends.

10.17.2 Flow of Perfect Gases with Friction (Fanno Flow)

This flow is based upon the conservation of mass (*i.e.*, continuity equation) and energy equations. Combining the conservation of mass and energy equations into a single equation and plotting it on the h–s diagram yield a curve called the Fanno line.

Assumptions:

(*a*) One-dimensional flow with friction

(*b*) Constant cross-sectional area of duct.

(*c*) Perfect gas with constant specific heat and molecular weight.

(*d*) Adiabatic flow.

Governing equations:

(*i*) Conservation of mass:

$$\rho_1 AV_1 = \rho_2 AV_2$$

or $\boxed{\rho_1 V_1 = \rho_2 V_2}$ $\mid \because A_1 = A_2$

(*ii*) Conservation of energy :

$$\boxed{h_1 + \frac{V_1^2}{2} = h_2 + \frac{V_2^2}{2}}$$

(*iii*) Increase in entropy:

$$\boxed{ds = c_v \log_e \frac{p_2}{p_1} + c_p \log_e \frac{v_2}{v_1}}$$

The momentum Eq. (10.17.2) has to be modified to take into account frictional force at the wall and so

$$p_1 - p_2 = \rho \ (V_2^2 - V_1^2) + \frac{P}{A} \int_0^l \tau_0 dx$$

where P is the wetted perimeter

          l is the length of the duct

and τ_0 is the wall share stress

The Darcy friction factor f' is related to the wall share stress as:

$$f' = \frac{\tau_o}{\frac{1}{2}\rho V^2}$$

$\therefore$ $$\boxed{p_1 - p_2 = \rho\left(V_2^2 - V_1^2\right) + \frac{P}{A} \int_0^l \frac{f'\rho V^2}{2} dx}$$

By simplifying above equations, we get

$$dM = \left(1 + \frac{\gamma - 1}{2} M^2\right) M \frac{dV}{V}$$

$$ds = R(1 - M^2)\frac{dV}{V}$$

where R = gas constant

$$dp = -[1 + (\gamma - 1)M^2]p\frac{dV}{V}$$

$$dT = -(\gamma - 1)M^2 T \frac{dV}{V}$$

$$d\rho = -\rho \frac{dV}{V}$$

Table 10.4 Change in Properties for a Given Change in Velocity V

	$V\uparrow$	$V\downarrow$
$M < 1$	$\rho\downarrow \ p\downarrow \ T\downarrow \ s\uparrow \ M\uparrow$	Not allowed
$M > 1$	Not allowed	$\rho\uparrow \ p\uparrow \ T\uparrow \ s\uparrow \ M\downarrow$

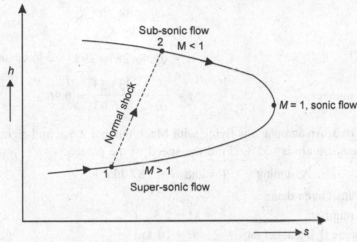

Fig. 10.23 Normal shock and Fanno line.

The observation in Table 10.4 can be summarised conveniently as follows. The effect of friction on a sub-sonic flow is to increase the velocity, Mach number and decrease the static temperature and static pressure. On the other hand, effect of friction on a supersonic flow, is to increase the static temperature and static pressure, while velocity and Mach number decrease. But the entropy is increase in both the cases because of the flow with friction and adiabatic.

Problem 10.1: Find the speed of the sound wave in air where the pressure and temperature are 101 kPa and 15°C respectively.

Solution: Given data:

Pressure: $p = 101$ kPa

Temperature: $T = 15°C = (273 + 15)$ K $= 288$ K

For adiabatic process:

Velocity of sound: $a = \sqrt{\gamma RT}$

where $\gamma = 1.4$ for air

                                                            $R = 287$ J/kg K

∴ $a = \sqrt{1.4 \times 287 \times 288}$ $= \mathbf{340.17 \ m/s}$

Problem 10.2: Determine the Mach number at a point on aircraft, which is flying at 1200 km/h at sea level where air temperature 25°C.

Solution: Given data:

Speed of air craft: $V = 1200$ km/h $= \dfrac{1200 \times 1000}{3600} = 333.33$ m/s

Temperature: $T = 25°C = (273 + 25)$ K $= 298$ K

Velocity of sound: $a = \sqrt{\gamma RT}$ for adiabatic process

where $\gamma = 1.4$ for air
                                             $R = 287$ J/kg K

∴ $a = \sqrt{1.4 \times 287 \times 298} = 346.03$ m/s

Mach number: $M = \dfrac{V}{a} = \dfrac{333.33}{346.03} = \mathbf{0.96}$

Problem 10.3: An aeroplane is flying with Mach number 2.5 at an height of 10 km where the temperature is –35°C. Find the speed of the plane.

Assuming $\gamma = 1.4$ and $R = 287$ J/kg K

Solution: Given data:

Mach number: $M = 2.5$
Aeroplane flying at height: $h = 10$ km
Temperature: $T = -35°C$
                                             $= (273 - 35)$ K $= 238$ K

Velocity of sound: $a = \sqrt{\gamma RT}$ for adiabatic process

where $\gamma = 1.4$ for air
                                             $R = 287$ J/kg K

∴ $a = \sqrt{1.4 \times 287 \times 238} = 309.23$ m/s

Mach number : $M = \dfrac{V}{a}$

                                             $2.5 = \dfrac{V}{309.23}$

or $V = 773.07$ m/s $= \mathbf{2783.05}$ **km/h**

Problem 10.4:   Determine the Mach number when an aeroplane is flying at 1000 km/h through still air having a pressure of 70 kPa and temperature –15°C. Determine also the pressure, temperature and density of air at the stagnation point on the nose of the aeroplane.

Solution: Given data:

Speed of aeroplane: $V = 1000$ km/h

                                             $= \dfrac{1000 \times 1000}{3600}$ m/s $= 277.77$ m/s

Pressure of air:	$p_1 = 70$ kPa
Temperature:	$T_1 = -15°C = (273 - 15)$ K $= 258$ K
Velocity of sound:	$a = \sqrt{\gamma RT}$ for adiabatic process
where	$\gamma = 1.4$ for air
	$R = 287$ J/kg K

$$\therefore \quad a = \sqrt{1.4 \times 287 \times 258} = 321.97 \text{ m/s}$$

$\therefore$ Mach number:
$$M = \frac{V}{a} = \frac{277.77}{321.97} = 0.86 \text{ km/h}$$

Stagnation pressure:
$$p_0 = p_1 \left[1 + \left(\frac{\gamma - 1}{2} \right) M^2 \right]^{\frac{\gamma}{\gamma - 1}}$$

$$= 70 \left[1 + \left(\frac{1.4 - 1}{2} \right) \times (0.86)^2 \right]^{\frac{1.4}{1.4 - 1}}$$

$$= 70 [1 + 0.148]^{3.5} = 113.47 \text{ kPa}$$

Stagnation temperature:
$$T_0 = T_1 \left[1 + \left(\frac{\gamma - 1}{2} \right) M^2 \right]$$

$$= 258 \left[1 + \left(\frac{1.4 - 1}{2} \right) (0.86)^2 \right]$$

$$= 296.18 \text{ K} = 23.18°C$$

Stagnation density:
$$\rho_0 = \frac{p_0}{RT_0} \text{ kg/m}^3$$

where	$p_0 = 113.47$ kPa $= 113.47 \times 10^3$ Pa
	$R = 287$ J/kg K
and	$T = 258$ K

$$\therefore \quad \rho_0 = \frac{113.47 \times 10^3}{287 \times 258} = 1.53 \text{ kg/m}^3$$

Problem 10.5: A supersonic aircraft flies at an altitude of 2000 m where the air temperature is 5°C. Determine the speed of the aircraft if its sound is heard 4.5 s after its passage over the head of an observer.

Solution: Given data:

Altitude:	$h = 2000$ m
Temperature of air:	$T = 5°C = (5 + 273) = 278$ K

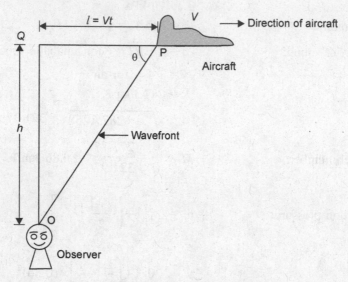

Fig. 10.24: Schematic for Problem 10.5

Sound heard after aircraft passage over the head of an observer:

$$t = 4.5 \text{ s}$$

Let O represent the observer and P in position of aircraft just over the observer. After 4.5 s, the aircraft reaches a position represented by Q. Line OQ represents the wave front and θ is the Mach angle.

Velocity of sound in air: $a = \sqrt{\gamma R T}$

where $\gamma = 1.4$

$R = 287$ J/kg K

$T = 278$ K

$\therefore$ $a = \sqrt{1.4 \times 287 \times 278} = 334.21$ m/s

From Fig. 16.24

$$\tan \theta = \frac{h}{l} = \frac{h}{Vt}$$

$$\tan \theta = \frac{2000}{V \times 4.5}$$

$$\frac{\sin \theta}{\cos \theta} = \frac{2000}{4.5V}$$

or $\sin \theta = \dfrac{2000 \cos \theta}{4.5V}$

also $\sin \theta = \dfrac{1}{M} = \dfrac{1}{\dfrac{V}{a}} = \dfrac{a}{V}$

$$\therefore \qquad \frac{a}{V} = \frac{2000 \cos\theta}{4.5V}$$

or
$$a = \frac{2000}{4.5}\cos\theta$$

$$334.21 = \frac{2000}{4.5}\times\cos\theta$$

$$\cos\theta = 0.7519$$

$$\theta = \cos^{-1}(0.7519) = 41.24°$$

$$\sin\theta = \frac{a}{V}$$

$$\sin 41.24° = \frac{334.21}{V}$$

or
$$V = 506.98 \text{ m/s}$$

The speed of aircraft is **506.98 m/s**

Problem 10.6: Determine the velocity of air flowing at the exit of the nozzle, fitted to a large container which contains air at a pressure of 30 bar (abs) and at a temperature of 20°C. The pressure at the exit of the nozzle is 21 bar (abs). Assuming $\gamma = 1.4$ and $R = 287$ J/kg K.

Solution: Given data:

Pressure of air in the container: $p_1 = 30$ bar $= 30 \times 10^5$ N/m²

Temperature of air in the container:
$$T_1 = 20°C = (20 + 273) \text{ K} = 293 \text{ K}$$

Pressure at the exit of the nozzle: $p_2 = 21$ bar $= 21 \times 10^5$ N/m²

Velocity at the exit of the nozzle: $V_2 = \sqrt{\dfrac{2\gamma}{\gamma-1}p_1v_1\left[1-\left(\dfrac{p_2}{p_1}\right)^{\frac{\gamma-1}{\gamma}}\right]}$

From the equation of state,
$$p_1v_1 = RT_1$$
$$30 \times 10^5 \times v_1 = 287 \times 293$$
$$\therefore \qquad v_1 = 0.0280 \text{ m}^3/\text{kg}$$

$$\therefore \qquad V_2 = \sqrt{\frac{2\times1.4}{1.4-1}\times30\times10^5\times0.0280\left[1-\left(\frac{21\times10^5}{30\times10^5}\right)^{\frac{1.4-1}{1.4}}\right]}$$

$$= \sqrt{588000\times[1-0.903]} = \sqrt{57036} = \textbf{238.82 m/s}$$

Problem 10.7:   A nozzle of exit diameter 20 mm is fitted to a large vessel which contains air at 20°C. The air flows from the vessel into atmosphere through the nozzle. For adiabatic flow, determine the mass rate of flow of air through the nozzle when pressure of air in vessel is (*i*) 58 kPa (gauge), and (*ii*) 295 kPa (gauge). Take $\gamma = 1.4$, $R = 287$ J/kg K and atmospheric pressure : $p_{atm} = 101$ kPa.

Solution: Given data:

Diameter at the nozzle exit: $d_1 = 20$ mm $= 0.02$ m

∴ Cross-sectional area: $A_2 = \dfrac{\pi}{4} d_2^2 = \dfrac{3.14}{4} \times (0.02)^2 = 3.14 \times 10^{-4}$ m^2

Temperature of air in vessel: $T_1 = 20°C = (20 + 273)$ K $= 293$ K

Adiabatic index: $\gamma = 1.4$

Gas constant: $R = 287$ J/kg K

Atmospheric pressure: $p_{atm} = 101$ kPa

Case I: Mass rate of flow of air when pressure in vessel is 58 kPa (gauge)

∴ $p_{g1} = 58$ kPa

Absolute pressure of air in vessel:

$$p_1 = p_{g1} + p_{atm} = 58 + 101 = 159 \text{ kPa}$$

Absolute pressure at nozzle exit:

$$p_2 = p_{atm} = 101 \text{ kPa}$$

∴ Pressure ratio : $\dfrac{p_2}{p_1} = \dfrac{101}{159} = 0.6352$

We know that the critical pressure ratio:

$$\frac{p^*}{p_o} = 0.528$$

As pressure ratio $\dfrac{p_2}{p_1}$ is more than the critical pressure ratio $\dfrac{p^*}{p_o}$. Then

Mass rate of flow of air: $m = A_2 \sqrt{\dfrac{2\gamma}{\gamma - 1} p_1 v_1 \left[\left(\dfrac{p_2}{p_1} \right)^{\frac{2}{\gamma}} - \left(\dfrac{p_2}{p_1} \right)^{\frac{\gamma+1}{\gamma}} \right]}$

where $A_2 = 3.14 \times 10^{-4}$ m^2
                $p_1 = 159 \times 10^3$ Pa
                $p_2 = 101 \times 10^3$ Pa
                v_1 in m^3/kg

From the equation of state,

$$p_1 v_1 = RT_1$$
$$159 \times 10^3 \times v_1 = 287 \times 293$$
∴ $v_1 = 0.5288$ m^3/kg

$$\therefore \; m = 3.14 \times 10^{-4} \sqrt{\frac{2 \times 1.4}{1.4-1} \times 159 \times 10^{3} \times 0.5288 \left[\left(\frac{101 \times 10^{3}}{159 \times 10^{3}} \right)^{\frac{2}{1.4}} - \left(\frac{101 \times 10^{3}}{159 \times 10^{3}} \right)^{\frac{1.4+1}{1.4}} \right]}$$

$$= 3.14 \times 10^{-14} \sqrt{588554.4 \left[(0.6352)^{1.428} - (0.6352)^{1.714} \right]}$$

$$m = 3.14 \times 10^{-4} \sqrt{588554.4 \left[0.5230 - 0.4594 \right]} = \mathbf{0.06075 \; kg/s}$$

Case II: Mass rate of flow of air when pressure in vessel is 295 kPa (gauge)

$$\therefore \qquad\qquad p_{g1} = 295 \; kPa$$

$\therefore$ Absolute pressure of air in vessel:

$$p_1 = p_{g1} + p_{atm} = 295 + 101 = 396 \; kPa$$

Absolute pressure at nozzle exit: $p_2 = p_{atm} = 101 \; kPa$

$\therefore$ Pressure ratio:

$$\frac{p_2}{p_1} = \frac{101}{396} = 0.255$$

As the pressure ratio $\dfrac{p_2}{p_1}$ is less than critical pressure ratio $\dfrac{p^*}{p_o}$, the mass rate of flow is constant and is equal to the mass rate of flow corresponding to critical pressure ratio $\dfrac{p^*}{p_o} = 0.528$.

$$\therefore \qquad m = A_2 \sqrt{\frac{2\gamma}{\gamma - 1} p_1 v_1 \left[\left(\frac{p^*}{p_o} \right)^{\frac{2}{\gamma}} - \left(\frac{p^*}{p_o} \right)^{\frac{\gamma+1}{\gamma}} \right]}$$

where

$$A_2 = 3.14 \times 10^{-4} \; m^2$$
$$p_1 = 396 \times 10^{3} \; Pa$$

$$\frac{p^*}{p_o} = 0.528$$

From equation of state,

$$p_1 v_1 = RT_1$$
$$396 \times 10^{3} \times v_1 = 287 \times 293$$

or

$$v_1 = 0.2123 \; m^3/kg$$

$$\therefore \; m = 3.14 \times 10^{-4} \sqrt{\frac{2 \times 1.4}{1.4-1} \times 396 \times 10^{3} \times 0.2123 \left[(0.528)^{\frac{2}{1.4}} - (0.528)^{\frac{1.4+1}{1.4}} \right]}$$

$$= 3.14 \times 10^{-4} \sqrt{588495.6 \left[(0.528)^{1.428} - (0.528)^{1.714} \right]}$$

$$= 3.14 \times 10^{-4} \sqrt{588495.6 \left[0.4017 - 0.3346 \right]} = \mathbf{0.06239 \; kg/s}$$

SUMMARY

1. **Equation of state:** The equation in temperature, pressure and volume (or specific volume or density) which used to describe the condition or state of the system, is called an equation of state.

   For a perfect gas, the equation of state is

   $$pV = mRT$$

   $$\frac{pV}{m} = RT$$

   $$pv = RT$$

   $$\frac{p}{\rho} = RT$$

   $$p = \rho RT$$

2. **Gas constant:** R. It is defined as the ratio of the universal gas constant ($\overline{R}$) to the molecular weight of a perfect gas.

   Mathematically,

   $$\text{Gas constant: } R = \frac{\text{Universal gas constant: } \overline{R}}{\text{Molecular weight: } M}$$

   $$R = \frac{\overline{R}}{M}$$

   where $R = 8.314 \text{ kJ/k mol K}$

3. **Thermodynamics processes:**

   (*i*) Isothermal process [$T = C$]

   Isothermal process takes place at constant temperature. The volume (or specific volume) of the gas varies inversely with pressure at constant temperature in this type of process.

   Mathematically,

   $$v \propto \frac{1}{p} \quad \text{at} \quad T = C$$

   or $pv = C$

   Isothermal process follows the Boyle's law which state that the specific volume of a perfect gas in inversely proportional to the absolute pressure when the temperature is kept constant.

   (*ii*) Isobaric process [$p = C$]

   Isobaric process takes place at constant pressure. The volume (or specific volume) of the gas varies directly with temperature at constant pressure in this type of process.

   Mathematically,

Contd...

$$v \propto T \quad \text{at} \quad p = C$$

$$\frac{v}{T} = C$$

Isobaric process follows the (Charles' law which state that the specific volume of a perfect gas is directly proportional to the absolute temperature when the process is kept constant.

(*iii*) Isochoric process (or isometric process) [$V = C$]

Isochoric process takes place at constant volume. The pressure of the gas varries directly with temperature at constant volume in this type of process.

Mathematically,

$$p \propto T \quad \text{at} \quad V = C$$

or $$\frac{p}{T} = C$$

This process follows the Gay-Lussac's law which state that the absolute pressure of a perfect gas is directly proportional to the absolute temperature when the volume is kept constant.

(*iv*) Adiabatic process [$pv^\gamma = C$]

When there is no heat transfer between the system and surroundings during a process, it is known as an adiabatic process. The properties, pressure, temperature and volume of the system will vary during this process.

This process is represented mathematically as
$$pv^\gamma = C$$

where $$\gamma = \frac{c_p}{c_v}, \text{ adiabatic index}$$

$$= 1.4 \text{ for air}$$

$p - v - T$ relationship:

$$\frac{T_2}{T_1} = \left(\frac{p_2}{p_1}\right)^{\frac{\gamma-1}{\gamma}} = \left(\frac{v_1}{v_2}\right)^{\gamma-1}$$

(*v*) Polytropic process [$pv^n = C$]:

Polytropic process is a real process followed by compression and expansion process. For example: Actual expansion in the nozzles, turbines, IC engines and actual compression in the compressors and IC engines are followed the polytropic process. This process can be represented by the equation

$$pv^n = C$$

where $$n = \text{polytropic index}$$

Contd...

p–v–T relationship:

$$\frac{T_2}{T_1} = \left(\frac{p_2}{p_1}\right)^{\frac{n-1}{n}} = \left(\frac{v_1}{v_2}\right)^{n-1}$$

4. **Steady and unsteady flow:**

   A flow is considered to be steady if fluid flow parameter such as velocity (V), pressure (p), temperature (T) etc. at any point do not change with time. If any one of these parameters changes with time at particular point, the flow is said to be unsteady.

   Mathematically, $\dfrac{\partial V}{\partial t} = 0$, $\dfrac{\partial p}{\partial t} = 0$ for steady flow

   $\dfrac{\partial V}{\partial t} \neq 0$, $\dfrac{\partial p}{\partial t} \neq 0$ for unsteady flow

5. **Uniform and non-uniform flow:**

   A type of flow in which velocity (V), pressure (p), density (ρ), temperature (T) etc. at any given time do not change with respect to space (i.e., length of direction of the flow) is called uniform flow.
   Mathematically,

   $$\frac{\partial V}{\partial s} = 0, \quad \frac{\partial p}{\partial s} = 0 \quad \text{for steady flow}$$

   In case of non-uniform flow, velocity, pressure, density etc at given time change with respect to space.
   Mathematically,

   $$\frac{\partial V}{\partial s} \neq 0, \quad \frac{\partial p}{\partial s} \neq 0 \quad \text{for non-uniform flow}$$

6. **Compressible and incompressible flow:**

   (*i*) Compressible flow [$\rho \neq C$]

   If the density of the fluid changes from point to point in the fluid flow, is referred to as compressible flow

   Mathematically,

   Density: $\rho \neq C$

   For example: Gases (like air, carbon dioxide etc) are compressible.

   (*ii*) Incompressible flow [$\rho = C$]

   If the density of the fluid remains constant at every point in the fluid flow, it is referred to as an incompressible flow.

   Mathematically,

   Density: $\rho = C$

Contd...

For example: Liquid are generally incompressible in nature (solids are also incompressible in nature).

Match number: $M = \dfrac{V}{a}$

where V = velocity of fluid

a = speed of sound in the flowing fluid

If $M > 0.2$, the gases are considered to be compressible.

If $M < 0.2$, the gases are considered to be incompressible.

7. **Rate of flow:**

   (*i*) Volume of the fluid flowing across the reaction per second is called discharge: Q

   $$Q = AV = mv \text{ m}^3/s$$

   where A = cross-sectional area of pipe

   V = average velocity of fluid across the

section.

   v = specific volume

   $$= \dfrac{1}{\rho}$$

   (*ii*) Mass of the fluid flowing across the section per second: m

   $$AV = mv$$

   or $$m = \dfrac{AV}{v} = \rho AV \text{ kg/s}$$

   (*iii*) Weight of the fluid flowing across the section per second: $\dot{W}$

   $$\dot{W} = mg \text{ N/s}$$

   $$= \rho AVg \text{ N/s}$$

8. **Continuity equation:** It is based on the law of conservation of mass. It means mass of fluid can neither be created nor be destroyed.

   $$m = \rho AV = \text{constant}$$

   $$\rho_1 A_1 V_1 = \rho_2 A_2 V_2 \quad \text{for compressible fluid}$$

   $$A_1 V_1 = A_2 V_2 \quad \text{for incompressible fluid}$$

9. **Steady flow energy equation (SFEE):**

   $$m\left[h_1 + \dfrac{V_1^2}{2} + gz_1 \right] + Q_{1-2} = m_2\left[h_2 + \dfrac{V_2^2}{2} + gz_2 \right] + W_{1-2}$$

Contd...

$$h_1 + \frac{V_1^2}{2} + gz_1 + \underset{1-2}{q} = h_2\frac{V_2^2}{2} + gz_2 + \underset{1-2}{w}$$

where $\underset{1-2}{q} = \dfrac{\underset{1-2}{Q}}{m}$, specific heat transfer, which is defined as the rate of

heat transfer per unit mass flow rate. Its SI unit is kJ/kg and $\underset{1-2}{w} = \dfrac{\underset{1-2}{W}}{m}$,

specific work transfer, which is defined as the rate of work done per unit mass flow rate. Its SI unit is kJ/kg.

10. **Stagnation state:**

Stagnation state is the state of flowing fluid at which the flow come to stop. This state is achieved by decelerating the fluid adiabatic isentropically to zero velocity. The enthalpy, temperature, pressure, density at stagnation state are called stagnation enthalpy h_0, stagnation temperature T_0, stagnation pressure p_0 and stagnation density ρ_0 respectively.

Stagnation enthalpy: $\quad h_0 = h_1 + \dfrac{V_1^2}{2}$

$$\frac{T_0}{T_1} = 1 + \frac{(\gamma-1)M^2}{2}$$

$$\frac{p_0}{p_1} = \left[1 + \frac{(\gamma-1)}{2}M^2\right]^{\frac{\gamma}{\gamma-1}}$$

$$\frac{\rho_0}{\rho} = \left[1 + \frac{(\gamma-1)}{2}M^2\right]^{\frac{1}{\gamma-1}}$$

11. **Velocity of sound:** $\quad a = \sqrt{\dfrac{dp}{d\rho}} = \sqrt{\dfrac{K}{\rho}}$

where $\qquad\qquad K$ = Bulk modulus of elasticity

ρ = density of fluid.

$a = \sqrt{\gamma RT}$ for adiabatic process

$a = \sqrt{RT}$ for isothermal process

12. **Mach number:** M. It is defined as the ratio velocity of the fluid to the velocity of sound in same fluid. It is denoted by letter M.

Mathematically,

Contd...

Mach number:
$$M = \frac{V}{a}$$

If
$0.9 < M < 1.1$, the flow is transonic
$M < 1$, the flow is subsonic
$1 < M < 7$, the flow is supersonic
$7 \leq M < 10$, the flow is hypersonic, and
$M \geq 10$, the flow is high hypersonic

13. Mach angle: $\quad \sin \theta = \dfrac{a}{V} = \dfrac{1}{M}$

$$\theta = \sin^{-1} \left(\frac{1}{M} \right)$$

14. **Nozzle.** It is a device which increases the velocity of the fluid and simultaneously it decreases the pressure of fluid. As the fluid flows through the nozzle it expand to a lower pressure and in this process the pressure falls and the velocity increases continuously from the entrance to exit of the nozzle.

   Types of nozzle:

   (*i*) Converging nozzle (or subsonic nozzle)

   (*ii*) Diverging nozzle (or supersonic nozzle)

   (*iii*) Converging – diverging nozzle.

15. Area-velocity relationship for compressible fluid is used to design the nozzle.

$$\frac{dA}{A} = \frac{dV}{V} \left[M^2 - 1 \right]$$

   where $\quad M = $ Mach number

16. Velocity at exit of nozzle: V_2

$$V_2 = 44.72 \sqrt{h_1 - h_2} \text{ m/s}$$

   where h_1 and h_2 in kJ/kg.

   and inlet velocity V_1 is neglected.

17. Critical pressure ratio: $\dfrac{p^*}{p_o} = \left(\dfrac{2}{\gamma + 1} \right)^{\frac{\gamma}{\gamma - 1}}$

18. **Normal shocks.**

   Normal shock waves are compression waves that seen in converging-diverging nozzles and turbomachinery blade passages. It usually occurs under off–design operating condition. The term 'normal' is used to denote the fact that the shock wave is perpendicular to the flow direction, before and after passage through the shock waves.

19. Flow of perfect gases with heat transfer [Rayleigh flow]

This flow is based upon the conservation of mass and momentum equations. Combining the conservation of mass and momentum equations into a single equation and plotting it on the $h - s$ diagram yield a curved called the Rayleigh line.

Assumptions:

(a) One dimensional flow with heat transfer

(b) Constant cross-sectional area and frictionless duct.

(c) Perfect gas with constant specific heat and molecular weight.

20. Flow of perfect gases with friction [Fanno flow]

This flow is based upon the conservation of mass (i.e., continuity equation) and energy equation. Combining the conservation of mass and energy equation into a single equation and plotting it on the $h–s$ diagram yield a curve called the Fanno line.

Assumptions:

(a) One-dimensional flow with friction.

(b) Constant cross-sectional area of duct.

(c) Perfect gas with constant specific heat and molecular weight.

(d) Adiabatic flow

ASSIGNMENT - 1

1. What is equation of state? Write its three different form.
2. Explain the following thermodynamics processes.

   (i) Isothermal process   (ii) Isochoric process

   (iii) Adiabatic process   (iv) Polytropic process
3. Explain briefly steady and unsteady flow.
4. Define compressible and incompressible flow.
5. State the equation of continuity. Write down it for compressible and incompressible flow.
6. State the steady flow energy equation.
7. What do you understand by stagnation state.
8. Prove that the following stagnation parameters.

   (i) Stagnation temperature : $T_0 = T_1\left[1+\left(\frac{\gamma-1}{2}\right)M^2\right]$

   (ii) Stagnation pressure : $p_0 = p_1\left[1+\left(\frac{\gamma-1}{2}\right)M^2\right]^{\frac{\gamma}{\gamma-1}}$

(iii) Stagnation density : $\rho_0 = \rho_1 \left[1 + \left(\dfrac{\gamma - 1}{2} \right) M^2 \right]^{\frac{1}{\gamma - 1}}$

9. Derive an expression for velocity of the sound wave in a compressible fluid in terms of change of pressure and change of density.

10. Show that the velocity of propagation of the pressure wave in a compressible fluid is given by

$$a = \sqrt{\dfrac{K}{\rho}}$$

where K is bulk modulus of elasticity.

11. Derive an expression for velocity of sound for an adiabatic process. Show that the velocity of sound wave:

(i) $a = \sqrt{\gamma RT}$ for adiabatic isentropic process

(ii) $a = \sqrt{RT}$ for isothermal process.

12. Define Mach number. What is the significance of Mach number in compressible flow?

13. Define the following terms:
   (i) Transonic flow   (ii) Sub-sonic flow
   (iii) Sonic flow   (iv) Super-sonic flow
   (v) Hyper-sonic flow   (vi) High hyper-sonic flow

14. Define the following terms:
   (i) Mach angle   (ii) Mach cone

15. With neat sketches, explain the process of propagation of sound wave due to disturbance in a compressible fluid for different Mach numbers.

16. Show by means of diagrams the nature of propagation of disturbance in compressible of propagation of disturbance in compressible flow when Mach number is less than one, is equal to one and is more than one.

17. What are a nozzle and a diffuser?

18. Explain the effect of area change in sub-sonic and super-sonic flow.

19. What do you understand by choking in nozzle flows?

20. What do you understand by critical pressure ratio? What is its value for air?

21. What is a shock? Where does it occur in a nozzle?

22. What is a Fanno line?

23. What is a Rayleigh line?

1. Find the speed of the sound wave in air where the pressure and temperature are 100 kPa and 14°C respectively. **Ans.** 339.58 m/s

2. Determine the Mach number at a point on aircraft, which is flying at 850 km/h at sea-level where air temperature is 15°C. Take $\gamma = 1.4$ and $R = 287$ J/kg K. **Ans.** 0.69

3. An aeroplane is flying at an height of 18 km, where the temperature is −50°C. The speed of the plane is corresponding to $M = 1.5$. Take $\gamma = 1.4$ and $R = 0.287$ kJ/kg K. **Ans.** 449 m/s

4. Find the Mach number when an aeroplane is flying at 900 km/h through still air having a pressure of 80 kPa and temperature −15°C. Find also the pressure, temperature and density of air at the stagnation point on the nose of the plane. **Ans.** 0.77, 118.42 kPa, 15.59°C, 1.43 kg/m^3.

5. Determine the velocity of air flowing at the exit of the nozzle, fitted to a large vessel which contains air at a pressure of 32 bar (abs) and at a temperature of 30°C. The pressure at the exit of the nozzle is 18 bar (abs). Assuming $\gamma = 1.4$ and $R = 287$ J/kg K. **Ans.** 303.35 m/s

6. A vessel contains air at a temperature of 30°C. Air flows from the vessel into atmosphere through a sub-sonic nozzle. The diameter at the exit of the nozzle is 25 mm. Assuming the adiabatic flow, determine the mass rate of flow of air through the nozzle when the pressure of air in tank is (*i*) 39 kPa (gauge), and (*ii*) 330 kPa (gauge). Assuming $\gamma = 1.4$, $R = 287$ J/kg K and atmospheric pressure = 101 kPa. **Ans.** 0.0905 kg/s (*ii*) 0.0991 kg/s

7. A supersonic nozzle is to be designed for air flow with Mach number 3 at the nozzle exit which is 200 mm in diameter. The pressure and temperature of air at the nozzle exit to be 7.85 kPa and 200 K respectively. Determine the reservoir pressure and temperature. Take $\gamma = 1.4$ **Ans.** 287.5 kPa, 560 K

   Hints : Since the velocity, in the reservoir is zero, the temperature and pressure there correspond to the stagnation condition,

$$p_0 = p\left[1 + \left(\frac{\gamma - 1}{2}\right)M^2\right]^{\frac{\gamma}{\gamma - 1}}$$

$$T_0 = T\left[1 + \left(\frac{\gamma - 1}{2}\right)M^2\right]$$

◻◻◻

References

1. Yunus A. Cengel and John M. Cimbala; *Fluid Mechanics*, Tata McGraw Hill Publishing Company Limited, New Delhi, 2006.

2. A.K. Jain; *Fluid Mechanics*, Khanna Publishers, Delhi, 2002.

3. Shiv Kumar; *Fluid Systems*, Satya Prakashan, New Delhi, 2008.

4. H. Schlichting; *Boundary Layer Theory*, McGraw Hill, New York, 1987.

5. F.M. White; *Fluid Mechanics*, McGraw - Hill, New York, 2003.

6. S.K. Som and C. Biswas; *Fluid Mechanics and Fluid Machines*, Tata McGraw Hill, New Delhi, 2006.

7. R.K. Bansal; *Fluid Mechanics and Hydraulic Machines*, Laxmi Publications (P) Ltd. New Delhi, 2002.

8. Waren L. McCabe, Julian C. Smith, Peter Harriott; *Unit Operations of Chemical Engineering*, McGraw-Hill, New York, 2005.

9. A.K. Mohanty; *Fluid Mechanics*, Prentice-Hall of India Private Limited, New Delhi, 1994.

10. R.W. Fox and A.T. McDonald; *Introduction of Fluid Mechanics*, Wiley, New York, 1999.

11. D. Rama Durgaiah; *Fluid Mechanics and Machinery*, New Age International (P) Limited, New Delhi, 2002.

12. C.T. Crowe, J.A. Roberson and D.F. *Engineering Fluid Mechanics*, Wiley, New York, 2001.

13. B.R. Munson, D.F. Young and T. Okiishi; *Fundamentals of Fluid Mechanics*, Wiley, New York, 2002.

© The Author(s) 2023
S. Kumar, *Fluid Mechanics (Vol. 2)*,
https://doi.org/10.1007/978-3-030-99754-0

Appendices

Standard Prefixes in SI Units

Prefix	Symbol	Multiple
exa	E	10^{18}
peta	P	10^{15}
tera	T	10^{12}
giga	G	10^{9}
mega	M	10^{6}
kilo	k	10^{3}
hecto	h	10^{2}
deka	da	10^{1}
deci	d	10^{-1}
centi	c	10^{-2}
milli	m	10^{-3}
micro	μ	10^{-6}
nano	n	10^{-9}
pico	p	10^{-12}
femto	f	10^{-15}
atto	a	10^{-18}

© The Author(s) 2023
S. Kumar, *Fluid Mechanics (Vol. 2)*,
https://doi.org/10.1007/978-3-030-99754-0

CONVERSION FACTORS

Length

$$1\text{m} = 1000 \text{ mm}$$
$$= 100 \text{ cm}$$
$$= 39.37 \text{ in}$$
$$= 3.281 \text{ ft}$$
$$= 1.0936 \text{ yd}$$
$$1 \text{ cm} = 10 \text{ mm}$$
$$1 \text{ in} = 2.54 \text{ cm}$$
$$1 \text{ ft} = 30.48 \text{ cm}$$
$$1 \text{ mile} = 1.609 \text{ km}$$
$$1 \text{ yd} = 3 \text{ ft}$$

Mass

$$1 \text{ kg} = 2.204 \text{ lb}$$
$$1 \text{ lb} = 0.4537 \text{ kg}$$
$$1 \text{ kg} = 1000 \text{ g}$$
$$1 \text{ tonne} = 1000 \text{ kg}$$
$$1 \text{ tonne} = 0.984 \text{ ton}$$
$$1 \text{ ton} = 1016.26 \text{ kg}$$
$$1 \text{ slug} = 14.59 \text{ kg}$$

Area

$$1 \text{ m}^2 = 10^4 \text{ cm}^2$$
$$1 \text{ km}^2 = 0.3862 \text{ mi}^2$$
$$1 \text{ ft}^2 = 0.09289 \text{ m}^2$$
$$1 \text{ in}^2 = 6.4516 \text{ cm}^2$$

Density

$$1 \text{ kg/m}^3 = 10^{-3} \text{ g/cm}^3$$
$$= 0.0624 \text{ lb/ft}^3$$
$$1 \text{ lb/ft}^3 = 16.025 \text{ kg/m}^3$$

Volume

$$1 \text{ litre} = 1000 \text{ cc or cm}^3$$
$$= 10^{-3} \text{ m}^3$$
$$= 0.0353 \text{ ft}^3$$
$$1 \text{ m}^3 = 35.32 \text{ ft}^3$$
$$1 \text{ m}^3 = 1000 \text{ litres}$$
$$= 10^6 \text{ cm}^3$$
$$1 \text{ gallon} = 4.546 \text{ litres}$$
$$= 4.546 \times 10^{-3} \text{ m}^3$$
$$= 8 \text{ pints}$$
$$1 \text{ pint} = 568.25 \text{ cc}$$
$$= 568.25 \times 10^{-6} \text{ m}^3$$

Force

$$1 \text{ N} = 1 \text{ kg m/s}^2$$
$$= 0.102 \text{ kgf}$$
$$= 10^5 \text{ dyne}$$
$$1 \text{ dyne} = 1 \text{ gm cm/s}^2$$
$$1 \text{ kgf} = 9.807 \text{ N}$$
$$= 2.204 \text{ lbf}$$

Discharge

$$1 \text{ litre/s} = 10^{-3} \text{m}^3/\text{s}$$
$$1 \text{m}^3/\text{s} = 10^3 \text{ litre/s}$$
$$= 35.32 \text{ ft}^3/\text{s}$$
$$1 \text{ cusecs} = 0.02831 \text{ m}^3/\text{s}$$

Energy and Work

$$1 \text{ Nm} = 1 \text{ J} = 10^7 \text{ erg}$$
$$1 \text{ erg} = 1 \text{ dyn. cm}$$
$$= 10^{-5} \times 10^{-2} \text{ Nm}$$
$$= 10^{-7} \text{ J}$$
$$1 \text{ kgfm} = 9.807 \text{ Nm}$$
$$= 7.229 \text{ ft lbf}$$
$$1 \text{ ft lbf} = 0.1383 \text{ kgfm}$$
$$= 1.356 \text{ Nm}$$
$$= 1.356 \text{ J}$$

$$1 \text{ kWh} = 3600 \text{ kJ}$$
$$= 860 \text{ kcal}$$
$$= 3.6 \text{ MJ}$$

Heat

$$1 \text{ kJ} = 0.2388 \text{ kcal}$$
$$= 0.9478 \text{ Btu}$$
$$1 \text{ Btu} = 1.055 \text{ kJ}$$
$$= 0.252 \text{ kcal}$$
$$= 778 \text{ ft lbf}$$
$$1 \text{ kcal} = 427 \text{ kgf m}$$
$$= 4.187 \text{ kJ}$$
$$= 3.968 \text{ Btu}$$
$$1 \text{ cal} = 4.187 \text{ J}$$

Btu = British thermal units

Power

$$1 \text{ Nm/s} = 1 \text{ J/s} = 1 \text{ W}$$
$$1 \text{ kW} = 1000 \text{ W}$$
$$= 860 \text{ kcal/h}$$
$$= 102 \text{ kgf m/s}$$
$$= 737.5 \text{ ft lbf/s}$$
$$= 1.359 \text{ hp (metric)}$$
$$= 1.341 \text{ hp (FPS)}$$
$$1 \text{ hp (metric)} = 75 \text{ kgf m/s}^2 \text{ (MKS)}$$
$$= 75g \text{ watt}$$
$$= 735.75 \text{ W}$$
$$1 \text{ hp (FPS)} = 745.70 \text{ W}$$
$$1 \text{ hp (metric)} = 4500 \text{ kgf m/min}$$

Pressure

$$1 \ N/m^2 = 1 \ Pa$$
$$1 \ bar = 10^5 \ N/m^2$$
$$= 10^2 \ kPa$$
$$= 0.1 \ M \ N/m^2$$
$$= 1.0197 \ kgf/cm^2$$
$$1 \ mbar = 10^{-3} \ bar$$
$$= 100 \ N/m^2$$
$$= 100 \ Pa$$
$$= 10.2 \ mm$$
$$1 \ atm = 101.325 \ kPa$$
$$= 1.01325 \ bar$$
$$= 760 \ mm \ of \ Hg$$
$$= 10.33 \ m \ of \ water$$
$$= 760 \ torr$$

$$1 \ torr = 1 \ mm \ of \ Hg$$
$$= 13.6 \ mm \ of \ water$$
$$= 1.334 \ mbar$$
$$= 133.38 \ Pa$$
$$1 \ mm \ of \ water = 1 \ kgf/m^2$$
$$= 9.807 \ N/m^2$$
$$= 0.0981 \ mbar$$
$$1 \ ata = 1 \ kgf/cm^2$$
$$= 9.807 \times 10^4 \ N/m^2$$
$$= 0.981 \times 10^5 \ N/m^2$$
$$= 0.981 \ bar$$

Temperature

$$T(K) = T \ °C + 273.15$$
$$1 \ °F = 1.8 \ °C + 32$$

Dynamic Viscosity

$$1 \ Ns/m^2 = 1 \ Pa.s$$
$$= 1 \ kg/ms$$
$$= 10 \ poise$$

$$1 \ poise = \frac{1}{10} \ Ns/m^2 = 0.1 \ Ns/m^2$$
$$= dyne\text{-}s/cm^2$$

$$1 \ centipoise = 10^{-2} \ poise$$
$$= 10^{-3} \ Ns/m^2$$
$$1 \ kgf \ s/m^2 = 98.1 \ dyn\text{-}s/cm^2$$
$$= 98.1 \ poise$$
$$1 \ lbf \ s/ft = 47.847 \ Ns/m^2$$

Kinematic Viscosity

$$1 \ m^2/s = 10^4 \ cm^2/s$$
$$= 10^4 \ stokes$$
$$1 \ stokes = 1 \ cm^2/s$$
$$1 \ centistokes = 10^{-2} \ stokes$$
$$= 10^{-6} \ m^2/s$$
$$1 \ ft^2/s = 0.0929 \ m^2/s$$

Mathematical Formulae

1. TRIGONOMETRICAL FORMULAE

$$\operatorname{cosec} A = \frac{1}{\sin A} \qquad\qquad \sec A = \frac{1}{\cos A}$$

$$\tan A = \frac{\sin A}{\cos A} = \frac{1}{\cot A} \qquad \cot A = \frac{\cos A}{\sin A} = \frac{1}{\tan A}$$

$$\sin^2 A + \cos^2 A = 1$$

$$\sec^2 A = 1 + \tan^2 A \qquad\qquad \operatorname{cosec}^2 A = 1 + \cot^2 A$$

$$\sin(90° + A) = \cos A \qquad\qquad \sin(90° - A) = \cos A$$

$$\cos(90° + A) = -\sin A \qquad\qquad \cos(90° - A) = \sin A$$

$$\tan(90° + A) = -\cot A \qquad\qquad \tan(90° - A) = \cot A$$

$$\sin(180° + A) = -\sin A \qquad\qquad \sin(180° - A) = \sin A$$

$$\cos(180° + A) = -\cos A \qquad\qquad \cos(180° - A) = -\cos A$$

$$\tan(180° + A) = \tan A \qquad\qquad \tan(180° - A) = -\tan A$$

$$\sin(360° + A) = \sin A \qquad\qquad \sin(360° - A) = -\sin A$$

$$\cos(360° + A) = \cos A \qquad\qquad \cos(360° - A) = \cos A$$

$$\cos(360° + A) = \tan A \qquad\qquad \tan(360° - A) = -\tan A$$

$$\sin(A + B) = \sin A \cos B + \cos A \sin B$$

$$\sin(A - B) = \sin A \cos B - \cos A \sin B$$

$$\cos(A + B) = \cos A \cos B - \sin A \sin B$$

$$\cos(A - B) = \cos A \cos B - \sin A \sin B$$

$$\tan(A + B) = \frac{\tan A + \tan B}{1 - \tan A \tan B}$$

$$\tan(A - B) = \frac{\tan A - \tan B}{1 + \tan A \tan B}$$

$$\sin 2A = 2 \sin A \cos A$$

$$\cos 2A = \cos^2 A - \sin^2 A = 2\cos^2 A - 1 = 1 - 2\sin^2 A$$

$$\tan 2A = \frac{2 \tan A}{1 - \tan^2 A}$$

$$\sin 2A = \frac{2\tan A}{1+\tan^2 A}$$

$$\cos 2A = \frac{1-\tan^2 A}{1+\tan^2 A}$$

$$\sin 3A = 3 \sin A - 4 \sin^3 A$$

$$\cos 3A = 4 \cos^3 A - 3 \cos A$$

$$\tan 3A = \frac{3\tan A - \tan^3 A}{1-3\tan^2 A}$$

$$\sin (A + B) \sin (A - B) = \sin^2 A - \sin^2 B = \cos^2 B - \cos^2 A$$

$$\cos (A + B) \cos (A - B) = \cos^2 A - \sin^2 B = \cos^2 B - \sin^2 A$$

$$\sin \frac{A}{2} = \pm\sqrt{\frac{1}{2}(1-\cos A)}$$

$$\cos \frac{A}{2} = \pm\sqrt{\frac{1}{2}(1+\cos A)}$$

$$\tan \frac{A}{2} = \frac{\sin A}{1+\cos A}$$

$$\sin \frac{A}{2} + \cos \frac{A}{2} = \pm\sqrt{1+\sin A}$$

$$\sin \frac{A}{2} - \cos \frac{A}{2} = \pm\sqrt{1-\sin A}$$

$$\sin (A + B) + \sin (A - B) = 2 \sin A \cos B$$

$$\sin (A + B) - \sin (A - B) = 2 \cos A \sin B$$

$$\cos (A + B) + \cos (A - B) = 2\cos A \cos B$$

$$\cos (A + B) - \cos (A - B) = -2 \sin A \sin B$$

$$\sin A + \sin B = 2 \sin \frac{A+B}{2} \cos \frac{A-B}{2}$$

$$\sin A - \sin B = 2 \cos \frac{A+B}{2} \sin \frac{A-B}{2}$$

$$\cos A + \cos B = 2 \cos \frac{A+B}{2} \cos \frac{A-B}{2}$$

$$\cos A - \cos B = -2 \sin \frac{A+B}{2} \sin \frac{A-B}{2}$$

2. DIFFERENTIAL CALCULUS FORMULAE

(a) $\dfrac{d(x)^n}{dx} = nx^{n-1}; \quad \dfrac{dx^5}{dx} = 5x^4, \quad \dfrac{d(x)}{dx} = 1$

(b) $\dfrac{d}{dx}(ax + b)^n = n(ax + b)^{n-1} \times a$

(c) $\dfrac{d(C)}{dx} = 0$

(d) $\dfrac{d(u \times v)}{dx} = u\dfrac{dv}{dx} + v\dfrac{du}{dx}$

(e) $\dfrac{d}{dx}\left(\dfrac{u}{v}\right) = \dfrac{v\dfrac{du}{dx} + u\dfrac{dv}{dx}}{v^2}$

(f) $\dfrac{d(\sin x)}{dx} = \cos x, \quad \dfrac{d(\cos x)}{dx} = -\sin x.$

(g) $\dfrac{d(\tan x)}{dx} = \sec^2 x, \quad \dfrac{d(\cot x)}{dx} = -\operatorname{cosec}^2 x$

(h) $\dfrac{d(\sec x)}{dx} = \sec x \tan x, \quad \dfrac{d(\operatorname{cosec} x)}{dx} = -\operatorname{cosec} x \cot x$

3. INTEGRAL CALCULUS FORMULAE

(a) $\displaystyle\int x^n \, dx = \dfrac{x^{n+1}}{n+1}$

(b) $\displaystyle\int 7 \, dx = 7x, \quad \int C \, dx = Cx$

(c) $\displaystyle\int (ax+b)^n \, dx = \dfrac{(ax+b)^{n+1}}{(n+1) \times n}$

(d) $\displaystyle\int_o^l x^5 \, dx = \left[\dfrac{x^4}{4}\right]_0^l = \dfrac{l^4}{4} - \dfrac{0}{4} = \dfrac{l^4}{4}$

(e) $\displaystyle\int \log_e x \, dx = \dfrac{1}{x}$

Index

□□□

Printed in the United States
by Baker & Taylor Publisher Services